Simulation and Optimization of
Internal Combustion Engines

Simulation and Optimization of Internal Combustion Engines

ZHIYU HAN

INTERNATIONAL

Warrendale, Pennsylvania, USA

400 Commonwealth Drive
Warrendale, PA 15096-0001 USA
E-mail: CustomerService@sae.org
Phone: 877-606-7323 (inside USA and Canada)
              724-776-4970 (outside USA)
FAX: 724-776-0790

Library of Congress Catalog Number
http://dx.doi.org/10.4271/ 9781468604016

Information contained in this work has been obtained by SAE International from sources believed to be reliable. However, neither SAE International nor its authors guarantee the accuracy or completeness of any information published herein and neither SAE In ternational nor its authors shall be responsible for any errors, omissions, or damages arising out of use of this information. This work is published with the understanding that SAE International and its authors are supplying information but are not attempting to render engineering or other professional services. If such services are required, the assistance of an appropriate professional should be sought.

ISBN-Print 978-1-4686-0400-9
ISBN-PDF 978-1-4686-0401-6
ISBN-ePub 978-1-4686-0402-3

To purchase bulk quantities, please contact: SAE Customer Service

E-mail: CustomerService@sae.org
Phone: 877-606-7323 (inside USA and Canada)
              724-776-4970 (outside USA)
Fax: 724-776-0790

Visit the SAE International Bookstore at books.sae.org

Chief Growth Officer
Frank Menchaca

Publisher
Sherry Dickinson Nigam

Director of Content Management
Kelli Zilko

Production and Manufacturing Associate
Erin Mendicino

Table of Contents

CHAPTER 4

In-Cylinder Turbulence   75

CHAPTER 5

Fuel Sprays   119

CHAPTER 6

Combustion and Pollutant Emissions 193

Preface

Internal combustion engines are the dominating power sources in many applications including automotive, marine, industrial equipment, agriculture equipment, general machinery, power station, and so forth. In the transportation sector, there are about 1.4 billion passenger vehicles and about 380 million commercial vehicles in use worldwide today, consuming about 35% of the total petroleum consumed. Pollutant and greenhouse gas emissions from internal combustion engines have unfriendly impacts on the environment and are of great concern.

Driven by increasingly stringent government regulations and elevated customer demands, reducing fuel consumption and emissions has been the goal of the internal combustion engine R&D community and manufacturers for years. Major advances have been achieved with the application of new technologies and innovation in engine combustion and tail-pipe emissions aftertreatment. Consequently, the regulated pollutant emissions of particulate matters (PM), nitroxides (NOx), carbon monoxide (CO), and unburned hydrocarbons from newly produced road vehicles were reduced by about one order of magnitude, and carbon dioxide (CO_2) emissions decreased by about 30% due to improvement in vehicle fuel economy in the past 30 years.

Although battery-driven electric vehicles (EVs) have taken global attention in the automotive sector, a wide adaptation of EVs hinges upon battery safety, energy storage, and charging infrastructure, which remain to be the challenges. In the foreseeable future, diversification in road-vehicle power sources is inevitable, but internal combustion engines will continue to take the major share. Internal combustion engines, battery-electric propulsion systems, and hybrid powertrain will all share space in the world market according to geographic region, energy resource availability, and end-usage. With the advancement of powertrain electrification, automotive engines will not die; rather, they will be reborn in a new life featuring high fuel conversion efficiency, near-zero emissions, and design simplicity.

The complexity of internal combustion engines has increased greatly. New technologies often introduce more design and operating variables, at the same time bringing about benefits, which makes engine optimization even more difficult. The goal is that engines deliver multiple optimal function targets at acceptable costs. The traditional experiment-based trial-and-error method does not work anymore. Engine simulation technology has developed and is applied to help achieve the goal.

Multidimension engine combustion simulation is the extension of computational fluid dynamics (CFD) with the addition of physical and chemical models of sprays, combustion, and emissions formation that are appropriate to internal combustion engines. This technology originated in the 1970s and is now widely used in many major engine and automotive companies worldwide backed by the dramatic increases in computer capability. With years of research and development, advances in predictive physical models and computational efficiency warrant their roles in the design and optimization of engine combustion systems. Multidimension engine simulation has proven to be effective in the development of new engine products and research exploratory of more efficient and cleaner new combustion concepts.

In this book, the fundamentals and up-to-date progress in multidimensional simulation and optimization of internal combustion engines are presented. The main contents are divided into three parts. The first part (Chapters 1 to 3) presents the progress and outlook of automotive engines and engine combustion simulation; provides overviews and basic theories of combustion in spark-ignition gasoline engines, diesel engines, and advanced low-temperature combustion concepts; and describes mathematical governing equations, numerical methods, and boundary conditions for engine simulation. The second part (Chapters 4 to 6) focuses on the physical and chemical models that describe the phenomena of turbulence, sprays, combustion, and emissions formation in internal combustion engines. Both Reynolds-averaged Navier–Stokes (RANS) and large-eddy simulation (LES) turbulence models are discussed; often-used models for fuel spray atomization, drop dynamics, evaporation and spray-wall interactions are given; various ignition and combustion models with elevated complexities are presented; soot and NOx formation models are introduced; and simulations of diesel, spark-ignition premixed and homogeneous charge compression ignition (HCCI) combustion are discussed.

The last part of this book (Chapters 7 and 8) provides some examples of combustion optimization with the use of engine simulation techniques. In Chapter 7, a model-driven combustion development methodology is described. Sprays, mixture formation, and spray wall impingement, which are among the key problems in direct-injection (DI) gasoline engines, are addressed for both wall-guided DI and spray-guided DI systems. In Chapter 8, modeling and optimization of combustion and emissions in diesel and alternative fuel (natural gas and biodiesel) engines are presented. The advantages of reactivity controlled compression ignition (RCCI) combustion of natural gas and diesel dual-fuel are demonstrated by simulation.

The goal of this book is to reflect major advances in physical models and applications in multidimensional engine simulation technology and to provide an up-to-date reference. Many models and their variations have been proposed in the past years. Each one makes incremental improvement either in prediction accuracy or in numerical efficiency. It is impossible to include all the models in a single book and this book intends to introduce the pioneer and/or the often-used models and the physics behind them, and provide readers with the fundamentals and ready-to-use knowledge. Furthermore, many application examples are provided, which focus on the key issues in those engines that impact combustion and performance significantly. Useful modeling methodology and techniques, as well as instructive results, are discussed through the examples. Hopefully, readers can understand the fundaments of these examples and be inspired to explore new ideas and means for better solutions in their studies and work.

My career has been associated with automotive engine research and development for many years. The time I spent studying for my Ph.D. at the Engine Research Center (ERC) of the University of Wisconsin-Madison in the mid-1990s was very delightful and fruitful. This is where I started my research on multidimensional engine combustion simulation. I would like to express my gratitude to Prof. Rolf Reitz at the ERC for his inspiration in my research. Prof. Reitz and I jointly published a number of technical papers in engine models including the widely-used compressible-flow RNG k-ε model. Many of our works are cited in this book. I am grateful also to the Ford Research Laboratory (FRL) of the Ford Motor Company where I had the opportunity to develop the in-house CFD code MESIM with my colleagues and to apply it to several advanced engine systems. My colleagues at FRL are appreciated for their collaboration and co-authorship of the technical publications, many of which are cited as well. The graduate students in my research group at Hunan University in China have made invaluable contributions in the engine simulation field. Some of their results are also presented. In addition, I want to thank research staff Dr. Zhenkuo Wu and Yong Wang, graduate student Mingqing Li, Jian Feng, Mengyang Lyu, Shuo Meng, and Yongzheng Sun in my research group at Tongji University in China for their assistance in preparing the material for this book.

My wife Yun Tang, daughter Amy Han, and son Andrew Han have strongly supported my career. I am truly thankful for their love, understanding, and support.

Zhiyu (Allen) Han
May 2021, Shanghai

Abbreviations

1D	One-dimensional
3D	Three-dimensional
ABCD	After bottom dead center
A/F	Air/fuel ratio
AI	Artificial intelligence
AMC	Adaptive multigrid chemistry
ALE	Arbitrary Lagrangian–Eulerian
ARMOGA	Adaptive range multi-objective genetic algorithm
ASI	After the start of injection
ATDC	After top dead center
BDC	Bottom dead center
BFSA	Brute-force sensitivity analysis
BMEP	Brake mean effective pressure
BML	Bray–Moss–Libby
BTDC	Before top dead center
CA	Crank-angle degrees
CCD	Charge coupled device
CDCI	Conventional diesel compression-ignition
CDM	Continuous droplet model
CI	Compression ignition
CID	Charge injection devices
CFD	Computational fluid dynamics
CFM	Continuous formulation model
CFM	Coherent flame model
CMC	Continuous multicomponent model
CMC	Conditional moment closure
COM	Control-oriented-model
CR	Compression ratio
CSAE	Chinese Society of Automotive Engineers
CTC	Characteristic time combustion
CTR	Cross-tumble ratio
CVI	Close-valve injection
DDM	Discrete droplet model
DI	Direct injection
DMC	Discrete multicomponent model
DME	Dimethyl ether
DNGDF	Diesel and natural gas dual-fuel
DNS	Direct numerical simulation
DOC	Diesel oxidation catalyst

DPF	Diesel particulate filter
DPI	Diesel pilot ignition
DPIK	Discrete particle ignition kernel
DRG	Directed relation graph
DRGEP	Directed relation graph with error propagation
DRGASA	Directed relation graph-aided sensitivity analysis
EBU	Eddy breakup
EDM	Eddy dissipation model
EDM	Equivalent diesel-fuel mass
EGR	Exhaust gas recirculation
EOI	End of injection
EPA	Environmental Protection Agency
ER	Expansion ratio
ERC	Engine Research Center
EV	Electric vehicle
EVO	Exhaust-valve opening
FRL	Ford Research Laboratory
FSN	Filter smoke number
GA	Genetic algorithm
GDICI	Gasoline direct injection compression ignition
GISFC	Gross indicated specific fuel consumption
GSE	Group species elimination
GRNG	Generalized Renormalization Group
GUI	Graphic user interface
HCCI	Homogeneous-Charge Compression Ignition
HCDI	Homogeneous-charge direct injection
HRR	Heat release rate
HSDI	High-speed direct injection
HWA	Hot wire anemometry
IC	Internal combustion
IMEP	Indicated mean effective pressure
ITE	Indicated thermal efficiency
IVC	Intake valve closure
i-VCT	Independent-variable camshaft timing
IVO	Intake valve opening
KH	Kelvin–Helmholtz
KH-RT	Kelvin–Helmholtz/Rayleigh–Taylor
KLSA	Knock limited spark advance
LANL	Los Alamos National Laboratory
LDA	Laser Doppler anemometry
LDV	Laser Doppler velocimeter
LEA	Laser extinction/absorption
LES	Large-eddy simulation
LHV	Laser homodyne velocimetry
LIEF	Laser induced exciplex fluorescence
LIF	Laser induced fluorescence
LISA	Linearized instability sheet atomization
LLNL	Lawrence Livermore National Laboratory
LRS	Laser Rayleigh scattering

L-S	Launder–Spalding
LSC	Light stratified-charge
LSD	Laser sheet droplet
LTC	Low-temperature combustion
MBT	Maximum brake torque
MD	Methyl decanoate
NEDC	New European Driving Cycle
MESIM	Multidimensional Engine Simulation
MRC	Mechanism Reduction Code
NGV	Natural gas vehicle
NMEP	Net mean effective pressure
NSC	Nagle and Strickland-Constable
NSGA	Non-dominated sorting genetic algorithm
NSFC	Net specific fuel consumption
NTC	Negative temperature coefficient
NVH	Noise, vibration, harshness
OVI	Open-valve injection
PAH	Polycyclic aromatic hydrocarbon
PCCI	Premixed-charge compression ignition
PDE	Partial differential equations
PDF	Probability density function
PDPA	Phase Doppler particle analyzer
PES	Percent energy substitution
PFI	Port fuel injection
PHEV	Plug-in hybrid electric vehicle
PISO	Pressure implicit with splitting of operators
PIV	Particle image velocimetry
PLIF	Planar laser induced fluorescence
PM	Particulate matter
PN	Particle number
PSO	Particle swarm optimization
QSOU	Quasi-second-order upwind
RANS	Reynolds-averaged Navier–Stokes
RCCI	Reactivity-Controlled Compression Ignition
RDE	Real-driving emissions
RIF	Representative interactive flamelet
RNG	Renormalization Group
ROI	Radius-of-influence
RT	Rayleigh–Taylor
SA	Spark-advance
SCDI	Stratified-charge direct injection
SCR	Selective catalytic reduction
SCV	Swirl control valve
SGDC	Sub-grid direct chemistry
SGDI	Spray-guided direct injection
SGS	Sub-grid scale
SI	Spark ignition
SIMPLE	Semi-implicit method for pressure-linked equations
SING	Spark-ignition natural gas

SMD	Sauter mean diameter
SMR	Sauter mean radius
SOC	Start of combustion
SOF	Soluble organic fraction
SOI	Start of injection
SR	Swirl ratio
SRS	Spontaneous Raman scattering
TAB	Taylor Analogy Breakup
TDC	Top dead center
TKE	Turbulence kinetic energy
TR	Tumble ratio
US	United States
VCR	Variable compression ratio
VCT	Variable cam timing
VISC	Vortex Induced Stratification Combustion
VVL	Variable valve lift
VVT	Variable valve timing
WLTC	Worldwide Harmonized Light Vehicles Test Cycle
WGDI	Wall-guided direct injection
WOT	Wide-open throttle

Nomenclature

B	Spalding mass transfer number
C_D	Drag coefficient
c_p	Specific constant-pressure heat capacity
c_v	Specific constant-volume heat capacity
D	Diffusion coefficient
D_{32}	Sauter mean diameter
d	Diameter
E	Activation energy
e	Specific internal energy
f	Drop distribution function
H^0	Enthalpy of formation
H_{cr}	Threshold for splash
h	Sheet thickness
K	Heat transfer coefficient
K_f	Rate of forward reaction
K_r	Rate of reverse reaction
K_c	Equilibrium constant
k	Turbulence kinetic energy
k	Specific heat ratio
L	Breakup Length
L_K	Kolmogorov length scale
L_I	Turbulence integral length scale
L_M	Turbulence micro length scale
L_ε	Turbulence macro length scale
m	Mass
Nu	Nusselt heat transfer number
n	Engine speed
$\mathbf{n}$	Unit normal vector
P	Engine power
Pr	Prandtl number
p	Pressure
Q	Heat
Q_f	Heat release
Q_{LHV}	Fuel lower heating value
Q_w	Wall heat transfer
q_w	Wall heat flux
R	Gas constant
Re	Reynolds number
r	Radius

r_{32}	Sauter mean radius
S^0	Entropy of formation
Sc	Schmidt number
Sh	Sherwood mass transfer number
S_l	Laminar flame speed
S_t	Turbulent burning velocity
T	Temperature
T	Taylor number
T_s	Surface temperature
t	Time
U	Internal Energy
$\mathbf{u}$	Gas velocity vector
$u^\star$	Shear speed
u'	Turbulence intensity
V	Volume
$\mathbf{v}$	Drop velocity vector
W	Molecular weight
We	Weber number
$\mathbf{x}$	Position vector
x_b	Fuel mass burned fraction
Y	Mass fraction
y	Distortion from sphericity
Z	Ohnesorge number
Δt	Time step
ε	Dissipation rate of turbulence kinetic energy
ε	Compression ratio
ε_e	Effective compression ratio
ϕ	Equivalence ratio
φ	Crank angle
η	Engine thermal efficiency
η_c	Combustion efficiency
η_i	Indicated thermal efficiency
η_m	Mechanical efficiency
η_v	Volumetric efficiency
κ	Von Karman constant
λ	Thermal conductivity
μ	Dynamic viscosity
μ_{SGS}	Sub-scale turbulent viscosity
μ_t	Turbulent viscosity
v	Kinematic viscosity
v_0	Laminar kinematic viscosity
v'	Forward molar stoichiometric coefficient
v''	Reverse molar stoichiometric coefficient
θ	Spray cone angle
ρ	Density
σ	Surface tension
$\boldsymbol{\sigma}$	Stress tensor
τ	Viscous stress tensor

Superscript

$\cdot$	Time rate of change
$\sim$	Mass (Favre) averaged
$-$	Time averaged
$+$	Non-dimensional parameters in law-of-the-wall
c	Combustion
s	Spray

Subscript

d	Droplet
g	Gas
KH	KH model
k	Species
l	Liquid
n	Normal direction
RT	RT model
t	Turbulence, Tangent direction
w	Wall

1

Introduction

Internal combustion (IC) engines play an indispensable role in our society. Tremendous advances had been achieved in automotive IC engines in the past few decades, including powertrain electrification. However, newer regulatory requirements keep pushing IC engines to deliver the next level of fuel economy and reduction in CO_2 emissions. For the automotive industry in China, the goals have been set up by the Chinese Society of Automotive Engineers [1]. By 2035, the average fuel consumption of new conventional passenger vehicles (powered by IC engines or hybrid powertrain) will be 4 L/100 km under the worldwide harmonized light vehicles test cycle (WLTC); while that of commercial vehicles like trucks and buses will be reduced by 15–20% and 20–25%, respectively, from the base year, 2019. These goals undoubtedly bring great challenges to the IC engines research and development community and industry.

The phenomena of flow, combustion, and emissions formation in IC engines are extremely complex and not yet fully understood. Complicated interactions within subsystems take place in an engine. Multiple design parameters need to be optimized for reliable and durable functions such as fuel economy, performance, emissions, and noise, vibration, and harshness (NVH) along with cost-effectiveness. Innovation and improvement in engine combustion, thus, will continue to be the key topic in engine research.

Multidimensional engine combustion simulation (also called combustion simulation or engine simulation) based on computational fluid dynamics (CFD) solves the conservation equations that describe the multi-component gas flow interacting with vaporizing fuel sprays, mixing, ignition, chemical reactions, effects of turbulence, and wall heat transfer in an engine. It can be used to optimize engine design and operating parameters with optimization algorithms [2]. It has also been a key factor in the development of new combustion concepts [3,4]. Furthermore, model-driven design methodologies have been effective in delivering optimal product designs to meet multiple goals in a short period of time [5–7]. Nowadays, combustion simulation has become an indispensable tool in engine research and development. Virtual design and performance optimization before the actual manufacture of hardware is the future of engine development. A large portion of engine calibration will be also done by computer modeling.

1.1 Recent Progress and Outlook of Automotive Engines

1.1.1 Achievement in Engine Performance and Emissions

Remarkable strides have been taken in technology innovation and advancement in automotive engines over the past 30 years. Key technologies of turbocharging, gasoline direct-injection (DI), enhanced engine control, variable devices (variable valve timing (VVT), variable valve lift (VVL), variable compression ratio (VCR), cylinder-deactivation, etc.), and smart thermal management have enabled more efficient and cleaner combustion with the use of new cycle (Atkinson/Miller cycle), high geometric compression ratio (up to 16), high tumble gas motion, and high exhaust gas recirculation (EGR) rate. As a result, engine performance, fuel economy, pollutants emissions and CO_2 emissions have been improved substantially.

Progress in automotive engine thermal efficiency is illustrated in Figure 1.1, which compiles recently reported maximum brake thermal efficiencies of engines in mass production. Historical data of Toyota Motor Corporation before 2015 are also cited in Figure 1.1 [8]. New-generation engines with a maximum thermal efficiency greater than 39% are common, which is an improvement of more than 10% over that of the last generation. In addition, laboratory engines with indicated thermal efficiency higher than 52% [9,10] and close to 60% [11] are also reported with innovative combustion strategies.

FIGURE 1.1 Evolution of the maximum thermal efficiency of automotive gasoline engines in mass production.

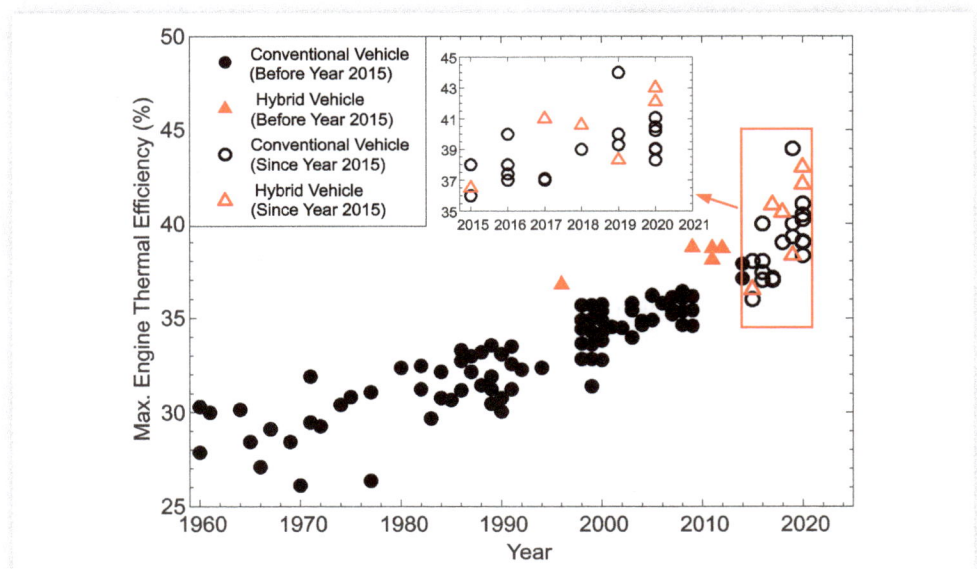

Courtesy of Zhiyu Han

Driven by government regulations and customer demands, the fuel economy of new vehicles has decreased steadily worldwide with the improvement of engine thermal efficiency and the advancement of vehicle technology. For instance, the annual nationwide-average fuel consumption of new passenger vehicles in China has reduced by 24.45% from 2012 to 2019 as shown in Figure 1.2. This is despite the fact that during the same period, the average vehicle weight had increased by 149 kg. If vehicular weight had been maintained at the level of 2012, there would have been an additional 10% reduction in fuel consumption.

FIGURE 1.2 Yearly-averaged fuel consumption of new passenger vehicles in China.

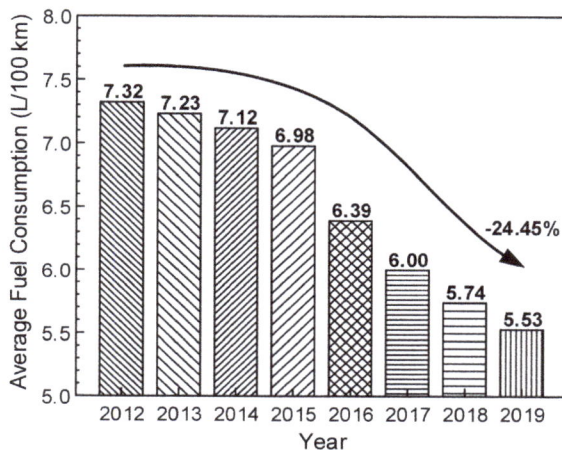

Courtesy of Zhiyu Han

The yearly changes in engine performance are illustrated in Figure 1.3. The data are from the "Wards 10 Best Engines" by the *Wards AutoWorld Magazine* [12] in the United States (US) and from the "Ten Best Engines" by the *Autosports Magazine* [13] in China, which started in 2016. Only data are used of four-cylinder gasoline engines. The data in China are those of the Chinese domestic brands to show their advances. It is seen that although the power density (power output per engine displacement) and torque density (torque output per engine displacement) of the naturally-aspirated engines are increased incrementally, turbocharging technology has doubled the engine performance in the United States and surged by about 80% in China in the past 20 years.

The first automotive pollutant emissions standard was enacted in 1963 in the United States, mainly as a response to the Los Angeles' smog problems. Since then, many countries and regions have imposed emissions standards. Consequently, tremendous reductions in pollutant emissions from on-road vehicles have been achieved worldwide. For example, the Stage 1 standards for light-duty vehicles were put in place in 2000 in China, and Stage 6a was enacted in 2020. Reinforced by which, the total emissions of hydrocarbon and CO for new vehicles launched in a year are reduced by a factor of 4, and NO_x by a factor of 10 as shown in Figure 1.4. In addition, particulate matter (PM) emissions of gasoline engines are regulated, and a more stringent test cycle, the WLTC is to replace the New European Driving Cycle (NEDC) at Stage 6. It is expected that mandatory Real-Driving Emissions (RDE) test procedure in the Stage 6b by 2023 will reduce the gap between the cycle-approval vehicle emissions and real-world emissions.

FIGURE 1.3 Performance of I4 gasoline engines. (a) power density, in the United States (b) power density, in China (domestic brands); (c) torque density, in the United States; (d) torque density, in China (domestic brands).

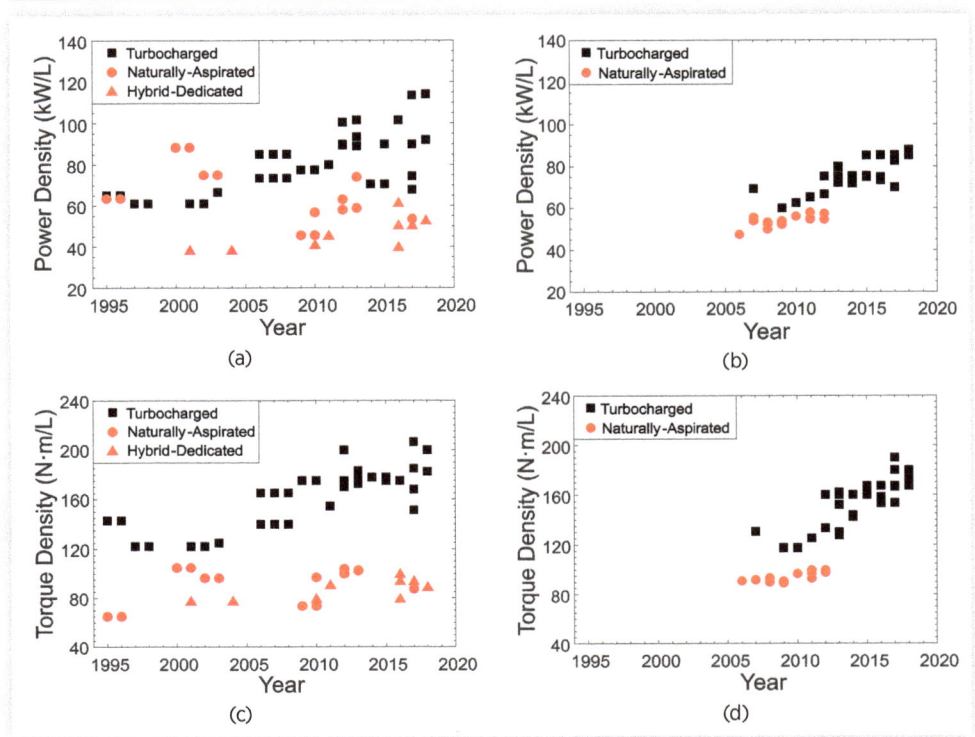

FIGURE 1.4 Evolution of emissions regulation for light-duty vehicles in China.

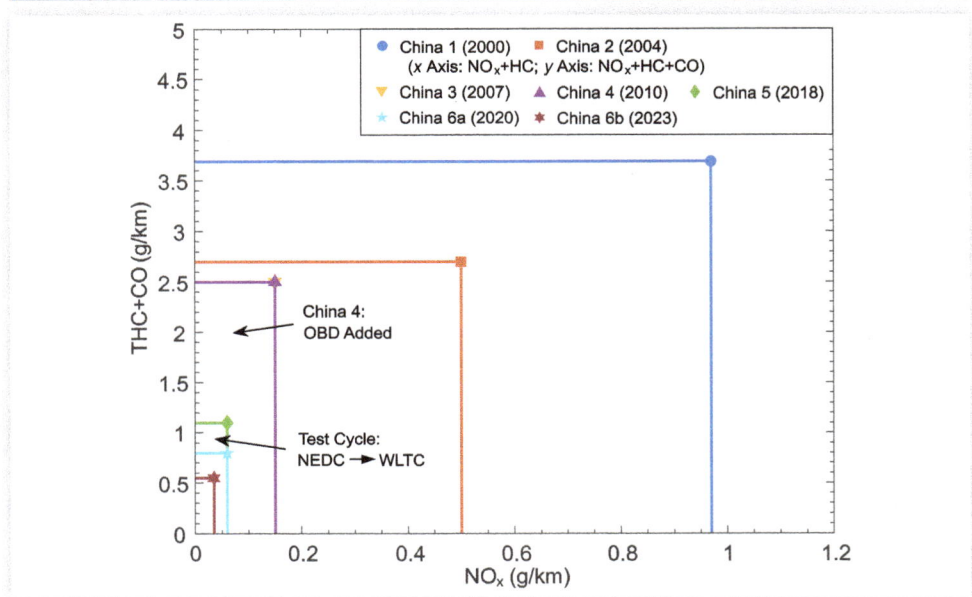

There are no global automotive CO_2-emission dedicated legislative regulations as of date. The reduction of CO_2 emissions has come from improvement in vehicle fuel economy. Legislation on vehicle CO_2 emissions on a life-cycle basis will be beneficial to promote utilization of low-carbon fuels and renewable energies. Utilization of alternative fuels such as natural gas, alcohol (methanol and ethanol), biodiesel, dimethyl ether (DME), hydrogen, etc., in IC engines can not only reduce petroleum consumption, but also lower CO_2 emissions [14]. For example, it was shown that stoichiometrically burning natural gas (consisting mainly of methane) can reduce 20% more CO_2 than the expected reduction in burning gasoline while meeting the China 6b emissions standards [15]. With abundant natural resources of methane in conventional natural gas, shale gas, and methane hydrate, large-scale usage of methane fuels in the transportation sector may be a realistic and economic way to reduce CO_2 emissions substantially in the near future as the infrastructure and vehicle technology are ready. Keeping this in mind, is it acceptable to produce electricity at the cost of high CO_2 emissions by burning coals in order to power battery-driven electric vehicles (EVs) before green electricity becomes reality?

1.1.2 Future Development of IC Engines

Although EVs have their merits in reducing pollutions in major cities, issues of battery safety, energy storage, and charging infrastructure remain unsolved. In the foreseeable future, diversification in road-vehicle power sources is inevitable. But, as concluded by engine and automotive experts worldwide that internal combustion engines will continue to take the major share in transportations [16]. Battery-electric propulsion systems, IC engines, and hybrid powertrain will all continue to have a share in the automobile market, depending on geographic region, energy resource availability, and end-use.

It is aggressively forecasted that by 2035, 50% of the newly produced light-duty vehicles in China will be EVs and the other 50% will be hybrid vehicles [1]. Since a hybrid vehicle is powered by an electric motor and an IC engines, the role of an IC engines will be changed from "Solo" (single power-source) to "Duo" (working together with a traction motor). Thus, there are needs to re-optimize an IC engines to better fit it into the new role. In addition, the development of an optimized energy-management control strategy for maximizing vehicular fuel economy based on driving destination and road conditions is of particular interest. Applications of artificial intelligence (AI) and cloud computing for hybrid powertrain are very useful.

In fact, powertrain hybridization provides a great opportunity for IC engines. Conventional engines will be transformed with features like high thermal (fuel conversion) efficiency, near-zero emissions, and design simplicity [17]. The new engines may be named dedicated hybrid engines. Since the operating ranges in engine speeds and loads become smaller in a dedicated hybrid engine (in particular, in a series-hybrid or range-extender system), some performance requirements, for example, low-end torque, rated power at high speed (e.g., at 6000 rpm), and low idle-speed, can be removed or relaxed as illustrated in Figure 1.5.

Thus, combustion optimization can be focused on either high-to-full loads or low-to-medium loads in the middle-speed range depending on the engine displacement (size) selected to meet the power demand. In the former case, knock combustion is the main challenge, while in the latter case, pumping loss and combustion efficiency are the major concerns. New combustion concepts of Homogeneous Charge Compression Ignition (HCCI) or Reactivity-Controlled Compression Ignition (RCCI) with high thermal efficiency and ultra-low pollutant emissions can be then adopted in the middle operating range. Issues

FIGURE 1.5 A reduced operating range of a dedicated hybrid engine.

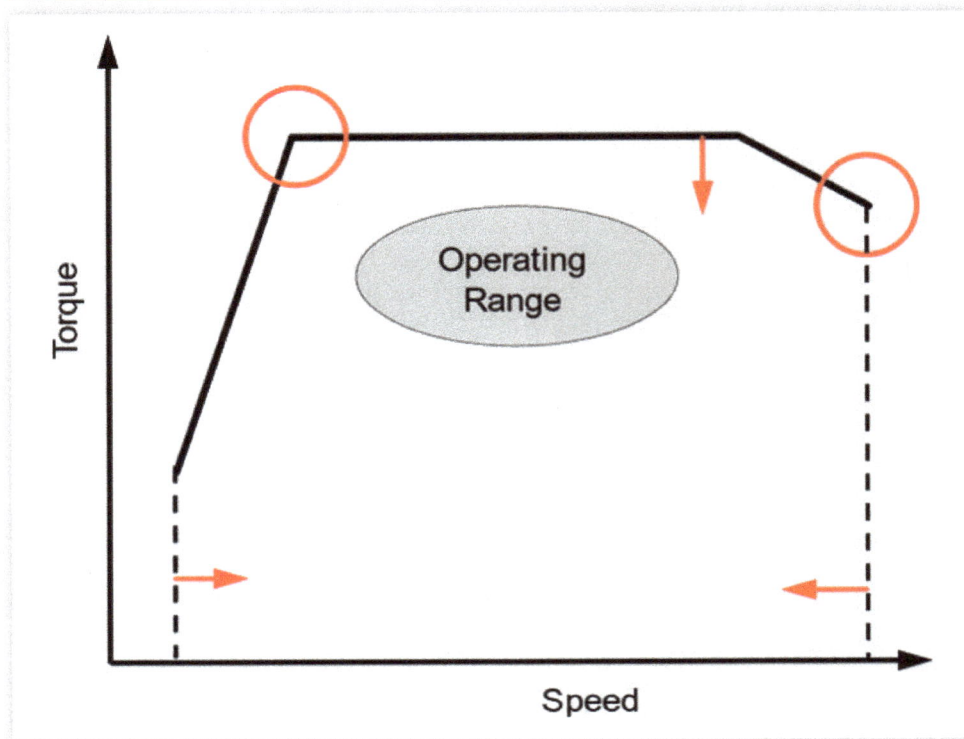

Courtesy of Zhiyu Han

in conventional engines can be then addressed with dedicated methods and technologies. For instance, in the case that a small displacement engine is selected due to the limitation of vehicle installation space, it should be focused on mitigation of knock combustion and improvement of fuel economy in the high-load region [18]. In addition, some functions that are provided by engine accessories (e.g., compressor for air conditioning and engine starter-generator) now can be provided by the electric part of the hybrid system and these accessory parts can be hence removed. Complicated technology such as turbocharging may not be needed in a dedicated hybrid engine since demanded engine power is greatly reduced. As a result, an IC engine in a hybrid vehicle becomes much simpler compared with its counterpart in a conventional engine-powered vehicle.

Electronically-controlled, high-pressure common-rail injection technology has enabled mass production of modern DI gasoline engines since the middle 1990s. Since gasoline fuels are directly injected into the cylinder, fueling can be controlled more accurately. Consequently, in-cylinder air-fuel mixing can be improved and cylinder-to-cylinder fueling variation can be reduced. In-cylinder charging cooling due to fuel vaporization helps to reduce knock tendency so that engine compression ratio can be increased or knock limited spark advance (KLSA) can be set to be more optimal. As a result, engine thermal efficiency can be raised.

Early-injection homogeneous combustion and late-injection stratified-charge combustion has been enabled by DI. Its good dynamic feature facilitates turbocharging to deliver superior low-end torque performance. Turbocharging DI downsized engines have replaced large-sized port fuel injection (PFI) engines enabling better vehicular fuel economy. Multi-hole injectors provide some flexibilities in the design of individual nozzle and spray-plume orientation.

Increasing injection pressure to current 35MPa level can help combustion in two ways. First, fuel spray droplets become smaller and more uniform [19], which leads to faster vaporization and mixing, and less wall wetting so that engine PM (or soot) emissions can be reduced. Second, higher injection pressure provides enhanced flexibility in using a multiple-injection strategy. Multiple injections can improve air-fuel mixing and mitigate knock tendency at high engine loads [6] and improve combustion stability and emissions at cold start [20]. Other spray atomization mechanisms and their applications, e.g., superheat flash-boiling generated sprays [21] and methods to control spray penetration should be further studied.

The thermal efficiency of the current gasoline engines in production has been over 40% as shown in Figure 1.1. Combustion in these engines is in stoichiometric, homogeneous, premixed regime in general. Fuel-lean combustion is needed to achieve 45–50% engine thermal efficiency [9,10] in which lean combustion stability may be the concern. Knock combustion is the biggest obstacle in increasing engine compression ratio and accurate prediction of KLSA in engine design and performance development phase is crucial. New methods with practical feasibility to mitigate knock tendency are called for.

For diesel engines, emissions control is still the key issue. In-cylinder formation of NO_x and soot can be mitigated through optimization of combustion chamber shape, injector nozzle design, flow motion, fuel injection strategy, and EGR. Aftertreatment devices of diesel oxidation catalyst (DOC), diesel particulate filter (DPF), and selective catalytic reduction (SCR) must be used to meet the stringent emissions standards. Diesel low-temperature combustion with significant in-cylinder NO_x reduction [22] may offer an opportunity to remove SCR.

New combustion concepts such as RCCI, HCCI, and Premixed-Charge Compression Ignition (PCCI) are under investigation. These initiate fuel-lean combustion in the low temperature-equivalence ratio region in which formations of soot and NO_x are low, offering exciting potentials of high thermal efficiency and ultra-low emissions [23,24]. However, the challenge is control over combustion phasing and rate, load expansion, combustion-mode switching, and transient operation. Among these three concepts, RCCI is the most promising since it offers practical means to control combustion [25,26]. Based on the fundamentals of RCCI, diesel and natural gas dual-fuel combustion may be the best application scenario since it is easy to be implement in the already existing dual-fuel application environment.

On the other hand, it has been realized that traditional commercial fuels or sole-component fuels cannot meet the requirements for HCCI combustion. To control autoignition timing and combustion phasing, and expand the load range in an HCCI engine, there have been researches on reformation or design of fuel molecular structure, fuel components and composition, and fuel physical-chemical properties to make them more suitable for practical engine combustion [27]. The cetane number, octane number, molecular structure, oxygen content, latent heat of evaporation, and boiling point and distillation are among the properties considered for alternates in the form of additives, fuel blending, and dual fuels.

Research on design and optimization of fuel and engines has been active in recent years. For instance, US national laboratories are conducting research in this area through the Co-Optimization of Fuels and Engines (Co-Optima) initiative from the US Department of Energy, which aims to co-develop fuels and engines in an effort to maximize energy efficiency and the utilization of renewable fuels [28]. The outlook of co-optimization of fuels and engines is to find solutions with potential for near-term improvements to the types of fuels and engines found in most vehicles currently on the road, as well as to the development of revolutionary engine technologies for a longer-term, higher-impact series of solutions. Progress has been made in identifying detailed effects of fuel property on gasoline engine thermal efficiency [29] and in mapping the relationships between production pathways, molecular structure, fuel chemical/physical properties, and engine performance [28].

In the fundamental research area, in addition to achievements in enriching engine fundamentals and in understanding novel combustion concepts, many advanced numerical models and experimental methods have been invented. CFD modeling results can be displayed three-dimensionally and in-cylinder images can be obtained by optical diagnostics techniques. Arguably, visualization of in-cylinder phenomena in an optically or numerically "transparent" engine can be said to be the most remarkable achievement of engine research in the past 30 years. It enables seeing the processes of interest inside an engine and helps to nail down issues.

Figure 1.6 is an example of computed soot formation in a diesel engine. The high soot concentration regions in the combustion chamber are displayed as the colored clouds at the tips of the reacting spray plumes. Spray droplets are shown by the colored particles in which different colors represent different drop sizes. With CFD visualization, reduction of diesel soot emissions by multiple injections can be seen [30].

FIGURE 1.6 Soot formation in a diesel engine.

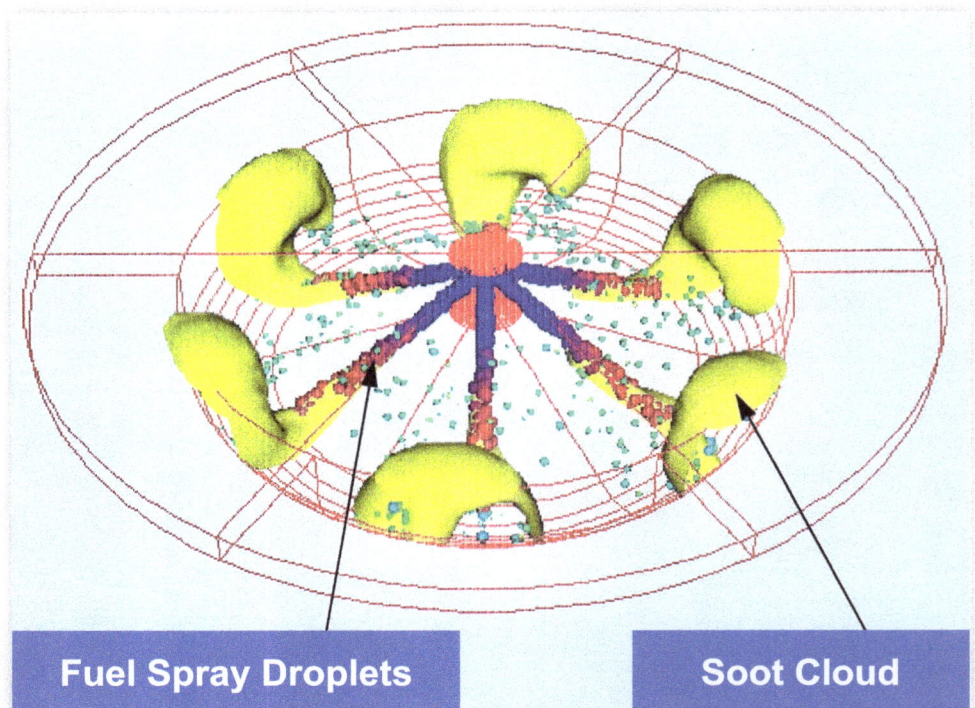

Fuel Spray Droplets   Soot Cloud

Courtesy of Zhiyu Han

Another example is given in Figure 1.7. In this case, the root cause of the high-level smoke emissions in an early design of a DI gasoline engine operated at homogeneous-charge combustion mode was pinned down by CFD modeling and optical imaging. The issue resulted from the fuel droplets impinged on the intake valves due to intake flow impacting on the sprays as seen in the CFD display, and the soot emissions were produced from the pool fires of the deposited fuels as seen. The design was then revised and the issue quickly resolved without compromise of other functions. A traditional trial-and-error method would take much more time and money to find out the problem and fix it.

FIGURE 1.7 Smoke from an early design of a DI gasoline engine. (a) simulation result; (b) optical engine image.

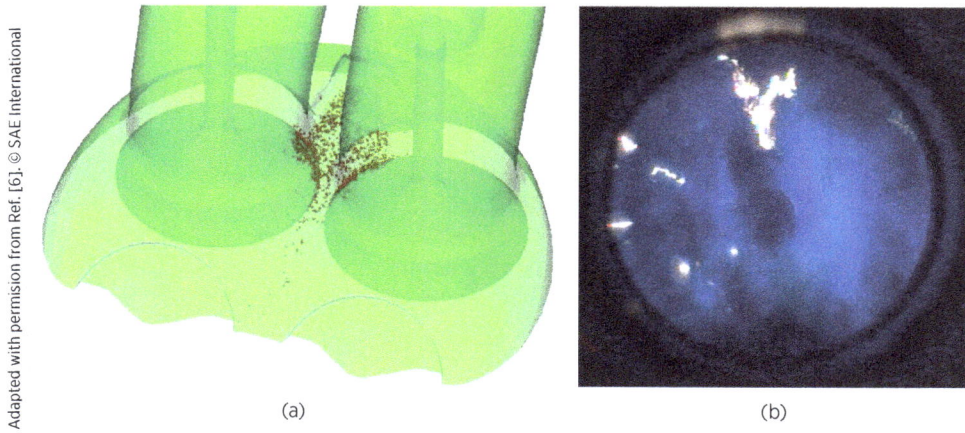

(a)   (b)

1.2 Roles of Multidimensional Engine Simulation

Historically, the early cycle simulation of internal combustion engines by analysis of some air-standard cycles was made at the early 20th century [31]. It was not until the late 1960s that significant advances in IC engines modeling were achieved. Since then, a number of combustion models have been developed with the rapid evolution of computers [32].

The mathematical models describing the engine in-cylinder processes are generally divided into two main groups: thermodynamics and multidimensional models, depending on whether the governing equations that give the model its predominant structure are based on energy conservation or on a full analysis of flow motion.

In thermodynamic models, the cylinder charge is generally assumed to have uniform pressure, temperature, and composition in one or several regions of the combustion chamber and only temporal variations are accounted for. These models can be used as diagnostic (heat release analysis) or predictive [33–35]. In heat release analyses, experimentally determined pressure diagrams are used as an input to calculate the heat release rate, which reflects the evolution of combustion globally. The heat release rate from a thermodynamic model is also frequently used in multidimensional computations to verify the rate of energy release from the overall system. Since these types of models are useful in combustion analysis (after careful model calibration) and are inexpensive in computer resources, they have been widely used in engine research, design, and testing to provide a certain level of information [36]. Thermodynamic models are not the focus of this book. Readers who are interested in these models can refer to a recent book by Caton [37].

Multidimensional models are based on the conservation equations of mass, momentum, energy, and species. They give a full analysis of the gas motion and account for the spatial and temporal variations of the dynamic flow fields and thermodynamic parameters within an engine. They are the extensions of CFD with added physical spray and combustion models applicable to IC engines. These models can predict details of engine processes, and the related wall heat transfer, sprays dynamics, combustion, and emissions production.

The flow field of an IC engine is three-dimensional, unsteady, and turbulent with moving boundaries. It comprises many time and length scales. In addition, small liquid fuel droplets in the flow, which interact with each other and with the gas, complicate the problem even more. It has been estimated that at least 10^{12} grid points are needed to be possible to resolve the flow-field around 10 μm diameter drops within a typical engine chamber domain and that is far beyond the current computer capability. Hence, the strategy used is to introduce sub-models that describe sub-processes on sub-grid scales, and engine processes are simulated by solving the governing equations coupled with the sub-models. The use of sub-models to describe the unsolved physical processes necessarily introduces empiricism into computations. However, the compromise between accuracy and feasibility of computation is justified by the insight that model calculations offer. Confidence in the model predictions and knowledge of their limitations is gained by comparison with experiments. Advances in sub-models when they become available are included to improve the overall prediction accuracy of the integrated model and to add model capability.

The roles of multidimensional engine simulation are:

1. To optimize engine design parameters to achieve multiple optimal function goals in upstream of development process before a prototype is built for test verification. The parameters usually include intake port shape (for volumetric efficiency and gas motion strength), combustion chamber/piston shape, compression ratio, injector position, spray specification, etc.

2. To evaluate the effects of key operating variables on engine functions (torque output, thermal efficiency, emissions, combustion stability, cold-start performance, etc.). The key parameters are engine operating conditions (speed and load), injection strategy (scheme, timing, pressure), valve timing, spark timing, EGR rate, gas motion (e.g., through opening of the butterfly valve in an intake port), etc.

3. To identify key controlling variables that provide guidelines for better organized and hence more efficient experimental development efforts.

4. To provide physical insights to help engineers understand why a design does or does not work.

5. To learn, educate, and communicate more intuitively due to the "see-through" features.

6. To inspire creativity and innovation.

Each of these roles is valuable. However, it is important to note that the multiple tasks listed are often performed at the same time. For example, task 1 and 2 should be performed together for optimal solutions. Comprehensive simulations should be carried out under multiple key operating conditions. When CFD models cannot directly predict an engine function due to the limitation of model accuracy and computer power, a combined deterministic and empirical approach can be used in which CFD modeling predicts an intermediate variable that can be correlated with the engine function empirically (see Sec. 7.1.1).

Engine CFD started in the early 1970s, but engine modeling activities were limited to cold flows and turbulence model development [32,38]. There were a number of factors that limited the activities, which included computer capacity, availability of a general CFD code, and incomplete spray models. The KIVA software [39] developed at the Los Alamos

National Laboratory (LANL) in the United States was released to the public in 1985. The commercial CFD software STAR-CD originated by a research group in the Imperial College London became available in the late 1980s and soon had the second release in the early 1990s [40]. Since KIVA opened its source code, it soon became dominant in the engine research community.

Simulation of diesel sprays in three-dimensional space was not plausible and feasible until the stochastic particle model by Dukowicz [41], instability-based spray atomization and breakup models by Reitz [42,43], and drop dynamic models for collisions and coalescences by O'Rourke [44] were established in the 1980s. These models have provided the framework of modeling sprays in IC engines, and have been proven to be of good computational efficiency and adequate prediction accuracy, and are widely used in engine combustion simulations.

The KIVA family codes were developed by a group of researchers at LANL. The history of KIVA was summarized by Amsden and Amsden [45]. The word KIVA has its origin in southwest United States; it is a Pueblo ceremonial chamber that is usually round and set underground. It is entered from above by means of a ladder through the roof. The analogy is made with a typical engine cylinder, in which the entrance and egress of gases is through valves set in the cylinder head.

The KIVA2 code [46] was released to public in 1989 and soon became the most popular software in the United States, spreading to other countries in a short period of time. KIVA2 added new features including computational efficiency improvements, numerical accuracy improvements, new or improved physical sub-models, and improvements in ease-of-use and versatility. KIVA3 [47], which is an extension of KIVA2, was released in 1993. KIVA3 improved the numerical treatments in KIVA2 by employing a block-structured mesh that eliminates the necessity of maintaining large regions of deactivated cells and results in a reduction of both storage and computer time for complex geometries. The block-structure meshing technique in KIVA3 allows deal with intake flows in realistic intake-port geometries. With some modifications in KIVA3, the code becomes possible to model a diesel engine with vertical valve motions [48] and a gasoline engine with canted valve motions [49].

The moving-valve modeling capability is very important, since it can simulate engine processes in realistic engine geometries with intake flows, allowing people to evaluate the effects of intake-charge motion and engine design parameters on engine combustion performance. The moving-valve capability of KIVA3, together with the improved and new physical models developed at the Engine Research Center (ERC) of the University of Wisconsin-Madison (Madison, Wisconsin), enabled the real application of KIVA in engine R&D activities in the automotive industry in the late 1990s. In response to the need to simulate intake flows, KIVA3V [50], an extended version of KIVA3, was released in 1997, which could model any number of vertical or canted valves in the cylinder head of an engine. Two years later, KIVA3V2 [51] replaced KIAV3V with add-on some features to enhance the robustness, efficiency, and usefulness of the overall program. During the same period, more commercial CFD software capable of modeling internal combustion engines with realistic engine geometries became available.

On the other hand, extensive improvements have been made in physical models, mainly driven by the ERC. These advances included advanced physical models (Renormalization Group (RNG) k- model, large-eddy simulation (LES) model, wall function model, Kelvin–Helmholtz/Rayleigh–Taylor (KH-RT) breakup model, spray impingement model, linearized instability sheet atomization (LISA) model, radius-of-influence (ROI) collision model, evaporation model, characteristic time combustion (CTC), and G-equation combustion models, KIVA–CHEMKIN model, soot model, etc.), code parallelization, and applications for automated engine optimization. The model status at the ERC in 1995 is presented in Table 1.1 [52]. Many of the models, e.g., models for combustion, heat transfer, wall impingement, drop breakup and atomization, soot, and turbulence are still being adopted in some in-house codes and commercial software. As a snapshot of the evolution in model development, an update of the ERC models in 2011 is shown in Table 1.2 [2] in which the progress in drop breakup, detailed kinetic ignition and combustion, and soot and NO_x formation models can be seen.

TABLE 1.1 Status of ERC KIVA2 models in 1995

	Original KIVA	Updated models
Intake flow	Assumed initial flow	KIVA-3 computed intake flow
Combustion	Arrhenius	Laminar-turbulent char time Flamelet model
Unburned hydrocarbon	Arrhenius	Laminar-turbulent char time
NO_x	Extended Zeldo'vich	Extended Zeldo'vich
Heat transfer	Law-of-the-wall	Compressible, unsteady
Wall impinge	None	Rebound-slide with breaking-up drops
Drop drag	Spherical drop	Distorting drops
Drop breakup and atomization	Taylor analogy breakup	Surface-wave-growth
Crevice flow	None	MIT model
Ignition	None	Shell multistep model
Vaporization	Single component	Multicomponent
Soot	None	Soot formation/Nagle-Strickland-Constable oxidation
Turbulence	Standard k-ε	Modified RNG k-ε

TABLE 1.2 Status of ERC KIVA3V models in 2011

Functions	Models
Turbulence	Modified RNG k-ε
Spray development	KH-RT Model, Gas-jet and ROI collision models
Spray impingement	Standard KIVA model
Ignition and combustion	ERC PRF mechanism (n-heptane part, 39 species and 141 reactions including NOx chemistry)
Soot	Two-step model with C_2H_2 as precursor
NOx	Twelve-step kinetics model (extend from GRI 3.0 mechanism embedded in the fuel mechanism)

As the capability of KIVA codes improved, interests in engine CFD modeling from the United States automotive makers increased. In the mid-1990s, there was a need to understand the complicated mixing behaviors in DI gasoline engines. Considering the open-source feature and prediction accuracy of fuel sprays, the KIVA3V code with the ERC models was transferred to the Ford Research Laboratory (FRL) of Ford Motor Company. Based on this, a group of researchers at FRL developed an in-house code called Multidimensional Engine Simulation (MESIM). MESIM is basically an extension of KIVA3V with enhanced capability, which integrates MESIM physics-based 3D solver with commercially available software for mesh generation and data post-processing. By 2004, MESIM had featured with a cutting-edge dynamic mesh algorithm [53,54], state-of-the-art physical models [55–62]. A user-friendly graphic user interface (GUI), the lack of which had made KIVA codes difficult for non-programmers, had also been developed.. MESIM was extensively used in developing a number of DI gasoline engines at FRL. The success of MESIM at FRL got it transferred to Ford's European and Japanese subsidiaries in 2003 [63].

While the early application of multidimensional engine simulation technology in the United States automotive and engine makers helped product R&D in these companies, industrial applications too had a great impact on the development and advancement of the models. Nowadays, engine simulations are routinely used in many engine and automotive companies worldwide, which in turn, have promoted the maturity and prosperity of the simulation technology.

Large-eddy simulations of engine turbulence and combustion in realistic geometry started to ramp up in the early 2000s [64]. Although the Reynolds-averaged Navier–Stokes (RANS) based turbulence models (e.g., k–ε models) have been widely used in industry, they have limited ability to resolve the detailed flow structures. This major issue of RANS simulation can be addressed by LES model. LES can be used to study cycle-to-cycle variability; provide more design sensitivity for investigating both geometrical and operational changes, and produce more detailed and accurate results in engine simulations.

Further development in engine models is anticipated. Again, there is a present need for knock prediction in gasoline engines to aid high compression-ratio engine development. Soot formation models that can predict particulate mass and number as regulated by current legislation are needed. An "universal combustion model" may be of interest that can be applicable for all combustion regimes (e.g., autoignition, premixed, partially premixed, and non-premixed). From a modeling point of view, the key role of a combustion model is to provide the right time and rate at which the reactants in the flow field are converted into products. It is in the reactant-to-product conversion that chemical energy is released in the computational cell(s), leading to the rise of gas temperature and pressure. Thus, the combined CFD hydrodynamics and detailed chemistry at the sub-grid level may be the candidate of the universal combustion model (see Sec. 6.4). As more flow structures are resolved with reduction in the cell sizes and more detailed chemical mechanisms are introduced into the computation, this model may reproduce the combustion phenomena more accurately.

High-fidelity combustion modeling with LES is the direction for the next level of engine simulation. LES can lead to a better understanding of lean combustion instability phenomena and variations of engine outputs that are associated with engine cyclic variability. There are, however, still some issues regarding LES engine simulations. Although computational costs are decreasing due to the rapid advances of computer power, performing LES simulation still requires a lot more time than performing a RANS simulation.

To conclude, IC engines will exist for a long time. Continuously improving the fuel consumption, pollutants and CO_2 emissions of an IC engine is essential. The goal is to achieve zero-impact emissions and high thermal efficiency. High-fidelity combustion models will help engineers to optimize engine design in rapid turn-round times, obtaining the results with high quality and low cost. In the rest of this book, we will review the physical models and modeling methodology for IC engine combustion. We will also discuss some examples of model-driven optimization to achieve improved engine designs in details. These examples will provide insights to the flow and combustion phenomena in an engine, and demonstrate the capabilities of engine combustion CFD simulation.

References

1. Chinese Society of Automotive Engineers. 2020. *Technology Roadmap for Energy Saving and New Energy Vehicles 2.0* [in Chinese]. Beijing: China Machine Press

2. Shi, Y., Ge, H.W., and Reitz, R.D. 2011. *Computational Optimization of Internal Combustion Engines.* London: Springer-Verlag

3. VanDerWege, B.A., Han, Z., Iyer, C.O., Muñoz, R.H., and Yi, J. 2003. "Development And Analysis of a Spray-Guided DISI Combustion System Concept." *SAE Transactions* 112, no. 4: 2135–53. DOI: 10.4271/2003-01-3105

4. Kokjohn, S.L., Hanson, R.M., Splitter, D., and Reitz, R. 2011. "Fuel Reactivity Controlled Compression Ignition (RCCI): A Pathway to Controlled High-Efficiency Clean Combustion." *International Journal of Engine Research* 12, no. 3 (June): 209–26. DOI: 10.1177/1468087411401548

5. Drake, M.C., and Haworth, D.C. 2007. "Advanced Gasoline Engine Development Using Optical Diagnostics And Numerical Modeling." *Proceedings of the Combustion Institute* 31, no. 1: 99–124. DOI: 10.1016/j.proci.2006.08.120

6. Han, Z., Weaver, C., Wooldridge, S., Alger, T., Hilditch, J., McGee, J., Westrate, B., Xu, Z., Yi, J., Chen, X., Trigui, N., and Davis, G. 2004. "Development of a New Light Stratified-Charge DISI Combustion System for a Family of Engines With Upfront CFD Coupling With Thermal And Optical Engine Experiments." *SAE Transactions* 113, no. 3 (March): 269–293. DOI: 10.4271/2004-01-0545

7. Yi, J., Wooldridge, S., Coulson, G., Hilditch, J., Iyer, C.O., Moilanen, P., Papaioannou, G., Reiche, D., Shelby, M., and VanDerWege, B. 2009. "Development And Optimization of the Ford 3.5 L V6 Ecoboost Combustion System." *SAE International Journal of Engines* 2, no. 1 (April): 1388–1407. DOI: 10.4271/2009-01-1494

8. Nakata, K., Nogawa, S., Takahashi, D., Yoshihara, Y., Kumagai, A., and Suzuki, T. 2016. "Engine Technologies for Achieving 45% Thermal Efficiency of SI Engine." SAE International Journal of Engines 9, no. 1 (September): 179–192. DOI: 10.4271/2015-01-1896

9. Hirose, I. 2019. "Our Way Toward the Ideal Internal Combustion Engine for Sustainable Future." Paper presented at the *28th Aachen Colloquium Automobile and Engine Technology, Eurogress Conference Center Aachen, Germany, October 7–9, 2019.*

10. Nagasawa, T., Okura, Y., Yamada, R., Sato, S., Kosaka, H., Yokomori, T., and Iida, N. 2020. "Thermal Efficiency Improvement of Super-Lean Burn Spark Ignition Engine by Stratified Water Insulation on Piston Top Surface." *International Journal of Engine Research* 22, no. 5 (April): 1421–1439. DOI: 10.1177/1468087420908164

11. Splitter, D., Wissink, M., DelVescovo, D., and Reitz, R.D. 2013. "RCCI Engine Operation Towards 60% Thermal Efficiency." SAE Technical Paper 2013-01-0279. DOI: 10.4271/2013-01-0279

12. 2018. "2019 Wards 10 Best Engines." *WARDSAUTO.* https://www.wardsauto.com

13. 2018. "Ten Best Engines" [in Chinese]. *China-Engine.* https://www.china-engine.net

14. Jiang, D., Huang, Z., Wu, D., Gao, J., Liao, S., Jiang, Y., and Zhang, J., eds. 2007. *Combustion of Alternative Fuels in Internal Combustion Engines* [in Chinese]. Xi'an Jiaotong University Press.

15. Han, Z., Wu, Z., Huang, Y., Shi, Y., and Liu, W. [In Press]. "Impact of Natural Gas Fuel Characteristics on the Design And Combustion Performance of a New Light-Duty CNG Engine." *International Journal of Automotive Technology* 22, no. 6.

16. Reitz, R.D., Ogawa, H., Payri, R., Fansler, T., Kokjohn, S., Moriyoshi, Y., Agarwal, A. et al. 2020. "IJER Editorial: The Future of the Internal Combustion Engine." *International Journal of Engine Research* 21, no. 1: 3–10. DOI: 10.1177/1468087419877990

17. Han, Z., Wu, Z., and Gao, X. 2019. "Development Trend of Internal Combustion Engines in the Revolution of Automotive Powertrain" [In Chinese]. *Journal of Automotive Safety and Energy* 10, no. 2: 146–160. DOI: 10.3969/j.issn.1674-8484.2019.02.002

18. Wu, Z., Han, Z., Shi, Y., Liu, W., Zhang, J., Huang, Y., and Meng, S. 2021. "Combustion Optimization for Fuel Economy Improvement of a Dedicated Range-Extender Engine." *Proceedings of the Institution of Mechanical Engineers, Part D: Journal of Automobile Engineering* 235, no. 9: 2525–2539. DOI: 10.1177/0954407021993620

19. Hoffmann, G., Befrui, B., Berndorfer, A., Piock, W.F., and Varble, D.L. 2014. "Fuel System Pressure Increase for Enhanced Performance of GDI Multi-Hole Injection Systems." *SAE International Journal of Engines* 7, no. 1 (April): 519–527. DOI: 10.4271/2014-01-1209

20. Xu, Z., Yi, J., Wooldridge, S., Reiche, D., Curtis, E., and Papaioannou, G. 2009. "Modeling the Cold Start of the Ford 3.5L V6 Ecoboost Engine." *SAE International Journal of Engines* 2, no. 1 (April): 1367–1387. DOI: 10.4271/2009-01-1493

21. Xu, M., Hung, D., Yang, J., and Wu, S. 2015. "Flash-Boiling Spray Behavior And Combustion in a Direct Injection Gasoline Engine." Paper presented at the *Australian Combustion Symposium, Melbourne, December 7–9, 2015.*

22. Dec, J.E. 2014. "Advanced Compression-Ignition Combustion for High Efficiency And Ultra-Low No$_x$ And Soot." In *Encyclopedia of Automotive Engineering*, edited by D. Crolla, D.E. Foster, T. Kobayashi, and N. Vaughan, 1–40. Chichester, West Sussex: John Wiley & Sons

23. Su, W., Zhao, H., Wang, J., and Huang, H. 2010. *Theory And Technology of HCCI Low-Temperature Combustion Engines* [in Chinese]. Beijing: Science Press.

24. Reitz, R.D. 2013. "Directions in Internal Combustion Engine Research." *Combustion and Flame* 160, no. 1: 1–8. DOI: 10.1016/j.combustflame.2012.11.002

25. Reitz, R.D., and Duraisamy, G. 2015. "Review of High Efficiency And Clean Reactivity Controlled Compression Ignition (RCCI) Combustion in Internal Combustion Engines." *Progress in Energy and Combustion Science* 46: 12–71. DOI: 10.1016/j.pecs.2014.05.003

26. Paykani, A., Garcia, A., Shahbakhti, M., Rahnama, P., and Reitz, R.D. 2021. "Reactivity Controlled Compression Ignition Engine: Pathways Towards Commercial Viability." *Applied Energy* 282: article 116174. DOI: 10.1016/j.apenergy.2020.116174

27. Lu, X., Han, D., and Huang, Z. 2011. "Fuel Design And Management for The Control of Advanced Compression-Ignition Combustion Modes." *Progress in Energy and Combustion Science* 37, no. 6 (December): 741–783. DOI: 10.1016/j.pecs.2011.03.003

28. Fouts, L., Fioroni, G.M., Christensen, E., Ratcliff, M.A., McCormick, R.L., Zigler, B.T., Sluder, C.S., et al. 2018. *Properties of Co-Optima Core Research Gasolines* (NREL/TP-5400-71341). Golden, Colorado: National Renewable Energy Laboratory.

29. Szybist, J.P., Busch, S., McCormick, R.L., Pihl, J.A., Splitter, D.A., Ratcliff, M.A., Kolodziej, C.P., et al. 2021. "What Fuel Properties Enable Higher Thermal Efficiency in Spark-Ignited Engines?" *Progress in Energy and Combustion Science* 82 (January): Article 100876. DOI: 10.1016/j.pecs.2020.100876

30. Han, Z., and Reitz, R.D. 1998. "Seeing Reduced Diesel Emissions." *Mechanical Engineering* 120, no. 1 (January): 62–63. DOI: 10.1115/1.1998-JAN-2

31. Hershey, R.L., Eberhardt, J.E., and Hottel, H.C. 1936. "Thermodynamic Properties of the Working Fluid in Internal-Combustion Engines." *SAE Transactions* 31 (January): 409–424. DOI: 10.4271/360140

32. Mattavi, J.N., and Amann, C.A. 1980. *Combustion Modeling in Reciprocating Engines*. New York/London: Plenum Press.

33. Krieger, R., and Borman, G. 1966. "The Computation of Apparent Heat Release for Internal Combustion Engines." *ASME Paper*. Article 66-WA/DGP-4.

34. Heywood, J.B., Higgins, J.M., Watts, P.A., and Tabaczynski, R.J. 1979. "Development And Use of a Cycle Simulation to Predict SI Engine Efficiency And No$_x$ Emissions." SAE Technical Paper No. 790291. DOI: 10.4271/790291

35. Blumberg, P.N., Lavoie, G.A., and Tabaczynski, R.J. 1979. "Phenomenological Models for Reciprocating Internal Combustion Engines." *Progress in Energy and Combustion Science* 5, no. 2: 123–167. DOI: 10.1016/0360-1285(79)90015-7

36. Davis, G.C., and Tabaczynski, R.J. 1988. "The Effect of Inlet Velocity Distribution And Magnitude on In-Cylinder Turbulence Intensity And Burn Rate—Model Versus Experiment." *Journal of Engineering for Gas Turbines and Power* 110, no. 3 (July): 509–514. DOI: 10.1115/1.3240164

37. Caton, J.A. 2016. *An Introduction to Thermodynamic Cycle Simulations for Internal Combustion Engines.* Chichester: Wiley.

38. El Tahry, S.H. 1983. "K-Epsilon Equation for Compressible Reciprocating Engine Flows." *Journal of Energy* 7: 345–353. DOI: 10.2514/3.48086

39. Amsden, A.A., Ramshaw, J.D., O'Rourke, P.J., and Dukowicz, J.K. 1985. *KIVA: A Computer Program for Two- And Three-Dimensional Fluid Flows With Chemical Reactions And Fuel Sprays* (LA-10245-MS). Los Alamos, New Mexico: Los Alamos National Laboratory.

40. CD-Adapto. 1993. *STAR-CD Version 2.2 Manual.* London, United Kingdom: Computational Dynamics Ltd.

41. Dukowicz, J.K. 1980. "A Particle-Fluid Numerical Model for Liquid Sprays." *Journal of Computational Physics* 35, no. 2: 229–253. DOI: 10.1016/0021-9991(80)90087-X

42. Reitz, R.D. 1978. Atomization And Other Breakup Regimes of a Liquid Jet (Publication No. 7907964). Princeton, New Jersey: Princeton University. ProQuest Dissertations and Theses.

43. Reitz, R.D. 1987. "Modeling Atomization Processes in High-Pressure Vaporizing Sprays." *Atomisation Spray Technology* 3, no. 4: 309–337.

44. O'Rourke, P.J. 1981. Collective Drop Effects on Vaporizing Liquid Sprays (Publication No. 8203229). Princeton, New Jersey: Princeton University. ProQuest Dissertations and Theses.

45. Amsden, D.C., and Amsden, A.A. 1993. "The KIVA Story: A Paradigm of Technology Transfer." *IEEE Transactions on Professional Communication* 36, no. 4: 190–195. DOI: 10.1109/47.259956

46. Amsden, A.A., O'Rourke, P.J., and Butler, T.D. 1989. *KIVA-II: A Computer Program for Chemically Reactive Flows With Sprays* (LA-11560-MS). Los Alamos, New Mexico: Los Alamos National Laboratory.

47. Amsden, A.A. 1993. *KIVA-3: A KIVA Program With Block-Structured Mesh for Complex Geometries* (LA-12503-MS). Los Alamos, New Mexico: Los Alamos National Laboratory.

48. Hessel, R.P. 1993. Numerical Simulation of Valved Intake Port And In-Cylinder Flows Using KIVA3 (Publication No. 9318621). Madison, Wisconsin: University of Wisconsin-Madison]. ProQuest Dissertations and Theses.

49. Han, Z. 1996. Numerical Study of Air-Fuel Mixing In Direct-Injection Spark-Ignition And Diesel Engines (Publication No. 9634937). Madison, Wisconsin: University of Wisconsin-Madison. ProQuest Dissertations and Theses.

50. Amsden, A.A. 1997. *KIVA-3V: A Block-Structured KIVA Program for Engines With Vertical or Canted Valves* (LA-13313-MS). Los Alamos, New Mexico: Los Alamos National Laboratory.

51. Amsden, A.A. 1999. *KIVA-3V, Release 2: Improvements to KIVA-3V* (LA-13608-MS). Los Alamos, New Mexico: Los Alamos National Laboratory.

52. Reitz, R.D., and Rutland, C.J. 1995. Development And Testing of Diesel Engine CFD Models. *Progress in Energy and Combustion Science* 21, no. 2: 173–196. DOI: 10.1016/0360-1285(95)00003-Z

53. Yi, J., Han, Z., Yang, J., Anderson, R., Trigui, N., and Boussarsar, R. 2000. "Modeling of the Interaction of Intake Flow And Fuel Spray in DISI Engines." SAE Technical Paper No. 2000-01-0656. DOI: 10.4271/2000-01-0656

54. Yi, J., Han, Z., and Trigui, N. (2002). "Fuel-Air Mixing Homogeneity And Performance Improvements of a Stratified-Charge DISI Combustion System." *SAE Transactions* 111, no. 4 (October): 965–975. DOI: 10.4271/2002-01-2656

55. Han, Z., and Reitz, R.D. 1995. "Turbulence Modeling of Internal Combustion Engines Using RNG K-E Models." *Combustion Science and Technology* 106, no. 4–6: 267–295. DOI: 10.1080/00102209508907782

56. Han, Z., and Reitz, R.D. 1997. "A Temperature Wall Function Formulation for Variable-Density Turbulent Flows With Application to Engine Convective Heat Transfer Modeling." *International Journal of Heat and Mass Transfer* 40, no. 3 (February): 613–625. DOI: 10.1016/0017-9310(96)00117-2

57. Han, Z., Xu, Z., and Trigui, N. 2000. "Spray/Wall Interaction Models for Multidimensional Engine Simulation." *International Journal of Engine Research* 1, no. 1: 127–146. DOI: 10.1243/1468087001545308

58. Han, Z., Xu, Z., Wooldridge, S.T., Yi, J., and Lavoie, G. 2001. "Modeling of DISI Engine Sprays With Comparison to Experimental In-Cylinder Spray Images." *SAE Transactions* 110, no. 3 (September): 2376–2386. DOI: 10.4271/2001-01-3667

59. Zeng, Y., and Han, Z. 2001. *Implementation of Multicomponent Droplet And Film Vaporization Models Into The KIVA-3V Code* (SRR-2001-0165). Dearborn, Michigan: Ford Research Laboratory.

60. Hilditch, J., Han, Z., and Chea, T. 2003. "Unburned Hydrocarbon Emissions From Stratified Charge Direct Injection Engines." SAE Technical Paper No. 2003-01-3099. DOI: 10.4271/2003-01-3099

61. Han, Z., and Xu, Z. 2004. "Wall Film Dynamics Modeling for Impinging Sprays in Engines." SAE Technical Paper No. 2004-01-0099. DOI: 10.4271/2004-01-0099

62. Iyer, C.O., Han, Z., and Yi, J. 2004. "CFD Modeling of a Vortex Induced Stratification Combustion (VISC) System." SAE Technical Paper No. 2004-01-0550. DOI: 10.4271/2004-01-0550

63. Han, Z., Yi, J., Hilditch, J., Xu, Z., Iyer, C., Fan, L., Curtis, E., and Davis, G. 2003. *Lecture Notes at the MESIM Technology Transfer Workshop*. Dearborn, Michigan.

64. Rutland, C.J. 2011. "Large-Eddy Simulations for Internal Combustion Engines – A Review." *International Journal of Engine Research* 12, no. 5: 421–451. DOI: 10.1177/1468087411407248

2

Combustion Basis of Internal Combustion Engines

An internal combustion (IC) engine is a thermal device that produces mechanical power from the chemical energy contained in the fuel. Fuel is mixed with air in an IC engine to form a flammable mixture and combustion occurs to release the chemical energy from the fuel. Based on the differences in fuel property, IC engines can be classified as diesel engines that burn diesel fuel and gasoline engines that burn gasoline fuel. The processes of mixture formation and subsequent ignition and combustion are very different in these two types of engines.

Low-temperature combustion (LTC) concepts take advantage of gasoline and diesel engines so that can dramatically reduce pollutant emissions and increase thermal efficiency at the same time. The knowledge of thermodynamics and the basis of mixture formation and combustion in engines are very useful in understanding the computer modeling technology discussed in this book. In this chapter, engine thermodynamic analysis and mechanisms to improve engine thermal efficiency and torque output is presented. The features of mixture formation and combustion in gasoline and diesel engines are then discussed in two separate sections. Concepts of advanced LTC are overviewed. This chapter will be an introduction to these subjects, outlining the basic and important aspects. Systematic descriptions of IC engine principles can be found in a few recommended books [1,2,3].

2.1 Thermodynamic Analysis

The processes taking place in an IC engine obey the laws of thermodynamics. Based on the boundaries of the studied system, equations of mass and energy conservation are formulated differently. In thermodynamic analysis, some important assumptions are made, including:

1. Thermodynamic properties (e.g., pressure and temperature) vary only with time and are spatially uniform in the studied system.

2. The fuel is instantaneously and completely vaporized and mixed with the air.

3. The mixture contents obey the ideal gas law.

4. The blow-by is neglected.

In the engine system, after the intake valves are closed, applying the first law to the in-cylinder system with pressure p, temperature T, mass m, and volume V, the energy balance equation is given as:

$$\frac{dU}{dt} = \frac{dQ}{dt} + m_f h_f - p\frac{dV}{dt}$$

(2.1)

The term dU/dt is the change rate of sensible internal energy U; and $p\,dV/dt$ is the rate of work done by the system due to system boundary displacement; $m_f h_f$ is the sensible enthalpy of the injected fuel mass m_f; Q is the difference between the chemical energy Q_f or "heat" released by combustion of the fuel and the heat transfer Q_w to the cylinder walls. Since the specific enthalpy $h_f \approx 0$, we have

$$\frac{dQ}{dt} = \frac{dQ_f}{dt} - \frac{dQ_w}{dt} = p\frac{dV}{dt} + \frac{dU}{dt}$$

(2.2)

The term dQ/dt is also called the net heat release rate (or apparent heat release rate), dQ_f/dt is the gross heat release rate. With ideal gas assumption, then

$$\frac{dQ}{dt} = p\frac{dV}{dt} + mc_v\frac{dT}{dt}$$

(2.3)

From the ideal gas law, $pV = mRT$, R is the gas constant. Since the mass of the in-cylinder mixture, m, remains the same before the exhaust vales are open, it follows that

$$\frac{1}{p}\frac{dp}{dt} + \frac{1}{V}\frac{dV}{dt} = \frac{1}{T}\frac{dT}{dt}$$

(2.4)

Then, the heat release rate from the fuel is given as:

$$\frac{dQ_f}{dt} = \left(1 + \frac{c_v}{R}\right)p\frac{dV}{dt} + \frac{c_v}{R}V\frac{dp}{dt} + \frac{dQ_w}{dt}$$

(2.5)

or

$$\frac{dQ_f}{dt} = \frac{k}{k-1} p \frac{dV}{dt} + \frac{1}{k-1} V \frac{dp}{dt} + \frac{dQ_w}{dt} \qquad (2.6)$$

Here k is the ratio of specific heat c_p/c_v. To evaluate energy balance on the base of engine crank angle (CA), φ, for a four-stroke engine $\varphi = 6nt$, n is engine speed. Equation (2.6) becomes

$$\frac{dQ_f}{d\varphi} = \frac{k}{k-1} p \frac{dV}{d\varphi} + \frac{1}{k-1} V \frac{dp}{d\varphi} + \frac{dQ_w}{d\varphi} \qquad (2.7)$$

where the change of the cylinder volume $dV/d\varphi$ can be obtained from engine kinetics once the engine structure and speed are given. Heat transfer $dQ_w/d\varphi$ can be calculated using an empirical convective correlation, e.g., Woschni's correlation [4].

Note that cylinder pressure, p, is unknown in Equation (2.7). It can be measured in an engine by using a pressure transducer or it can be calculated with the use of computer modeling. CFD-based multi-dimensional combustion simulation can provide a solution with spatial and temporal variations to reflect the non-uniformity reality (see Chapter 6). Other simplified thermodynamic models including non- and quasi-dimensional models can be also employed [5].

If we define the fuel mass burned fraction $x_b = m_{fb}/m_{fc}$, and $Q_{fb} = m_{fb} Q_{LHV}$, where Q_{LHV} is the fuel's lower heating value, m_{fb} is the fuel mass burned (a function of crank angle), and m_{fc} is the cyclic total fuel mass, we then obtain the change rate of the fuel mass burned fraction as:

$$\frac{dx_b}{d\varphi} = \frac{1}{m_{fc} Q_{LHV}} \left(\frac{k}{k-1} p \frac{dV}{d\varphi} + \frac{1}{k-1} V \frac{dp}{d\varphi} + \frac{dQ_w}{d\varphi} \right) \qquad (2.8)$$

The instantaneous fuel mass burned fraction as a function of φ can be calculated by integrating the above equation as:

$$x_b(\varphi) = \frac{1}{m_{fc} Q_{LHV}} \int_{\varphi_0}^{\varphi} \left(\frac{k}{k-1} p \frac{dV}{d\varphi} + \frac{1}{k-1} V \frac{dp}{d\varphi} + \frac{dQ_w}{d\varphi} \right) d\varphi \qquad (2.9)$$

here φ_0 is the crank angle at which combustion starts.

The fuel mass burned fraction $x_b(\varphi)$ is very useful to characterize the combustion process as shown schematically in Figure 2.1. For instance, several parameters can be defined to represent some important combustion concepts for a spark ignition (SI) engine:

1. Flame-development angle $\Delta\varphi_d$ or $\Delta\varphi_{10}$: The crank-angle interval between the spark discharge and the time when a small but significant fraction of the cylinder mass has burned or fuel chemical energy has been released. Usually, this fraction is 10%.

FIGURE 2.1 Definition of various combustion angles on mass fraction burned versus crank angle.

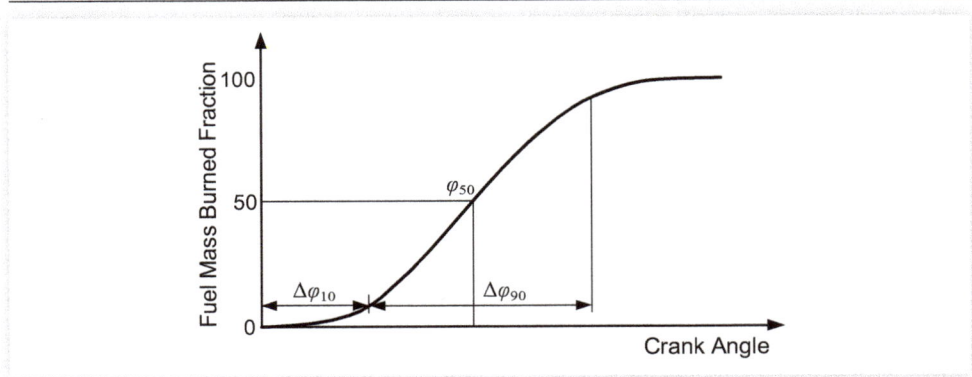

Courtesy of Zhiyu Han

2. Rapid-burning angle $\Delta\varphi_b$ or $\Delta\varphi_{90}$: The crank-angle interval required to burn the bulk of the charge. It is defined as the interval between the end of the flame-development stage and the end of the flame-propagation process (usually mass fraction burned or energy-release fraction of 90%).

3. Overall burning angle $\Delta\varphi_o$: The duration of the overall burning process. It is the sum of $\Delta\varphi_d$, $\Delta\varphi_b$, and the duration of the last 10% fuel burned.

4. φ_{50} angle: The crank angle at which 50% of the in-cylinder fuel has burned. It is used to characterize the phasing of the combustion process and determine a combustion retard parameter. For a wide range of engines and their operating conditions, φ_{50} at maximum brake torque (MBT) timing occurs at 5° to 7° after top dead center (ATDC).

In general, for a combustion system with a main combustion chamber and a pre-chamber (e.g., swirl chamber in an in-direct injection diesel engine and pre-chamber in a gasoline engine), the fuel burned mass fraction in the pre-chamber can be derived [6].

Based on energy conservation, the fuel heat release rate in each chamber can be given as:

$$\frac{dQ_{fi}}{d\varphi} = \frac{dU_i}{d\varphi} + \frac{dH}{d\varphi} + P_i \frac{dV_i}{d\varphi} + \frac{dQ_{wi}}{d\varphi} \qquad (2.10)$$

where $i = 1$ for the main chamber, $i = 2$ for the pre-chamber. H is the enthalpy of the gas flowing through the passage connecting the main chamber and pre-chamber. The mass change rate in each chamber due to combustion and the flow passing through the passage is

$$\frac{dm_i}{d\varphi} = m_{fc} \frac{dx_{bi}}{d\varphi} - (-1)^{i+\delta} \frac{dm}{d\varphi} \qquad (2.11)$$

and the gas temperature change rate in each chamber is

$$\frac{dT_i}{d\varphi} = T_i \left(\frac{1}{p_i} \frac{dp_i}{d\varphi} + \frac{1}{V_i} \frac{dV_i}{d\varphi} - \frac{1}{m_i} \frac{dm_i}{d\varphi} \right) \tag{2.12}$$

The mass flow rate of the gas passing through the passage is given as:

$$\frac{dm}{d\varphi} = \begin{cases} \dfrac{\mu F_p}{6n} \left(\dfrac{p_j m_j}{V_j} \right)^{0.5} \sqrt{\dfrac{2k}{k+1} \left(\dfrac{2}{k+1} \right)^{\frac{2}{k+1}}} & \pi \le \left(\dfrac{2}{k+1} \right)^{\frac{k}{k-1}} \\[4ex] \dfrac{\mu F_p}{6n} \left(\dfrac{p_j m_j}{V_j} \right)^{0.5} \sqrt{\dfrac{2k}{k+1} \left(\pi^{\frac{2}{k}} - \pi^{\frac{k+1}{k}} \right)} & \pi > \left(\dfrac{2}{k+1} \right)^{\frac{k}{k-1}} \end{cases} \tag{2.13}$$

where μ is the discharge coefficient of the passage connecting the two chambers, F_p is the section area of the connecting passage, and

$$\begin{cases} p_1 > p_2 : \delta = 1, & j = 1, & \pi = \dfrac{p_2}{p_1} \\[2ex] p_2 > p_1 : \delta = 0, & j = 2, & \pi = \dfrac{p_1}{p_2} \end{cases} \tag{2.14}$$

Thus, the fuel mass burned fraction in each chamber can be derived as:

$$\frac{dx_i}{d\varphi} = \frac{(-1)^{i+\delta}[c_{pj}T_j + (A_i - B_i)T_i]\dfrac{dm}{d\varphi} + \left(1 + \dfrac{A_i}{R}\right)\dfrac{p_i dV_i}{d\varphi} + \dfrac{A_i V_i}{R}\dfrac{dp_i}{d\varphi} + \dfrac{dQ_{wi}}{d\varphi}}{m_{fc}[C_i + (A_i - B_i)T_i]} \tag{2.15}$$

where A_i, B_i, C_i are functions of gas properties varying with gas temperature, which are given by Zhou et al. [6]. With these equations and measured pressure in each chamber, the fuel burned rate at each chamber can be obtained. Note that the discharge coefficient, μ, of the passage can be given as an empirical value obtained from a steady-state flow test. However, it was argued that μ is not a constant and a method to calculate it based on the measured pressure of the motoring engine was given by Zhou et al. [7].

The specific fuel consumption of an engine is defined as the fuel flow rate per unit power output with the unit of g/kWh. It measures how efficiently an engine is using the fuel supplied to produce work at a specific operating condition. Engine thermal efficiency or fuel conversion efficiency is defined as:

$$\eta = \frac{P}{\dot{m}_f Q_{LHV}} \tag{2.16}$$

where P is the engine power, $\dot{m}_f$ is fuel mass flow rate. It is clear seen that engine thermal efficiency is a normalized parameter (i.e., independent of engine size). P and $\dot{m}_f$ can be precisely measured in engine tests or can be obtained from engine computer simulations. Note that the engine thermal efficiency is also related to the fuel's heating value. Thus, when comparing the engine thermal efficiency of different engines or of engines with different fuels, the heating value of the fuel used must be clearly given.

The term "thermal efficiency" is preferred in this book as it can be related directly to the indicated thermal efficiency η_i from thermodynamic cycle analysis. Considering the combustion efficiency η_c and mechanical efficiency η_m, which includes the pumping loss of an engine, we can derive that

$$\eta = \eta_i \cdot \eta_c \cdot \eta_m \qquad (2.17)$$

For a constant-volume cycle (also called Otto Cycle), which is the ideal cycle for an SI engine, the indicated thermal efficiency is given as:

$$\eta_i = 1 - \frac{1}{\varepsilon^{k-1}} \qquad (2.18)$$

where ε is the geometric compression ratio (CR), which is defined as the ratio of the cylinder volume when the piston at the bottom dead center (BDC) and the cylinder volume when the piston at the top dead center (TDC) for a reciprocating engine. From Equations (2.17) and (2.18), it becomes very clear that the strategy to improve the thermal efficiency or fuel consumption of an SI engine include:

1. Increase of the geometric compression ratio ε so as to increase the indicated thermal efficiency η_i, although this can be limited by engine knock combustion and has an adverse effect on frictions.

2. Reduction of heat transfer through the cylinder walls to increase the specific ratio k so as to increase the indicated thermal efficiency η_i, e.g., to adopt lean combustion for higher k.

3. Improvement of combustion completeness so as to raise the combustion efficiency η_c.

4. Reduction of pumping loss so as to increase the mechanical efficiency η_m.

5. Reduction of the friction loss of moving parts (e.g., piston rings, bearings, valves, etc.) so as to increase the mechanical efficiency η_m.

6. Reduction of the energy consumption of the accessory parts (e.g., water pump, oil pump, etc.) so as to increase the mechanical efficiency η_m.

Combustion efficiency η_c is the fraction of the fuel energy supplied that is released in the combustion process. Figure 2.2 shows combustion efficiency varies with the fuel/air equivalence ratio [1]. For SI engines, when the mixtures are at lean equivalence ratios not far from the stoichiometric ratio, the combustion efficiency is usually in the range of 95% to 98%. For mixtures richer than stoichiometric, lack of oxygen prevents complete

FIGURE 2.2 Variation of engine combustion efficiency with fuel/air equivalence ratio.

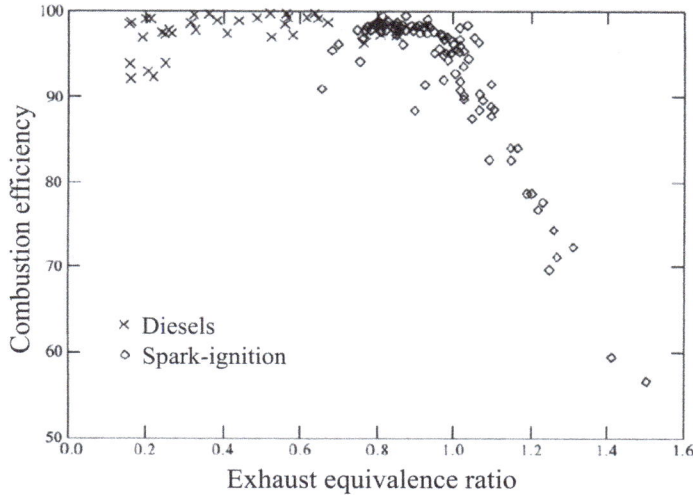

combustion of the fuel carbon and hydrogen, and the combustion efficiency steadily decreases as the mixture becomes richer. Combustion efficiency is little affected by other engine operating and design variables provided the engine combustion process remains stable. For diesel engines, which always operate lean, the combustion efficiency is about 98%.

Although the mechanisms for improving the thermal efficiency discussed earlier are for an SI engine, they are also applicable to a diesel engine and other alternative-fuel engines. In a diesel engine, pumping loss is generally low due to its unthrottled operations. Since a diesel engine has a higher CR and higher specific heat ratio due to lean combustion, it is 20–30% more efficient than an SI engine.

Similarly, the brake mean effective pressure (BMEP), which is a relative engine performance measure can be given as:

$$BMEP = \eta_v \cdot \eta_i \cdot \eta_c \cdot \eta_m \cdot \frac{1}{A/F} \cdot \frac{p_a}{RT_a} \cdot Q_{LHV} \qquad (2.19)$$

where η_v is the volumetric efficiency which measures the effectiveness of an engine's induction process and is affected by intake port design details; A/F is the air–fuel ratio; p_a and T_a are the pressure and temperature of the air at the inlets (intake ports). This relationship shows that any mechanism that improves thermal efficiency will also improve engine performance. Other important mechanisms to improve engine performance include:

1. Raising volumetric efficiency and intake air density to maximize amount of air inducted into the cylinder (e.g., super- or turbo-charging).

   2. Achieving minimum air–fuel ratio that can be burned efficiently to fully utilize induced air.

All technologies to improve the fuel economy and performance of an engine fall fundamentally into one or more mean(s) discussed above. It is important to note that many technologies improve the thermal efficiency of an engine based on the same or similar fundamentals. For example, VVT, VVL, EGR, DI stratified-charge operation, and cylinder deactivation can reduce pumping loss of an SI engine operated at part loads. However, the pumping-loss reduction benefits from these technologies when applied to an engine simultaneously cannot be added together.

In an overexpansion case, the thermal efficiency is a direct function of the expansion ratio (ER) rather than CR if the former is greater than the latter in an SI gasoline engine. The effective CR is defined as the ratio of the cylinder volume when the piston is at the position at the intake valve closure (IVC) time and the cylinder volume when the piston is at TDC. The effective CR is the one affecting gas compression in a practical SI engine. The ER can be increased with the use of the Atkinson cycle in which the volumetric change during the compression process is less than that during the expansion process. It can be achieved in a conventional four-stroke cycle engine by suitable choice of exhaust valve opening and intake valve closing positions relative to BDC. If the crank angle between exhaust valve opening and BDC on the expansion stroke is less than the crank angle between BDC and intake valve closing on the compression stroke, then the actual geometric ER is greater than the actual geometric CR.

When the IVC is retarded, the effective CR reduces quickly. Figure 2.3 gives an example of the effective CR versus IVC of a 1.5-liter 4-cylinder SI engine. The reduced effective CRs from different geometric CRs converge when the IVC is overly retarded. In the example in Figure 2.3, the maximum difference among the geometric CRs is 6 units; it becomes less than 2 units when the IVC is retarded to 120° after bottom dead center (ABDC). It is the effective CR that is limited by knock combustion. Thus, retard IVC is used to reduce the effective CR to avoid knock combustion.

The indicated thermal efficiency of the Atkinson cycle is given as:

$$\eta_i = 1 - \frac{1}{\varepsilon_e^{k-1}} + \frac{c_v T_{IVC}}{Q^*} \left(\frac{\varepsilon_e}{\varepsilon}\right)^{1-k} \left[k\left(\frac{\varepsilon_e}{\varepsilon}\right)^{k-1} - (k-1)\left(\frac{\varepsilon_e}{\varepsilon}\right)^{k} - 1 \right] \qquad (2.20)$$

where ε_e is the expansion ratio, Q^* is the heat added to the system per unit mass of working fluid. It is seen that the sum of first two terms in Equation (2.20) is the same as that given in Equation (2.18) if the ER is equal to the CR, since the third term is always greater than or equal to zero when ε_e and ε are the same, it is an addition to the indicated thermal efficiency of the conventional constant-volume cycle. The calculated indicated thermal efficiency of the same 1.5-liter SI engine versus the effective CR and geometric CR (the same as ER for the Atkinson cycle) is shown in Figure 2.4. As can be seen, it is the ER that can improve the indicated thermal efficiency, and when the ER is fixed, the change in the effective CR by adjusting the IVC does very little to alter the indicated thermal efficiency. This means that, in the Atkinson cycle with a high geometric CR, a retarded IVC needs to be adopted to avoid knock combustion. The engine's indicated thermal efficiency does not deteriorate with the retarded IVC since the ER remains the same.

FIGURE 2.3 Effects of IVC on effective compression ratio.

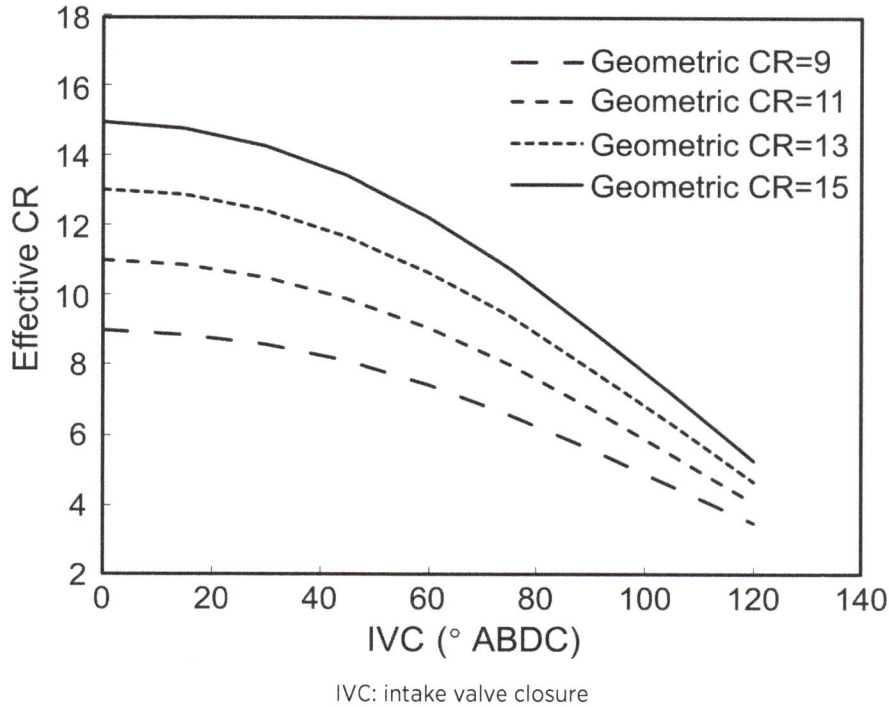

IVC: intake valve closure

FIGURE 2.4 Effects of expansion ratio (ER) and effective compression ratio (CR) on indicated thermal efficiency.

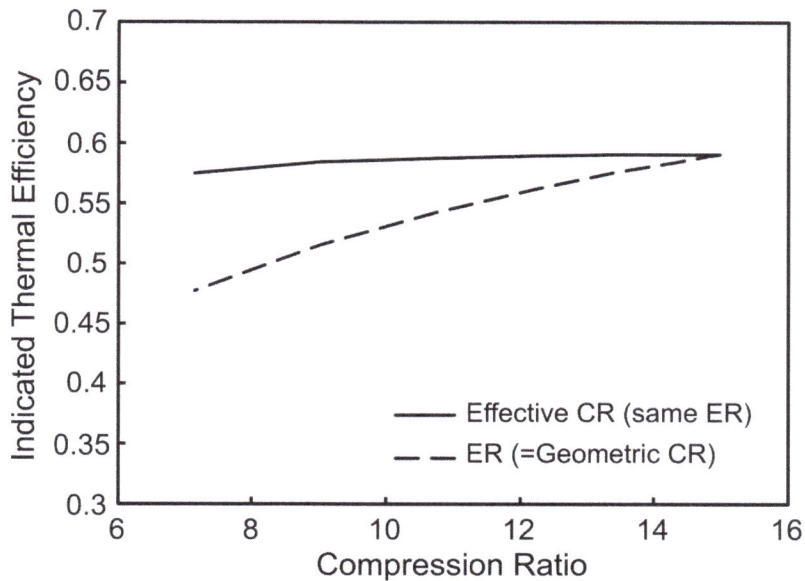

Unfortunately, the Atkinson cycle causes deterioration in engine performance since the inducted charge is pushed back into the intake ports by the piston due to the late IVC. This is a fundamental disadvantage of the Atkinson cycle. The indicated mean effective pressure (IMEP) of the Atkinson cycle can be evaluated as:

$$IMEP = \eta_i \cdot p_{IVC} \left(\frac{Q^\star}{c_v T_{IVC}} \right) \left(\frac{1}{k-1} \right) \left(\frac{\varepsilon_e}{\varepsilon} - \frac{1}{\varepsilon} \right)^{-1} \tag{2.21}$$

Since the quantity in the last bracket is always less than unit, the greater the ER, the lesser the IMEP.

2.2 Mixture Formation and Combustion in Spark-Ignition Gasoline Engines

In most conventional SI engines, combustion is initiated by a spark after the air–fuel mixing process completes and flammable mixture forms. Thus, mixture formation and combustion can be managed in two separated processes in analysis or practice. There are two approaches to form mixture in SI engines. One is to deliver fuel into the intake ports, which then mixes with air in the ports. Another one is to directly inject fuel into the cylinder. In the former approach, electronically-controlled PFI replaced carburetors in the late 1970s, in which injectors mounted on the intake manifold of the engine inject fuel in the intake ports. Usually, one injector is used for one cylinder.

In a PFI engine, fuel is injected under the injection pressure of about 0.3 MPa into the intake ports when the intake valves are either closed (close-valve injection (CVI)) or open (open-valve injection (OVI)). These processes are illustrated in Figure 2.5 for both CVI and OVI, under 20°C engine conditions [8]. In the case of CVI, fuel sprays are targeted on the intake-valve back surfaces where the temperature is the highest in the intake system. Fuel spray droplets and films formed on the valve/port surfaces evaporate and the fuel vapors mix with the surrounding air. When the intake valves are open, the air–fuel mixture and remaining liquid fuel enter the cylinder. The evaporation and mixing processes continue during the induction and compression stroke, and the charge is well mixed before spark ignition occurs near the end of compression. At the early stage of the intake-valve opening, high-speed in-cylinder gases first flow out of the cylinder into the ports that peel the fuel films or puddles off the metal surfaces. This back-flow phenomenon helps the mixing substantially. In the case of OVI, some fuel droplets penetrate through the valve-openings and enter the cylinder, and some of these droplets impinge on the cylinder head and liner surfaces due to the combined effects of spray momentum and gas motion momentum.

For emission-control purpose, air–fuel mixture is prepared at stoichiometric ratio in a PFI SI engine, which is not difficult at engine warm-up conditions. However, during engine cold-start operations, since the engine is cold, fuel evaporation becomes difficult and only the volatile components of gasoline fuel vaporize. Hence, excessive fuels are injected to form the ignitable mixture. The problems of mixture formation at cold-start lead to the very high engine-out hydrocarbon and CO emissions that contribute the most in an emission test cycle. Fuel enrichment is also used at high to full loads to maximize torque output, prevent knocking and reduce exhaust temperature.

FIGURE 2.5 CFD computed images to show (a) close-valve-injection (CVI), and (b) open-valve-injection (OVI) in a PFI gasoline engine. Fuel injection starts during the expansion stroke for CVI and during the intake stroke for OVI. Images are shown in a transparent and cut-away geometry fashion, and fuel droplets are represented by the particles.

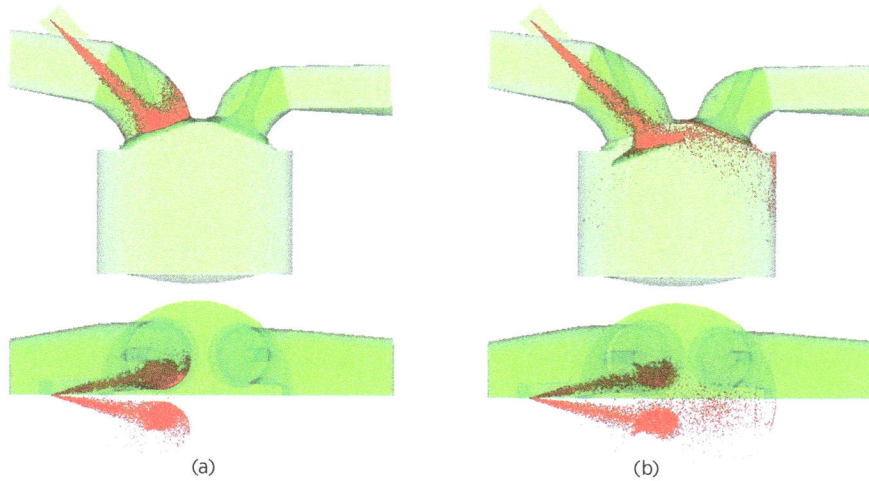

(a)   (b)

CFD: computational fluid dynamics; PFI: port fuel injection

It was not until the late 1990s that many DI gasoline concepts were implemented in mass production [9–13]; although several trials of mass production of DI gasoline engines had been undertaken and terminated prior to this. Advances in fuel injection technology and R&D tools (CFD modeling and optical diagnostics) enabled the development and implementation of modern DI engines.

In DI engines, gasoline is directly injected into the cylinder to form desired air–fuel mixture. DI combustion systems can be broadly divided into homogeneous-charge DI (HCDI) and combined stratified/homogeneous charge or stratified-charge DI (SCDI). In HCDI systems, fuel injection takes place during the induction stroke to form stoichiometric homogeneous mixture before spark ignition. These systems have demonstrated engine performance advantages over conventional PFI due to improved volumetric efficiency and knock tolerance as a thermodynamic consequence of charge cooling due to fuel vaporization inside the cylinder [14]. An increased CR can be also used to improve engine full-load output and part-load fuel consumption.

In SCDI systems, homogeneous charge operation is used at high engine loads. At part loads, stratified-charge operations are used in which a locally-rich mixture cloud is formed in the vicinity of the spray plug by fuel injection during the late compression stroke. SCDI engines have demonstrated the advantages over conventional PFI part-load fuel consumption improvement due primarily to pumping reduction, CR increase, and overall lean operation.

SCDI systems may be further divided into wall-guided, air-guided, and spray-guided; wall-guided SCDI can be further divided into standard tumble, reverse tumble, or swirl, and spray-guided SCDI into close or far spacing [15]. These may be further divided based on injection technology and chamber design. While these nomenclatures have been created to describe the main air–fuel mixture formation features of these systems in a rather simple way, they reflect the complexity of DI systems and the steady evolution of injection technology.

FIGURE 2.6   Layout of a wall-guided stratified-charge DI combustion system.

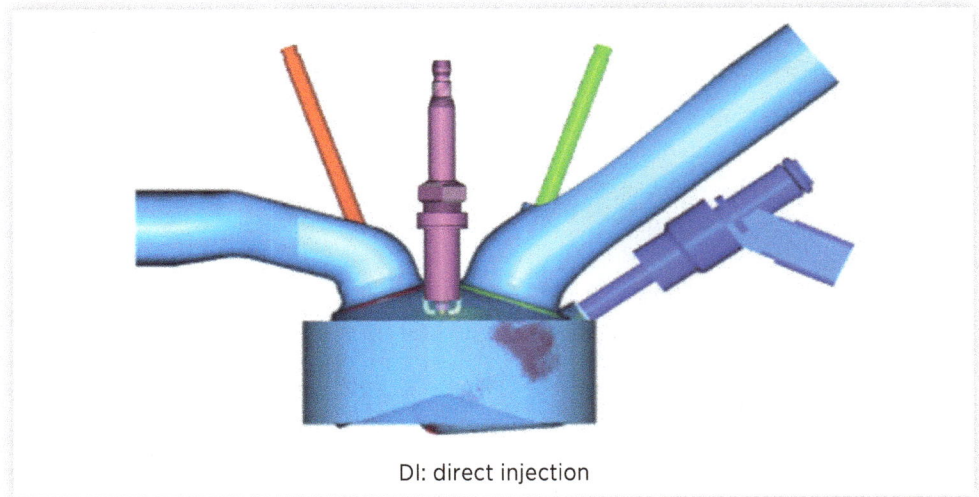

DI: direct injection

The layout of a light stratified-charge (LSC) DI combustion system [16] is shown in Figure 2.6 as an example of wall-guided systems. The system uses a high-pressure swirl injector, which is mounted below the intake ports, a centrally located spark plug, and a shallow bowl-in-piston combustion chamber. This system demonstrated an improvement of 10.1% in vehicle fuel economy compared to the baseline PFI engine over the NEDC cycle. Sprays are injected into the shallow bowl on top of the piston. Fuel droplets and vapor, as well as the entrained air flow move toward the piston bowl, and the mixture cloud is redirected by the wall of the piston bowl and spray-induced flow.

Figure 2.7 illustrates the stratified-mixture formation process by the evolution of computed fuel spray, air–fuel ratio distribution, and gas vector fields in a wall-guided SCDI engine [17]. The spray generates strong vortex gas motion due to air entrainment and the spray-generated vortex flow moves the air–fuel mixture cloud towards the spark plug. The spark plug is located near the center of the chamber with its gap protruded into the chamber. When fuel is injected into the cylinder at the start of injection (SOI) of 70° before top dead center (BTDC) in this case, liquid fuel vaporizes. Very rich mixtures form inside the spray and the mixtures become leaner at the periphery of the spray region. As the spray impinges on the piston bowl, a structure is formed such that the rich mixtures are formed near the wall and the lean mixtures are at the periphery far from the wall. The gases near the wall are continuously replenished with fuel vapors due to vaporization of the fuel films on the wall and maintain a low air–fuel ratio. At the mixture cloud periphery far from the wall, the vapor continuously mixes out with fresh air as a result of convection and diffusion. At the late stage of compression, the vortex gas motion moves the air–fuel mixture cloud toward the spark plug region. To ensure stable ignition, the design must be optimized so that the mixture air–fuel ratio near the spark plug gap region is within flammability limits (e.g., from 9 to 23) at the time of ignition.

In a spray-guided SCDI, the fuel injector is centrally located close to the spark plug in the engine cylinder head. A wide-angle fuel spray is injected during the late compression stroke. Locally rich mixtures are formed around the spray. The spark plug gap is placed at

FIGURE 2.7 Formation of stratified mixture in a wall-guided DI engine as demonstrated by (a) CFD computed evolution of spray; (b) air–fuel ratio distribution; (c) gas velocity vector fields. Time sequence from the top row: 55°, 45°, 35°, and 25° BTDC.

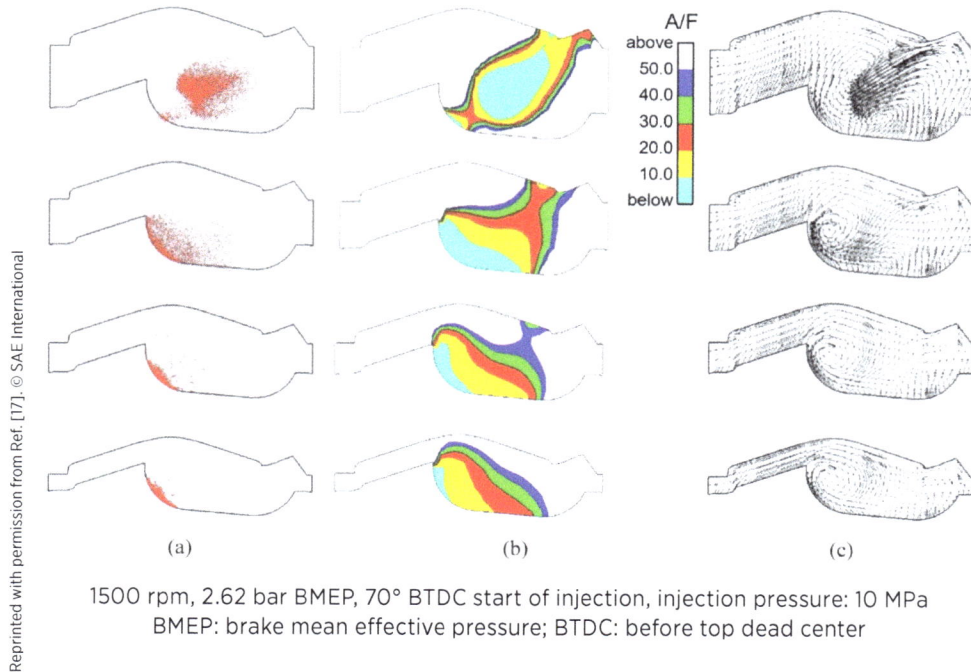

1500 rpm, 2.62 bar BMEP, 70° BTDC start of injection, injection pressure: 10 MPa
BMEP: brake mean effective pressure; BTDC: before top dead center

the periphery of the spray where flammable mixtures present. An example of spray-guided SCDI systems is shown in Figure 2.8. This particular DI concept is called vortex induced stratification combustion (VISC) system [18,19]. It uses an outward-opening pintle injector. The spark plug gap is located at the periphery of the spray where flow recirculation is induced by the spray. A very shallow piston crown design enables the VISC concept, which is used to maintain the mixture cloud without dissipation and hold it near the spark plug. With the piston crown, the fuel mixture can stay longer at the spark plug and the spark can be initiated as many as ten or more crank angle degrees after the end of injection (EOI). More details of VISC are described in Sec. 7.6.

CFD computed air–fuel ratio distributions and superimposed spray and gas velocity vector fields are shown in Figure 2.9 to demonstrate mixture formation in the spray-guided VISC engine [19]. In this case, an 80° spray is injected when the piston is near the TDC. At 24° BTDC, fuel injection ends. It is seen that as the spray penetrates the chamber environment, it vaporizes quickly in the hot air environment, and large-scale vortices are generated at the tip of the spray due to the viscous shear between the high velocity spray droplets and the low velocity surrounding gas. The gas vortices promote air entrainment and thus mix air with fuel vapors. At 16° BTDC, flammable mixture cloud is formed around the spark plug gap, which is ready to be ignited.

FIGURE 2.8 Layout of the VISC system as an example of spray-guided SCDI systems.

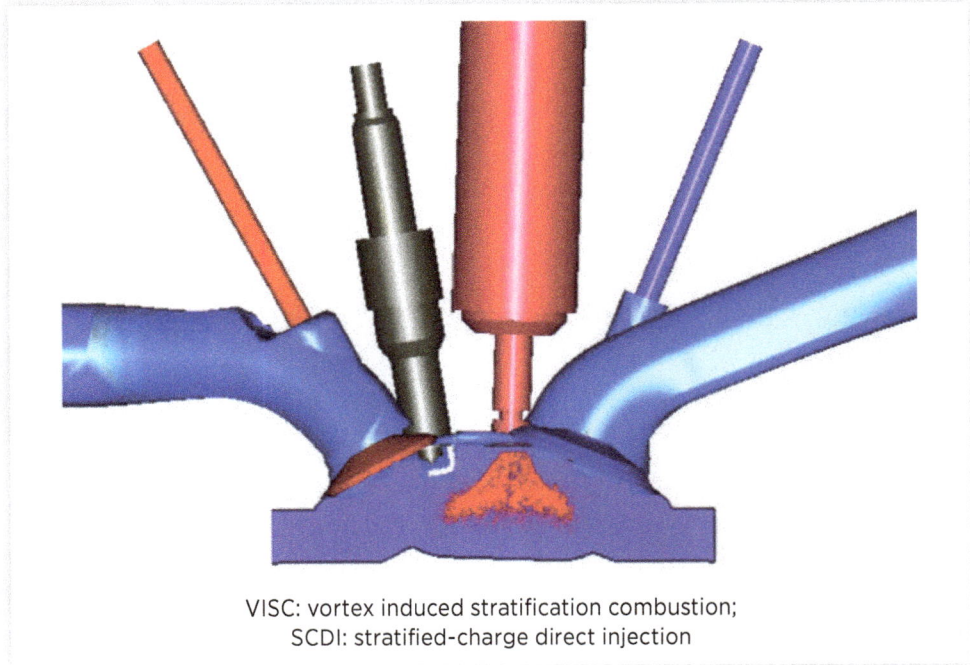

VISC: vortex induced stratification combustion;
SCDI: stratified-charge direct injection

Reprinted with permission from Ref. [19]. © SAE International

FIGURE 2.9 Formation of stratified mixture in a spray-guided DI engine as demonstrated by (a) CFD computed air–fuel ratio distributions and (b) superimposed spray and gas velocity vector fields. Time sequence from the top row: 24°, 22°, and 16° BTDC. The location of the spark plug gap is marked by the cross.

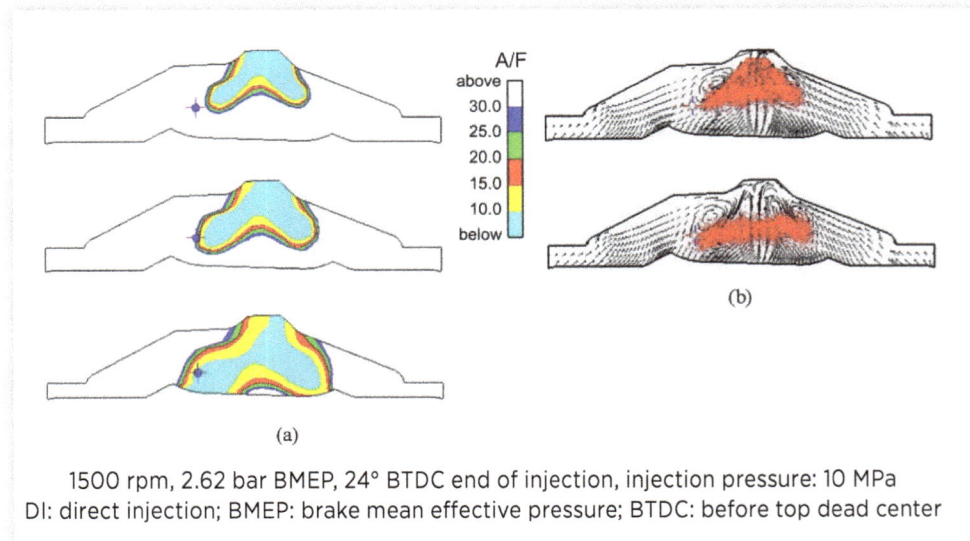

1500 rpm, 2.62 bar BMEP, 24° BTDC end of injection, injection pressure: 10 MPa
DI: direct injection; BMEP: brake mean effective pressure; BTDC: before top dead center

Adapted with permission from Ref. [19]. © SAE International

Precise control over the fuel injection and air–fuel mixing process to form the desired in-cylinder mixture is one of the key factors for a successful DI gasoline engine design (see Secs. 7.3 to 7.6). Injection technology that can deliver suitable fuel sprays is the critical enabler. Generally, a wide spray angle and small spray droplets are desired. Pressure-swirl injectors were used in the DI engines launched in the late 1990s, which could produce 60°~80° hollow-cone sprays due to liquid sheet atomization (see Sec. 5.2.3). The Sauter mean diameter (SMD) of the droplets is about 20 mm under 10 MPa injection pressure. Outward-opening pintle injectors as used in the VISC DI engine can produce 90° hollow-cone sprays with an SMD of about 14 mm under 10 MPa injection pressure. In recent years, multi-hole injectors have been widely adopted [20]. Driven by the reduction of pollutant emissions, particularly soot emissions, the injection pressure has been escalated from 10 MPa to 35 MPa in production engines in the past 20 years, aiming to give greater multiple-injection flexibility and reduce wall wettings. The main advantage of multi-hole injectors is the enhanced flexibility in the geometry and spatial orientation of the nozzle holes so that it is possible to design the orientation of individual plumes of the sprays to better fit the combustion chamber setups. Sprays images from these three types of injectors taken under room conditions are shown in Figure 2.10. The overall structure and details are quite different.

FIGURE 2.10 Comparison of sprays generated by (a) a multi-hole injector; (b) an outward-opening pintle injector; (c) a pressure-swirl injector.

(a)   (b)   (c)

Reprinted with permission from Ref. [19]. © Springer Nature

As the injection pressure is increased, smaller spray droplets are atomized. The SMD of the spray droplets from multi-hole injectors decreases from about 20 μm to 13 μm, and then to 9 μm as the injection pressure is raised from 10 MPa to 20 MPa, and then to 35 MPa under room conditions. This change in droplet size has a significant impact on drop vaporization. Since the vaporization time of a liquid droplet is approximately proportional to the square of its diameter (see Sec. 5.4), the smaller droplets need much less time to vaporize, which means there is more time available for mixing before spark ignition and hence the flexibility of late injections and multiple injections is greater.

In conventional PFI SI engines, combustion is initiated by the electrical discharge of the spark plug (see Sec. 6.6.1). Ignition timing is an important variable for controlling the combustion of SI engines. It influences engine performance, fuel consumption, and exhaust gas emissions. It takes up to approximately 2 msec from the time of ignition until the time of maximum combustion rate. Therefore, spark advance must be set according to engine speed, load, air–fuel ratio, etc. For high combustion efficiency, it is necessary that approximately 50% of the fuel is consumed shortly (5°~7° CA) after TDC. In this way, an approximation to the ideal constant volume combustion is achievable.

After ignition has been initiated, the flame starts to propagate outward from the ignition kernel through the combustion chamber (see Sec. 6.3). The flame front is distorted by the shape of the combustion chamber and the in-cylinder flows. In addition, the temperature distribution in the combustion chamber, the temperatures of the combustion chamber walls, and the homogeneity of the air–fuel mixture during combustion influence the contour of the flame front during its propagation. Figure 2.11 shows the flame propagation of an individual combustion cycle in an SI engine operating at 800 rpm, measured with the laser-aided high-speed imaging technique [22]. The intake flow was seeded with silicone-oil droplets as markers to identify burned gas regions. Spark ignition starts at 16 crank angle degree (cad) BTDC and a flame kernel is clearly visible (the dark area) at the spark plug gap location 6 cad later. The flame propagates away from the spark gap with an irregular front, and by 16 cad it passes out the observation window.

FIGURE 2.11 Images illustrating the flame propagation in an SI engine.

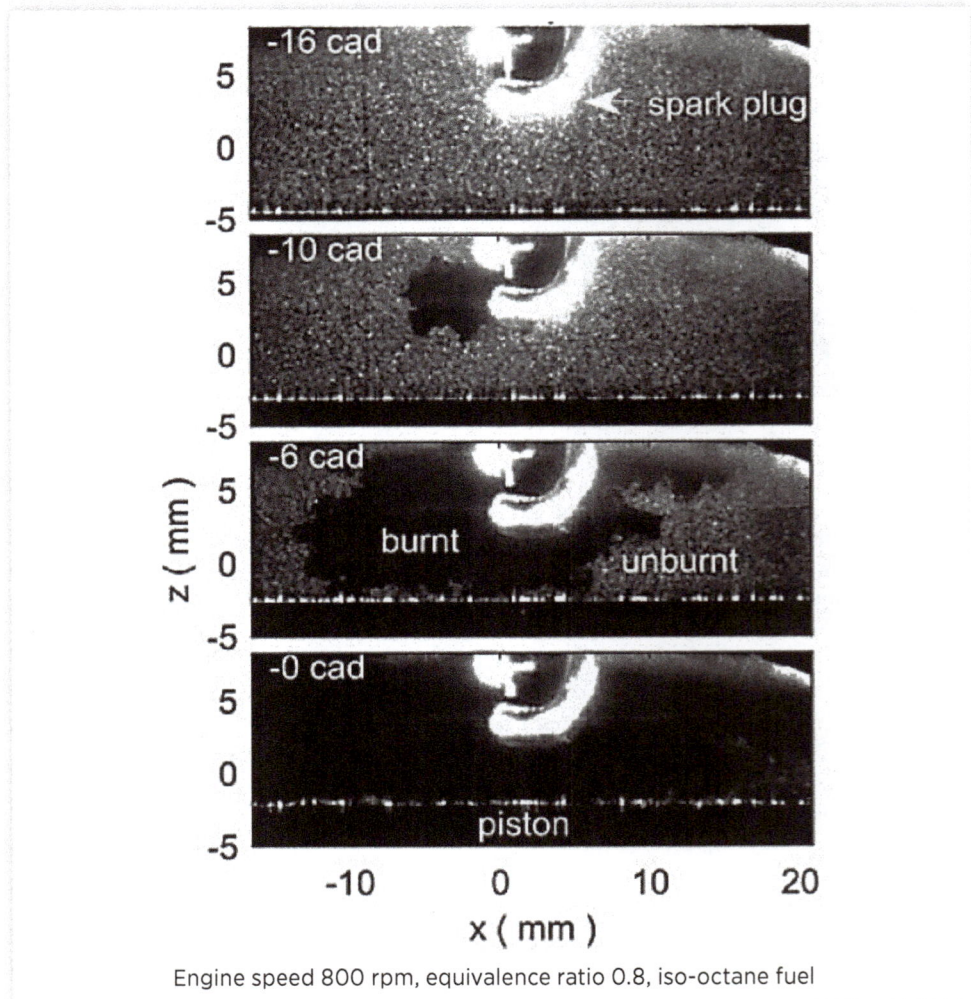

Engine speed 800 rpm, equivalence ratio 0.8, iso-octane fuel

Thermal efficiency and power output can both be improved by increasing the CR of SI engines. However, for high engine loads, achieved (for example) by boosting or downsizing SI engines, the increased CR can lead to engine knocking (abnormal combustion) and then to severe engine damage. It has been known that knocking combustion is a result of auto-ignition of end-gas pockets in the region of unburned air–fuel mixture in front of the propagating flame front during the late phase of normal combustion. Owing to the local auto-ignition in the end-gas region, pressure waves are induced and the amplitude of the pressure waves depends on the pressure, temperature, mixture composition, and the volume of the pockets in the end gas where auto-ignition occurs. The pressure waves propagate through the combustion chamber and are reflected at the combustion chamber walls. The characteristic pressure oscillations can be recognized in the measured pressure traces. Normal knock combustion can be mitigated by retarding spark ignition timing with fuel consumption penalty as it is well reproducible and, thus, controllable.

In some highly turbocharged DI engines, a more extreme type of knocking combustion can occur, especially under low-speed and high-load conditions. This phenomenon is referred to as super-knock or mega-knock. It is characterized by pressure amplitudes that are significantly higher than in the case of normal knock. In addition, super-knock occurs stochastically so it is not controllable in the same manner as a normal knock. It is generally accepted that super-knock originates from pre-ignition [1,23].

Fuel consumption reduction in an SI engine depends highly on mitigating knocking. Many methods have been demonstrated to be effective including retarding spark timing, charge enrichment, cooled EGR, turbulence enhancement in the combustion chamber, and multiple injections in DI engines. Eliminating pre-ignition is an effective approach to suppress super-knock. The main origins of pre-ignition include oil droplets and deposit particles. The subject of knock combustion has been investigated probably since the birth of gasoline engines, and there has been a large number of achievements [24,25].

In SCDI engines, combustion is usually organized in such a way that homogeneous-charge and stratified-charge combustion mode switches under certain engine load and speed conditions to take advantage of each mode. Figure 2.12 illustrates conceptually the engine operational modes on the engine load and speed map for the LSC DI engine as shown in Figure 2.6. The relative air/fuel ratio λ (fuel lean mixture $\lambda > 1$, fuel rich mixture $\lambda < 1$) is shown in each region. The engine operates at stoichiometric homogeneous-charge mode at high to full loads for performance. The engine runs at stoichiometric homogeneous-charge mode with high EGR in the middle load and speed range for fuel economy benefits. At low load and speed, the engine runs at stratified-charge mode to further improve fuel economy. The LSC DI concept uses a reduced stratified-charge operation window ranging from the idle operation to low speed, light load. The reduced stratified-charge window allows a significant relaxation in the requirements for the lean aftertreatment system, but still enables significant fuel economy gains over non-stratified-charge operation.

Combustion in a homogeneous-charge DI engine (or combustion mode) behaves similarly to that in a PFI engine, which is characterized as premixed flame propagation. However, stratified-charge combustion leads to increased unburned hydrocarbon and soot emissions caused by local rich combustion and pool fire from fuel film due to spray wall impingement [17,26,27]. Minimizing spray wall impingement can significantly reduce hydrocarbon and soot emissions in stratified-charge combustion and in engine cold-start operation. Increasing injection pressure can reduce soot emissions, since both the mean size (SMD) and the size to measure large drops (D_{V90}) of the fuel sprays are lowered [28]. The lowered spray SMD and D_{V90} are beneficial for evaporation and mixing, and hence good for diminishing local-rich combustion.

FIGURE 2.12 Operation map of a stratified-charge DI engine.

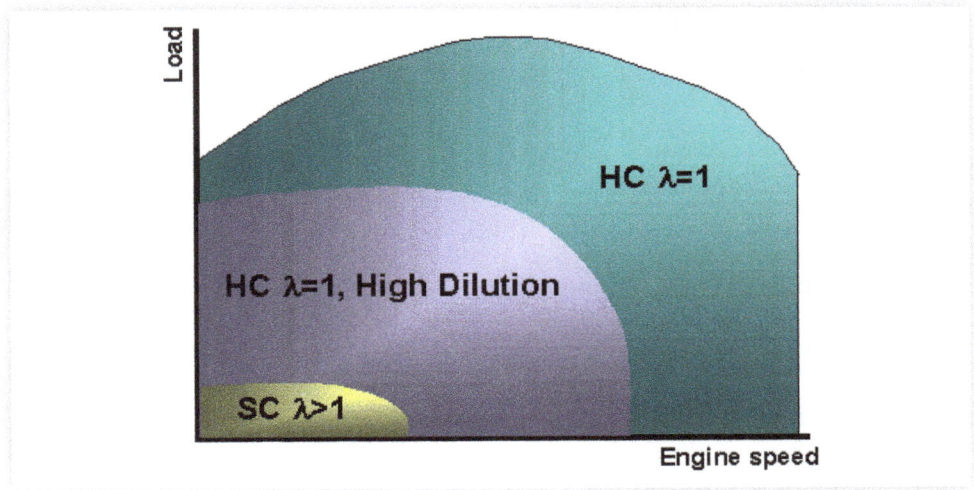

FIGURE 2.13 Comparison of fuel consumption of a PFI, wall-guided DI, and spray-guided VISC system over a mini-map.

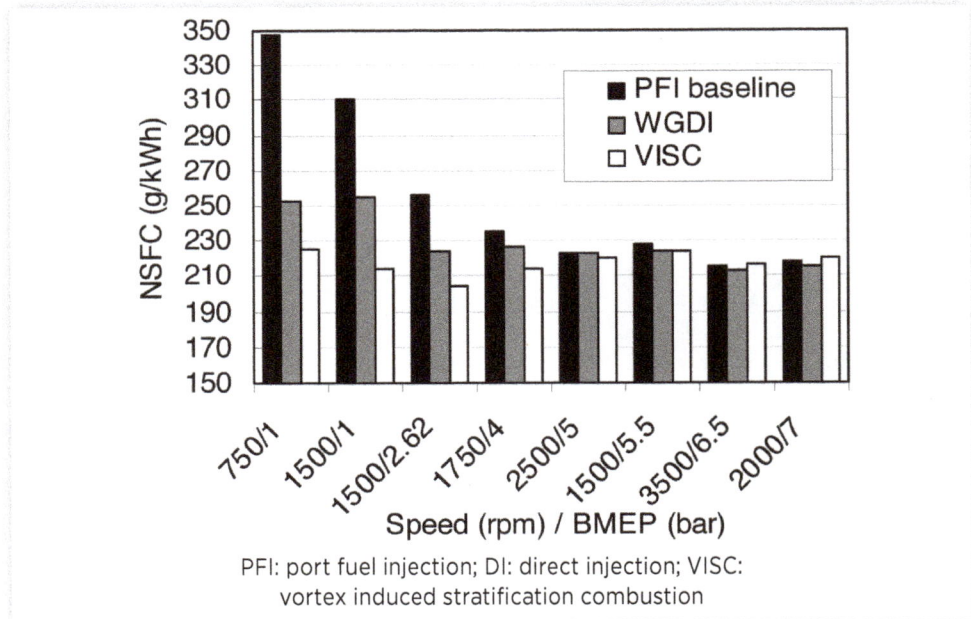

PFI: port fuel injection; DI: direct injection; VISC: vortex induced stratification combustion

It is interesting to compare the fuel consumption of PFI, wall-guided DI, and spray-guided DI engines, although a fair comparison is often difficult due to a lack of data on the same base. Such a comparison was made by VanDerWege et al. [18] with the data from the same engine configuration under the same operating conditions. Figure 2.13 shows the net specific fuel consumption (NSFC) at eight engine operation points measured in a 0.5-liter single-cylinder engine for a PFI combustion system with independent-Variable Camshaft

Timing (i-VCT), a wall-guided DI system with i-VCT, and the spray-guided VISC system with fixed cam timing. The CR for the PFI, wall-guided DI, and VISC is 10.5, 11.0, and 10.6, respectively. NSFC is calculated based on the net indicated mean effective pressure, which is the full-cycle (720°) integration of the measured cylinder pressure trace, and NSFC accounts for the pumping work by definition.

For the first three points from left in Figure 2.13, both the wall-guided DI and VISC are running in stratified-charge mode. The improvements of wall-guided DI over PFI and of VISC over wall-guided DI are obvious and are mainly due to reduction of pumping and heat transfer losses with fuel-lean stratified-charge combustion. The additional benefits of VISC over wall-guided DI mainly come from the improved combustion efficiency due to reduced hydrocarbon and CO emissions, and improved combustion phasing in the VISC system. At the fourth and fifth points, only the VISC is operating stratified as it can operate stratified at higher loads than wall-guided DI. As the load increases, the benefit for the stratified charge becomes smaller. For the last three points, all three systems are stoichiometric, homogeneous charges. As expected, the fuel consumption results are very similar. Vehicle fuel economy with the use of the three engines was also projected by modeling, but not including transient effects such as cold-start and NO_x-trap purges. The results show a roughly 18% reduction in fuel consumption for VISC compared with PFI, and a 6% reduction compared with wall-guided DI over a vehicle test cycle.

2.3 Combustion in Diesel Engines

Unlike in an SI engine, mixture formation and combustion phenomena cannot be separated in a conventional diesel engine. Diesel sprays resulting from jet atomization (see Sec. 5.2.2) are injected from several nozzle holes of the injector into the combustion chamber towards the end of the compression stroke, just before the desired SOC. The liquid fuel sprays penetrate into the high-temperature, high-pressure air in the combustion chamber. The fuel vaporizes and the fuel vapor mixes with the air. In a few crank-angle-degree delays after fuel injection, auto-ignition spontaneously occurs at the peripheries of the sprays that initialize combustion (see Sec. 6.6.2). The consequent high-pressure compresses the unburned portion of the charge and shortens the delay before ignition for the air–fuel mixture, which then burns rapidly. Injection continues until the desired amount of fuel has entered the cylinder. Atomization, vaporization, air–fuel mixing, and combustion continue until essentially all the fuel has passed through each process. In addition, the air remaining in the cylinder continues to mix with the burning and already burned gases throughout the combustion and expansion processes.

Diesel combustion is termed non-premixed combustion or diffusion combustion because the gaseous fuel and the air must diffuse into each other and then react. Since combustion in diesel engines is initialized by auto-ignition, it is also called compression ignition combustion. For much of the diesel combustion process, the fuel burning rate is controlled by the rate at which fuel mixes with the air. Spontaneous ignition, resulting from the high air temperature and pressure in the cylinder when injection commences, and the fact that air–fuel mixing controls the burning rate, are the distinguishing features of the diesel combustion process. As described, the compression-ignition combustion process is extremely complex. The details of the process depend on the characteristics of the fuel, on the design of the combustion chamber and fuel-injection system, and the engine's operating conditions. It is an unsteady and heterogeneous combustion process with complex chemistry. However, the fact that mixing controls the reaction rate can be used to model the combustion process (see Sec. 6.2).

There is no knock limit in a diesel engine as in an SI engine resulting from spontaneous ignition of the premixed fuel and air in the end-gas. Hence, a higher engine CR can be used in the compression-ignition engine, improving its thermal efficiency relative to an SI engine. The nature of heterogeneous diffusion combustion in diesel engines leads to the formation of excessive amounts of soot, which is discharged out of the engine as black smoke. High-temperature reactions also result in excessive NO_x emissions.

Figure 2.14 illustrates the auto-ignition and subsequent diffusive flame process. The time-resolved images of the natural light emission are taken from an optically accessible DI heavy-duty diesel engine with the use of the chemiluminescence imaging technique [29]. The engine has a bore of 139.7 mm, a combustion chamber diameter of 97.8 mm, and a CR of 16. The images show diesel auto-ignition to be a progressive process that occurs simultaneously over the downstream region of all the fuel jets. The first chemiluminescence is detected shortly after the start of injection (1.0° to 2.5° after the start of injection (ASI)). And it grows brighter and shifts downstream as the jet penetrates across the chamber until soot luminosity dominates the emission. The liquid-phase fuel reaches a maximum penetration of about 24 to 25 mm from the injector (along the nominal fuel-jet axis) at 3.5° to 4.0° ASI. At this point, enough hot in-cylinder air has been entrained into the jet to vaporize all the fuel. The liquid length then remains fairly constant as the vapor-phase fuel and air mixture penetrates across the chamber. Later, after the premixed burn is well underway, the diffusion flame forms (at about 6.5° ASI), and the liquid length becomes about 4–5 mm shorter, apparently due to localized heating by the diffusion flame.

The early stages of weak chemiluminescence are primarily the result of contributions from formaldehyde (CH_2O) and benzene (CH) emissions, corresponding to the "cool-flame" phenomena in rich mixtures. And the leading regions of the diesel sprays contain a relatively well-mixed rich vapor–air mixture in the fuel–air equivalence ratio range of 2–4. The much brighter emission images at 4° to 4.5° ASI result from soot particle radiation. By 5.0° to 5.5° ASI, the emission from the leading portion of the jet (downstream of the liquid fuel) is much brighter indicating that soot formation is beginning in this region. This indicates that exothermic chemical reactions are already well underway in the large fuel vapor region in the leading portion of all the jets; the fuel vapor has broken down and polyaromatic hydrocarbons (PAHs) have formed throughout the entire jet cross-section in the main chemiluminescent region downstream of the liquid fuel, indicating that hot-combustion, premixed-burn reactions have already begun in the main jet by the time the soot spots appear.

In the final two images, at 6.5° to 7° ASI, the luminous emission becomes significantly brighter as soot concentrations increase. These images show the development of diffusion flames around the individual sprays. These diffusion flames, located at the spray boundaries with a very rich mixture consisting of fuel vapor, partially reacted fuel molecules, PAHs, soot particles, CO, and H_2 on the inside of the flame and air on the outside, constitute the second mixing-controlled phase of diesel combustion in which most of the heat releases out.

Research processes have led to a conceptual model of DI diesel combustion. Based on the work of Dec [30], a more complete understanding of diesel combustion, soot, and NO_x formation is illustrated in Figure 2.15. When liquid fuel is injected into the cylinder, it atomizes and the resulted sprays rapidly enter the hot air or mixture. A liquid length exists, beyond which the fuel vapor–air mixture continues to entrain hot gases, and eventually undergoes a premixed combustion process, generating a large amount of soot precursors (PAHs and acetylene (C_2H_2)) in the hot products of the fuel-rich combustion. These soot precursors go on to form soot particles. Meanwhile, a diffusion flame surrounds the hot products and extends upstream. This liquid length determines the reach of the fuel spray

FIGURE 2.14   Temporal sequence of natural flame emission images to show auto-ignition and flames in a diesel engine. The injector location is shown by the small white spot. The field of view is 74 mm by 55 mm.

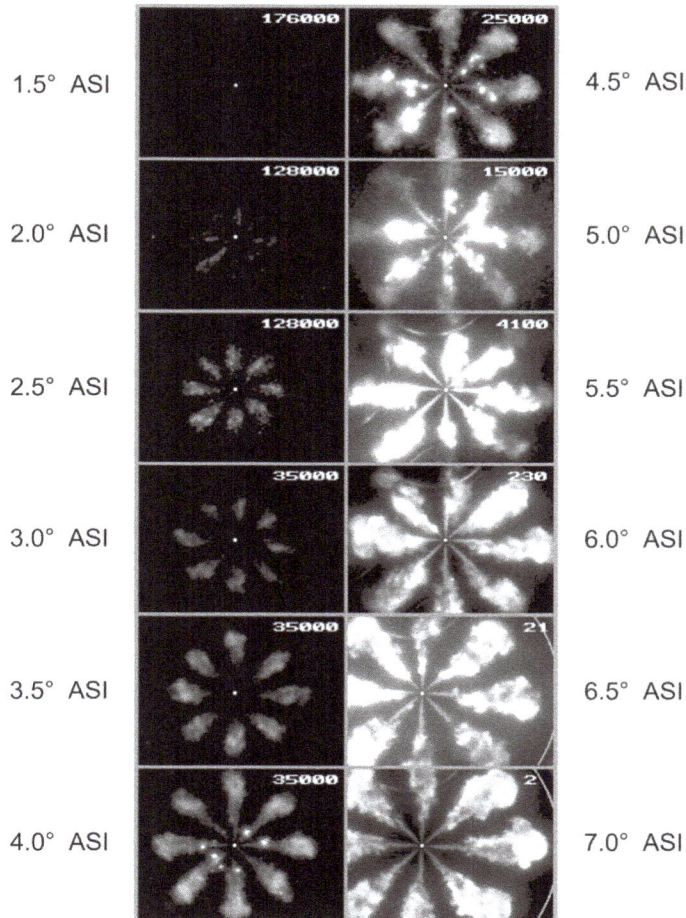

1200 rpm, intake temperature: 160°C, pressure: 206 kPa,
water and lubricant oil temperature: 95°C, 11.5°
BTDC SOI, peak injection pressure 68 MPa. BTDC:
before top dead center; SOI: start of injection

jet and, hence, the mixture entrainment. Beyond that, diffusion combustion continues at the outer region of the combustion zone. Soot formation continues in the downstream regions where oxygen concentration is low. Since the temperature and oxygen concentration are high in the diffusion flame regions, NO_x formation occurs mainly in these regions. In summary, the fuel and air first react in a fuel-rich mixture, leading to soot formation; this rich mixture then burns out in a high-temperature diffusion flame at the jet periphery, leading to NO_x formation. Note that this conceptual illustration is based on quasi-steady diesel combustion in large-bore heavy-duty diesel engines, and the picture may change

FIGURE 2.15 A conceptual diesel combustion model to show key features of spray jet liquid length, premixed flames, soot formation region, diffusion-flame location, and high NO$_x$ concentration locations.

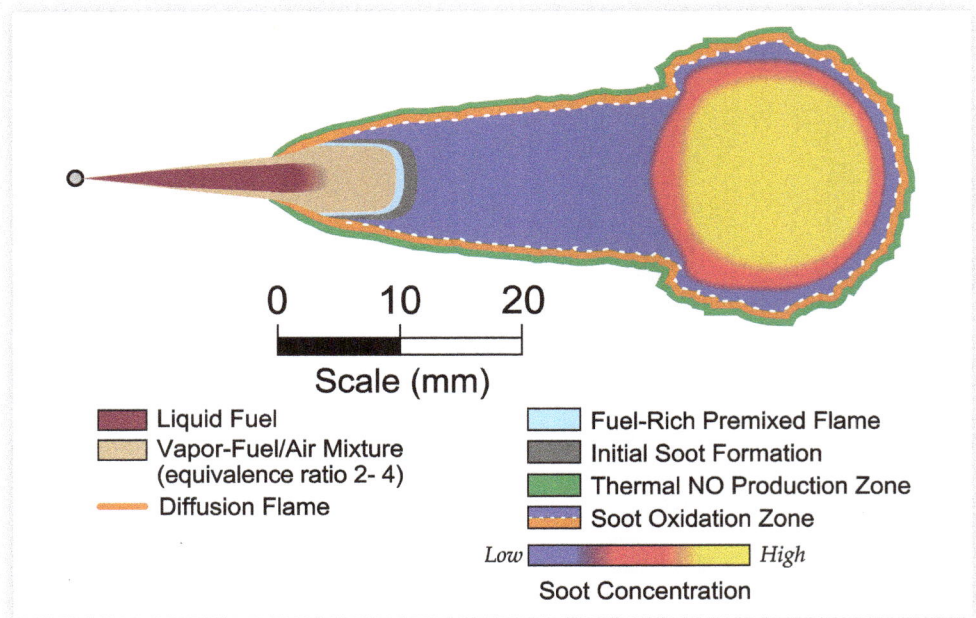

for automotive high-speed diesel engines, which use strong swirl flows that influence the distribution of the physical and chemical fields.

Since diesel combustion occurs spontaneously at multiple sites and there is no knock combustion limit, diesel engine size spans with the bore diameter ranging from 70 mm to 900 mm. Different injection systems and combustion chamber designs fit for different-size engines as shown in Figure 2.16. For automotive high-speed diesel engines, it is particularly challenging to have the combustion chamber shape, intake gas flow, and injection strategies work together to achieve high efficiency and clean combustion [31] (see Sec. 8.1). Strong swirl motion is often induced by using a helical intake port and re-entrant combustion chamber is proposed to generate turbulence for better mixing and combustion as shown in Figure 2.16c.

One of the important aspects in diesel combustion is to optimize injection strategy for improved engine functions. Multiple injections, enabled by electronically-controlled common-rail injection technology, have many advantages. Pilot injection is effective for engine combustion noise control and for NO$_x$ reduction [32]. Split injections or multiple injections can reduce both soot and NO$_x$ emissions simultaneously [33], which is a challenge due to the soot −NO$_x$ trade-off characteristics.

The mechanism to reduce soot and NO emissions using split injections was studied computationally by Han et al. [34] (see Sec. 8.1.1). The NO reduction mechanism is similar to that of a single injection with retarded injection timing. Soot formation is reduced after the first injection pause. Soot formation is reduced because the soot-producing rich regions at the spray tips are no longer replenished with a rich mixture. Since the subsequent injection takes place in a high-temperature environment left from the combustion products of the first injection, the injected fuel burns more rapidly, soot formation rates are decreased, and the net soot production can be reduced dramatically.

FIGURE 2.16 Common types of direct-injection diesel engine combustion systems: (a) quiescent chamber with multi-hole nozzle for larger engines; (b) bowl-in-piston chamber with moderate swirl and multi-hole nozzle for medium engines; (c) re-entrant bowl-in-piston chamber with high swirl and multi-hole nozzle for small-bore engines.

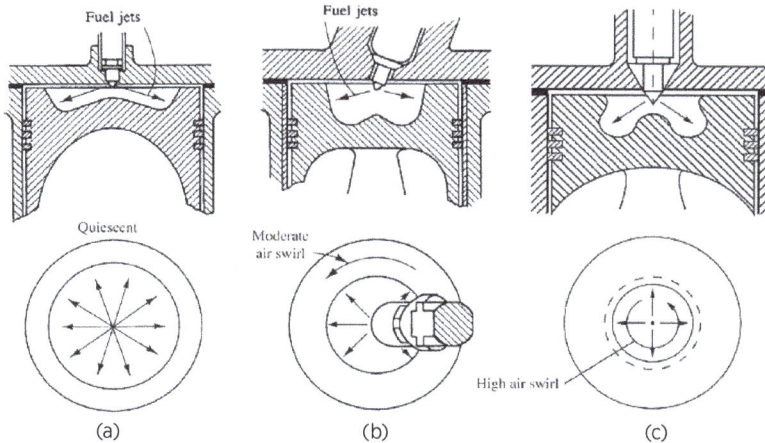

More recently, multiple injection strategies with up to five injection pulses have been implemented in production diesel engines. The objective of each injection pulse is different. The fuel quantities in each injection pulse and the dwells between pulses must be carefully adjusted through experimental calibration. Among the injection pulses, the main injection occurs around TDC for generating torque, the pilot injection before the main is to reduce premixed CR so as to control engine noise and NO_x, and the post-injection is aimed at reducing soot emissions. After the post-injection, there are two late-injection pulses to assist the engine-out emission aftertreatment devices. The early late injection takes place around the middle of the expansion stroke aimed to raise the exhaust gas temperature to initiate the particulate regeneration process in DPF. The last one, the late injection pulse, occurring near the exhaust-valve open (EVO) time, is used when a reducing agent (hydrocarbons) is needed to reduce NO_x in a SCR system.

2.4 Advanced Concepts of Low-Temperature Combustion

In the past few decades, research on diesel combustion is largely focused on the reduction of soot and NO_x emissions using in-cylinder formation control and engine-out emissions aftertreatment to meet the increasingly strict emissions regulations. Although advanced combustion systems with increased injection pressure (up to 240 MPa in production), EGR, optimized piston-bowl geometries, and improved in-cylinder flows have resulted in substantial reductions in emissions, it appears unlikely that conventional mixing-controlled diesel combustion can meet current and future emission requirements without expensive aftertreatment systems. On the other hand, government regulation and market demand have driven technology advances to reduce fuel consumption (increase thermal efficiency) further in SI gasoline engines. The goal is to produce engines with thermal efficiency greater than 50% for automotive engines under ultra-low NO_x and particulate emissions in the next 10~15 years.

Advanced combustion concepts with compression-ignition, LTC such as HCCI, RCCI, and PCCI have demonstrated promising results to reach that goal. For example, RCCI combustion has demonstrated up to 20% improved fuel efficiency over conventional diesel and 40–50% over conventional SI gasoline engines [35]. The common features of these LTC concepts and strategies are compression-ignited combustion, premixed or partially premixed charge preparation, highly diluted mixture by air and/or EGR, high CR, and unthrottled operation. The advantages are high thermal efficiency due to high CR, high specific heat ratio and no pumping loss, and low in-cylinder NO_x and soot formations.

LTC has been first demonstrated in HCCI combustion in the early 1980s [36]. Renewed intensive research on HCCI has begun since the middle of the 1990s. In HCCI or controlled auto-ignition combustion, fuel is injected into the intake ports, which is then mixed with air to form a premixed homogeneous mixture. The mixture is compressed by the piston and auto-ignited as the piston moves towards the end of compression. The mixture is made very dilute either with a fuel/air equivalence ratio typically less than 0.45, or with high levels of EGR for the equivalence ratio up to stoichiometric. The mixture reacts and burns volumetrically as it is compressed to auto-ignition temperatures. Because of the high dilution, combustion temperatures are low, resulting in very low in-cylinder NO_x formation, and the charge is sufficiently well-mixed to prevent soot formation. Thermal efficiencies are comparable to those of a diesel engine [37,38]. A variety of fuels including gasoline, diesel, ethanol, natural gas, and others have been studied for HCCI combustion.

The nature of HCCI combustion is auto-ignition, which is controlled by hydrocarbon oxidation kinetics and occurs volumetrically in the cylinder, and therefore it is greatly affected by the in-cylinder mixture temperature. Figure 2.17 shows HCCI combustion images captured with a high-speed camera [39]. The images were taken from one cycle in a 0.5-liter (93-mm Bore and 73.3-mm Stroke) pend-roof-chamber 4-stroke PFI single-cylinder engine. Ignition at the 2'o clock and 8'o clock locations occurred by 6.8° ATDC. Scattered flames, initiated by ignition, grow and fast chemical reactions take place in the entire volume of the chamber. At 16.4° ATDC, reactions continue but the main combustion stage (represented by intensive blue lights) comes to an end. Large-scale flame propagation is not seen from these images, which supports the fact that HCCI combustion is not mixing-controlled. Fast chemical reactions lead to a rapid cylinder pressure raise in HCCI, which can generate noise and vibration problems.

FIGURE 2.17 Sequential combustion images obtained by a highspeed color camera. The time sequence from the left image is 5.2°, 6.8°, 8.4°, 10.0°, 11.6°, 13.2°, 14.8°, and 16.4° BTDC.

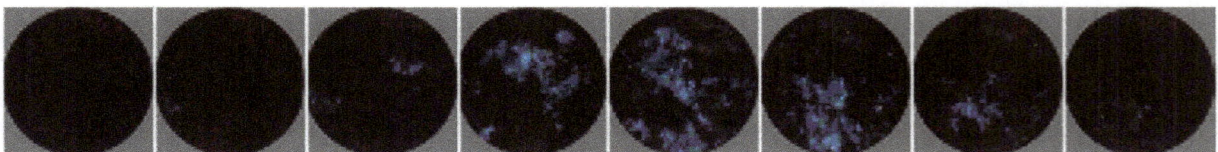

800 rpm, air–fuel ratio: 32, intake air temperature: 150°C.
BTDC: Before top dead center

Another method to achieve LTC is PCCI combustion [40]. This combustion method lengthens the period between the SOI and the SOC thereby enhancing the air–fuel mixing before the SOC. In this case, it is intended to separate mixing and combustion so that the mixture charge is premixed to some extent depending on the separation time. Fuel (diesel or gasoline) is injected into the cylinder during the compression stroke, fuel evaporates and mixes with air (and EGR gases), and the mixtures are then auto ignited. PCCI combustion is primarily governed by chemical kinetics and not by turbulent flame propagation. Figure 2.18 compares the combustion images, taken by a high-speed video camera, of conventional diesel compression-ignition (CDCI) (high temperature) combustion and PCCI (low-temperature) combustion in the same engine with diesel fuel [41]. The engine bore is 100 mm, stroke is 125 mm, and CR is 17.5. In the CDCI combustion mode as shown in the top row, the SOI is −10° ATDC, and combustion can be first observed near −1.5° ATDC through the start of the premixed burn. At 6.5° ATDC, combustion reaches a peak rate (reflected by peak natural luminosity), the flaming jet impinging on and spreading out along the circumference of the piston bowl rim. The yellowish colors of the flames indicate high-temperature rich diffusive burning. The bottom row shows the PCCI combustion mode in which the SOI is −32° ATDC. With a delay time of about 15°, ignition is observed by −15° ATDC. Then volumetric reactions take place in the spray regions and continue in that way. Blue luminosity indicates low-temperature combustion in the PCCI. In this case, the occurrence of pool fire is observed by the luminous diffusion flame along the circumference of the piston bowl in the images from −12.7° to 0.78° ATDC. The location where the pool fire is observed is in accord with that where the spray jet impinged.

FIGURE 2.18 Sequential combustion images taken by high-speed video camera showing CDCI combustion (top row) and PCCI combustion (bottom). The number by each image is the crank angle for that image.

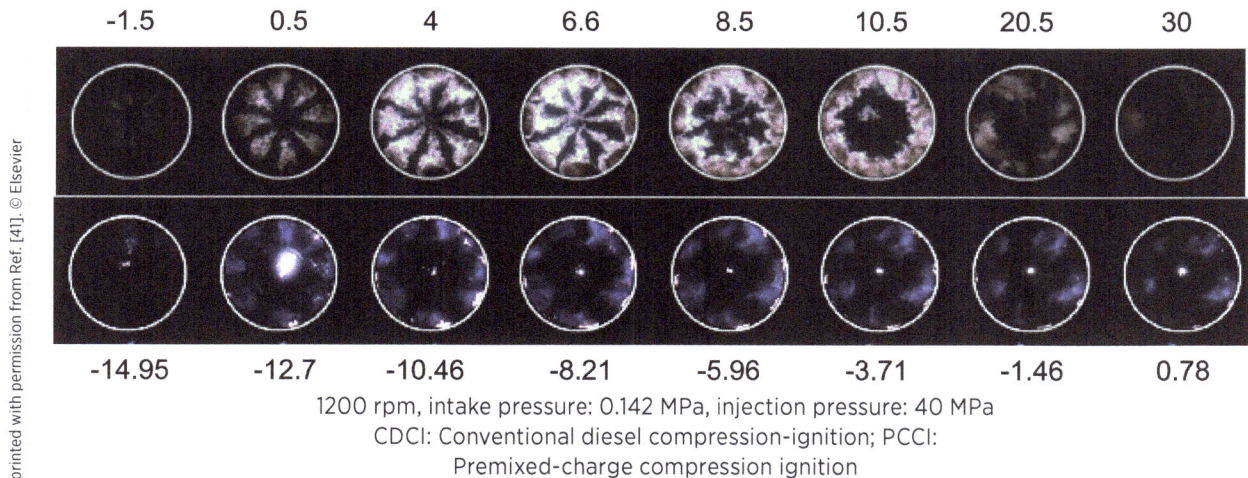

1200 rpm, intake pressure: 0.142 MPa, injection pressure: 40 MPa
CDCI: Conventional diesel compression-ignition; PCCI: Premixed-charge compression ignition

Since HCCI and PCCI are chemical kinetics-controlled reactions, it is not difficult to understand they are sensitive to the temperature and composition conditions, which is affected by intake gas temperature and pressure, and EGR rate as well [42]. This comes to a combustion controllability issue. Excessive formation of hydrocarbon and CO emissions at low engine loads is another issue that deteriorates combustion efficiency greatly. For example, the combustion efficiency reduces from 95% at $\phi>0.24$ to 55% at $\phi=0.08$ as reported by Dec [38].

As summarized by Reitz [35], despite the advantages of HCCI, several problems limit its adoption. The first difficulty is the achievement of a homogeneous charge in the engine. Many methods of fuel preparation have been proposed, such as PFI or early DI with a narrow spray included angle to avoid spray-wall impingement. A second difficulty is the limited operating range of HCCI. Due to the nature of volumetric combustion, the cylinder contents virtually ignite simultaneously and this can produce unacceptable noise and knocking at high loads. Attempts have been made to increase the load by injecting water, applying high EGR rates, or by introducing temperature stratification in the cylinder.

An additional difficulty with HCCI is control of the SOC timing, which means managing the engine's operation in an expected smooth way. In conventional CI engines, the SOC is controlled by the fuel injection timing, while in SI engines the spark initiates combustion. On the other hand, HCCI ignition is controlled by the charge mixture temperature, composition, and compression history. More control is possible with a low reactivity fuel, such as gasoline whose high resistance to auto-ignition allows more mixing time prior to combustion, producing a lower local equivalence ratio and thus lower NO_x and soot emissions. Indeed, the best fuel for HCCI operation would have auto-ignition qualities between those of diesel and gasoline. The high load operation is favored with a low cetane number (e.g., high octane number gasoline), while low load operation favors a high cetane number (e.g., diesel). PCCI offers some degrees of combustion timing controllability by adjusting SOI or multiple injections, which, however, are not sufficient.

The load limitation issues may be bypassed by using a mode switching strategy like the one used in SCDI, that is with HCCI at low and medium loads and conventional combustion at high loads. The extended operating range has been demonstrated using spark-assisted HCCI combined with multiple injections/multiple ignitions and valve-timing strategies. In Gasoline Direct Injection Compression Ignition (GDICI) operation [43], more control over the SOC timing can be achieved with suitably timed multiple injections to create partially premixed (or "premixed enough") in-cylinder equivalence-ratio stratification. Since higher equivalence-ratio regions are more reactive, this leads to sequential auto-ignition at different spatial locations in the chamber, and pressure rise rates are reduced, with the result that higher loads can be achieved.

A novel LTC combustion concept of RCCI was invented by Dr. Reitz's group at the Engine Research Center of the University of Wisconsin in 2010. In the original RCCI concept, in-cylinder fuel blending is arranged using PFI of a lower reactivity fuel (e.g., gasoline) coupled with optimized in-cylinder multiple injections of more reactive fuel (e.g., diesel) [35]. This dual-fuel strategy provides a feasible mean to greatly improve combustion controllability. Thus, intensive research on RCCI has been carried out at the Engine Research Center and many other centers across the world. The list of technical publications on RCCI is very long and readers are recommended to first refer to the papers by Reitz et al. [44], Paykani et al. [45], Harari [46], and Paykani et al. [47].

RCCI combustion has been demonstrated ultra-low in-cylinder NO_x and soot emissions while achieving gross indicated thermal efficiencies close to 60% [48]. This efficiency is about 10% higher than that of conventional diesel combustion (i.e., about 20% less fuel consumed). In addition, RCCI does not require high fuel injection pressures since the injections occur well before TDC, thus providing time for fuel vaporization and mixing. RCCI extends the engine's operable load range beyond that of HCCI or PCCI since fuel blending allows the

FIGURE 2.19 Effects of port-injected gasoline amount on RCCI phasing combustion.

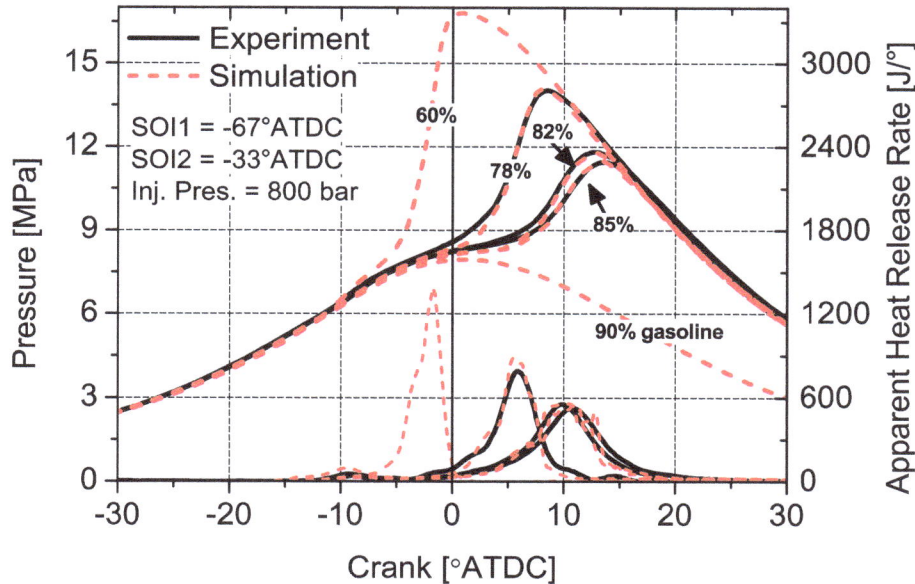

RCCI: reactivity controlled compression ignition

combustion to be tailored. Indeed, gasoline–diesel RCCI has been demonstrated up to 23 bars gross IMEP in a heavy-duty diesel engine with a CR of approximately 12:1.

Both equivalence ratio and reactivity stratifications in the chamber are generated in RCCI. Combustion progresses sequentially from the high reactivity regions to low reactivity regions, thereby effectively lowering pressure rise rates. This capability of RCCI is shown in Figure 2.19. As the amount of port-injected gasoline is varied, combustion phasing is alternated as indicated by both measured and CFD computed cylinder pressure and heat release rates [49]. The heat release rate curves show evidence of a cool flame reaction from the diesel fuel, followed by two distinct bumps on the high temperature heat release. It was thought that the first bump corresponds to high temperature oxidation of CO formed from oxidation of the diesel fuel and the second bump corresponds to oxidation of CO formed from the gasoline breakdown.

In contrast to other LTC concepts, the combustion phasing of RCCI can be altered by adjusting the proportions of the more and less reactive fuels on a same or next-cycle basis with appropriate combustion feedback control. RCCI can operate with a wide range of fuels, including gasoline, natural gas, and diesel. It can be also used with a single fuel. For example, in the work of Splitter et al. [50], the engine operates with port injection of gasoline, plus a small quantity of cetane improver such as di-tert-butyl peroxide (DTBP) added to the direct-injected fuel (also gasoline). DTBP can be replenished at oil change intervals.

RCCI combustion has been demonstrated to be able to run at high engine loads with high thermal efficiency and low emissions as shown in Figure 2.20. Operation with 14.6 bar IMEP load with 56% gross indicated thermal efficiency is achieved, ringing intensity is within the acceptable level, and the in-cylinder NO_x and soot levels are under the United States Environmental Protection Agency 2010 (US EPA 2010) on-highway truck emissions

FIGURE 2.20 RCCI engine performance data versus engine loads (IMEP). 2.44-liter single-cylinder engine, 1300 rpm. The solid horizontal lines show the US 2010 EPA. on-highway truck emissions limits for NO_x and soot.

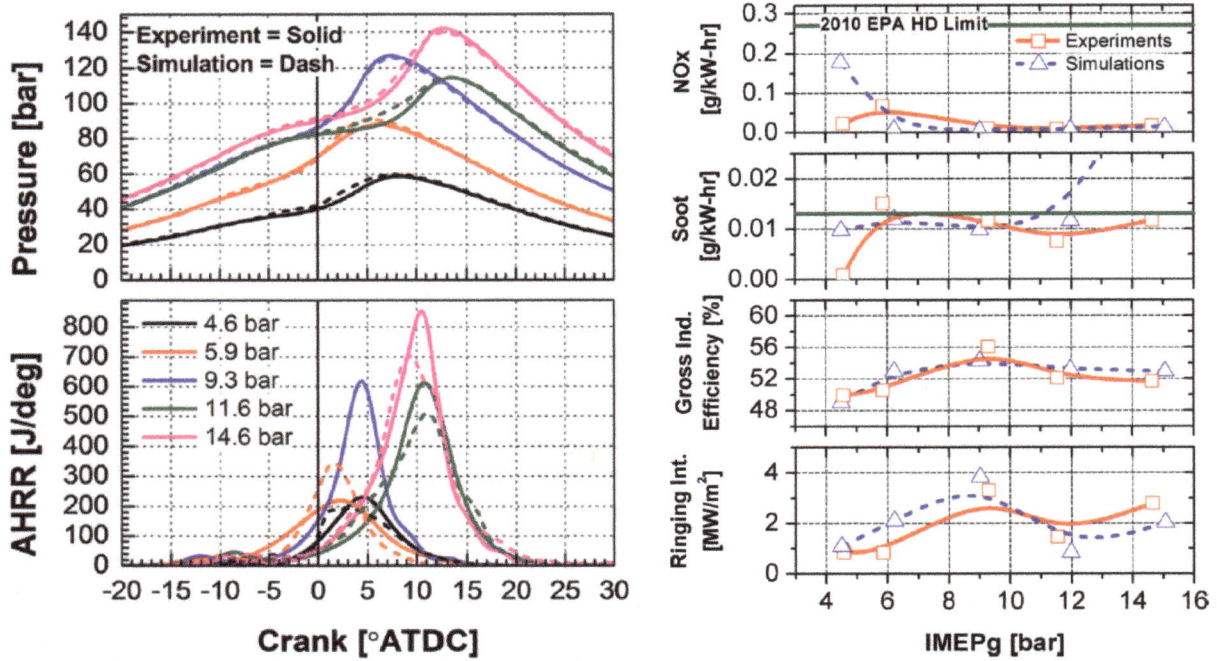

RCCI: reactivity controlled compression ignition; IMEP: indicated mean effective pressure; US EPA: United States Environmental Protection Agency

Reprinted with permission from Ref. [51]. © SAGE Publications

limits [51]. The fuel flexibility of RCCI has been demonstrated in Figure 2.21 [52]. PFI gasoline/DI gasoline with DTBP, E85/diesel, and gasoline/diesel are tested. It is seen that the E85/diesel combination delivered about 59% thermal efficiency, which indicates the good alternative-fuel adaptability of RCCI.

Due to the potential of RCCI, recent research has been focused on the implementation of RCCI concepts to commercial vehicle applications. Efforts have been made about engine transit behavior over entire load conditions, engine control methods, conventional and hybrid vehicle applications, etc. The latest results have been reviewed by Paykani et al. [47], and they indicate that control-oriented-model (COM) based models can control CA50 (φ_{50}) for different RCCI engine transient conditions; single-fuel RCCI operation with biofuels is a promising way to commercialize RCCI combustion; combustion mode-switching with COM-based models is required to capture transient RCCI emissions; and plug-in hybrid electric vehicles (PHEVs) with RCCI engine have the potential to reduce local and global emissions.

Mixture preparation is an essential aspect of all LTC concepts. Arguably, fuel injection strategy is the distinctive feature and key enabler of specific LTC methods. Thus, it is important to understand the basis of the fuel injection strategy used in each LTC. This

FIGURE 2.21 Performance of RCCI with different fuels. 2.44-liter single cylinder engine, 1300 rpm.

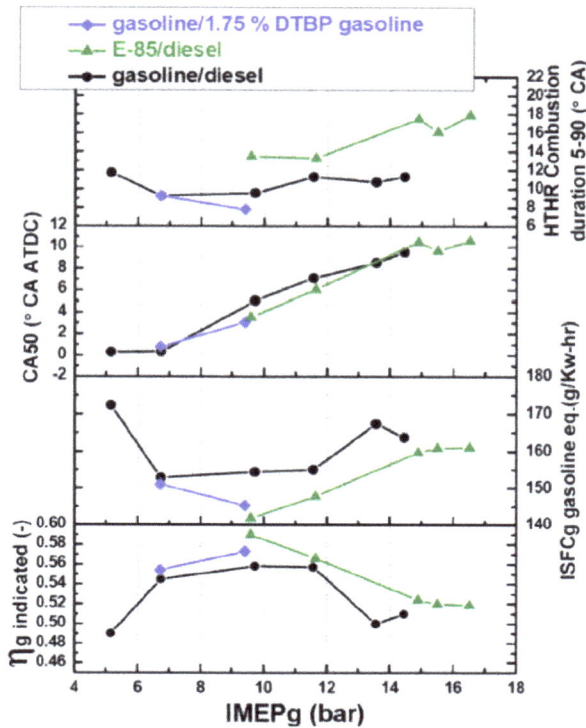

RCCI: reactivity controlled compression ignition

is illustrated in Figure 2.22. In CDCI combustion, fuel is always directly injected into the combustion chamber. HCCI uses external (port) injection. PCCI employs in-cylinder fuel injection as defined in this book, although some researchers called their external fuel injection strategy, PCCI. For RCCI, one fuel with lower reactivity is injected in the intake ports and the other more reactive one is internally injected. Note that, in-cylinder fuel injection can be single or multiple in all LTC concepts.

In conclusion, the emissions reduction potential of new combustion concepts is summarized in Figure 2.23 in which contours of the fuel–air equivalence ratio ϕ and combustion temperature T at which soot and NO_x formation occur are plotted [1,38]. CDCI combustion is in the soot and NO_x formation regions, leading to high levels of emissions. SI combustion also produces considerable engine-out NO_x emissions, but they are readily removed by modern three-way catalysts due to stoichiometric combustion. LTC combustion concepts such as HCCI, PCCI, and RCCI are in the low NO_x and soot formation regions.

Other recent research and development efforts have resulted in new promising high-efficiency engine concepts as well. For example, a high indicated thermal efficiency of 56% in a gasoline engine was reported with the use of spark-controlled compression ignition, which was proposed by Mazda engineers to used spark to control HCCI combustion [53]. On

FIGURE 2.22   Basic injection strategy used in CDCI, HCCI, PCCI, and RCCI combustion.

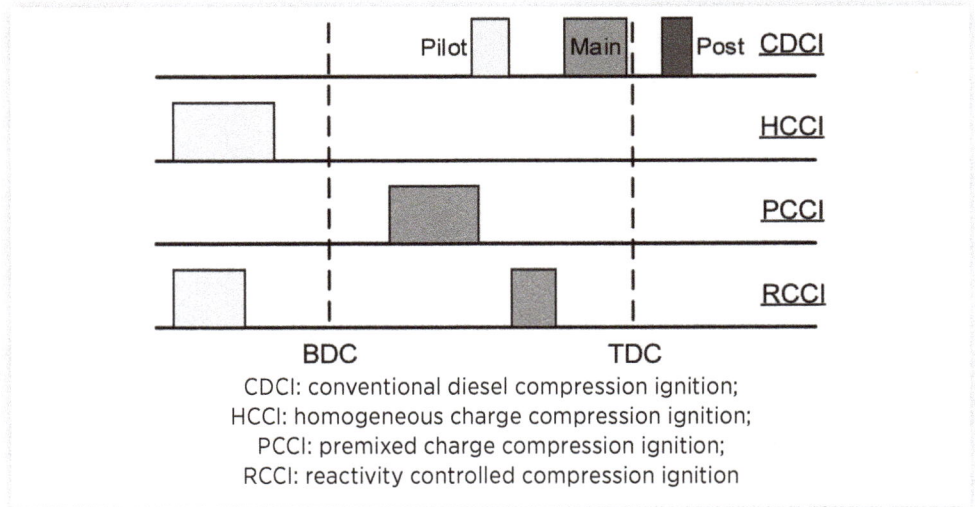

CDCI: conventional diesel compression ignition;
HCCI: homogeneous charge compression ignition;
PCCI: premixed charge compression ignition;
RCCI: reactivity controlled compression ignition

FIGURE 2.23   Diagram showing the ranges for soot and NO_x formation and the regions for CDCI, SI, HCCI, PCCI, RCCI engines. The darkest regions represent the highest concentration of soot and NO_x.

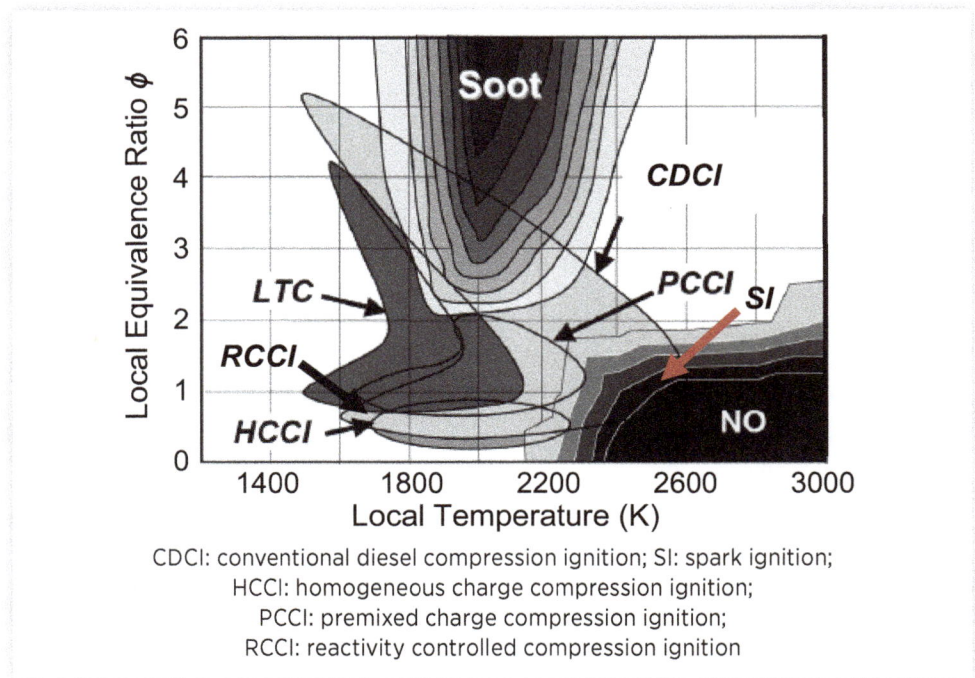

CDCI: conventional diesel compression ignition; SI: spark ignition;
HCCI: homogeneous charge compression ignition;
PCCI: premixed charge compression ignition;
RCCI: reactivity controlled compression ignition

the other hand, some hybrid vehicle configurations require the engines to run in narrower speed-load ranges compared to conventional engine-powered vehicles. This offers a design opportunity to optimize combustion for improved thermal efficiency for the hybrid dedicated engines [54].

References

1. Heywood, J.B. 2018. *Internal Combustion Engine Fundamentals*. 2nd Ed. New York, USA: McGraw-Hill Education.

2. Zhou, L., Liu, Y., and Gao, Z. 2011. *Internal Combustion Engines* [In Chinese]. 3rd ed. Beijing, China: China Machine Press.

3. Crolla, D., Foster, D.E., Kobayashi, T., and Vaughan, N, eds. 2014. *Encyclopedia of Automotive Engineering*. Chichester, UK: John Wiley & Sons.

4. Borman, G., and Nishiwaki, K. 1987. "Internal-combustion engine heat transfer." *Progress in Energy and Combustion Science* 13, no. 1: 1–46. DOI: 10.1016/0360-1285(87)90005-0

5. Caton, J.A. 2016. *An Introduction to Thermodynamic Cycle Simulations for Internal Combustion Engines*. Chichester, UK: John Wiley & Sons.

6. Zhou, L., Han, Z., and Xu, B. 1990. "Study of Computation Methods for Rates of Heat Release in Swirl Chamber Diesel Engines" [In Chinese]. *Journal of Xi'an Jiaotong University* 24, no. 5: 35–44.

7. Zhou, L., Song, S., Han, Z., Xu, B., Chang, Y., and Du, J. 1991. "Evaluation and Study on the Heat Release Rate of Swirl Chamber Diesel Engine." SAE Technical Paper No. 911786. DOI: 10.4271/911786

8. Han, Z., and Xu, Z. 2004. "Wall Film Dynamics Modeling for Impinging Sprays in Engines." SAE Technical Paper No. 2004-01-0099. DOI: 10.4271/2004-01-0099

9. Kume, T., Iwamoto, Y., Iida, K., Murakami, M., Akishino, K., and Ando, H. 1996. "Combustion Control Technologies for Direct Injection SI Engine." *SAE Transactions* 105, no. 3(February): 704–717. DOI: 10.4271/960600

10. Harada, J., Tomita, T., Mizuno, H., Mashiki, Z., and Ito, Y. 1997. "Development of Direct Injection Gasoline Engine." *SAE Transactions* 106, no. 3(February): 767–776. DOI: 10.4271/970540

11. Takagi, Y., Itoh, T., Muranaka, S., Iiyama, A., Iwakiri, Y., Urushihara, T., and Naitoh, K. 1998. "Simultaneous Attainment of Low Fuel Consumption High Output Power and Low Exhaust Emissions in Direct Injection SI Engines." *SAE Transactions* 107, no. 3(February): 215–225. DOI: 10.4271/980149

12. Koike, M., Saito, A., Tomoda, T., and Yamamoto, Y. 2000. "Research and Development of a New Direct Injection Gasoline Engine." *SAE Transactions* 109, no. 3(March): 543–552. DOI: 10.4271/2000-01-0530

13. Stiebels, B., Krebs, R., and Pott, E. 2001. "FSI-Gasoline Direct Injection Engines from Volkswagen." In: *Proceedings of the 2001 Global Powertrain Congress: Vol. A. Advanced Engine Design & Development*. Detroit, MI.

14. Anderson, R., Yang, J., Brehob, D., Vallance, J., and Whiteaker, R. 1996. "Understanding The Thermodynamics of Direct Injection Spark Ignition (DISI) Combustion Systems: An Analytical and Experimental Investigation." *SAE Transactions* 105, no. 3(October): 2195–2204. DOI: 10.4271/962018

15. Zhao, F., Lai, M.C., and Harrington, D.L. 1999. "Automotive Spark-Ignited Direct-Injection Gasoline Engines". *Progress in Energy and Combustion Science* 25, no. 5(October): 437–562. DOI: 10.1016/S0360-1285(99)00004-0

16. Han, Z., Weaver, C., Wooldridge, S., Alger, T., Hilditch, J., McGee, J., Westrate, B., Xu, Z., Yi, J., Chen, X., Trigui, N., and Davis, G. 2004. "Development of a New Light Stratified-Charge DISI Combustion System for a Family of Engines With Upfront CFD Coupling with Thermal and Optical Engine Experiments." *SAE Transactions* 113, no. 3(March): 269–293. DOI: 10.4271/2004-01-0545

17. Han, Z., Yi, J., and Trigui, N. 2002. "Stratified Mixture Formation and Piston Surface Wetting in a DISI Engine." SAE Technical Paper No. 2002-01-2655. DOI: 10.4271/2002-01-2655

18. VanDerWege, B.A., Han, Z., Iyer, C.O., Muñoz, R.H., and Yi, J. 2003. "Development and Analysis of a Spray-Guided DISI Combustion System Concept." *SAE Transactions* 112, no. 4(October): 2135–2153. DOI: 10.4271/2003-01-3105

19. Iyer, C.O., Han, Z., and Yi, J. 2004. "CFD Modeling of a Vortex Induced Stratification Combustion (VISC) System." SAE Technical Paper No. 2004-01-0550. DOI: 10.4271/2004-01-0550

20. Yi, J., Wooldridge, S., Coulson, G., Hilditch, J., Iyer, C.O., Moilanen, P., Papaioannou, G., Reiche, D., Shelby, M., and VanDerWege, B. 2009. "Development and Optimization of the Ford 3.5 L V6 EcoBoost Combustion System." *SAE International Journal of Engines* 2, no. 1(April): 1388–1407. DOI: 10.4271/2009-01-1494

21. Ando, H., and Arcoumanis, C.D. 2009. "Flow, Mixture Preparation and Combustion in Four-Stroke Direct-Injection Gasoline Engines." In *Flow and Combustion in Reciprocating Engines*, edited by C.D. Arcoumanis and T. Kamimoto, 137–171. Berlin, Germany: Springer-Verlag.

22. He, C., Kuenne, G., Yildar, E., van Oijen, J., Di Mare, F., Sadiki, A., Ding, C.P., Baum, E., Peterson, B., and Böhm, B. 2017. "Evaluation of the Flame Propagation Within an SI Engine Using Flame Imaging and LES." *Combustion Theory and Modelling* 21, no. 6: 1080–1113. DOI: 10.1080/13647830.2017.1343498

23. Wang, Z., Liu, H., Song, T., Qi, Y., He, X., Shuai, S., and Wang, J. 2015. "Relationship Between Super-Knock and Pre-ignition." *International Journal of Engine Research* 16, no. 2(April): 166–180. DOI: 10.1177/1468087414530388

24. Wang, Z., Liu, H., and Reitz, R.D. 2017. "Knocking Combustion in Spark-Ignition Engines." *Progress in Energy Combustion Science* 61: 78–112. DOI: 10.1016/j.pecs.2017.03.004

25. Kalghatgi, G. 2018. "Knock Onset, Knock Intensity, Superknock and Preignition in Spark Ignition Engines." *International Journal of Engine Research* 19, no. 1(October): 7–20. DOI: 10.1177/1468087417736430

26. Wooldridge, S., Lavoie, G., and Weaver, C. 2003. "Convection Path for Soot and Hydrocarbon Emissions from the Piston Bowl of a Stratified Charge Direct Injection Engine." In the *Proceedings of the Third Joint Meeting of the US Sections of the Combustion Institute*. Chicago, Illinois: The University of Illinois.

27. Hilditch, J., Han, Z., and Chea, T. 2003. "Unburned Hydrocarbon Emissions from Stratified Charge Direct Injection Engines." SAE Technical Paper No. 2003-01-3099. DOI: 10.4271/2003-01-3099

28. Lee, Z., Kim, T., Park, S., and Park, S. 2020. Review on Spray, Combustion, and Emission Characteristics of Recent Developed Direct-Injection Spark Ignition (DISI) Engine System With Multi-Hole Type Injector." *Fuel* 259, article 116209. DOI: 10.1016/j.fuel.2019.116209

29. Dec, J.E., and Espey, C. 1998. "Chemiluminescence Imaging of Autoignition in a DI Diesel Engine." *SAE Transactions* 107, no. 3(October): 2230–2254. DOI: 10.4271/982685

30. Dec, J.E. 1997. "A Conceptual Model of DL Diesel Combustion Based on Laser-Sheet Imaging." *SAE Transactions* 106, no. 3(February): 1319–1348. DOI: 10.4271/970873

31. Miles, P.C., and Andersson, Ö. 2016. "A Review of Design Considerations for Light-Duty Diesel Combustion Systems." *International Journal of Engine Research* 17, no. 1 (October), 6–15. DOI: 10.1177/1468087415604754

32. Shundoh, S., Komori, M., Tsujimura, K., and Kobayashi, S. 1992. "NO_x Reduction From Diesel Combustion Using Pilot Injection With High Pressure Fuel Injection." SAE Technical Paper No. 920461. DOI: 10.4271/920461

33. Tow, T.C., Pierpont, D., and Reitz, R.D. 1994. "Reducing Particulate and NO_x Emissions by Using Multiple Injections in a Heavy Duty D.I. Diesel Engine. *SAE Transactions* 103, no. 3(March): 1403–1417. DOI: 10.4271/940897

34. Han, Z., Uludogan, A., Hampson, G.J., and Reitz, R.D. 1996. "Mechanism of Soot and NO_x Emission Reduction Using Multiple-Injection in a Diesel engine." *SAE Transactions* 105, no. 3(February): 837–852. DOI: 10.4271/960633

35. Reitz, R.D. 2013. "Directions in Internal Combustion Engine Research." *Combustion and Flame* 160, no. 1(January): 1–8. DOI: 10.1016/j.combustflame.2012.11.002

36. Najt, P.M., and Foster, D.E. 1983. "Compression-Ignited Homogeneous Charge Combustion." SAE Technical Paper No. 830264. DOI: 10.4271/830264

37. Zhao, F., Asmus, T.N., Assanis, D.N., Dec, J.E., Eng, J.A., and Najt, P.M. 2003. *Homogeneous Charge Compression Ignition (HCCI) Engines: Key Research and Development Issues.* Warrendale, USA: SAE International.

38. Dec, J.E. 2014. "Advanced Compression-Ignition Combustion for High Efficiency and Ultra-Low NO_x and Soot." In *Encyclopedia of Automotive Engineering*, edited by D. Crolla, D.E. Foster, T. Kobayashi, and N. Vaughan, 1–40. Chichester, UK: John Wiley & Sons.

39. Kakuho, A., Nagamine, M., Amenomori, Y., Urushihara, T., and Itoh, T. 2006. "In-Cylinder Temperature Distribution Measurement and its Application to HCCI Combustion." SAE Technical Paper No. 2006-01-1202. DOI: 10.4271/2006-01-1202

40. Iwabuchi, Y., Kawai, K., Shoji, T., and Takeda, Y. 1999. "Trial of New Concept Diesel Combustion System - Premixed Compression-Ignited Combustion." SAE Technical Paper No. 1999-01-0185. DOI: 10.4271/1999-01-0185

41. Kim, K., Kim, D., Jung, Y., and Bae, C. 2013. "Spray and Combustion Characteristics of Gasoline and Diesel in a Direct Injection Compression Ignition Engine." *Fuel* 19, no. 1(October): 616–626. DOI: 10.1177/1468087417736430

42. Dempsey, A.B., Walker, N.R., Gingrich, E., and Reitz, R.D. 2014. "Comparison of Low Temperature Combustion Strategies for Advanced Compression Ignition Engines With a Focus on Controllability." *Combustion Science and Technology* 186, no. 2: 210–241. DOI: 10.1080/00102202.2013.858137

43. Ra, Y., Loeper, P., Reitz, R., Andrie, M., Krieger, R., Foster, D.E., Durrett, R., et al. 2011. "Study of High Speed Gasoline Direct Injection Compression Ignition (GDICI) Engine Operation in the LTC Regime." *SAE International Journal of Engines* 4, no. 1(April): 1412–1430. DOI: 10.4271/2011-01-1182

44. Reitz, R.D., and Duraisamy, G. 2015. "Review of High Efficiency and Clean Reactivity Controlled Compression Ignition (RCCI) Combustion in Internal Combustion Engines." *Progress in Energy and Combustion Science* 46: 12–71. DOI: 10.1016/j.pecs.2014.05.003

45. Paykani, A., Kakaee, A.H., Rahnama, P., and Reitz, R.D. 2016. "Progress and Recent Trends in Reactivity-Controlled Compression Ignition Engines." *International Journal of Engine Research* 17, no. 5(July): 481–524. DOI: 10.1177/1468087415593013

46. Harari, P.A. 2018. "Comprehensive Review on Enabling Reactivity Controlled Compression Ignition (RCCI) in Diesel Engines." *Integrated Research Advances* 5, no. 1: 5–19.

47. Paykani, A., Garcia, A., Shahbakhti, M., Rahnama, P., and Reitz, R.D. 2021. "Reactivity Controlled Compression Ignition Engine: Pathways Towards Commercial Viability." *Applied Energy* 282, article 116174. DOI: 10.1016/j.apenergy.2020.116174

48. Splitter, D., Wissink, M., DelVescovo, D., and Reitz, R.D. 2013. "RCCI Engine Operation Towards 60% Thermal Efficiency." SAE Technical Paper No. 2013-01-0279. DOI: 10.4271/2013-01-0279

49. Kokjohn, S.L., Hanson, R.M., Splitter, D.A., and Reitz, R.D. 2009. "Experiments and Modeling of Dual-Fuel HCCI and PCCI Combustion Using In-Cylinder Fuel Blending." SAE Technical Paper No. 2009-01-2647. DOI: 10.4271/2009-01-2647

50. Splitter, D., Reitz, R., and Hanson, R. 2010. "High Efficiency, Low Emissions RCCI Combustion by Use of a Fuel Additive." *SAE International Journal of Fuels and Lubricants* 3, no. 2(October): 742–756. DOI: 10.4271/2010-01-2167

51. Kokjohn, S.L., Hanson, R.M., Splitter, D., and Reitz, R. 2011. "Fuel Reactivity Controlled Compression Ignition (RCCI): A Pathway to Controlled High-Efficiency Clean Combustion." *International Journal of Engine Research* 12, no. 3(June): 209–226. DOI: 10.1177/1468087411401548

52. Splitter, D., Hanson, R., Kokjohn, S., and Reitz, R.D. 2011. "Reactivity Controlled Compression Ignition (RCCI) Heavy-Duty Engine Operation at Mid-and High-Loads With Conventional and Alternative Fuels." SAE Technical Paper No. 2011-01-0363. DOI: 10.4271/2011-01-0363

53. Hirose, I. 2019. *Our Way Toward the Ideal Internal Combustion Engine for Sustainable Future*. 28th Aachen Colloquium Automobile and Engine Technology, Aachen, Germany.

54. Wu, Z., Han, Z., Shi, Y., Liu, W., Zhang, J., Huang, Y., and Meng, S. 2021. "Combustion Optimization for Fuel Economy Improvement of a Dedicated Range-Extender Engine." *Proceedings of the Institution of Mechanical Engineers, Part D: Journal of Automobile Engineering* 235, no. 9(February): 2525–2539. DOI: 10.1177/0954407021993620

3

Mathematical Description of Reactive Flow with Sprays

Combustion in IC engines comprises complicated strongly-coupled physical and chemical processes of air intake-charging and compression, liquid spray injection and vaporization, air–fuel mixing, chemical reactions, and emissions formation. The processes occur as the blend of gas mixture and liquid droplets flows within a complex geometry. These processes are governed by the conservation laws of fluid dynamics and are subject to the thermodynamics principles. In order to mathematically describe these processes in the CFD approach, they are usually decomposed into three parts: gas-phase fluid dynamics, gas-phase chemical kinetics, and liquid-phase spray dynamics. Each part is described by the corresponding mathematical formulations. After applying appropriate physical models, the governing equations are solved using proper numerical methods. In this chapter, the governing equations for the gas-phase and liquid-phase flows are presented initially. The commonly used numerical methods are described next. The boundary conditions necessary for solving the governing equations are discussed. The gas-phase thermal boundary models, which are used to predict the heat transfer through the cylinder walls, are described in detail at the end.

3.1 Governing and Spray Equations

3.1.1 Governing Equations of Gas Phase

In this section, we give the governing equations of motion for the gas phase in the Cartesian coordinate system. These equations are also called the Navier–Stokes equations. To make it easy to follow, we will first give the governing equations for compressible, viscous, heat-conducting ideal gas flow. Similar but more complicated equations for flows involving multiple species, sprays, and chemical reactions will follow. Derivation of these equations can be found in many technical books [1]. The mass conservation equation is

$$\frac{\partial \rho}{\partial t} + \frac{\partial(\rho u_j)}{\partial x_j} = 0 \tag{3.1}$$

where ρ is the mass concentration or mass density of the fluid, t is the time.

The momentum conservation equation is

$$\frac{\partial(\rho u_i)}{\partial t} + \frac{\partial(\rho u_i u_j)}{\partial x_j} = -\frac{\partial p}{\partial x_i} + \frac{\partial \sigma_{ij}}{\partial x_j} + \rho g_i \tag{3.2}$$

where p is the pressure of the flow, σ_{ij} is the Stokesian stress tensor, and g_i is the gravity. The Stokesian stress tensor is defined as:

$$\sigma_{ij} = \mu\left(\frac{\partial u_i}{\partial x_j} + \frac{\partial u_j}{\partial x_i} - \frac{2}{3}\frac{\partial u_k}{\partial x_k}\delta_{ij}\right) \tag{3.3}$$

where μ is the dynamic viscosity and δ_{ij} is the tensorial Kronecker symbol, which is

$$\delta_{ij} = \begin{cases} 1: i = j \\ 0: i \neq j \end{cases}$$

Equation (3.3) represents the relation between the stress and the velocity gradient for a Newtonian fluid. The strain-rate tensor, which is often referred to, is defined as:

$$S_{ij} = \frac{1}{2}\left(\frac{\partial u_i}{\partial x_j} + \frac{\partial u_j}{\partial x_i}\right) \tag{3.4}$$

The internal energy conversation equation is

$$\frac{\partial(\rho e)}{\partial t} + \frac{\partial(\rho u_j e)}{\partial x_j} = \frac{\partial p}{\partial t} + \frac{\partial}{\partial x_j}\left(u_i \sigma_{ij} - q_j\right) \tag{3.5}$$

where e is the internal energy and q_j is the heat flux due to heat conduction, which is

$$q_j = -\lambda\frac{\partial T}{\partial x_j} \tag{3.6}$$

where T is the temperature of the fluid and λ is the thermal conductivity.

These equations can be applied to compressible cold flows in engines (e.g., those in the induction process, or those in motored engines). However, they must be extended in order to describe the processes involving multiple species, sprays, and chemical reactions. In these flows, the mass conservation equation for species k is given as:

$$\frac{\partial(\rho_k)}{\partial t} + \frac{\partial(\rho_k u_j)}{\partial x_j} = \frac{\partial}{\partial x_j}\left(\rho D \frac{\partial(\rho_k / \rho)}{\partial x_j}\right) + \dot{\rho}_k^s + \dot{\rho}_k^c \tag{3.7}$$

where ρ_k is the mass density of species k, and it relates to the total mass density of the mixture as $\rho = \Sigma \rho_k$. The source terms of $\dot{\rho}_k^c$ and $\dot{\rho}_k^s$ are due to chemical reaction and spray evaporation/condensation, respectively. D is the single diffusion coefficient and based on Fick's Law of diffusion, it is defined as:

$$D = \frac{\mu}{\rho Sc} \tag{3.8}$$

where Sc is the Schmidt number. By summing Equation (3.7) over all species, the mass conversation equation for the whole gas flow is given as:

$$\frac{\partial \rho}{\partial t} + \frac{\partial(\rho u_j)}{\partial x_j} = \dot{\rho}^s \tag{3.9}$$

The momentum conservation equation for the whole flow is

$$\frac{\partial(\rho u_i)}{\partial t} + \frac{\partial(\rho u_i u_j)}{\partial x_j} = -\frac{\partial p}{\partial x_i} + \frac{\partial \sigma_{ij}}{\partial x_j} + F_i^s + \rho g_i \tag{3.10}$$

where F_i^s is a spray-induced source term that will be defined in the next section.

The energy conservation equation is

$$\frac{\partial(\rho e)}{\partial t} + \frac{\partial(\rho e u_j)}{\partial x_j} = -p\frac{\partial u_j}{\partial x_j} + \frac{\partial J_j}{\partial x_j} + \dot{Q}^s + \dot{Q}^c \tag{3.11}$$

where J_j is the heat flux, which is the sum of contributions due to heat conduction and enthalpy diffusion.

$$J_j = -\lambda \frac{\partial T}{\partial x_j} - \rho D \sum_{k=1}^{N} h_k \frac{\partial(\rho_k / \rho)}{\partial x_j} \tag{3.12}$$

where $\dot{Q}^s$ and $\dot{Q}^c$ are the source terms due to spray and chemical reaction, respectively. N is the total number of species.

The state equation is given by assuming the gas mixture is an ideal gas. This assumption is very reasonable considering the pressure and temperature range inside an IC engine.

$$p = RT \sum_{k=1}^{N} \frac{\rho_k}{W_k} \tag{3.13}$$

Here, R is the universal gas constant and W_k is the molecular weight of species k.

3.1.2 **Spray Equation**

Solving the essential dynamics of a spray and its interactions with the gas phase is an extremely complicated problem. To calculate the mass, momentum, and energy exchange between the gas-phase and the liquid-phase sprays, one must account for a distribution of the spray drop sizes, velocities, and temperatures. In many engine sprays, drop breakup, collisions, and coalescence must also be considered. Theoretically, the liquid phase can also be described using the Navier–Stokes equations in a detailed way. However, its interactions with the gas-phase flow are extremely complicated due to the large differences in timescales and length scales. Thus, the separated flow model has been proposed to consider the effects of the finite rate transport between the two phases.

In general, there are three different approaches in the separated flow model: the discrete droplet model (DDM), the continuous droplet model (CDM), and the continuous formulation model (CFM). In DDM, the spray is represented by a finite number of droplet groups. The motion and transport of these droplet groups are tracked through the flow field using a Lagrangian formulation. Hence, DDM is also referred to as the Lagrangian approach [2]. The mean quantities of the liquid phase are computed through statistical methods. The effects of the liquid phase on the gas phase are considered by introducing appropriate spray source terms into the governing equations of the gas phase (i.e., Equations (3.9) to (3.11)). It is convenient to construct physical models and numerical algorithms in DDM. Thus, the Lagrangian approach dominates current CFD simulations of two-phase flows.

CDM is applicable only when few phenomena must be considered. Otherwise, the computational cost will be very high. CFM treats the gas phase and liquid phase as continuous phases and solves both with an Eulerian formulation. It is thus referred to as an Eulerian approach in mathematics, distinguishing it from the Lagrangian approach. It takes the dispersed phase (liquid sprays) as a continuous fluid and introduces several continuous scalar fields to represent the dispersed phase. Quantities relevant to the dispersed phase are defined at nodes, which are generally coincident with those used for the continuous-phase grid, and the mean-field equations are derived for both phases. Therefore, the dispersed phase is modeled at the macroscopic level with this approach. This method leads to significant difficulties in modeling complex phenomena such as droplet breakup, droplet interaction, and droplet evaporation, which are essential in engine applications. It is also very difficult to establish the representation of the turbulent stresses and transport in the liquid phase. Lecture notes of Reitz (1996) give detailed descriptions of these approaches [3].

Since DDM is the model most commonly used in engine simulations, we discuss only this approach in this book and all the computational examples are based on it. The details of the DDM method are discussed in Sec. 5.2.1. The DDM approach assumes that, after the primary breakup, the formed droplets are small enough to be viewed as point sources. Thus, spray dynamics can be described by the Williams spray equation [4] in which the spray is represented by a droplet distribution function f that has ten independent variables in addition to time. These are the three droplet position components $\mathbf{x}$, three velocity components $\mathbf{v}$, drop radius r, temperature T_d, drop distortion from sphericity y, and its time rate of change $\dot{y} = dy/dt$. The dimensionless quantity y is proportional to the displacement of the droplet surface from its equilibrium position divided by the droplet radius r (see Sec. 5.3.1). The droplet distribution function is

$$f = f(\mathbf{x}, \mathbf{v}, r, T_d, y, \dot{y}, t) \tag{3.14}$$

It is defined in such a way that $f(\mathbf{x}, \mathbf{v}, r, T_d, y, \dot{y}, t)\, d\mathbf{v}\, dr\, dT_d\, dy\, d\dot{y}$ is the probable number of droplets per unit volume at position $\mathbf{x}$ and time t with velocity in the interval $(\mathbf{v}, \mathbf{v} + d\mathbf{v})$, radii in the interval $(r, r + dr)$, temperatures in the interval $(\dot{T}_d, T_d + d\dot{T}_d)$, displacement parameters in the intervals $(y, y + dy)$ and $(\dot{y}, \dot{y} + d\dot{y})$ [5].

The first moment of f is the number density of the droplets:

$$n = \int f \cdot d\mathbf{v}\,dr\,dT_d\,dy\,d\dot{y} \tag{3.15}$$

The second moment about radius r relates to the liquid volume fraction θ and liquid macroscopic density ρ_l':

$$\theta = \int \frac{4}{3}\pi r^3 f \cdot d\mathbf{v}\,dr\,dT_d\,dy\,d\dot{y} \tag{3.16}$$

$$\rho_l' = \rho_d \theta = \int \frac{4}{3}\pi r^3 \rho_d f \cdot d\mathbf{v}\,dr\,dT_d\,dy\,d\dot{y} \tag{3.17}$$

where ρ_d is the liquid microscopic density (drop density). Note that ρ_l' can nevertheless be comparable to or larger than the gas density because of the large ratio of liquid macroscopic density to gas density.

The time evolution of f is obtained by solving a form of the spray equation as:

$$\frac{\partial f}{\partial t} + \nabla_x \cdot (f\mathbf{v}) + \nabla_v \cdot \left(f\frac{d\mathbf{v}}{dt} \right) + \frac{\partial}{\partial r}\left(f\frac{dr}{dt} \right) + \frac{\partial}{\partial T_d}\left(f\frac{dT_d}{dt} \right) + \frac{\partial}{\partial y}(f\dot{y}) + \frac{\partial}{\partial \dot{y}}\left(f\frac{d\dot{y}}{dt} \right)$$
$$= \dot{f}_{coll} + \dot{f}_{bu} \tag{3.18}$$

The terms $\dot{f}_{coll}$ and $\dot{f}_{bu}$ arise from droplet collisions and breakups, respectively (see Sec. 5.3.1 and 5.3.2). By solving Equation (3.18), the spray exchange terms in Equations (3.9) to (3.11) are given as:

$$\dot{\rho}^s = -\int f \rho_d 4\pi r^2 \frac{dr}{dt} d\mathbf{v}\,dr\,dT_d\,dy\,d\dot{y} \tag{3.19}$$

$$\mathbf{F}^s = -\int f \rho_d \left[\frac{4}{3}\pi r^3 \mathbf{F}' + 4\pi r^2 \frac{dr}{dt}\mathbf{v} \right] d\mathbf{v}\,dr\,dT_d\,dy\,d\dot{y} \tag{3.20}$$

$$\dot{Q}^s = -\int f \rho_d \left\{ \begin{array}{l} 4\pi r^2 \dfrac{dr}{dt}\left[I_l + \dfrac{1}{2}(\mathbf{v} - \mathbf{u})^2 \right] + \\[2mm] \dfrac{4}{3}\pi r^3 \left[c_l \dot{T}_d + \mathbf{F}' \cdot (\mathbf{v} - \mathbf{u} - \mathbf{u}') \right] \end{array} \right\} d\mathbf{v}\,dr\,dT_d\,dy\,d\dot{y} \tag{3.21}$$

And the source term in the turbulence k-ε models (see Sec. 4.2 and 4.3) is given as:

$$\dot{W}^s = -\int f \rho_d \frac{4}{3} \pi r^3 \mathbf{F}' \cdot \mathbf{u}' \, d\mathbf{v} \, dr \, dT_d \, dy \, d\dot{y} \qquad (3.22)$$

where $F' = d\mathbf{x}/dt - \mathbf{g}$. And $(\mathbf{v} - \mathbf{u})$ is the relative velocity between the drop and the gas. $\boldsymbol{u}'$ is the gas-phase turbulence intensity, I_l and c_l are the internal energy and specific heat of the liquid drops, respectively.

Equation (3.18) describes the evolution of the droplet distribution function. However, sub-models for drop breakup, distortion, collisions, drag force, evaporation, and drop/wall interaction must be provided, which will be addressed later (see Chapter 5).

Models are also needed to define the source terms associated with chemical reactions. They will also be discussed later. With all the sub-models given, supplemented with properly treated initial and boundary conditions and fluid properties, the above system of equations can be numerically solved to describe an engine flow.

3.2 Numerical Methods

In order to numerically solve the governing partial differential equations (PDEs), approximations to the partial differentials are introduced. These approximations convert the partial derivatives to finite-difference expressions, which are used to rewrite the PDEs as algebraic equations. The approximate algebraic equations are subsequently solved at discrete points within the domain of interest with the use of computers. There are many numerical techniques and schemes are available with their own advantages and disadvantages in terms of computational accuracy, efficiency, and stability [6]. Instead of examining the vast contents of numerical methods of CFD, we will explain the main features of the numerical treatments in the often-used engine simulation code KIVA and CONVERGE.

3.2.1 The KIVA Code

In the KIVA family codes [5,7–9], the gas-phase solution procedure is based on a finite volume method called the arbitrary Lagrangian–Eulerian (ALE) method [10,11]. Spatial differences are formed on a finite-difference mesh with arbitrary hexahedrons that subdivides the computational region into a number of small hexahedron cells. The use of an arbitrary mesh shape is a great advantage to engine simulation since the arbitrary mesh can conform to curved boundaries and can move to follow changes in combustion chamber geometry. In this method, except for the velocity, all gas flow quantities are stored in the cells. The gas velocities are stored in momentum cells, which are centered about the cell vertices. Thus, the velocities are located at the vertices while other quantities are located at the cell centers.

The temporal difference scheme in the KIVA codes is largely implicit. Implicit differencing is used for all the diffusion terms and the terms associated with pressure wave propagation. The coupled implicit equations are solved by using a semi-implicit method for pressure-linked equations (SIMPLE) algorithm [12]. The transient solution is marched out in a sequence of finite time increments called cycles or timesteps. On each cycle, the values of the dependent variables are calculated from those on the previous cycle. In the ALE method, each cycle is divided into two phases: a Lagrangian phase and a rezone phase. In the Lagrangian phase, the vertices move with the fluid velocity, and there is no convection across cell boundaries. In the rezone phase, the flow field is frozen, the vertices are moved to new user-specified positions, and the flow field is remapped or rezoned onto the new computational mesh. This remapping is accomplished by convecting variables across the boundaries of the computational cells, which are regarded as moving relative to the flow field. A quasi-second-order upwind (QSOU) scheme [5] is used to calculate convection in the rezone phase.

The timesteps are calculated based on accuracy, not stability, criteria in KIVA. This feature makes it different from some commercial CFD software in which stability takes the priority to avoid customer complaints due to computation crashes. An automatic timestep control method is used in the KIVA codes, which requires the timestep Δt to satisfy the criteria. The first accuracy condition on Δt is

$$\Delta t_{acc} < \sqrt{f_a \Delta x \left| \frac{D\mathbf{u}}{Dt} \right|^{-1}} \qquad (3.23)$$

where the typical value of f_a is 0.5 and Δx is an average cell dimension. The second accuracy condition on Δt is

$$\Delta t_{res} < \frac{f_r}{|\lambda|} \qquad (3.24)$$

where $f_r = 1/\sqrt{3}$ and λ is an eigenvalue of the strain-rate tensor. This criterion limits the amount of cell distortion that can occur due to mesh movement.

Two other accuracy criteria for Δt arise from the need to couple accurately the flow field and source terms due to chemical heat release and mass and energy exchange with the spray. They are given as:

$$\Delta t_{ch} < f_{ch} \left(\frac{\rho_i e_i}{\dot{Q}_i^c} \right) \qquad (3.25)$$

$$\Delta t_{sp} < f_{sp} \left(\frac{\dot{\rho}_i^s}{\rho_i} \right) \qquad (3.26)$$

where the typical value for f_{ch} and f_{sp} are taken to be 0.1. In addition, the timestep Δt_{gr} is used to limit the amount by which the timestep can grow. That is

$$\Delta t_{gr} = 1.02\Delta t \qquad (3.27)$$

The Courant stability condition are also used to limit the convection timestep Δt_c. The criterion is given as:

$$\Delta t_c \leq f_{con} \Delta t_c \, min \left(\frac{V_i}{\delta V_i} \right) \qquad (3.28)$$

where V_i is the volume of cell i, and δV_i is the flux volume calculated for the cell. The typical value of f_{con} is taken to be 0.2 to reduce the timestep for computational accuracy.

Then, the main timestep used for the computation for cycle $n + 1$ is given by

$$\Delta t^{n+1} = min(\Delta t^{n+1}_{acc}, \Delta t^{n+1}_{rst}, \Delta t^{n+1}_{ch}, \Delta t^{n+1}_{sp}, \Delta t^{n+1}_{gr}, \Delta t^{n+1}_{c}, \Delta t_{mx}, \Delta t_{mxca}) \tag{3.29}$$

Timesteps Δt_{mx} and Δt_{mxca} are the input maximum timestep and the maximum timestep based on an input maximum crank angle, respectively.

The spray dynamics equation is solved with the stochastic particle technique, which is based on the Monte Carlo method and of discrete particle methods. This method has been proven very efficient and accurate. In the discrete particle method, the continuous droplet probability distribution function f is approximated by a discrete distribution f':

$$f' = \sum_{p=1}^{N_p} N_p \delta(\mathbf{x} - \mathbf{x}_p) \delta(\mathbf{v} - \mathbf{v}_p) \delta(r - r_p) \delta(T_d - T_{d_p}) \delta(y - y_p) \delta(\dot{y} - \dot{y}_p) \tag{3.30}$$

Each particle p is composed of a number of droplets N_p having equal location $\mathbf{x}_p$, velocity $\mathbf{v}_p$, size r_p, temperature T_{d_p}, and oscillation parameters y_p and $\dot{y}_p$. Particle and droplet trajectories coincide, and the particles exchange mass, momentum, and energy with the gas in the computational cells in which they are located.

Difference expressions can be formulated to approximate the governing equations. The difference equations are then calculated using the aforementioned numerical method. A series of computational steps are needed to complete the calculation process. The details of the computational procedure have been written about by Amsden et al. [5].

In order to solve the finite difference equations, a set of grid points within the domain, as well as the boundaries of the domain, must be specified. The creation of such a grid system is called grid generation. Grid generation had been bottlenecks for engine simulations due to the complicated boundary geometries (e.g., curved piston shape, canted valves), which are in motion. It was not until the middle 1990s that pioneer simulations of engine intake flow with realistic valve-motion were reported for a DI diesel engine [13] and for a DI SI gasoline engine [14].

In KIVA3 [8], a block-structured mesh is used with connectivity defined through indirect addressing. This allows complex geometries to be modeled with high efficiency because large regions of deactivated cells are no longer necessary. KIVA3V [9] adds the flexibility to model any number of vertical or canted valves in the cylinder head of an engine as it treats the valves as solid objects that move through the mesh using the so-called snapping technique.

Although these advances in grid generation make simulations possible of a full engine cycle with intake and exhaust processes, the grid generation was still time consuming and tedious. A cutting-edge method was developed at the FRL in the early 2000s to generate a moving-valve engine mesh rapidly, reducing the time of mesh generation from 3–4 weeks to 2–3 days for a new engine geometry [15].

3.2.2 The CONVERGE Code

A breakthrough innovation of engine grid generation has been made by Senecal and co-workers [16]. They founded the commercial CFD code CONVERGE and released updated versions [17]. Since CONVERGE has been widely licensed for engine simulations worldwide, we briefly describe its methods of grid generation and numeric here.

CONVERGE uses a modified cut-cell Cartesian method that eliminates the need for the computational grid to be morphed with the geometry of interest while still properly representing the geometry. This approach has two significant advantages. First, the type of grid used is chosen for computational efficiency instead of geometry. This allows the use of simple orthogonal grids, which greatly simplifies the numeric of the code. Second, the grid generation time and complexity are significantly reduced, as the complex geometry only needs to be mapped onto the underlying orthogonal grid. The user is only required to provide a file containing the surface of the geometry, represented as a closed triangulated surface. This file is easily output in stereolithography (STL) format from most CAD packages.

CONVERGE performs the actual grid generation internally at runtime and uses the supplied triangulated surface to cut the cells that are intersected by the surface. Any cell can be cut by an arbitrary number of surface triangles, thus allowing the grid to represent the boundary accurately. The use of simple orthogonal grids allows for a variety of mesh manipulation options that can enhance the accuracy and increase the speed of simulation. The manipulation options include refinement of the overall mesh resolution, grid size mapping from coarse to fine cells during a simulation, fixed grid embedding by a user to add grid resolution locally in critical flow sections, and adaptive grid embedding based on a pre-specified critical value of the sub-grid scale. All these features of mesh generation make CONVERGE unique and advantageous in engine combustion simulations.

In CONVERGE, the conservation equations are solved using the finite volume method. All computed values are collocated at the center of the computational cell with the use of the Rhie and Chow algorithm [18] to maintain co-located variables and eliminate undesirable checkerboarding. The transport equations are solved using the pressure implicit with splitting of operators (PISO) method of Issa [19]. The PISO algorithm implemented in CONVERGE starts with a predictor step where the momentum equation is solved. After the predictor, a pressure equation is derived and solved, which leads to a correction, which is then applied to the momentum equation. This process of correcting the momentum equation and re-solving can be repeated as many times as necessary to achieve the desired accuracy. After the momentum predictor and first corrector step have been completed, the other transport equations are solved in series.

CONVERGE also features a multigrid pressure solver and a timestep control algorithm. It can be run in parallel on shared and distributed memory computers. More detailed numerical methods can be found in the CONVERGE manual [17].

3.3 Boundary Conditions

3.3.1 General Description

Boundary conditions must be needed for solving the conservation equations. Boundary conditions also provide necessary information to determine some important parameters of the simulated engines such as the airflow rate at the intake port openings and the heat transfer through the cylinder walls. There are typically three types of boundary conditions used in engine simulations: rigid wall boundary, inflow boundary, and outflow boundary. Generally, each quantity, denoted as q, in the conservation equations requires a value of V (Dirichlet condition)—the spatial change of q requires a value of V_i (Neumann condition)—at each boundary. Thus, boundary conditions can be given as:

$$q = V \qquad (3.31)$$

or

$$\frac{\partial q}{\partial x_i} = V_i \qquad (3.32)$$

where q can be a solved quantity (e.g., pressure, temperature, velocity, species mass fraction, energy, etc.). In many cases, V_i is equal to zero.

For the inflow boundary, the boundary conditions can be Dirichlet or Neumann. For the outflow boundary, the normal components of $\partial q/\partial x_i$ are set to zero. There are several types of rigid wall boundary conditions for velocity and temperature. The velocity boundary conditions on rigid walls can be free slip, no slip, or turbulent law-of-the-wall.

Temperature boundary conditions include adiabatic walls and fixed temperature walls. In engine simulations, turbulent law-of-the-wall velocity conditions with fixed wall temperatures are ordinarily used. The engine wall temperatures are usually estimated based on limited experimental data. However, conjugate heat transfer analysis can be performed that allows wall temperatures to be predicted with the use of finite element computation of the metal walls [20]. In this case, the wall temperatures and heat fluxes are solved iteratively by calculating the gas convective heat transfer near the wall and heat conduction inside the wall.

Velocity boundary conditions on rigid walls are introduced either by imposing the value of the velocity on the walls or the value of the wall stress $\boldsymbol{\sigma}_w = \boldsymbol{\sigma} \cdot \mathbf{n}$, where $\boldsymbol{n}$ is the unit normal to the wall. On no slip walls, the gas velocity is set equal to the wall velocity:

$$\boldsymbol{u} = w_{wall}\boldsymbol{k} \qquad (3.33)$$

where the wall is assumed to be moving with speed w_{wall} in the z-direction. On free-slip boundaries the normal gas velocity is set equal to the normal wall velocity

$$\boldsymbol{u} \cdot \boldsymbol{n} = w_{wall}\boldsymbol{k} \cdot \boldsymbol{n}$$

On the other hand, the velocity profile near the wall can be numerically provided. However, numerical solutions of the complete conservation equations are impractical because one cannot afford sufficient resolution (mesh size) in engine simulations. For turbulent engine flows, the law-of-the-wall functions are suggested to calculate the gas velocity and temperature in the wall region. This means the computed fluid velocities and temperatures at the grid points closest to walls are matched to the wall functions to determine the wall shear stresses and heat losses numerically. This approach is discussed in detail in the next sections.

When the flow field is assumed to have an N-fold periodicity about the z-axis (cylinder axis), periodic boundaries can be used. In this case, the computational region is composed of points in the pie-shaped sector $0 \leq \theta \leq 2\pi/N$. The periodic boundaries are those for which $\theta = 0$ and $\theta = 2\pi/N$. The conditions imposed on these boundaries can be inferred from the assumed N-fold periodicity. For a scalar quantity q, the requirement is that $q(r,\theta,z) = q(r,\theta + 2\pi/N,z)$. For a vector $\boldsymbol{v}$, the requirement is that $\boldsymbol{v}(r,\theta + 2\pi/N,z) = \boldsymbol{R} \cdot \boldsymbol{v}(r,\theta,z)$, where $\boldsymbol{R}$ is the rotation matrix corresponding to the angle $2\pi/N$. Simulation with periodic boundaries is often used for diesel engines if the combustion chamber is symmetric about the cylinder axis and the sprays are evenly distributed in the chamber. Thus, a sector mesh including one spray is adopted for computational efficiency. Examples will be given later.

Specifying the boundary conditions for engine simulations are often difficult and somewhat arbitrary, partially due to lack of the data of the simulated engine (e.g., the wall temperatures) and partially due to the inexplicit assumptions in the theory (e.g., turbulence kinetic energy (TKE) and length scale at the inflow boundary). Temporal changes of the conditions at the boundaries (e.g., air pressure waves at the intake port openings) add complications to the problem. Appropriate experimental data and good engineering knowledge certainly help to address this issue. On the other hand, some numerical methods and techniques have been developed to mitigate this issue and to improve simulation accuracy.

Quasi-dimensional or one-dimensional (1D) thermodynamic models can determine detailed evolution of gas velocity, pressure, and temperature, etc. at the engine inflow and outflow boundaries. They can be integrated with the three-dimensional (3D) combustion simulations to provide the latter with the needed boundary and initial conditions. The often-used approach is to integrate a 1D engine model and a CFD code for combined engine simulations [21–23]. Usually, 1D simulation model is used for modeling of hydraulic and gas dynamics systems (e.g., gas exchange processes in the intake and exhaust manifolds), and 3D CFD code for modeling of the in-cylinder combustion and emissions. For example, in the work of Millo et al. [23], 1D engine modeling by means of GT-SUITE [24] is combined with 3D CFD by means of CONVERGE for combustion simulations of a diesel engine executed within a specifically designed calculation methodology. The predicted results are more accurate in comparison with the experimental data.

Multiple-cycle simulation is another useful method to generate converged boundary and initial conditions in engine simulation. Wu and Han [25] assessed the effects of the number of simulation cycles on the convergence of the computed results in the simulation of an automotive turbocharged port-injection natural gas engine. Computations of the engine cylinder with intake ports/runners and exhaust ports were carried out with the use of the CONVERGE code, and the results from the previous cycle were used as the boundary and initial conditions for the current cycle simulation. By comparison of the computed parameters from each cycle, they concluded that three cycles were needed in order to obtain converged results.

Figure 3.1 shows the comparisons of the computed in-cylinder tumble and swirl ratios, overall fuel equivalence ratio and TKE averaged over the total in-cylinder mass from the first four simulation cycles. It is clear that one-cycle simulation cannot predict good results, in particular, the injected natural gas motion cannot be predicted correctly in the first cycle since the natural gas motion has not developed in the intake ports. After the third cycle, the computed parameters converged together and the results become repeatable cycle-by-cycle.

3.3.2 Velocity Law-of-the-Wall Function

According to the turbulent boundary layer theory, in the near-wall region, if the distance from the wall is small enough, the gas velocity can be approximated as in the logarithmic region or the laminar sublayer region of the turbulent boundary layer. Derivations of the turbulent velocity law-of-the-wall functions can be found in the literature [26] and the resultant formulations are very similar. Here we introduce the formulations used in the KIVA codes.

FIGURE 3.1 Evolution of (a) the tumble ratio and swirl ratio, (b) overall equivalence ratio (Φ), and (c) mass-averaged TKE in simulation of a turbocharged port-injection natural gas engine.

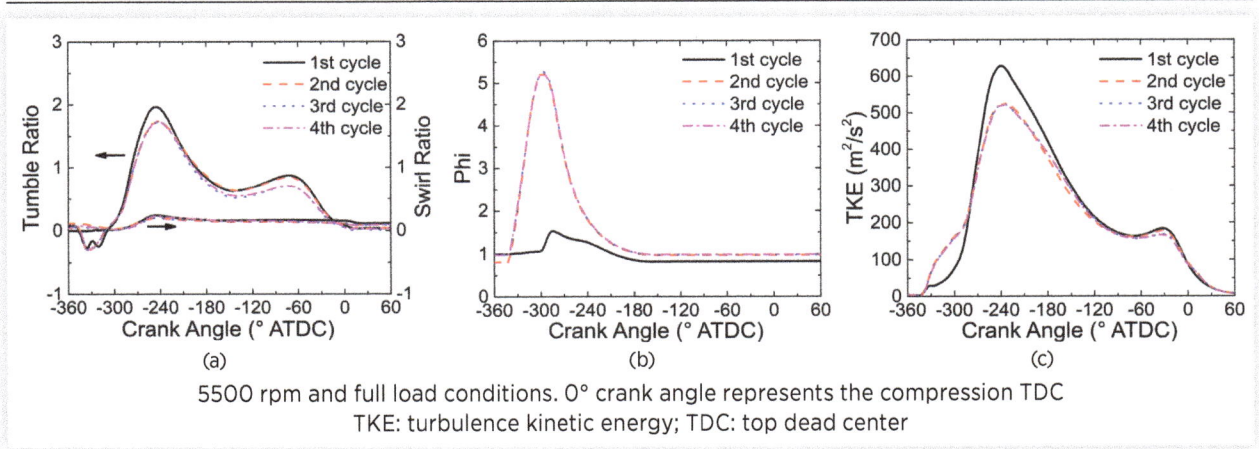

5500 rpm and full load conditions. 0° crank angle represents the compression TDC
TKE: turbulence kinetic energy; TDC: top dead center

In the law-of-the-wall model, it is assumed that:

1. Gradients normal to the wall are much greater than those parallel to the wall.
2. The fluid velocity is directed parallel to a flat wall.
3. Pressure gradients are neglected.
4. Viscous dissipation, and the Dufour and enthalpy diffusion effects on energy flux are neglected.
5. Radiation heat transfer is neglected.
6. No fuel film or spray impingement is on the considered wall surface.
7. The gas is ideal.

Hence, the wall velocity function can be derived from the conservation equations. Amsden et al. [5] made the derivation and give the velocity function as:

$$\frac{v}{u^*} = \begin{cases} \frac{1}{\kappa}\ln\left(c_{lm}R^{7/8}\right) + B & R > R_c \\ \\ R^{1/2} & R \le R_c \end{cases} \tag{3.34}$$

where $R = vy/v_0$ is the Reynolds number based on the gas velocity relative to the wall $v = |\boldsymbol{u} - w_{wall}\boldsymbol{k}|$ and the distance y from the wall. v_0 is the kinematic viscosity, u^* is called shear speed, R_c defines the boundary between the logarithmic region and the laminar sublayer region of the turbulent boundary layer. In Equation (3.34), the constants B, c_{lm}, κ, and R_c are given as $B = 5.5$, $c_{lm} = 0.15$, $\kappa = 0.4327$, and $R_c = 114$ for the classic k-ε turbulence model (see Sec. 4.2).

The classic velocity law-of-the-wall function proposed by Launder and Spalding [27] is:

$$\frac{v}{u^*} = \begin{cases} \dfrac{1}{\kappa}\ln(y^+) + B & y^+ > 10.18 \\\\ y^+ & y^+ \leq 10.18 \end{cases} \tag{3.35}$$

In Equation (3.35), $y^+ = u^* y / v_0$. The value of von Karman constant $\kappa = 0.4$ and the constant $B = 5.5$ is based on flow experiments. The values of these two constants are very close to those in Equation (3.34). Note that in Equation (3.34) no iterative calculations are needed to solve for the unknown shear speed u^*.

At the boundary, if k-ε turbulence models are used to simulate the gas turbulence, u^* is calculated from turbulent kinetic energy k as $u^* = k^{1/2} C_\mu^{1/4}$. At the wall, the boundary conditions for the turbulent kinetic energy k and its dissipation rate ε are given as:

$$\nabla k \cdot \boldsymbol{n} = 0 \tag{3.36}$$

$$\varepsilon = \frac{C_\mu^{3/4} k^{3/2}}{\kappa y} \tag{3.37}$$

where C_μ is a model constant and it is equal to 0.09 in the classic turbulence k-ε model, y is the physical distance from the nearest wall.

3.3.3 Temperature Wall Function and Wall Heat Transfer

Temperature boundary conditions on rigid wails are introduced by specifying either the wall temperature T_w, or the wall heat flux q_w. For adiabatic walls, q_w is set to be zero. For fixed temperature walls that are also either free slip or no slip, the wall temperature is prescribed. For fixed temperature walls using the turbulent law-of-the-wall conditions, q_w can be determined by the formulations discussed hereafter.

The temperature wall function provides a formulation to calculate the gas temperature in the wall cells. It can also give the expression to calculate heat flux transferring to or from the wall. We will first derive the temperature wall function following Han and Reitz [28], and then discuss the wall heat transfer modeling.

In the near wall region of the in-cylinder engine flows, with the assumptions of the flows given earlier, Equation (3.11) can be reduced as:

$$\frac{\partial q}{\partial y} = -\rho c_p \frac{\partial T}{\partial t} + \frac{dp}{dt} - \dot{Q}^c \tag{3.38}$$

where c_p is the specific heat, and the heat flux is

$$q = -(\lambda + \lambda_t)\frac{\partial T}{\partial y} \tag{3.39}$$

where λ_t is the turbulent conductivity. The first term of the right-hand-side of Equation (3.38) is the transient term and accounts for the energy change with time in the control volume. The second one is the pressure work term and the third one is the heat generation term due to chemical reactions.

It is known that engine wall heat transfer is unsteady in nature as described by Equation (3.38). However, we make an approximation of the process by invoking the quasi-steady assumption. Integration of Equation (3.38) from the wall ($y = 0$) gives

$$-c_p\left(\frac{\mu}{Pr}+\frac{\mu_t}{Pr_t}\right)\frac{dT}{dy}=q_w+Gy \tag{3.40}$$

The average chemical heat release $G = \overline{\dot{Q}^c}$. The Prandtl number $Pr = c_p\mu/\lambda$, turbulent Prandtl number $Pr_t = c_p\mu_t/\lambda_t$, with μ_t being the turbulent viscosity. We introduce the dimensionless quantities as:

$$v^+=\frac{v_t}{v_0};\quad y^+=\frac{u^*y}{v_0};\quad G^+=\frac{Gv_0}{q_wu^*} \tag{3.41}$$

then, we have

$$-\frac{\rho c_p u^*}{q_w}dT=\frac{1}{\left(\dfrac{1}{Pr}+\dfrac{v^+}{Pr_t}\right)}dy^++\frac{G^+y^+}{\left(\dfrac{1}{Pr}+\dfrac{v^+}{Pr_t}\right)}dy^+ \tag{3.42}$$

Equation (3.42) is integrated from 0 to y^+. The left-hand-side of Equation (3.42) then becomes

$$T^+=\int_{T_w}^{T}\frac{\rho c_p u^*}{q_w}dT=\frac{\rho c_p u^*T\ln\left(T/T_w\right)}{q_w} \tag{3.43}$$

In order to integrate the right-hand-side of Equation (3.42), correlations describing the changes of Pr_t and v_t ($v_t = \mu_t/\rho$) are needed. Based on analyses of formulations in the literature, a curve-fit correlation has been proposed by Han and Reitz [28]. This correlation is given as:

$$\begin{cases}\dfrac{v^+}{Pr_t}=a+by^++cy^{+2} & y^+\leqslant y_0^+ \\[4mm] \dfrac{v^+}{Pr_1}=my^+ & y^+>y_0^+\end{cases} \tag{3.44}$$

where the values of the constants a, b, c, and m are set to be 0.1, 0.025, 0.012, and 0.4767, respectively. The transition value of y_0^+ is chosen as 40.0, mainly due to mathematical consideration. Substituting Equation (3.44) into Equation (3.42) and splitting the integration into two parts, we get

$$T^+ = \int_0^{y_0^+} \frac{1}{Pr^{-1} + a + by^+ + cy^{+2}} dy^+ + \int_{y_0^+}^{y^+} \frac{1}{my^+} dy^+$$
$$+ \int_0^{y_0^+} \frac{G^+ y^+}{Pr^{-1} + a + by^+ + cy^{+2}} dy^+ + \int_{y_0^+}^{y^+} \frac{G^+}{m} dy^+$$

(3.45)

where Pr^{-1} is neglected in the second part of the integration, thus assuming the turbulent effect is dominant in this region of boundary layer. Finally, the temperature wall function equation is given as:

$$T^+ = 2.1 \ln(y^+) + 2.1 G^+ y^+ + 33.4 G^+ + 2.5$$

(3.46)

and the corresponding formulation for wall heat flux is given as:

$$q_w = \frac{\rho c_p u^* T \ln(T/T_w) - (2.1y^+ + 33.4)Gv_0/u^*}{2.1 \ln(y^+) + 2.5}$$

(3.47)

If the source term G can be neglected, Equation (3.47) then becomes

$$q_w = \frac{\rho c_p u^* T \ln(T/T_w)}{2.1 \ln(y^+) + 2.5}$$

(3.48)

For incompressible flows, Equation (3.46) is still valid, but Equation (3.43) becomes

$$T^+ = \int_{T_w}^{T} \frac{\rho c_p u^*}{q_w} dT = \frac{\rho c_p u^* (T - T_w)}{q_w}$$

(3.49)

Correspondingly, we have

$$q_w = \frac{\rho c_p u^* (T - T_w) - (2.1y^+ + 33.4)Gv_0/u^*}{2.1 \ \ln(y^+) + 2.5}$$

(3.50)

Again, if combustion is not considered, Equation (3.50) reduces to

$$q_w = \frac{\rho c_p u^* (T - T_w)}{2.1 \ln(y^+) + 2.5}$$

(3.51)

Comparing Equation (3.48) for compressible flows and Equation (3.51) for incompressible flows, it is immediately noticed that the impact of gas temperature on heat flux is quite different due to gas compressibility.

Now we discuss wall heat transfer modeling. Heat transfer in an engine is a classic subject of engine research. It is important because it influences engine efficiency, exhaust emissions, and component thermal stresses. Accurate prediction of wall heat transfer is not only needed for a better understanding of heat loss mechanism, but also necessary for improving the overall accuracy of engine simulation. Heat flux through combustion chamber walls is mainly due to gas-phase convection, fuel film conduction, and high-temperature gas and soot radiation. In many cases such as premixed-charge engines and on surfaces of DI engines without spray impingement, gas-phase convective heat transfer is the major concern.

Engine heat transfer phenomena have been studied extensively for many decades. Numerous mathematical models have been proposed. The traditional models (correlations) based on dimensional analysis are useful from the viewpoint of global analyses [29]. However, they cannot provide spatial resolution. Additionally, these models lack a sound theoretical basis and their predictions are often inaccurate when applied beyond the conditions under which their empirical constants are determined.

Since the boundary layer of an engine in-cylinder flow is thin, relative to practical computational grid sizes, velocity and temperature wall functions are often used in engine multidimensional computations to solve the near-wall shear stress and heat transfer as discussed earlier. Some models were proposed based on the turbulent boundary layer theory and incompressible flow assumptions [5,27]. However, predictions with the use of those models were not satisfactory. High underprediction of wall heat fluxes was found by Reitz [30].

The assumption of incompressibility for engine flows is inaccurate because the gas density varies significantly due to piston motion and combustion. From the derivation of the equations for wall heat flux, the effects of gas compressibility are apparent. The temperature wall function model for compressible flows as given in Equations (3.46) to (3.48) was demonstrated to have improved the prediction of engine heat transfer significantly [28].

Prediction of a premixed-charge SI engine is compared with experimental data to assess the heat transfer model as shown in Figure 3.2. The engine is operated at 1500 rpm and a middle-load. Simulations are made using the KIVA2 code. The computation monitoring locations are the same as the measurement locations labeled HT-1 to HT-5. Four radial locations of HT-1 to HT-4 are situated on the engine head, one (HT-5) on the cylinder liner. Alkidas (1980) gives more information on the measurement [31]. As seen in Figure 3.2, satisfactory prediction is obtained in terms of both the phase and magnitude of the heat flux. It is also seen that the heat flux peak values occur at later crank angle from HT-1 to HT-5 as the flame propagates across the chamber.

Gas compressibility influences wall heat transfer prediction significantly as indicated in Figure 3.3, in which the model using Equation (3.48) is referred to as the compressible model, the model using Equation (3.51) is referred to as the incompressible model. In addition, the Launder-Spalding (L-S) model for incompressible flows by Launder and Spalding [27] is also used for comparison. The L-S model is given as:

$$T^+ = \frac{Pr_t u^*}{(\tau_w/\rho U)} + Pr_t \frac{\pi/4}{\sin(\pi/4)}\left(\frac{A}{\kappa}\right)^{0.5}\left(\frac{Pr}{Pr_t}-1\right)\left(\frac{Pr_t}{Pr}\right)^{0.25} \tag{3.52}$$

FIGURE 3.2 Prediction of wall heat transfer flux in a premixed-charge SI engine using a temperature wall function model.

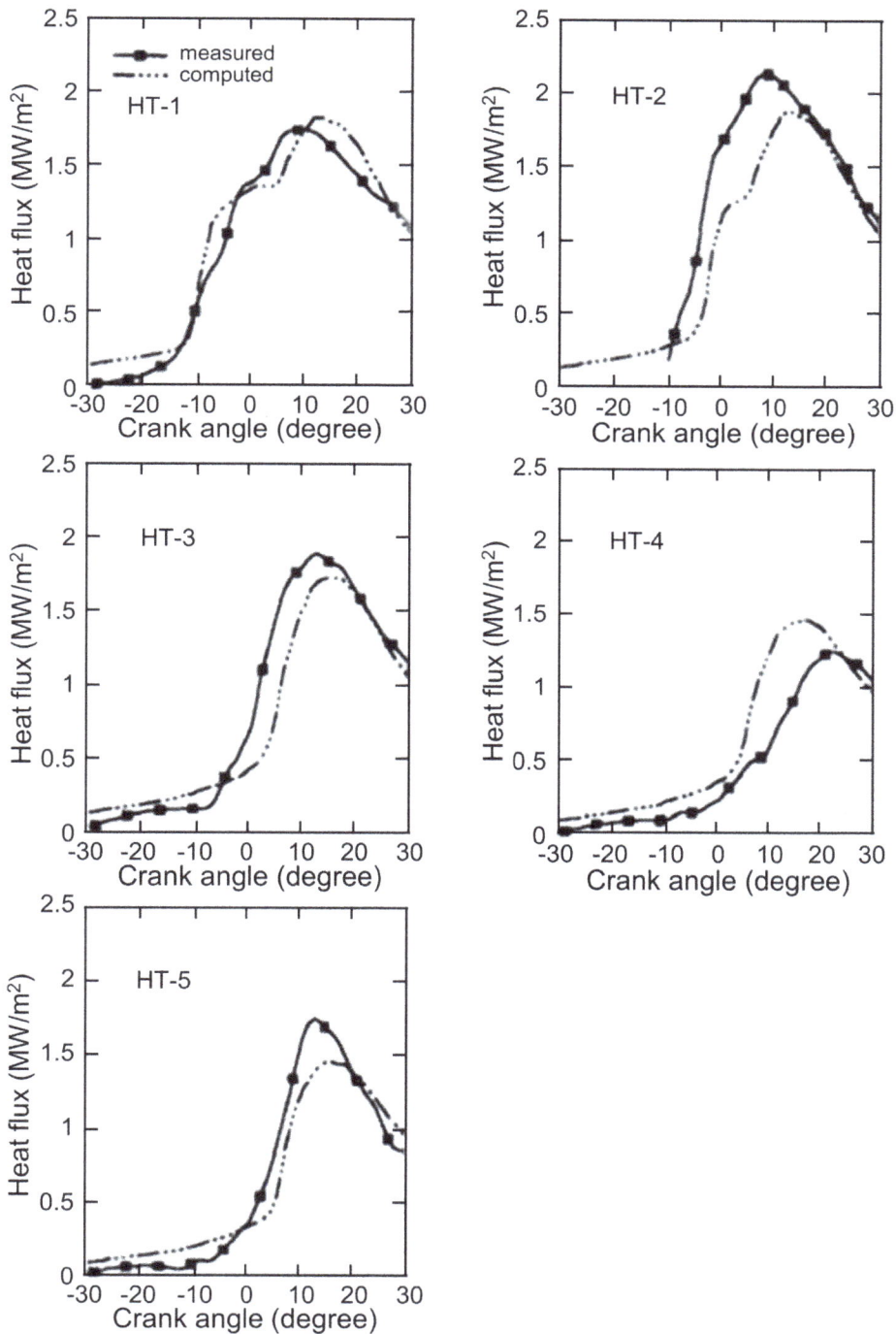

where T^+ is defined by Equation (3.49) and the Van Driest constant A is 26. The results at HT-1 in Figure 3.3 clearly show that the models excluding gas compressibility significantly underpredict heat transfer in the engine.

In the derivation of the present heat transfer model, the quasi-steady assumption is invoked as seen in Equation (3.40). As a result, the transient temperature term and the pressure work in the energy equation are absent in the model. The effects of unsteadiness were examined by adding an approximation term q_{ss} to the heat flux [28], and the results at HT-1 are illustrated in Figure 3.4. It is seen from Figure 3.4 (a) that the effect of unsteadiness is insignificant.

FIGURE 3.3   Comparison of predicted wall heat flux showing the effect of gas compressibility: (a) fired-engine case; (b) motored-engine case.

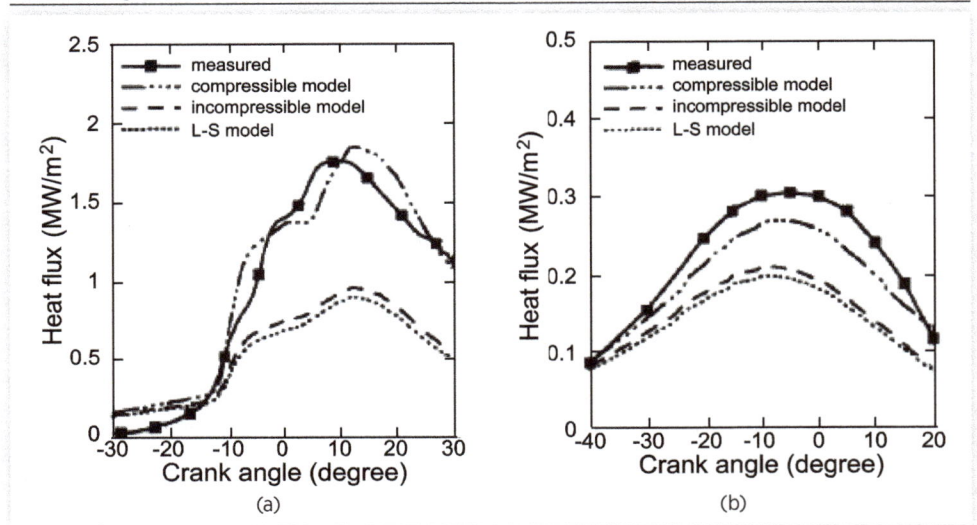

(a)   (b)

FIGURE 3.4   Comparison of predicted wall heat flux showing (a) the effects of unsteadiness and (b) chemical heat release.

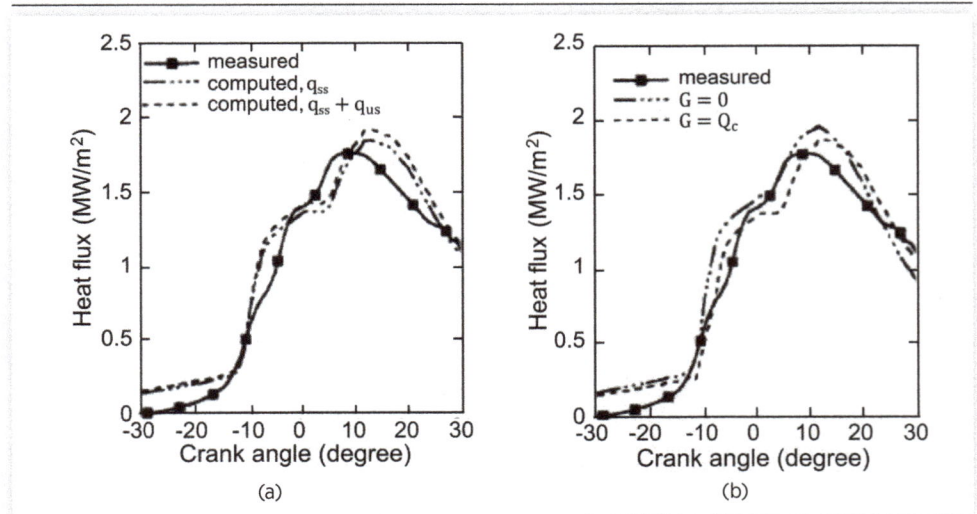

(a)   (b)

The effect of chemical heat release is included in the model. However, Figure 3.4 (b), in which the results of Equation (3.47) (including heat release) and Equation (3.48) (excluding heat release) are compared, indicates that overlooking the heat release source in the energy equation does not cause large errors in wall heat flux predictions.

Heat transfer in a Caterpillar diesel engine is also simulated [28]. In the simulation, three locations on the cylinder head, at 15, 45, and 60 mm from the cylinder axis, are chosen to monitor wall heat flux computations and are referred to as HT-1, HT-2, and HT-3. HT-1 is located near the injector, HT-2 is located over the piston bowl close to the edge of the bowl, and HT-3 is located over the squish area near the cylinder liner. In the Caterpillar engine, spray droplets impinge on the piston-bowl surface shortly after injection, and there is minimal spray wall interaction on the cylinder head. Hence, the neglect of spray effects in the present heat transfer model is valid for the analysis of the heat transfer process on the cylinder head.

The computed wall heat fluxes for the Caterpillar engine are shown in Figure 3.5 in which the heat release rate (HRR) is also given for reference. It is seen that, unlike the homogeneous-charge engine discussed in Figure 3.2, the wall heat-flux distribution in this diesel engine is highly uneven. Two heat-flux peaks are seen at HT-I. The first peak corresponds to the initial premixed combustion after ignition, as indicated by the HRR curve. The second peak is caused by the high gas temperature due to diffusion combustion. Heat flux at HT-2 increases rapidly at about 10° ATDC. At this time, both computations and experimental flame images indicate that the high-temperature flame reaches the edge of the piston bowl where HT-2 is located. Hence, the value of heat flux at HT-2 is as high as 10 MW/m². Later, at about 20° ATDC, flame extends into the squish region, and the wall heat flux at HT-3 increases. The predicted heat flux changes are consistent with the flame development process shown by the endoscope combustion images, and the magnitudes are also in the range of previously measured values in heavy-duty diesel engines [29].

FIGURE 3.5   Computed wall heat flux in a Caterpillar heavy-duty diesel engine. Heat release rate (HRR) is given to indicate flame development.

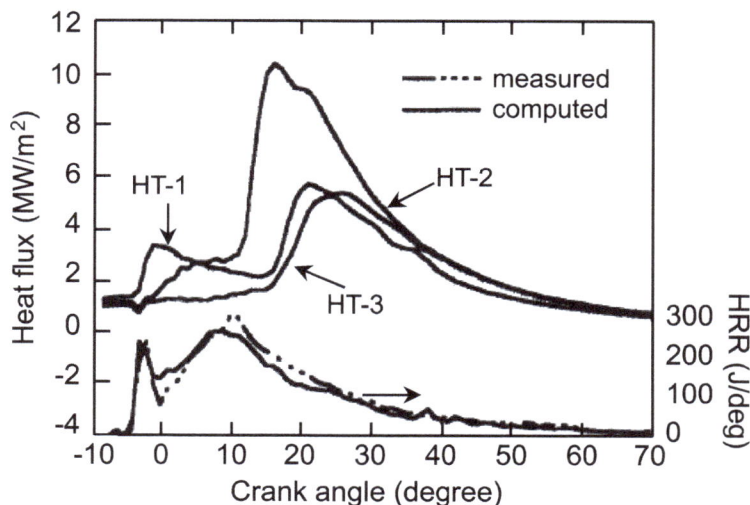

The compressible-flow wall function heat transfer model has been widely used due to its accuracy and simplicity. Rakopoulos et al. [32] performed a critical evaluation of serval heat transfer models through detailed comparisons with experimental data from three SI engines and three diesel engine cases with different CRs, swirl ratios and piston shapes. They have concluded that the model of Han and Reitz [28], although having a very simple formulation, performs satisfactorily. In most of the engines and operating conditions examined, the model captures the magnitude of the peak heat flux and its trend during the compression and expansion strokes, and the results are close to the experimental values. The model can be used in both SI and diesel engines, for low and high engine speeds and for variable CR and load.

References

1. Kuo, K.K. 1986. *Principles of Combustion.* 2nd Ed. Hoboken, New Jersey: Wiley-Interscience.

2. Dukowicz, J.K. 1980. "A particle-fluid numerical model for liquid sprays." *Journal of Computational Physics* 35, no. 2 (April): 229–253. DOI: 10.1016/0021-9991(80)90087-X

3. Reitz, R.D. 1996. "Computer Modeling of Sprays." Spray Technology Short Course Note. Pittsburgh, PA.

4. Williams, F.A. 1958. "Spray combustion and atomization." *Physics of Fluids* 1, no. 6: 541–545. DOI: 10.1063/1.1724379

5. Amsden, A.A., O'Rourke, P.J., and Butler, T.D. 1989. *KIVA-II: A Computer Program for Chemically Reactive Flows with Sprays* (LA-11560-MS). Los Alamos, New Mexico: Los Alamos National Laboratory.

6. Chung, T.J. 2010. *Computational Fluid Dynamics.* 2nd ed. New York, USA: Cambridge University Press.

7. Amsden, A.A., Ramshaw, J.D., O'Rourke, P.J., and Dukowicz, J.K. 1985. *KIVA: A Computer Program for Two- and Three-Dimensional Fluid Flows With Chemical Reactions and Fuel Sprays* (LA-10245-MS). Los Alamos, New Mexico: Los Alamos National Laboratory.

8. Amsden, A.A. 1993. *KIVA-3: A KIVA Program With Block-Structured Mesh for Complex Geometries* (LA-12503-MS). Los Alamos, New Mexico: Los Alamos National Laboratory.

9. Amsden, A.A. 1997. *KIVA-3V: A Block-Structured KIVA Program for Engines With Vertical or Canted Valves* (LA-13313-MS). Los Alamos, New Mexico: Los Alamos National Laboratory.

10. Hirt, C.W., Amsden, A.A., and Cook, J.L. 1974. "An Arbitrary Lagrangian-Eulerian Computing Method for All Flow Speeds." *Journal of Computational Physics* 14, no. 3 (March): 227–253. DOI: 10.1016/0021-9991(74)90051-5

11. Pracht, W.E. 1975. "Calculating Three-Dimensional Fluid Flows at All Speeds With an Eulerian-Lagrangian Computing Mesh." *Journal of Computational Physics* 17, no. 2 (February), 132–159. DOI: 10.1016/0021-9991(75)90033-9

12. Patankar, S.V., and Spalding, D.B. 1972. "A Calculation Procedure for Heat, Mass and Momentum Transfer in 3-D Parabolic Flows. *International Journal of Heat and Mass Transfer* 15: 1787–1806. DOI: 10.1016/B978-0-08-030937-8.50013-1

13. Hessel, R.P. 1993. Numerical Simulation of Valved Intake Port And In-Cylinder Flows Using KIVA3 (Publication No. 9318621). Madison, Wisconsin: University of Wisconsin-Madison. ProQuest Dissertations and Theses.

14. Han, Z. 1996. Numerical Study of Air-Fuel Mixing in Spark-Ignition and Diesel Engines (Publication No. 9634937). Madison, Wisconsin: University of Wisconsin-Madison. ProQuest Dissertations and Theses.

15. Yi, J., Han, Z., and Trigui, N. 2002. Fuel-Air Mixing Homogeneity and Performance Improvements of a Stratified-Charge DISI Combustion System. *SAE Transactions* 111, no. 4 (October): 965–975. DOI: 10.4271/2002-01-2656

16. Senecal, P.K., Richards, K.J., Pomraning, E., Yang, T., Dai, M.Z., McDavid, R.M., Patterson, M.A., Hou, S., and Shethaji, T. 2007. "A New Parallel Cut-Cell Cartesian CFD Code for Rapid Grid Generation Applied to In-Cylinder Diesel Engine Simulations." SAE Technical Paper No. 2007-01-0159. DOI: 10.4271/2007-01-0159

17. Richards, K.J., Senecal, P.K., and Pomraning, E. 2017. *CONVERGE Manual (v2.4)*. Madison, Wisconsin: Convergent Science Inc.

18. Rhie, C.M., and Chow, W.L. 1983. "Numerical Study of the Turbulent Flow Past an Airfoil With Trailing Edge Separation. *AIAA Journal* 21, no. 11: 1525–1532. DOI: 10.2514/3.8284

19. Issa, R.I. 1985. "Solution of the Implicitly Discretised Fluid Flow Equations by Operator-Splitting." *Journal of Computational Physics* 62, no. 1 (January): 40–65. DOI: 10.1016/0021-9991(86)90099-9

20. Wiedenhoefer, J.F., and Reitz, R.D. 2000. "Modeling the Effect of EGR And Multiple Injection Schemes on IC Engine Component Temperatures." *Numerical Heat Transfer, Part A: Applications* 37, no. 7: 673–694. DOI: 10.1080/104077800274028

21. Riegler, U.G., and Bargende, M. 2002. "Direct Coupled 1D/3D-CFD-Computation (GT-Power/Star-CD) of the Flow in the Switch-Over Intake System of an 8-Cylinder SI Engine With External Exhaust Gas Recirculation." *SAE Transactions* 111, no. 3 (March): 1554–1565. DOI: 10.4271/2002-01-0901

22. Kolade, B., Morel, T., and Kong, S. 2004. "Coupled 1-D/3-D Analysis of Fuel Injection And Diesel Engine Combustion." *SAE Transactions* 113, no. 3 (March): 515–524. DOI: 10.4271/2004-01-0928

23. Millo, F., Piano, A., Peiretti Paradisi, B., Marzano, M.R., Bianco, A., and Pesce, F.C. 2020. "Development And Assessment of an Integrated 1D-3D CFD Codes Coupling Methodology for Diesel Engine Combustion Simulation And Optimization. *Energies* 13, no. 7: 1612.

24. Gamma Technologies. 2015. *GTISE Help (Version 2016)*. Westmont, Illinois: Gamma Technologies LLC.

25. Wu, Z., and Han, Z. 2018. "Numerical Investigation on Mixture Formation in a Turbocharged Port-Injection Natural Gas Engine Using Multiple Cycle Simulation." *Journal of Engineering for Gas Turbines and Power* 140, no. 5 (May): article 051704. DOI: 10.1115/1.4039106

26. Warsi, Z.U.A. 2005. *Fluid Dynamics: Theoretical And Computational Approaches*. 3rd ed. CRC Press: Boca Raton, Florida.

27. Launder, B.E., and Spalding, D.B. 1974. "The Numerical Computation of Turbulent Flows." *Computer Methods in Applied Mechanics and Engineering* 3: 269–289. DOI: 10.1016/B978-0-08-030937-8.50016-7

28. Han, Z., and Reitz, R.D. 1997. "A Temperature Wall Function Formulation for Variable-Density Turbulent Flows With Application to Engine Convective Heat Transfer Modeling." *International Journal of Heat and Mass Transfer* 40, no. 3 (February): 613–625. DOI: 10.1016/0017-9310(96)00117-2

29. Borman, G., and Nishiwaki, K. 1987. "Internal-Combustion Engine Heat Transfer." *Progress in Energy and Combustion Science* 13, no. 1: 1–46. DOI: 10.1016/0360-1285(87)90005-0

30. Reitz, R.D. 1991. "Assessment of Wall Heat Transfer Models for Premixed-Charge Engine Combustion Computations." *SAE Transactions* 100, no. 3 (February); 397–413. DOI: 10.4271/910267

31. Alkidas, A.C. 1980. "Heat Transfer Characteristics of a Spark-Ignition Engine." *Journal of Heat Transfer* 102, no. 2 (May): 189–193. DOI: 10.1115/1.3244258

32. Rakopoulos, C.D., Kosmadakis, G.M., and Pariotis, E.G. 2010. "Critical Evaluation of Current Heat Transfer Models Used in CFD In-Cylinder Engine Simulations And Establishment of a Comprehensive Wall-Function Formulation. *Applied Energy* 87, no. 5 (May): 1612–1630. DOI: 10.1016/j.apenergy.2009.09.029

4

In-Cylinder Turbulence

Turbulence modeling is the most important and challenging part of IC engine simulation. Since the in-cylinder turbulent flows directly affect fuel spray processes, air–fuel mixing, and combustion in engines [1,2], adequate prediction of turbulence behavior is necessary for understanding these phenomena better for improving engine performance and emissions. This chapter starts off with the basic features of the in-cylinder turbulent flows. . After the introduction of the RANS theory, detailed descriptions of the widely-used classic k-ε model and the RNG k-ε model are given. Finally, the method and sub-grid models of LES are described along with several application examples of LES simulation of IC engines.

4.1 Turbulence Features in Reciprocating Engines

4.1.1 In-Cylinder Flows

In-cylinder flows in reciprocating IC engines are highly complex and turbulent. The flows are often expressed as the bulk (or mean) flows with turbulent eddies (small-scale fluctuations). To separate the in-cylinder flow structures into bulk motion and fluctuation is a matter of definition, which is important and useful. This concept lays the foundation for describing engine turbulent flows through experimental measurements and theoretical computations and the two parts can be compared to each other. It also supports engineering ideas to generate bulk rotating flows in engine designs.

During the induction process in a reciprocating engine, jet flows form when the gases pass through the intake valves. The interaction of the inflows with the cylinder wall and moving piston forms large-scale rotating motions, which are referred to as swirl and tumble as shown in Figure 4.1. In a 4-valve gasoline engine with a pent-roof dome, wall jet flows are generated, and the competition of the wall jets leads to the tumble motion in the cylinder as shown in in Figure 4.2. The tumble flows rotate about an axis orthogonal to the cylinder axis. The tumble motion structure in a DI gasoline engine is illustrated by the computed flow streamlines in Figure 4.3. It has been concluded that high level tumble motions lead to enhanced gas turbulence, which is in favor of improving engine thermal efficiency for DI gasoline engines [4] and PFI gasoline engines [5].

FIGURE 4.1 Schematic disgram of swirl and tumble in an IC engine.

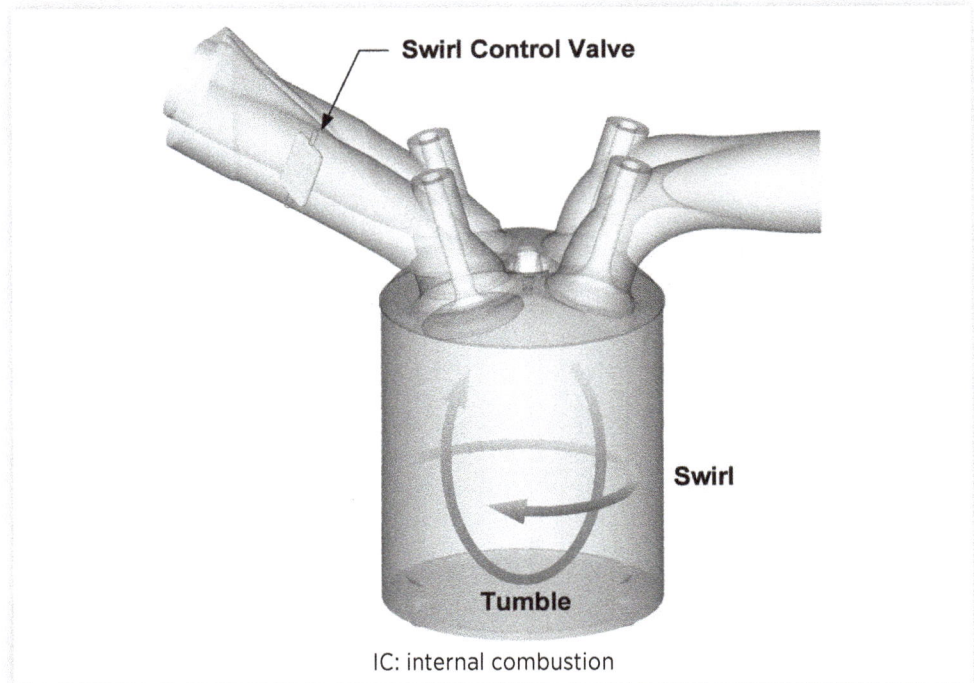

IC: internal combustion

Courtesy of Zhiyu Han

In a diesel engine, swirling motions about the cylinder axis are induced by the intake flow with the use of helical ports. While strong swirl motion is mostly seen in high-speed automotive diesel engines, it is also useful in DI gasoline engines, and improves the mixture homogeneity at full engine loads at low-speeds and combustion stability at part engine loads [6,7]. In this case, a swirl control valve is placed in one of the two intake ports as illustrated in Figure 4.1, the intake flows from the two ports are skewed when the swirl valve is partially or fully closed. Swirling flows are therefore induced and the strength of the swirl motion is determined by the openness of the swirl valve.

As the piston moves toward the TDC position, radial gas motions also occur in the piston bowls in diesel engines, or the domes of the cylinder heads usually adopted in gasoline engines. This type of flow is referred to as the squish flow. Squish also occurs when the piston moves away from the TDC, which is also called a reverse squish. The squish flow is found helpful to accelerate fuel–air mixing and flame diffusion in high-speed diesel engines with deep piston bowls. Theoretical correlations to calculate the squish velocity near the

FIGURE 4.2 Velocity contours on two cutting planes indicating the intake jet flows in a modeled engine. The stripes represent the high-velocity gas jets.

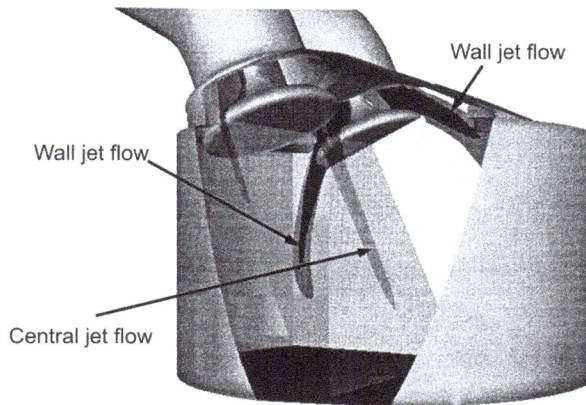

Wall jet flow

Wall jet flow

Central jet flow

FIGURE 4.3 Streamlines (ribbons) showing the tumble flow structure. The red color indicates higher velocity, and the blue lower velocity.

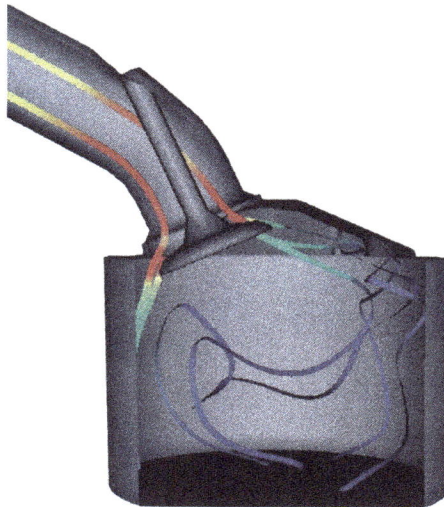

TDC are established with the assumption of incompressibility and mass conservation in the combustion chamber [2].

Strong interactions between the swirl flow and the squish flow result in dominant fluid mechanics in a combustion chamber of a bowl-in-piston diesel engine near the end of compression [8,9]. The squish flow significantly disrupts the radial distribution of the swirl velocity and causes large departures from a solid-body structure. The flow disruptions vary considerably with swirl level and bowl geometry and, in turn, change the transport of existing turbulence and generate additional turbulence.

In addition, the interactions between the fuel spray generated flow motion and the existing tumble, swirl, and squish flow add more complexity to the in-cylinder fluid mechanics. These interactions are of especial interest since combustion occurs after the fuel spray is injected. The superimposed fluid mechanics due to the spray–gas flow interactions ultimately affect the air–fuel mixing and flame development in the cylinder.

The enhancement of the flow turbulence by fuel injection is observed and qualified through measurements with a two-component, fiber-coupled laser Doppler velocimeter (LDV) in a four-valve diesel engine with re-entrant bowl-in-piston geometry [9]. The results in Figure 4.4 demonstrates the interaction of fuel sprays with the swirl and squish flows, resulting in enhanced local turbulence kinetic energy. The enhancement varies with the injection pressure as well as the swirl ratio (R_s). The effects of the fuel injection are more profound than that of the swirl. The local turbulence intensity is magnified by more than double.

FIGURE 4.4   Measured evolution of turbulent kinetic energy with fuel injection in a diesel engine.

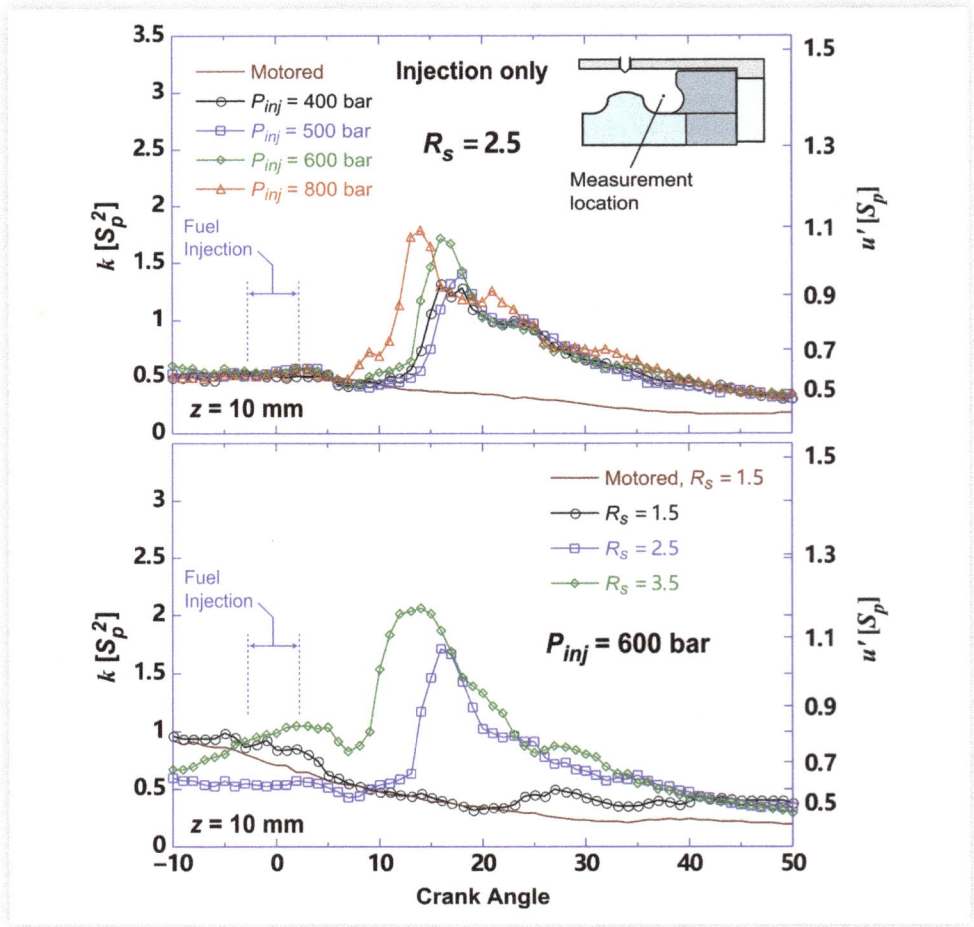

The effects of sprays on the gas turbulence in DI gasoline engines are investigated with the use of three-dimensional computations [3,10]. In the model engine, fuel sprays are vertically injected into the cylinders during the induction stroke to form a homogeneous mixture, where the injectors are centrally located on the top of the cylinder head. The engine operates at a wide-open throttle (WOT) and 1500 rpm condition with an injection-pressure of 4.76 MPa. When the fuel is injected, the intake-created tumble flow and the spray-induced flow interact with each other, which results in the enhanced central bulk flow. Also, the sprays cause gas entrainment forming strong vortexes at the periphery of the spray, which significantly suppresses the intake-created tumble motion of the gas flows shown in Figure 4.5. The degree of the suppression is dependent on SOI.

FIGURE 4.5 Effects of fuel injection on flow tumble ratio in a direct-injection gasoline engine.

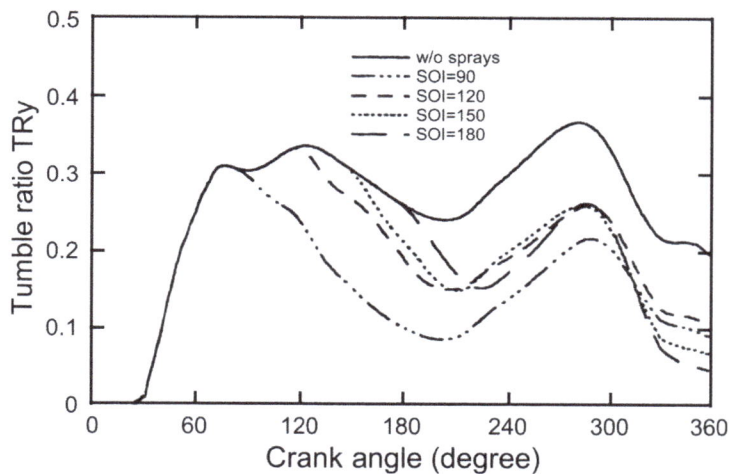

For the same engine, the mean velocity magnitudes along the line 30 mm below the head surface are plotted in Figure 4.6 in which zero-x represents the cylinder axis and the negative values of x-coordinate represent the distances away from the cylinder axis on the intake-valve side. This indicates that the local mean gas velocity is amplified by as much as 3.5 times by the injected spray in the core region of the cylinder for this engine configuration and injection scheme. Fuel injection also enhances the turbulence by the spray-induced velocity gradients in the spray region as indicated by Figure 4.7 in which the turbulence intensity is plotted on the same horizontal lines as those in Figure 4.6, although turbulence intensity is seen to be decreased along the cylinder axis.

The effects of the spray-induced flows are also examined by the global turbulence intensity, which is mass-averaged over the entire cylinder as shown in Figure 4.8. This computational quantity gauges the average level of the in-cylinder turbulence intensity and is useful in parametric comparisons. The spray-generated turbulence also decays during the compression process as the piston moves toward the TDC and the turbulence intensity near the TDC is substantially higher in the late injection cases.

Effects of fuel injection on gas mean velocity magnitude: (a) Monitoring line is located on the central plane between the two intake valves, and 30 mm below the head surface; (b) Monitoring line is located on the central plane orthogonal to the former plane, 30 mm below the head surface.

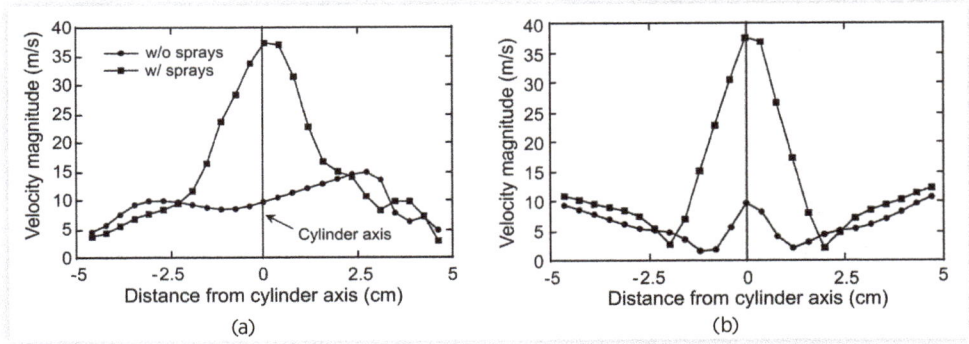

Reprinted with permission from Ref. [3]. © SAE International

FIGURE 4.7 Effects of fuel injection on gas turbulence intensity: (a) Monitoring line is located on the central plane between the two intake valves, and 30 mm below the head surface; (b) Monitoring line is located on the central plane orthogonal to the former plane, 30 mm below the head surface.

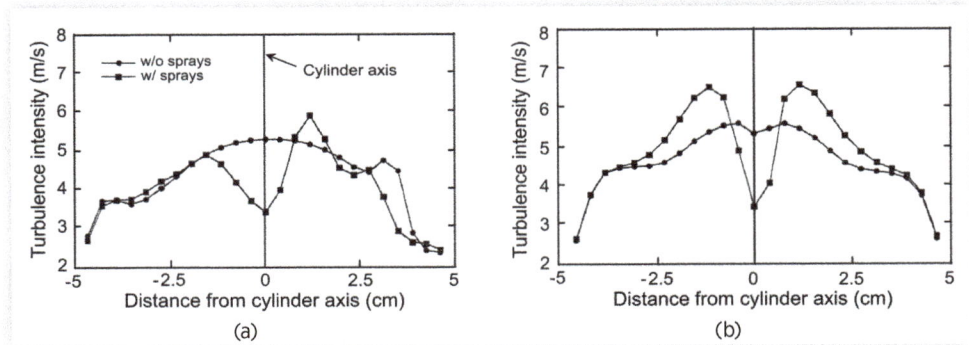

Reprinted with permission from Ref. [3]. © SAE International

FIGURE 4.8 Effects of fuel injection on cylinder-mass-averaged turbulence intensity in a direct-injection gasoline engine. The intensity is normalized by the mean piston speed.

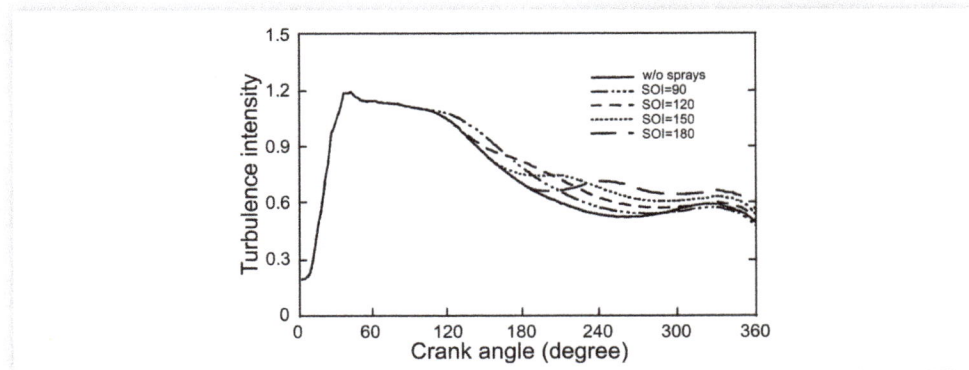

Reprinted with permission from Ref. [10]. © SAE International

©2022, SAE International

The general characteristics of the in-cylinder flows in reciprocating IC engines are summarized as:

1. Unsteady due to the piston motion.
2. Three-dimensional as a result of the engine geometry.
3. Turbulent at all engine speeds and for all inlet/cylinder dimensions.
4. Bulk flow in phase with the engine cycle.
5. Cycle-to-cycle variations in local flow properties.
6. Large-scale rotating flow persistent to the end of compression.
7. Additional turbulence generated due to fuel injection.

The large-scale structures of the in-cylinder flows (mean flows) generally decay during the compression process but maintain certain levels near the compression TDC, at which combustion takes place. The large-scale flows influence the transport of the spray droplets and vapors, helping air–fuel mixing. On the other hand, the small-scale fluctuations accelerate fuel vapor diffusion and mixing, and they are directly correlated with the turbulent flame speed. It has been found that the flow turbulence intensity near the compression TDC increases proportionally with the increase of the engine speed [2,11]. The value of turbulence intensity lies midway between half the mean piston speed and the mean piston speed as shown in Figure 4.9. Engines without significant swirl or tumble are at the lower end of this range. Small high-speed DI diesel engine chambers with deep, re-entrant-shaped bowl-in-piston geometries, with high swirl, are at the higher end. Gasoline engines with pent-roof combustion chamber designs have intermediate values of the ratio. Arguably, the proportionality of the in-cylinder turbulence intensity to engine speed is the most remarkable feature of engine in-cylinder flows. Since engine combustion rate is proportional to engine flow turbulence intensity, IC engines can run at speeds from several hundred revolutions per minute for marine engines to more than ten thousand revolutions per minute for racing engines.

FIGURE 4.9 Turbulence intensity at the compression TDC for several flow configurations and chamber geometries as a function of mean piston speed.

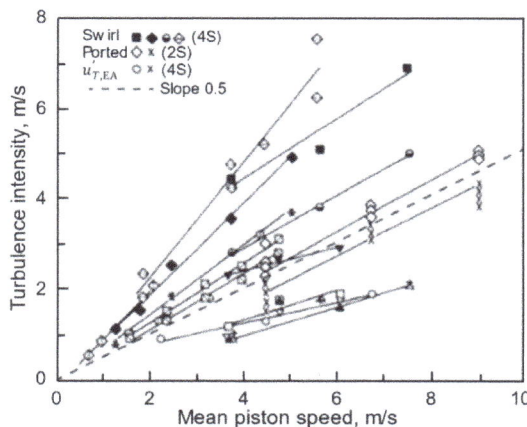

4.1.2 **Turbulence Scales**

The complexity of engine turbulence offers timely challenges to accurately describe it both experimentally and computationally since the turbulence process behaviors at a wide range of length and timescales in physical space. Widely accepted theories usually characterize turbulence by several length scales. Three types of turbulence-length scale are relevant in engine flows [12,13]. The integral scale L_I represents the extent of coherent motions and is a measure of the large eddies. The microscale or Taylor microscale L_M is believed to be a rough measure of the spacing of the very thin shear layers in which viscous dissipation occurs. The Kolmogorov scale L_K is a measure of the size of the smallest turbulence structures.

The integral length L_I and timescales τ_I are related by

$$L_I = \bar{U}\tau_I \tag{4.1}$$

where $\bar{U}$ is the mean velocity of the flow. In flows where the large-scale structures are convected, τ_I is a measure of the time it takes a large eddy to pass between two points. It is also an indication of the lifetime of a large eddy.

Turbulence energy dissipates the smallest flow structures, that is at the Kolmogorov scale level, and the kinetic energy in these smallest structures is dissipated through molecular viscosity into heat. Thus, the Kolmogorov length scale can be related by the energy dissipation rate ε and the gas molecular viscosity ν_0 as:

$$L_K = \left(\frac{\nu_0^3}{\varepsilon}\right)^{1/4} \tag{4.2}$$

and the corresponding Kolmogorov timescale is

$$\tau_K = \left(\frac{\nu_0}{\varepsilon}\right)^{1/2} \tag{4.3}$$

The microscale or Taylor microscale length scale L_M is defined by relating the fluctuating strain rate of the turbulent flow field to the turbulence intensity $u'/L_M \approx \partial u/\partial x$. For homogeneous, isotropic, and equilibrium turbulence, the microscale length scale L_M and timescale τ_M is related by

$$L_M = \bar{U}\tau_M \tag{4.4}$$

In this flow situation, the following relations are given as [14]:

$$\varepsilon = \frac{Au'^3}{L_I} = \frac{15\nu_0 u'^2}{L_M^2} \tag{4.5}$$

$$\frac{L_K}{L_I} \approx \left(\frac{u'L_I}{\nu_0}\right)^{-3/4} = Re_t^{-3/4} \tag{4.6}$$

$$\frac{L_M}{L_I} = \left(\frac{15}{A}\right)^{1/2} Re_t^{-1/2} \tag{4.7}$$

where $Re_t = u' L_I / v_0$ is the turbulent Reynolds number, and A is a constant of order one.

Equation (4.2) implies the assumption that the dissipation of turbulence energy is determined by the rate at which turbulence energy is provided at the integral scale level. Thus, evaluation of length scales and their evolution allows the examination of the turbulent dissipation rate. Hence, it is believed that the behavior of the in-cylinder turbulent flow can be characterized by monitoring the kinetic energy and the integral length-scale variation of the turbulent eddies that contribute to the turbulence production during the intake and compression processes [15]. The determination of turbulence length scales is, therefore, important in the study of in-cylinder flow processes.

Direct length-scale measurements have been made in engines. Collings et al. [16] and Dinsdale et al. [17] measured integral and micro length scales in a motored IC engine using LDA. Also, Ikegami et al. [18] measured the lateral integral length scale in an engine using a laser homodyne velocimetry (LHV) technique. The elongated-volume LDV technique was used by Fraser and Bracco [19] to measure the lateral fluctuation integral length scales for 64 crank-angle degrees (CA) near TDC. They showed that the length scale reached a minimum of 5-10° before TDC. Corcione and Valentino [14] measured the integral length scale in a medium-duty diesel engine equipped with a conventional toroidal combustion chamber using a two-probe volume LDV technique. They made measurements during the last 30° of the compression stroke and showed a monatomic decrease of the lateral integral scales until TDC. Miles et al. [20] recently measured the integral length scale in an engine with fuel injection in a re-entrant bowl-in-piston geometry.

Measurements of length scales reported in the literature are summarized by Miles [9]. The length scales reported are reasonably consistent and show little variation with the specific combustion chamber geometry employed. Most of the studies reporting the temporal evolution of the length scale indicate that it exhibits a broad minimum near TDC. The temporal evolution, as well as the magnitude of the length scale, has been shown to be reasonably insensitive to engine speed. This is because the spatial scale can be related to the mean velocity and a timescale, the mean motion scales with the engine speed while the timescale decreases, resulting in a spatial scale that is almost constant. Generally, the lateral length scales measured near TDC are in the range of 1–3 mm, while the longitudinal scales range from roughly 2–8 mm.

The length scales can be computed using turbulence models. The k-ε turbulence models (see Sec. 4.2) give for the turbulence macro length scale L_ε as:

$$L_\varepsilon = \frac{C_\mu^{3/4} k^{3/2}}{\kappa \varepsilon} \tag{4.8}$$

where κ is the von Karman constant.

Some studies [18,19] have been carried out to compare the measured integral length scale L_I with the predicted length scale L_ε on the assumption that L_ε is proportional (or equal) to L_I. However, the computations with this assumption have not been successful. An example is shown in Figure 4.10, where measured lateral integral length scale L_I is compared with the predicted length scale L_ε [21]. It is seen that both the commonly-used standard k-ε model and the RNG k-ε model (see Sec. 4.2) do not reproduce the measurements although the former appears to work better than the latter.

Reprinted with permission from Ref. [21]. © Elsevier

FIGURE 4.10 Comparison of the measured integral length scale and the k-ε model computed length scale under the assumption of equilibrium turbulence.

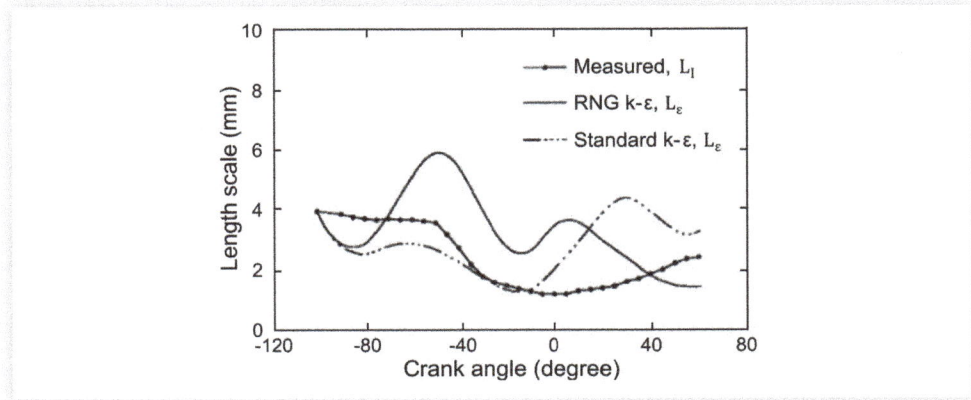

It is argued that the L_ε should not be directly proportional to the integral length scale L_I in engine flows considering the nonequilibrium nature of the engine turbulence flows [21]. Since the engine turbulence is nonequilibrium, the assumption that $L_I \sim L_\varepsilon$ for engine turbulence is quite questionable, and it may not be justified to directly compare the measured integral length scale with the predicted length scale.

Tennekes and Lumley [12] showed that the integral length scale is directly proportional to $k^{3/2}/\varepsilon$ for equilibrium turbulence in a steady, homogeneous, mean shear flow. In this case, the transport equation of turbulent kinetic energy reduces to

$$-\overline{u_i u_j} S_{ij} = 2\nu_0 \overline{s_{ij} s_{ij}} \tag{4.9}$$

where u_i is the fluctuating velocity, S_{ij} and s_{ij} are the mean and fluctuating rate of strain, respectively, and an overbar represents a Reynolds average. Equation (4.9) states that in this flow, the rate of production of turbulent energy by mean flow gradients equals the rate of viscous dissipation. If $S_{ij} \sim k^{1/2}/L_\varepsilon$ and $-\overline{u_i u_j} \sim u'^2$ are assumed and $\varepsilon = 2\nu_0 \overline{s_{ij} s_{ij}}$ is defined, then it follows that L_I is directly proportional to L_ε under the equilibrium turbulence condition.

In an engine, however, the flow undergoes rapid compression and expansion due to the piston movement, and engine turbulence is likely to be nonequilibrium in nature. The computed volume-averaged ratio of the turbulent-to-mean-strain timescale for the case in Figure 4.10 ranges mostly from two to three, which indicates the present engine flow is indeed evolving under rapid distortion conditions [21].

Wu et al. [22] performed direct numerical simulation (DNS) computations of flows under rapid compression conditions. They found that the Taylor length scale L_M and the integral length scale L_I remain proportional to each other during rapid compression. A similar finding was made experimentally by Dinsdale et al. in engine measurements [17]. Accordingly, $L_I \sim L_M$, and if it is assumed that the Taylor relation $\varepsilon = 10\nu_0 k/L_M^2$ holds also for isotropic compressed turbulence (which is reasonable because it follows from the definition of the Taylor length scale), we have $\varepsilon \sim \dfrac{k^{3/2}}{L_\varepsilon} \sim \dfrac{\nu_0 k}{L_M^2}$

therefore,

$$L_\varepsilon \sim L_I Re_t \ or \ L_\varepsilon = CL_I Re_t \tag{4.10}$$

where C is a constant, which can be shown to be

$$C = \left(\frac{L_\varepsilon}{L_I Re_t}\right)_{t=0} = \left(\frac{3}{2}\right)^{1/2} \frac{C_\mu^{3/4}}{\kappa} \left(\frac{v_0}{L_I^2} \frac{k}{\varepsilon}\right)_{t=0} \tag{4.11}$$

where the reference time $t=0$ is at the beginning of the computation.

The significance of Equation (4.10) is that it relates the integral length scale L_I (which can be measured) and the model length scale L_ε (which can be computed) for flows under rapid compression/expansion conditions. It shows that in a rapidly distorted isotropic turbulence, L_ε (or $k^{3/2}/\varepsilon$) is no longer directly proportional to L_I as it is in equilibrium turbulence. Instead, it is proportional to the product of L_I and the turbulent Reynolds number.

Figure 4.11 compares the measured integral length scale with the computed length scale deduced by using Equation (4.10). The value of the proportionality constant C in Equation (4.10) cannot be determined by using Equation (4.11) here, because ε at $t=0$ is not known. $C=0.005$ was chosen so that the measurement and computation could be compared in magnitude. As can be seen in Figure 4.11, the computed length scale using the standard and the RNG turbulence models now agrees with the measured scale very well in trend.

FIGURE 4.11 Integral length scales under the assumption of nonequilibrium and rapid distortion.

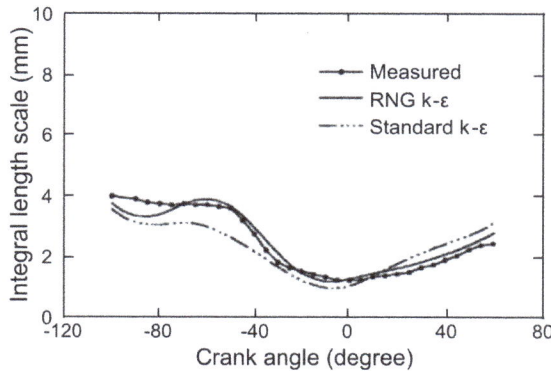

The physical interpretation of L_ε given by Equation (4.10) for rapidly compressed turbulence can also be used to assess the physical constraints on the length scales in engine flows. It is physically reasonable to expect that the size of large eddies should scale in some way with the cylinder dimension. Coleman and Mansour [23] show that the constraint satisfied during a rapid compression is $\rho L_M^3 = const$. Since $L_I \sim L_M$, we then have

$$\rho L_I^3 = const \tag{4.12}$$

Hence, for a rapidly compressed turbulence, the physical scaling postulate and its mathematical representation become consistent with each other. The size of large eddies, represented by L_p, decreases during the compression process and increases during the expansion process, as seen in Figure 4.11.

It is important to note that the non-equilibrium behavior of engine turbulence may lead to a reconstruction of the ε-equation. Although some modifications were introduced to the turbulent eddy-breakup timescale in the mixing-controlled combustion models for diesel engine combustion simulation [24,25], only marginal improvements were achieved. This may be due to the ad hoc way that was adopted.

4.2 RANS Methodology and Classical k-ε Models

4.2.1 RANS Methodology

The Navier–Stokes equations govern turbulent flows in theory. Limited by the computing capability in the foreseeable future, it is impossible to directly solve them for all the physical quantities in a time-dependent and three-dimensional space. As an approximation, the RANS method is used to describe the time averaged quantities of the flow field. The turbulent fluctuations are described through a Reynolds stress model or turbulent viscosity model. In the Reynolds averaging method, an instantaneous quantity q is decomposed into a mean $\bar{q}_i$ and a fluctuating quantity q'_i. That is:

$$q = \overline{q}_i + q'_i \tag{4.13}$$

Similarly, for the velocity and density:

$$u_i = \overline{u}_i + u'_i, \quad \rho = \overline{\rho}_i + \rho'_i \tag{4.14}$$

The time-averaged conservation equations can be obtained by applying the Reynolds averaging procedure. However, for compressible turbulent flows, density fluctuation will result in several quantities associated with density fluctuation, leading to some drawbacks [26]. Hence, mass–weight averaging (or Favre averaging) is used and the resulting equations are of simpler form. The mass-weighted mean velocity is defined as:

$$\tilde{u}_i = \frac{\overline{\rho u_i}}{\overline{\rho}} \tag{4.15}$$

The velocity may be then written as:

$$u_i = \tilde{u}_i + u''_i \tag{4.16}$$

Multiplying Equation (4.16) by ρ and decomposing the velocity, u_i, into two parts, we obtain

$$\rho u_i = \rho(\tilde{u}_i + u_i'') = \rho \tilde{u}_i + \rho u_i'' \tag{4.17}$$

By time-averaging the Equation (4.17), we have

$$\overline{\rho u_i} = \overline{\rho} \tilde{u}_i + \overline{\rho u_i''} \tag{4.18}$$

From the definition of $\tilde{u}_i$ given by Equation (4.16), we get

$$\overline{\rho u_i''} = 0 \tag{4.19}$$

Similarly, we can define other quantities including pressure, temperature, and internal energy.

By applying Eq. (4.19), it is straightforward to yield the relation for an instantaneous quantity q as:

$$\tilde{q} - \overline{q} = \frac{\overline{\rho' q''}}{\overline{\rho}} = \frac{\overline{\rho' q'}}{\overline{\rho}} \tag{4.20}$$

Equation (4.20) shows the difference between a time-averaged and a mass-weighted averaged quantity. For incompressible flows or if the density fluctuation can be ignored in compressible flows, the mass-weighted averaging approach becomes the same as the time averaging. In engine simulations, mass-weighted averaging is usually used to eliminate the density fluctuation terms.

Applying the Favre averaging to the governing equations of motion for a compressible, viscous, heat-conducting ideal gas flow as given in Chapter 3:

$$\frac{\partial \rho}{\partial t} + \frac{\partial (\rho u_j)}{\partial x_j} = 0 \tag{4.21}$$

$$\frac{\partial (\rho u_i)}{\partial t} + \frac{\partial (\rho u_i u_j)}{\partial x_j} = -\frac{\partial p}{\partial x_i} + \frac{\partial \sigma_{ij}}{\partial x_j} + \rho g_i \tag{4.22}$$

$$\frac{\partial (\rho e)}{\partial t} + \frac{\partial (\rho u_j e)}{\partial x_j} = \frac{\partial p}{\partial t} + \frac{\partial}{\partial x_j}(u_i \sigma_{ij} - q_j) \tag{4.23}$$

$$q_j = -\lambda \frac{\partial T}{\partial x_j} \tag{4.24}$$

We then obtain the conservation equations for compressible turbulent flows.

1. The mass conservation equation:

$$\frac{\partial \overline{\rho}}{\partial t} + \frac{\partial (\overline{\rho}\tilde{u}_j)}{\partial x_j} = 0 \tag{4.25}$$

2. The momentum conservation equation:

$$\frac{\partial (\overline{\rho}\tilde{u}_i)}{\partial t} + \frac{\partial (\overline{\rho}\tilde{u}_i\tilde{u}_j)}{\partial x_j} = -\frac{\partial \overline{p}}{\partial x_i} + \frac{\partial}{\partial x_j}\left(\overline{\sigma}_{ij} - \overline{\rho u_i'' u_j''}\right) + \overline{\rho}g_i \tag{4.26}$$

3. The energy conservation equation:

$$\frac{\partial (\overline{\rho}\tilde{e})}{\partial t} + \frac{\partial (\overline{\rho}\tilde{e}\tilde{u}_j)}{\partial x_j} = \frac{\partial \overline{p}}{\partial t} + \tilde{u}_j\frac{\partial \overline{p}}{\partial x_j} + \overline{u_j''\frac{\partial p}{\partial x_j}} + \frac{\partial}{\partial x_j}(-\overline{q}_j - \overline{\rho e'' u_j''}) + \overline{\sigma}_{ij}\frac{\partial \tilde{u}_i}{\partial x_j} + \overline{\sigma_{ij}\frac{\partial u_i''}{\partial x_j}} \tag{4.27}$$

In Equations (4.26), $\overline{\rho u_i'' u_j''}$ is the Reynolds stress tensor, and represents the turbulent stresses due to turbulent diffusion of momentum. This term is unknown and, thus, it makes these governing equations unclosed, and needs to be modeled.

In a multi-component gas mixture system, the mass and momentum conservation equations for compressible turbulent flows with sprays can be generalized as follows:

$$\frac{\partial \overline{\rho}}{\partial t} + \frac{\partial (\overline{\rho}\tilde{u}_j)}{\partial x_j} = \overline{\dot{\rho}^s} \tag{4.28}$$

$$\frac{\partial (\overline{\rho}\tilde{u}_i)}{\partial t} + \frac{\partial (\overline{\rho}\tilde{u}_i\tilde{u}_j)}{\partial x_j} = -\frac{\partial \overline{p}}{\partial x_i} + \frac{\partial}{\partial x_j}(\overline{\sigma}_{ij} - \overline{\rho u_i'' u_j''}) + \overline{\rho}g_i + \overline{F_i^s} \tag{4.29}$$

Various models have been proposed to approximate the solution of the Reynolds stress [27,28], which are usually referred to as turbulence models. The models can be divided into three categories or groups in general. In the first group, the models are based on the mixing length hypothesis in which the eddy or turbulent viscosity is assumed. The turbulent viscosity is then related to flow velocity gradients and a mixing length. These models are simple, requiring no additional differential equations to be solved. In the second group, turbulence kinetic energy k and its dissipation rate ε are modeled through two additional transport equations for k and ε, respectively. Models in this group are, therefore, often called the two-equation models. The transport equations of the Reynolds stress are established and solved for each component of the Reynolds stress in the third group, which are referred to as the Reynolds stress models.

The RANS methodology is widely used in the simulation of engineering flows due to their computation simplicity. For combustion simulation in IC engines, the RANS is still the dominant method, while other methods to solve for turbulence are also proposed including

LES [29,30] and the probability density function (PDF) method [31,32]. Two-equation models, particularly the k-ε models, are the most widely used turbulence models because of their simplicity and effectiveness, while the Reynolds stress models for engine simulation are rarely used [33]. Therefore, we will focus on the two-equation models, including the classic k-ε [34] and the RNG k-ε model [35], in the following section.

4.2.2 The Classical k-ε Model

In the early 1970s, the well-known classic (or standard) k-ε model was proposed for incompressible flows [34,36]. A linear relationship between the Reynolds stress tensor and rate of strain is assumed as:

$$-\overline{\rho u_i'' u_j''} = \mu_t \left(\frac{\partial \tilde{u}_i}{\partial x_j} + \frac{\partial \tilde{u}_j}{\partial x_i} \right) - \frac{2}{3} \delta_{ij} \left(\overline{\rho} k + \mu_t \frac{\partial \tilde{u}_k}{\partial x_k} \right) \tag{4.30}$$

where the turbulence kinetic energy, k, is defined as:

$$k = \frac{1}{2} \overline{u_i'' u_i''} \tag{4.31}$$

And the Reynolds stress tensor can be written as:

$$\overline{\tau}_{ij} = \overline{\rho u_i'' u_j''} = -2\mu_t \overline{S}_{ij} + \frac{2}{3} \overline{\rho} k \delta_{ij} \tag{4.32}$$

where $\overline{S}_{ij}$ is the strain-rate tensor. The turbulent (or eddy) dynamic viscosity μ_t is given by:

$$\mu_t = C_\mu \overline{\rho} \frac{k^2}{\varepsilon} \tag{4.33}$$

where ε is the turbulence dissipation rate and C_μ is a model constant. The transport equations for the turbulence kinetic energy k and turbulence dissipation rate ε can be derived as:

$$\frac{\partial(\overline{\rho}k)}{\partial t} + \frac{\partial(\overline{\rho}k\tilde{u}_j)}{\partial x_j} = \frac{\partial}{\partial x_j} \left[\left(\frac{\mu_t}{\sigma_k} + \mu \right) \frac{\partial k}{\partial x_j} \right] - \overline{\tau}_{ij} \frac{\partial \tilde{u}_i}{\partial x_j} - \overline{\rho}\varepsilon \tag{4.34}$$

$$\frac{\partial(\overline{\rho}\varepsilon)}{\partial t} + \frac{\partial(\overline{\rho}\varepsilon\tilde{u}_j)}{\partial x_j} = \frac{\partial}{\partial x_j} \left[\left(\frac{\mu_t}{\sigma_\varepsilon} + \mu \right) \frac{\partial \varepsilon}{\partial x_j} \right] - \frac{\varepsilon}{k} \left(C_1 \overline{\tau}_{ij} \frac{\partial \tilde{u}_i}{\partial x_j} - C_2 \overline{\rho}\varepsilon \right) \tag{4.35}$$

where σ_k and σ_ε are the turbulent Prandtl numbers for k and ε, respectively. The constants in Equations (4.34) and (4,35) are recommended by Launder and Spalding [34] as:

$$C_\mu = 0.09, \quad C_1 = 1.44, \quad C_2 = 1.92$$
$$\sigma_k = 1.0, \quad \sigma_\varepsilon = 1.30$$

The classic k-ε turbulence model has been developed for incompressible thin shear flows and it has been proven to be accurate and sufficient in a wide arrange of engineering computations. Since turbulence flows in engines are undergoing density changes due to piston motion and combustion, the fluid compressibility needs to be addressed. Gosman and Watkins [37] extended the k-ε model to an engine flow. The mass-weighted averaging technique is employed and there is little or no modification in the form of the models.

The standard k-ε model for engine turbulence flows is given in Equations (4.36) and (4.37) with the consideration of density variation and fuel sprays [38]. The vector differential operator ∇ is used and the overbar symbol representing the mass-average is omitted in the equations for brevity.

$$\frac{\partial \rho k}{\partial t} + \nabla \cdot (\rho \boldsymbol{u} k) = -\frac{2}{3}\rho k \nabla \cdot \boldsymbol{u} + \boldsymbol{\sigma} : \nabla \boldsymbol{u} + \nabla \cdot \left[\left(\frac{\mu_t}{\sigma_k} \right) \nabla k \right] - \rho \varepsilon + \dot{W}^s \tag{4.36}$$

$$\frac{\partial \rho \varepsilon}{\partial t} + \nabla \cdot (\rho \boldsymbol{u} \varepsilon) = -\left(\frac{2}{3}C_1 - C_3 \right) \rho \varepsilon \nabla \cdot \boldsymbol{u} + \nabla \cdot \left[\left(\frac{\mu_t}{\sigma_\varepsilon} \right) \nabla \varepsilon \right]$$
$$+ \frac{\varepsilon}{k}(C_1 \boldsymbol{\sigma} : \nabla \boldsymbol{u} - C_2 \rho \varepsilon + C_s \dot{W}^s) \tag{4.37}$$

where $\boldsymbol{u}$ is the gas velocity vector, and $\boldsymbol{\sigma}$ is the viscous stress tensor, which is given as:

$$\boldsymbol{\sigma} = 2\mu \left(\nabla \boldsymbol{u} - \frac{1}{3}\delta_{ij} \nabla \cdot \boldsymbol{u} \right) \tag{4.38}$$

Note that there are some terms added to the equations for incompressible flows as given in Equations (4.34) and (4.35). The term $(2C_1/3 - C_3)\nabla \cdot \boldsymbol{u}$ in the ε-equation accounts for length scale change when the velocity dilatation is not zero for compressible flows. $C_3 = -1.0$ is recommended by Amsden et al. [38] in the KIVA codes. Source term involving the quantity $\dot{W}^s$ arise due to interaction with the spray. It is suggested $C_s = 1.5$, based on the postulate of length scale conservation in interactions of spray and turbulence, and other model constants are the same as given previously.

To account for flow compressibility in engines, research was carried out on the effect of velocity dilatation on turbulence dissipation rate, that is, on the ε-equation, which led to various values for C_3. Reynolds [13] argued that the ε-equation of Gosman and Watkins [37] could not recover the rapid spherical distortion limit. Hence, a different C_3 constant for the velocity dilatation term was proposed from his rapid distortion analysis. Other values for C_3 were also suggested based on different assumptions [39–42]. The value of C_3 value in these assumptions ranged from –1.0 to 1.0. Since it is very difficult to validate turbulence models under engine conditions, there is still great uncertainty about the value for C_3 and its effects on engine combustion modeling remain unclear. Alternatively, a formulation to calculate C_3 is given in the RNG model in the next section.

4.3 RNG k-ε Models

A new k-ε turbulence model has been proposed by Yakhot and Orszag [43]. This model follows the two-equation framework and is derived from the fundamental governing equations for fluid flows using the RNG theory. In the high Reynolds number limit, the RNG k-ε model is almost identical in form to the standard k-ε model. One of the features of this model is that all the model constants can be explicitly evaluated based on certain assumptions and mathematical development [44]. An additional term has been added in the ε-equation [45]. This term changes dynamically with the rate of strain of the turbulence, providing more accurate predictions for flows with rapid distortion and anisotropic large-scale eddies. Accurate predictions have been found in the modeling of homogeneous shear flows [46]. Applications of the RNG k-ε model to a number of complex flows such as separated flows have also yielded excellent results in cases where the standard k-ε model predictions have been unsatisfactory [47]. The RNG k-ε has been extended to compressible two-phase flows by Han and Reitz [35] for engine combustion simulation. In the following sections, the RNG method as well as the RNG k-ε models are described.

4.3.1 RNG Methodology

Application of RNG methods to the analysis of turbulence can be traced back to the end of 1970s. Forster et al. [48] studied a randomly stirred flow governed by the forced Navier–Stokes equation using RNG methods. However, the early RNG work on turbulence [48,49] led only to scaled exponents and did not predict the amplitudes of scaled quantities. Yakhot and Orszag [43] introduced the renormalized perturbation theory combined with RNG calculations to analyze an incompressible flow subject to a random force. Their analysis quantitatively fixed the amplitudes and thereby flow constants, and gave primary evaluation of turbulence models such as two-equation transport approximations and sub-grid scale models for LES.

Derivation of the k-ε equation using RNG methods was originated by Yakhot and Orszag [43]. However, several problems including numerical errors in the derivation were indicated by Smith and Reynolds [50]. Yakhot and Smith [44] rederived the ε-equation in which the model constants were reevaluated. These constants are close to those commonly used in the standard k-ε model. More importantly, an extra term in the ε-equation was revealed, which was believed to be important for rapid distortion limits. This extra term has been closed by Yakhot et al. [45] via an infinite scale expansion in η, the ratio of the turbulent to mean strain timescale.

It should be mentioned that the RNG turbulence theory is based on certain assumptions and approximations. For example, a scale-invariant random force was introduced into the Navier–Strokes equation, and an ε-expansion technique was used as ε was set to 4 [43]. These assumptions, as well as the RNG procedure, are discussed in detail by several researchers, and different approaches have been proposed [51,52]. It is clear that more justifications are needed for the formulation of a complete RNG turbulence theory. However, our interest in this research is to study the performance of the established RNG k-ε model under engine flow conditions and issues related to the justification of the RNG theory itself are excluded from consideration here.

The main idea of the RNG methods and the high Reynolds number form of the RNG k-ε model are summarized here. An incompressible turbulent flow of unit density, velocity u, pressure p, and kinematic viscosity v_0 is assumed to be fully described by

$$\frac{\partial u_i}{\partial t} + u_j \frac{\partial u_i}{\partial x_j} = -\frac{\partial p}{\partial x_i} + v_0 \nabla^2 u_i + f_i \qquad (4.39)$$

and

$$\frac{\partial u_i}{\partial x_i} = 0 \tag{4.40}$$

where f_i is a Gaussian random force that is statistically isotropic and divergence-free. The significance of this force is discussed in detail by Lam [51] who shows that it arises from a filtering operation on high frequency eddies. In modeling of the turbulence stress (e.g., Reynolds stress) $\bar{\tau}_{ij} = \overline{u_i' u_j'}$, where u_i' is the fluctuating velocity and an overbar represents a Reynolds mean, the isotropic eddy viscosity model is used. Hence, $\bar{\tau}_{ij} = \frac{2}{3}k\delta_{ij} - \nu_t(\partial \bar{u}_i / \partial x_j + \partial \bar{u}_j / \partial x_i)$, and δ_{ij} is the Kronecker delta tensor.

The turbulent viscosity ν_t is modeled as:

$$\nu_t = C_\mu \frac{k^2}{\varepsilon} \tag{4.41}$$

where $k = \left(\overline{u_i' u_i'}\right)/2$ is the turbulent kinetic energy, $\varepsilon = \nu_0 \overline{\partial u_i' / \partial x_j \partial u_i' / \partial x_j}$ is the turbulence dissipation rate, and C_μ is a constant. The dynamic equations for the kinetic energy and the homogenous part of the instantaneous rate of kinetic energy dissipation can be obtained from Equations (4.39) and (4.40) as:

$$\frac{\partial k}{\partial t} + u_i \frac{\partial k}{\partial x_i} = -\nu_0 \frac{\partial u_i}{\partial u_j}\frac{\partial u_i}{\partial u_j} + \nu_0 \nabla^2 k - \frac{\partial u_i p}{\partial x_i} + u_i f_i \tag{4.42}$$

$$\frac{\partial \varepsilon}{\partial t} + u_i \frac{\partial \varepsilon}{\partial x_i} = 2\nu_0 \frac{\partial u_i}{\partial x_j}\frac{\partial f_i}{\partial x_j} - 2\nu_0 \frac{\partial u_i}{\partial x_j}\frac{\partial u_l}{\partial x_j}\frac{\partial u_i}{\partial x_l} - 2\nu_0^2 \left(\frac{\partial^2 u_i}{\partial x_j \partial x_j}\right)^2$$
$$- 2\nu_0 \frac{\partial u_i}{\partial x_j}\frac{\partial^2 p}{\partial x_i x_j} + \nu_0 \nabla^2 \varepsilon \tag{4.43}$$

However, the higher-order terms in Equations (4.42) and (4.43) make these transport equations unsolvable in the sense of modeling. These higher-order terms must be closed in terms of lower-order or mean flow quantities.

The way to close the instantaneous transport equations for turbulence parameters, as in Equations (4.42) and (4.43) constitutes one of the major differences between the RNG approach and the standard or classic approach. In the classic approach empirical reasoning, dimensional analysis, as well as invariance constraints are used, and the model constants are evaluated by reference to benchmark experiments. The RNG approach applies the dynamic RNG theory and the ε-expansion to deal with the fundamental equations. RNG is a type of mathematical operation that consists of a continuous family of transformations. Applying the RNG methods to Equations (4.42) and (4.43), and (4.39) and (4.40), bands of high wave numbers (i.e., small scales) are systematically removed. The effect of the small scales, which have been removed, is accounted for by the dynamic equations for the renormalized (large scale) velocity field through the presence of an eddy viscosity. In the limit, as successively larger scales are removed, the k-ε model is derived. The constants of the model are calculated explicitly. The resulting transport equations read as:

$$\frac{\partial k}{\partial t} + \overline{u}_i \frac{\partial k}{\partial x_i} = 2\nu_t S_{ij} S_{ij} - \varepsilon + \frac{\partial}{\partial x_i}\left(\alpha_k \nu \frac{\partial k}{\partial x_i}\right) \tag{4.44}$$

$$\frac{\partial \varepsilon}{\partial t} + \overline{u}_i \frac{\partial \varepsilon}{\partial x_i} = 2C_1 \frac{\varepsilon}{k} \nu_t S_{ij} S_{ij} - C_2 \frac{\varepsilon^2}{k} - R + \frac{\partial}{\partial x_i}\left(\alpha_\varepsilon \nu \frac{\partial k}{\partial x_i}\right) \tag{4.45}$$

where

$$R = \frac{\nu_t S^3 \left(1 - \eta/\eta_0\right)}{1 + \beta \eta^3} \tag{4.46}$$

or, alternatively

$$R = \frac{C_\mu \eta^3 (1 - \eta/\eta_0)}{1 + \beta \eta^3} \frac{\varepsilon^2}{k} \tag{4.47}$$

The ratio of the turbulent to mean strain-timescale η is given as:

$$\eta = S\frac{k}{\varepsilon} \tag{4.48}$$

where $S = \sqrt{2 S_{ij} S_{ij}}$ is the magnitude of the mean strain rate, $S_{ij} = (\partial \bar{u}_i/\partial x_j + \partial \bar{u}_j/\partial x_i)/2$, and $\nu = \nu_0 + \nu_t$. The model constants are given in Table 4.1. For comparison, the constants used in the KIVA standard k-ε model are also listed.

TABLE 4.1   Model Constants in the k-ε Models

	α_k	α_ε	C_μ	C_1	C_2	C_3	η_0	β	C_s
RNG k-ε [35]	1.39	1.39	0.0845	1.42	1.68	Eq. (4.64)	4.38	0.012	1.5
Standard k-ε [38]	1.0	1/1.3	0.09	1.44	1.92	-1.0	-	-	1.5

Courtesy of Zhiyu Han

Compared with the standard k-ε model, the k-equation remains the same in the RNG version. However, an extra term R appears in the ε-equation. It is closed by an approximation rather than by a rigorous RNG derivation. R is small for weakly strained turbulence, such as a homogenous shear flow, and is large in the rapid distortion limit when η approaches infinity. Choudhury et al. [47] explained how R changes dynamically with the mean-strain rate. In regions of large η, the sign of R is changed and the turbulent viscosity is decreased accordingly. Hence, they concluded that this feature of the RNG k-ε model was responsible for the improvement of their modeling of separated flows.

4.3.2 **The RNG k-ε Model for Variable-Density Flows**

Turbulence in IC engines is complicated. Although the Mach number of an in-cylinder flow is usually low, the moving piston makes the fluid undergo large density variations. With such flow behavior, the turbulence model discussed in Sec. 4.3.1 is not directly applicable. Flow compressibility must be considered. The RNG k-ε model for variable-density flows proposed by Han and Reitz [35] is thus applicable and is discussed here.

In the history of engine turbulence modeling, consideration of compressibility has mainly focused on the effect of velocity dilatation on the turbulence dissipation rate [13,40,41]. However, research with the aid of DNS has revealed that compressibility also influences the evaluation of the turbulent kinetic energy due, in part, to pressure dilatation [53]. The effect of compressibility depends on Mach number; rapidity of compression and mean strain modes (one-dimensional, planar, spherical, etc.) [54].

Since a model that can account for arbitrary strain and is independent of strain rate at low Mach number is unavailable, the present work closes the RNG ε-equation under the conditions of low Mach number by assuming rapid distortion and isotropic (spherical) mean strain.

The RNG k-ε model applied to a compressible turbulent flow is given as:

$$\rho \frac{Dk}{Dt} = P - \rho\varepsilon + \frac{\partial}{\partial x_i}\left(\alpha_k \mu \frac{\partial k}{\partial x_i}\right) \tag{4.49}$$

$$\rho \frac{D\varepsilon}{Dt} = \frac{\varepsilon}{k}(C_1 P - C_2 \rho\varepsilon) - \rho R + C_3 \rho\varepsilon \nabla \cdot \boldsymbol{u} + \frac{\partial}{\partial x_i}\left(\alpha_\varepsilon \mu \frac{\partial \varepsilon}{\partial x_i}\right) \tag{4.50}$$

where the production of k is modeled using the isotropic eddy viscosity assumption as:

$$P = 2C_\mu \rho \frac{k^2}{\varepsilon}\left(S_{ij}S_{ij} - \frac{1}{3}(\nabla \cdot \boldsymbol{u})^2\right) - \frac{2}{3}\rho k \nabla \cdot \boldsymbol{u} \tag{4.51}$$

The form of the k-equation given in Equation (4.49) is justified since the effects of both the pressure dilatation and dilatation dissipation are negligible under the assumption of isotropic distortion at very low Mach number [54]. Hence, no further work is needed for the k-equation. The formulation of the ε-equation is similar to that of the standard one. However, the constant C_3 is unknown. Here C_3 is determined through the analysis of rapid spherical distortion. This method was originated by Reynolds [13] and extended by Coleman and Mansour [23].

For a rapid isotropic compression at a low Mach number, turbulence can be described by

$$\frac{dk}{dt} = -\frac{2}{3}(\nabla \cdot \boldsymbol{u})k \tag{4.52}$$

$$\frac{d\omega^2}{dt} = -\frac{4}{3}(\nabla \cdot \boldsymbol{u})\omega^2 \tag{4.53}$$

where $\omega^2 = \overline{\omega_i \omega_i} = \varepsilon / v_0$. On the other hand, the ε-equation, Equation (4.50), can be applied to this flow as well. Under spherical distortion, it is obvious that $S = \sqrt{2/3} |\nabla \cdot \boldsymbol{u}|$, so that

$$P = -\frac{2}{3}\rho k (\nabla \cdot \boldsymbol{u}) \tag{4.54}$$

and

$$R = \sqrt{2/3}\, C_u C_\eta \eta \varepsilon |\nabla \cdot \boldsymbol{u}| \tag{4.55}$$

where

$$C_\eta = \frac{\eta(1 - \eta / \eta_0)}{1 + \beta \eta^3} \tag{4.56}$$

Thus, Equation (4.50), as the convection and diffusion terms vanish due to rapidity, reduces to

$$\frac{d\varepsilon}{dt} = \left(-\frac{2}{3}C_1 + C_3\right)\varepsilon \nabla \cdot \boldsymbol{u} - C_2 \frac{\varepsilon^2}{k} - \sqrt{2/3}\, C_u C_\eta \eta \varepsilon |\nabla \cdot \boldsymbol{u}| \tag{4.57}$$

For a homogeneous low Mach number flow, as mentioned earlier, the effect of dilatation dissipation is negligible. Hence $\varepsilon \equiv v_0 \omega^2$. Then, Equation (4.57) becomes

$$\frac{d\omega^2}{dt} = \left[\left(-\frac{2}{3}C_1 + C_3\right)\nabla \cdot \boldsymbol{u} - \frac{1}{v_0}\frac{dv_0}{dt} - \sqrt{2/3}\, C_u C_\eta \eta |\nabla \cdot \boldsymbol{u}| - C_2 \frac{v_0 \omega^2}{k}\right]\omega^2 \tag{4.58}$$

In the rapid distortion regime,

$$\left|S\frac{k}{\varepsilon}\right| \sim \left|(\nabla \cdot \boldsymbol{u})\frac{k}{\varepsilon}\right| \gg 1 \tag{4.59}$$

then

$$C_\eta \eta = -\frac{1}{\beta \eta_0} \tag{4.60}$$

and $v_0^{-1}\, dv_0/dt \sim 0(\nabla \cdot \boldsymbol{u})$ (see Eq. (4.63). The last term in the bracket on the right-hand side of Eq. (4.58) is relatively small, since it is of order ε/k while other terms are of order $\nabla \cdot \boldsymbol{u}$, and it can be neglected. Comparing Equations (4.58) and (4.53), for Equation (4.53) to be recovered by the ε-Equation (4.50), the model constants must satisfy

$$\left(-\frac{2}{3}C_1 + C_3\right)\nabla \cdot \boldsymbol{u} - \frac{1}{v_0}\frac{dv_0}{dt} - \sqrt{2/3}\, C_u C_\eta \eta |\nabla \cdot \boldsymbol{u}| = -\frac{4}{3}\nabla \cdot \boldsymbol{u} \tag{4.61}$$

Therefore

$$C_3 = \frac{-4 + 2C_1 + 3(v_0 \nabla \cdot \boldsymbol{u})^{-1}(dv_0/dt) + (-1)^\delta \sqrt{6C_\mu C_\eta}\eta}{3} \quad (4.62)$$

where the Kronecker delta is dependent on the sign of velocity dilatation that is

$$\begin{cases} \delta = 1, & \nabla \cdot \boldsymbol{u} < 0 \\ \delta = 0, & \nabla \cdot \boldsymbol{u} > 0 \end{cases}$$

For an ideal gas, the molecular viscosity $\mu_0 \propto T^m$, where T denotes temperature and $m = 0.5$ [55]. In a closed thermodynamic system, polytropic processes are described by $p/\rho^n = const$. A particular polytropic process is defined by a value of n, which ranges from zero to infinity. It is straightforward to show that

$$\left(v_o \nabla \cdot \boldsymbol{u}\right)^{-1} \frac{dv_0}{dt} = 1 - m(n-1) \quad (4.63)$$

Therefore, Equation (4.62) can be written as:

$$C_3 = \frac{-1 + 2C_1 - 3m(n-1) + (-1)^\delta \sqrt{6C_\mu C_\eta}\eta}{3} \quad (4.64)$$

Thus, the RNG ε-equation is closed with C_3 calculated by Equation (4.64). It is revealed by Equation (4.64) that, first, at the rapid distortion limit C_3 is dependent on the sign of velocity dilatation, that is, dependent on whether the distortion processes are compressing or expanding. Second, C_3 is also dependent on the exponent of a polytropic process, n, as well. The evaluation of turbulence dissipation is affected by the thermal details of compression/expansion processes. If the process is isothermal, i.e., $n=1$, the effect of viscosity–temperature variation on ε vanishes. In a limiting case, if the gas density does not change (a constant-volume process in a closed system), C_3 will approach infinity because of the infinite value of n. However, this effect will be canceled by the zero-velocity dilatation. In this case, the incompressible model is recovered. The value of n is determined by the details of heat transfer through boundaries in IC engines. Typically, n ranges between 1.3 to 1.4 depending on the engine design. Hence, the influence of the value used in this range is minor. For an adiabatic process with $n=1.4$ (assumed in the present study), as $C_\eta\eta$ is evaluated by Equation (4.60), C_3 is calculated to be -0.9 for positive velocity dilatation, and to be 1.726 for negative velocity dilatation.

Finally, we come to the high Reynolds number turbulence model formulation for compressible flows with sprays as:

$$\frac{\partial \rho k}{\partial t} + \nabla \cdot (\rho \boldsymbol{u} k) = -\frac{2}{3}\rho k \nabla \cdot \boldsymbol{u} + \boldsymbol{\sigma} : \nabla \boldsymbol{u} + \nabla \cdot (\alpha_k \mu \nabla k) - \rho \varepsilon + \dot{W}^s \quad (4.65)$$

$$\frac{\partial \rho \varepsilon}{\partial t} + \nabla \cdot (\rho \boldsymbol{u} \varepsilon) = -\left[\frac{2}{3}C_1 - C_3 + \frac{2}{3}C_\mu C_\eta \frac{k}{\varepsilon}\nabla \cdot \boldsymbol{u}\right]\rho \varepsilon \nabla \cdot \boldsymbol{u} + \nabla \cdot (\alpha_\varepsilon \mu \nabla \varepsilon)$$
$$+ \frac{\varepsilon}{k}\left[(C_1 - C_\eta)\boldsymbol{\sigma} : \nabla \boldsymbol{u} - C_2 \rho \varepsilon + C_s \dot{W}^s\right] \quad (4.66)$$

In Equation (4.66), C_η is calculated by Equation (4.56).

The low Reynolds number modifications are also given by the RNG theory [43] so that wall functions are not necessarily required. However, this approach would require very fine meshes (i.e., $y^+ < 5$). Such fine meshes are not practical for engine modeling. Hence, the wall functions as discussed in Chapter 3 are recommended for calculation of the near-wall turbulence and heat transfer in engine simulations.

It is assumed that turbulence production is equal to turbulence dissipation in the logarithmic layer of a turbulent boundary layer. Hence, $\eta = \sqrt{C_\eta}$ remains a constant. It can be derived that the von Karman constant is

$$\kappa = \left[\frac{\left(C_2 - C_1 + C_\eta \right)\sqrt{C_\mu}}{\alpha_\varepsilon} \right]^{1/2} \tag{4.67}$$

Solving for κ gives $\kappa = 0.4$ for the RNG k-ε model. While for the standard k-ε model, $\kappa = 0.4327$. The boundary conditions for k and ε are given by Equations (3.36) and (3.37).

The RNG k-ε model has been assessed against experimental data for engine flows [35]. The proposed evaluation of C_3 is shown to give an improved prediction of turbulence for engine compression and expansion flows. In the simulation of diesel combustion, it is found that $\eta < \eta_0 = 4.38$ during the entire combustion process and the R team in the ε-equation (4.50) is always positive; therefore, a lower turbulence dissipation rate and, hence, a higher turbulent diffusivity is produced when the R term is absent. The resulting higher turbulent diffusivity can suppress the premixing flame and enhance the diffusion combustion when R is excluded from the model.

The model was also used to simulate combustion in a diesel engine [56,57]. Figure 4.12 shows the computed gas temperature contours in the plane of the spray axis, in which the fuel droplet distributions are superimposed for reference. When the RNG model is used, ignition is predicted to occur in the central region of the chamber at early times (about 2° BTDC, not shown), and then, combustion quickly develops embracing the spray by TDC. By about 10° ATDC, the high-temperature flame moves over the edge of the piston bowl into the squish region under the influence of the spray and reverse squish flow. In comparison with the classic k-ε model results by Patterson et al. [58] who simulated the same engine under the same conditions, it is seen that, although the approximately same ignition is predicted, the flame regions are confined in the piston bowl and they expand with the increase of the chamber volume as the piston moved down.

The RNG-predicted flame structure is supported by the endoscope combustion visualization, which indicates that, at 6° ATDC, the high-temperature gases begin to move over the top of the bowl into the squish region. The consistency between the predicted high-temperature flows with combustion images suggests that the RNG k-ε model has resolved more large-scale flow structures during combustion properly. While much higher turbulent viscosity is predicted by using the classic k-ε model. Consequently, flow structures are suppressed and the combustion is predicted to be confined within the piston bowl, in disagreement with the endoscope images. It is also seen that the classic turbulence model predicts a much lower gas temperature (inconsistent with experimental data), which results in about ten times lower NO_x emissions compared to the measured data.

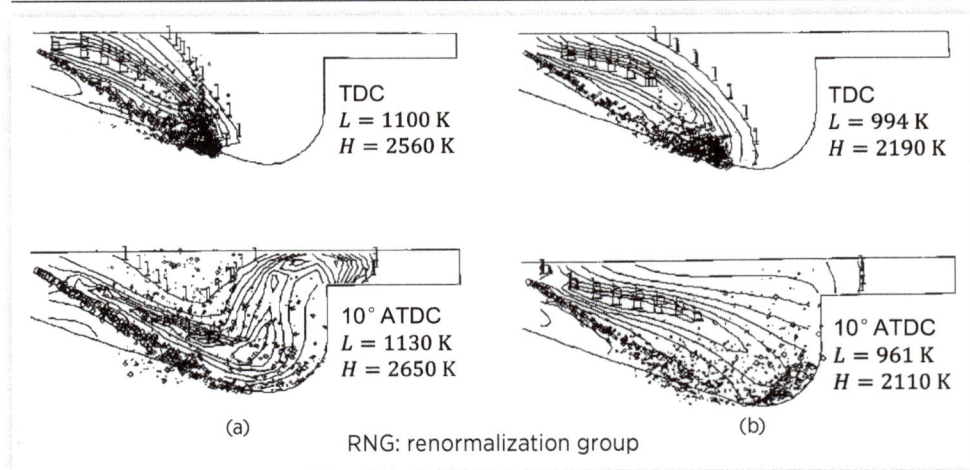

FIGURE 4.12 Computed temperature contours in the plane of the spray axis of a diesel engine. Fuel droplets are superimposed for reference (since all droplets are projected on the plane, some droplets appear outside of the piston outline): (a) the RNG k-ε model; (b) the classic k-ε model.

TDC
$L = 1100$ K
$H = 2560$ K

TDC
$L = 994$ K
$H = 2190$ K

10° ATDC
$L = 1130$ K
$H = 2650$ K

10° ATDC
$L = 961$ K
$H = 2110$ K

(a)

(b)

RNG: renormalization group

As more experimental data become available in a realistic engine geometry, the RNG k-ε model and other two-equation models have been assessed extensively. Recently, model validations have been continued in gasoline engines [59] and diesel engines [60]. The general conclusion is that the RNG k-ε and the classic k-ε model can reproduce the large-scale flow structure and cycle-averaged mean quantities reasonably well. More accurate prediction of the turbulence details needs more complicated modeling methods such as LES. The RNG k-ε model has been found to add beneficial improvements in the predictions in test cases including channel flow with a sudden expansion, gaseous jet impingement, high-pressure fuel spray injection in nonreactive and reactive environments, as well as in-cylinder flows in a light-duty optical diesel engine [60]. The RNG model has been recognized to be capable of providing enough accuracy with good computation efficiency and robustness for engine engineering simulations, and it is the de facto standard option in engine combustion simulations.

4.3.3 Other RNG k-ε Model Variants

Some model variations have been suggested aiming to improve the predictive accuracy of the original compressible RNG k-ε model as discussed. Considering one-dimensional compression along the piston axis or two-dimensional axisymmetric compression in an engine, an additional term is added to the ε-equation [61]. The model constants C' and C_3 are given for one-dimensional compression as:

$$C_3 = -2 + \frac{2}{3}C_1 + (-1)^\delta \sqrt{2} C_\mu C_\eta \eta$$

$$C' = 2 - \frac{4C_1}{3} \tag{4.68}$$

For two-dimensional axisymmetric compression, they are

$$C_3 = -\frac{3}{2} + \frac{2}{3}C_1 + (-1)^\delta C_\mu C_\eta \eta$$

$$C' = \frac{1}{2} - \frac{C_1}{3} \tag{4.69}$$

To introduce flow anisotropy effects to turbulence, Wang et al. [62] proposed a generalized closure for the RNG ε-equations, referred to as the generalized RNG (GRNG) k-ε model. The GRNG k-ε model incorporates the effects of a non-isotropic mean strain rate tensor through augmenting the RNG model coefficients by effective dimensionality coefficients n and a of the strain rate field:

$$a = 3\frac{S_{11}^{\ 2} + S_{22}^{\ 2} + S_{33}^{\ 2}}{(|S_{11}| + |S_{22}| + |S_{33}|)^2} - 1 \tag{4.70}$$

$$n = 3 - \sqrt{2a}$$

where n=1 for unidirectional axial compression, n=2 for cylindrical-radial compression (squish), n=3 for spherical compression. Based on validations from experimental and DNS data from various basic flow configurations, including homogeneous turbulence under adiabatic unidirectional and spherical compression, jet flows, and engine flows, polynomial fits for the RNG model constants were derived as:

$$C_2 = b_0 + b_1 n + b_2 n^2 \tag{4.71}$$

$$C_3 = -\frac{n+1}{n} + \frac{2}{3}C_1 + \sqrt{\frac{2(1+a)}{3}}C_\mu C_\eta \eta (-1)^\delta \tag{4.72}$$

with $b_0 = 2.496$, $b_1 = -0.686$, $b_2 = 0.11$. In this way, the GRNG k-ε equations can be solved in the same way as the RNG k-ε model, where the constants C_2 and C_3 are calculated from Equations (4.71) and (4.72).

4.4 Large-Eddy Simulation

LES is a numerical method for simulating turbulent flows. Although the RANS theory-based turbulence models (e.g., k-ε turbulence models) have been widely used in the industry, they have limited ability to resolve the detailed flow structures that are responsible for mixing and combustion in engines. This major issue of RANS simulation can be addressed by the LES model. For example, LES can be used to study cycle-to-cycle variability, provide more design sensitivity for investigating both geometrical and operational changes, and produce more detailed and accurate results in engine simulations.

The self-similarity theory of Kolmogorov [63] implies that the large eddies of a turbulent flow are dependent on the device geometry, while smaller eddies are self-similar and have a universal character. Therefore, the basic idea of LES is to solve only for large-scale eddies explicitly (i.e., those that can be resolved on the computational mesh), and to model the more universal small-scale (sub-grid scale (SGS)) eddies on large meshes using a sub-grid model. In contrast, in RANS, all fluctuations about the ensemble means are modeled. Currently, it is generally agreed that the next generation of turbulence modeling in CFD for many applications will be some form of LES. As computing power increases, the ability to use LES in engine-simulations is increasing, although the relative maturity and accumulated modeling experiences of RANS k-ε models will permit them to continue to play a dominant role in engine simulations in the industry.

The LES approach requires SGS models to close the equations. Many SGS models and modeling techniques have been proposed since the introduction of the first SGS stress model by Smagorinsky [64] and the early applications to engine simulations [65–67]. In this section, we will provide an introductory description of the LES methodology in the context of engine simulations. The most used sub-grid turbulence models are also introduced. Application examples of engine simulations using LES will be presented and the results will be compared with those from RANS modeling in some cases. The section does not include the increasing advances in LES or details descriptive equations of the models discussed. More background information is available for LES theory and for engine simulations in literature [28–30].

4.4.1 LES Methodology and Sub-Grid Models

LES is based on the concept of filtering (or spatially averaging) flow variables. For velocity u_i, its resolved component $\bar{u}_i$ is defined by the integral filtering operation

$$\overline{u(x)}_i = \int G(x, y)u(y)_i\, dy \qquad (4.73)$$

The filter function G is a normalized function with local support and a representative length scale, Δ, which is proportional to the computational grid size. The most common filter functions are Gaussian functions. For variable density flows as in engines, a density weighted or Favre filtering is defined as:

$$\tilde{u}_i = \frac{\overline{\rho u_i}}{\bar{\rho}} \qquad (4.74)$$

The instantaneous flow velocity can be decomposed into the resolved term $\tilde{u}_i$ and a reminder u_i' as:

$$u_i = \tilde{u}_i + u_i' \qquad (4.75)$$

This decomposition can be viewed as local filtering of the velocity field that leaves the large eddies and removes the SGS velocities. Then, $\tilde{u}_i$ represents the large eddies and u_i' represents the sub-grid velocities. Recall that in the RANS approach, the flow field is decomposed into an ensemble mean and a fluctuating component (see Equation 4.14). This decomposition method reflects a key difference between the LES and RANS.

The resolved terms are so named because they are the ones that are represented on the LES grid. Note that they represent locally filtered values and are not "mean" or "ensemble averaged" values. Thus, even though the notation is analogous to that used in RANS modeling, the properties of the LES terms are different. Most notably, the filtering operation results in the following properties

$$\bar{\bar{u}}_i \neq \bar{u}_i, \ \overline{u_i'} \neq 0 \tag{4.76}$$

For a compressible viscous flow, applying the filtering operation to the Navier–Stokes momentum equation results in the LES momentum equation:

$$\frac{\partial \left(\bar{\rho} \tilde{u}_i \right)}{\partial t} + \frac{\partial \left(\bar{\rho} \tilde{u}_i \tilde{u}_j \right)}{\partial x_j} = -\frac{\partial \bar{p}}{\partial x_i} + \frac{\partial}{\partial x_j} \left(\tilde{\sigma}_{ij} - \bar{\rho} (\widetilde{u_i u_j} - \tilde{u}_i \tilde{u}_j) \right) + \bar{\rho} g_i \tag{4.77}$$

Note that this LES momentum equation looks the same as the RANS equation (4.26). However, the physical interpretation between them is conceptually different. The term $\bar{\rho}(\widetilde{u_i u_j} - \tilde{u}_i \tilde{u}_j)$ is called sub-grid stress, which replaces the Reynolds stress $\overline{\rho u_i'' u_j''}$. It is unknown and requires modeling. We use the superscript SGS to denote the quantities related to the SGS when it is needed. Thus, the sub-grid stress tensor is written as:

$$\tilde{\tau}_{ij}^{\ SGS} = \bar{\rho}(\widetilde{u_i u_j} - \tilde{u}_i \tilde{u}_j) \tag{4.78}$$

There are many types of LES sub-grid stress models for $\tilde{\tau}_{ij}^{\ SGS}$ available. Rutland [30] has classified them into seven types based on whether turbulent viscosity is adopted and how many modeling transport equations are employed. The pros and cons of these types of models are also summarized.

4.4.1.1 Smagorinsky Model

The Smagorinsky model [64,68–70] is the most common among the LES sub-grid stress models, in which the sub-grid stresses is assumed to be

$$\tau_{ij}^{\ SGS} = 2\mu_{SGS}\tilde{S}_{ij} \tag{4.79}$$

where $\tilde{S}_{ij} = (\partial \tilde{u}_i / \partial x_j + \partial \tilde{u}_j / \partial x_i)/2$ is the strain-rate tensor. μ_{SGS} is the sub-grid scale turbulent viscosity and is modeled by

$$\mu_{SGS} = \bar{\rho} \ell_s^{\ 2} \left| \tilde{S} \right| = \bar{\rho}(C_s \Delta)^2 \left| \tilde{S} \right| \tag{4.80}$$

In this equation, ℓ_s is the Smagorinsky length scale (analogous to the mixing length), which, through the Smagorinsky coefficient C_s, is taken to be proportional to the filter width, Δ. The typical value of C_s is approximately 0.17, and $\tilde{S} = (2 \tilde{S}_{ij} \tilde{S}_{ij})^{1/2}$ is the filtered strain rate. The filter width or length scale, Δ, is commonly related to the local cell size. It could be, for example, $\Delta = \sqrt[3]{\Delta x \Delta y \Delta z}$.

The Smagorinsky model is an algebraic (e.g., zero-equation) turbulent viscosity model. The model coefficient C_s needs to be adjusted for each simulation situation. The model requires fine grids to obtain good results. Since the model is easy to implement, it often appears as an option in commercial CFD codes. The Smagorinsky model is fairly common in engine simulations, but the dense grid requirement is usually too restrictive.

4.4.1.2 Dynamic Smagorinsky Model

A major improvement in the Smagorinsky approach occurred when the dynamic approach was developed by Germano et al. [71], and the resulted model is often called as dynamic Smagorinsky model. In this approach, the adjustable coefficient Cs can be determined using the dynamic procedure along with some additional assumptions [72]. In the dynamic procedure, two different filter widths are introduced, $\tilde{\Delta}$ and $\bar{\tilde{\Delta}}$, where $\bar{\tilde{\Delta}} > \tilde{\Delta}$. Quantities filtered at the smaller scale are denoted by the squiggle notation while the overbar notation denotes filtering (or called as "test filter") at the larger scale. Thus $\bar{\tilde{u}}$ represents the LES-computed velocity field filtered at scale $\bar{\tilde{\Delta}}$.

An exact relationship can be derived between the sub-grid scale stress tensors at the two different filter widths, which is

$$C_s = \frac{L_{ij}M_{ij}}{2M_{kl}M_{kl}}\frac{1}{\tilde{\Delta}^2} \tag{4.81}$$

where L_{ij} is called as the Leonard tensor. L_{ij} and M_{kl} are computed from the LES resolved velocity field:

$$L_{ij} = \overline{\tilde{u}_i \tilde{u}_j} - \bar{\tilde{u}}_i \bar{\tilde{u}}_j \tag{4.82}$$

$$M_{kl} = \overline{|\tilde{S}|\tilde{S}_{kl}} - \left(\frac{\bar{\tilde{\Delta}}}{\tilde{\Delta}}\right)^2 |\bar{\tilde{S}}|\bar{\tilde{S}}_{kl} \tag{4.83}$$

The dynamic procedure is very powerful and can be used in many situations to find modeling coefficients. When used with the Smagorinsky model, the results are reasonably good for nonreacting flows. However, dense grids are required.

One of the main shortcomings of using the Smagorinsky model and the dynamic Smagorinsky model in engine combustion simulations is that they do not like the RANS k-ε turbulence models to provide applicable turbulent length scales and timescales that are often used in other spray and combustion physics models (see Chapters 5 and 6).

The simple form of the Smagorinsky model provides an important concept that can be analogized with the RANS-based k-ε turbulence model. From Equation (4.80), the sub-scale turbulence viscosity can be rewritten by using a length scale and velocity scale in the RANS as:

$$\mu_{SGS} \propto \bar{\rho}\ell u'' \tag{4.84}$$

On the other hand, the RANS-based k-ε approach describe the turbulent viscosity μ_t by Equation (4.33) as:

$$\mu_t = C_\mu \bar{\rho} \frac{k^2}{\varepsilon}$$

Thus, in the LES model, the length scale l and the velocity scale can be linked as $l \sim k^{1.5}/\varepsilon$. and $u'' \sim k^{1/2}$. This analysis leads us to understand the proposal of k-equation sub-model for simulating turbulent viscosity using turbulent kinetic energy at the sub-grid level.

4.4.1.3 k-Equation Model

The k-equation model [73,74] has similar expression as Equation (4.34), the sub-grid kinetic energy is given by

$$k = \frac{1}{2}(\widetilde{u_i u_i} - \tilde{u}_i \tilde{u}_i) \tag{4.85}$$

The model for the sub-grid stress tenor is

$$\tilde{\tau}_{ij} = -2\mu_t \tilde{S}_{ij} + \frac{2}{3} k \bar{\rho} \delta_{ij} \tag{4.86}$$

Note, we have used $\tilde{\tau}_{ij}$ to replace $\tilde{\tau}_{ij}^{SGS}$, and μ_t to replace μ_{SGS} for simplicity. The turbulent viscosity for the model is given as:

$$\mu_t = C_v k^{1/2} \Delta \tag{4.87}$$

The sub-grid dissipation rate is

$$\varepsilon = \frac{C_\varepsilon k^{3/2}}{\Delta} \tag{4.88}$$

where the constant $C_v = 0.1$ and $C_\varepsilon = 0.7$.

The k-equation model has several advantages. First, it incorporates more physical processes, such as the convection, production, and dissipation of sub-grid kinetic energy. Second, the sub-grid kinetic energy provides a velocity scaling that can be used in other models, such as combustion, scalar transport, and sprays. Third, models that use a sub-grid k-equation provide a better model for the sub-grid stresses and thus work better on the coarser grids commonly found in engine simulations. LES engine simulations with the use of the k-equation sub-model have been performed for engine flows with good results [75,76].

4.4.1.4 Dynamic Structure Model

The dynamic structure model has been proposed by the research group of Rutland at the Energy Resource Center (ERC) of the University of Wisconsin-Madison [67,77]. It does not use turbulent viscosity to model the sub-grid stress tensor; instead, a tensor coefficient is obtained directly from the dynamic procedure. This tensor coefficient is multiplied by the turbulent kinetic energy that is obtained from a transport equation. The modeled stress tensors are given by:

$$\tilde{\tau}_{ij} = C_{ij}k \tag{4.89}$$

and

$$C_{ij} = 2\frac{L_{ij}}{L_{kk}} \tag{4.90}$$

where L_{ij} is the Leonard tensor as defined by Equation (4.82). Thus, the sub-grid stresses are given via the dynamic procedure obtaining C_{ij} and the sub-grid kinetic energy k.

The most distinctive feature of the dynamic structure model is that no turbulent viscosity is used. Thus, the budget of turbulent kinetic energy is maintained between the grid scale velocity field and the sub-grid k-equation. In other words, energy removed from the grid scales goes into the sub-grid kinetic energy. Hence, the model is very robust and works well in a wide range of grid resolutions [30].

The dynamic structure model has been developed for practical applications of engines, in which the number of grid cells must remain reasonable. This model works very well in engine applications and provides a good model for the sub-grid turbulent kinetic energy for use in combustion, scalar mixing, and spray models. This model approach has been used for diesel engine simulations with good results [78–80].

Wall boundary conditions for LES sub-models are not very well developed. Currently, most LES simulations use one of two approaches for wall boundary conditions: (a) no special treatment of the wall, except for additional grid points, and (b) using the wall functions as developed in the RANS approach. This is especially true when wall heat transfer is calculated.

In order to apply the LES approach for engine combustion simulations, other physics models for fuel spray and combustion (chemical reactions) are required to be incorporated into the simulations. In almost all cases, the combustion models are essentially RANS models. RANS-based engine simulations use the Lagrangian parcel methodology to model fuel spray (see Chapter 5), which is adapted in the LES approach. The spray modeling issue of LES is representation of the sub-grid interaction of the Lagrangian spray particles with the continuous gas phase. RANS-based physics models of spray breakup and collision, drag forces, turbulent dispersion, and evaporation are used in the LES approach.

In summary, theoretically, a RANS methodology converges to an exact solution of the averaged Navier–Stokes equations. In LES, by contrast, smaller spatial/temporal scales are resolved as the grid spacing/computational time step is refined. An LES methodology converges to an exact solution of the unfiltered Navier–Stokes equations, that is, to DNS. This suggests that high-fidelity numerical methods are an important component of LES [28]. However, the LES approach should be viewed as an evolution of the RANS in the context of engine simulations so that the knowledge accumulated in RANS can be transferred to LES.

It is promising to use LES in simulation of IC engines. High-fidelity combustion models with LES may lead to a better understanding of lean combustion instability phenomena and variations of engine outputs that are associated with engine cyclic variability. There are, however, still some issues regarding LES engine simulations. Although the increased computational costs are being lightened due to the rapid advances of computer power, performing an LES simulation still requires a lot more time than performing a RANS simulation since the former usually needs computational results from multiple cycles (5~10 cycles or more) in order to obtain the assemble averages for comparison with experimental data (e.g., cylinder pressures and heat releases). More importantly, the LES method fundamentally implies that the results are individual cycle-based, and they do not represent ensemble averages. This can be a disadvantage if one is trying to compare to experimental results that are often averaged over many cycles. Even in the case of comparison with cyclic resolved measurements, one would have a problem nailing down the exact cycle on which the computational and experimental results should be compared since the randomness in conditions that causes the cyclic variations has not coincided in the simulation and experiment. Only the feature (or a typical cycle) of the model resolved cyclic results can be verified experimentally. Therefore, modelers should match the CFD tool to the problem at hand and use LES appropriately.

4.4.2 Engine Simulation Examples

In this section, several cases are described in which LES has been successfully used to study some aspects of IC engines. These examples help to demonstrate how LES is currently being used.

4.4.2.1 Intake and In-Cylinder Flows

A comparative cold flow analysis between RANS and LES cycle-averaged velocity and turbulence predictions was carried out for a single cylinder engine under motored conditions using high-speed particle image velocimetry (PIV) measurements as the reference data by researchers at the General Motor Company engine research group [59]. The modeled engine is a 0.57-liter single-cylinder engine with a two-valve head, simple intake and exhaust port geometries, and a pancake-shaped combustion chamber. A commercial CFD code, CONVERGE, is used for computation. The computational domain for full geometry simulations includes the intake and exhaust plenums, whereas that for partial geometry simulations only include the intake and exhaust ports. A base mesh of 1mm is used for both LES and RANS simulations in order to make the comparisons mesh independent. The standard k-ε turbulent model, RNG k-ε turbulent model for RANS, and a one-equation eddy viscosity LES model is used. The ensemble-averaged (mean) and root mean square (rms) velocities have been averaged over 56 cycles both for PIV and LES. To eliminate contamination by initial conditions, the first 10 cycles for LES have been ignored. All the RANS simulations were run for two cycles, and the results of the second cycle are used for the analysis.

The mean velocity results from various means are shown in Figure 4.13, and the corresponding turbulence velocity magnitudes are also shown. The observations are summarized as:

(a)   LES gives overall better predictions of mean velocity magnitudes as compared to RANS. The jet structure is better resolved for LES and the rightward curl is appropriately captured as opposed to RANS which wrongly predicts a leftward curl in the jet structure except for the RNG k-ε model partial geometry simulation. Furthermore, the RANS jet structure is irregular, whereas the LES jet structure is much more stable.

(b)   LES predictions match PIV measurements much better than RANS, which give high turbulence regions properly enclose the jet and the bottom stationary flow as in PIV. The standard k-ε model wrongly predicts the high turbulence regions rightward of the jet, and the RNG k-ε model gives a reasonable prediction. Furthermore, LES captures the local peaks and troughs in the contours, whereas RANS contours miss out on that and give a pretty smooth picture. Thus, it is clear that LES gives more accurate turbulence predictions.

FIGURE 4.13 Two-dimensional contours of mean velocity with overlaid velocity vectors (left) and turbulence velocity magnitude (right) at Y=0 plane (the center plane through the valves) and 100 CA in intake stroke: (a) PIV, (b) LES full, (c) LES partial, (d) RANS (k-ε) full, (e) RANS (k-ε) partial, (f) RANS (RNG k-ε) full, (g) RANS (RNG k-ε) partial.

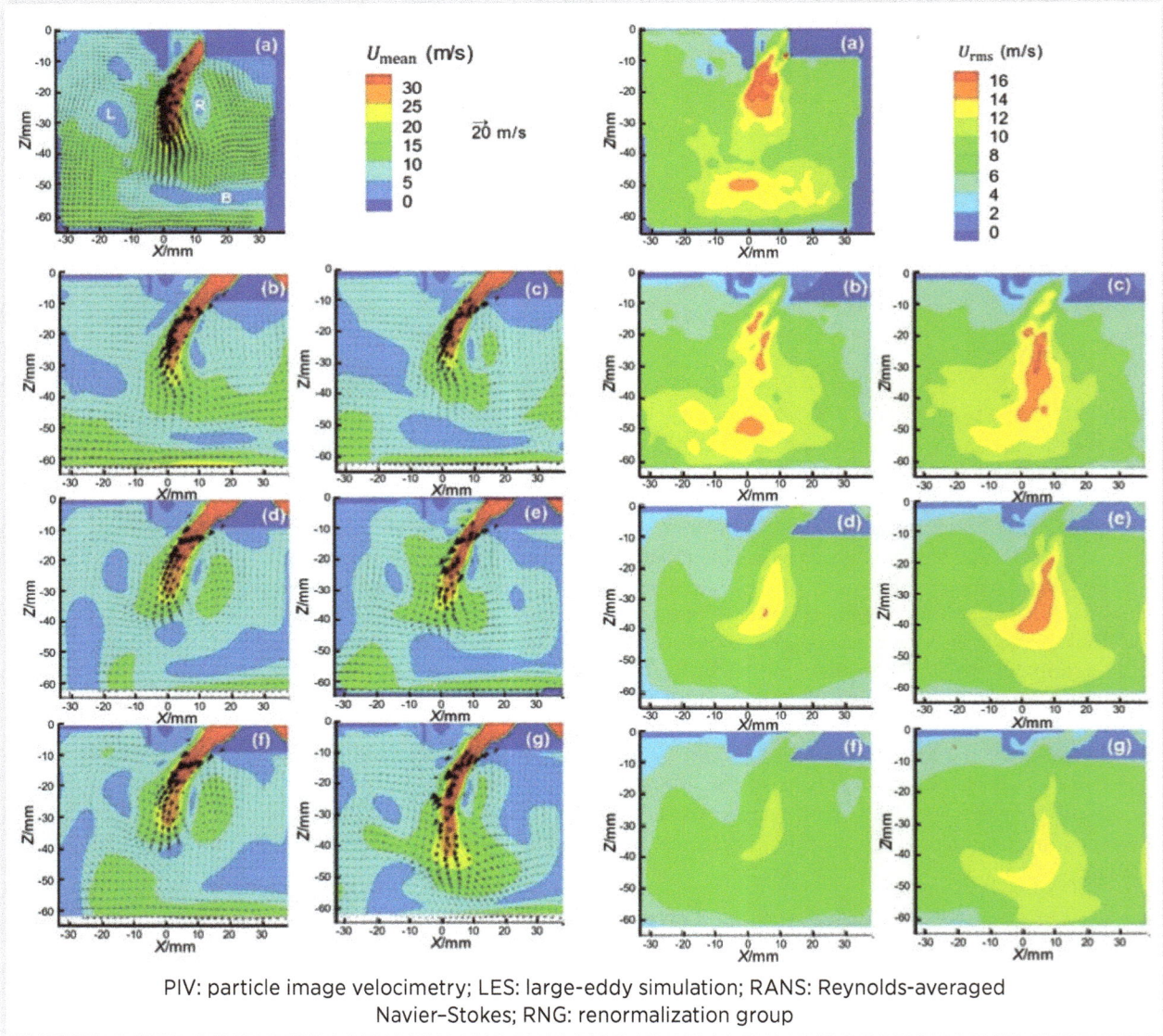

PIV: particle image velocimetry; LES: large-eddy simulation; RANS: Reynolds-averaged Navier–Stokes; RNG: renormalization group

Comparisons are also made near the end of the compression stroke at which time the local velocity and turbulence fields directly influence the combustion processes. After careful examination, they have concluded that:

(a) Both RANS and LES full and partial geometry simulations can qualitatively capture all the major PIV mean flow and turbulence structures. LES simulation with the inclusion of the runner plenums is the best one.

(b) RANS partial geometry simulations are better than full geometry, as RANS results correspond to the second cycle, which is important to practical simulation works. Between the two RANS turbulence models, the RNG k-ε model gives marginally better mean flow and turbulence predictions as compared to the standard k-ε model.

(c) The RANS RNG k-ε model simulations without the runner plenums (computationally least expensive) are good enough for capturing overall qualitative flow trends. However, if one is interested in getting more reasonably accurate estimates of the turbulence fields, LES simulations must be performed.

There are other studies too that use LES simulating engine intake and in-cylinder flows [65,81]. In the work of Yu et al. [81], a modified 0.48-liter Volvo D5 light-duty diesel engine with a ω-type bowl-in-piston combustion chamber was used for PIV measurement and LES modeling. The turbulence production in the intake and compression phases is mainly due to the large-scale flow motion. In the later stage of the compression phase, the flow is characterized by more isotropic turbulent eddies. The features of the flow structures were reproduced by the LES simulations.

4.4.2.2. Cycle-to-Cycle Combustion Variation

One of the main capabilities of LES is to reflect engine cycle-to-cycle variations. Vermorel et al. [82] studied these phenomena by LES simulation of nine consecutive complete engine cycles of a PSA XU10 4-valve PFI spark-ignition engine with burning premixed propane fuel. The results obtained were compared with experimental data on cycle-to-cycle cylinder pressures. An LES model with the Smagorinsky stress sub-grid model in the AVBP CFD code [83] was used to perform the simulations. Nine consecutive full engine cycles of 720 CA were simulated, covering all phases of the four-stroke cycles: intake, compression and combustion, expansion, and exhaust in the full engine geometry. In particular, flow in the intake and exhaust ports is solved, even when they are closed, to provide proper conditions upon valve opening. After the end of each cycle, the simulation carries on with the flow and conditions achieved at the end of the preceding cycle to simulate the real engine situation. The mesh size is about 0.2 mm and 1.5 mm, which gives a total number of 254,000 hexahedra at TDC and 628,000 at BDC. The return time for a complete cycle was 120 hours on 32 Xeon® processors of a Linux cluster.

Figure 4.14 presents the cylinder pressures for the nine consecutive simulated cycles, along with the cycle-averaged mean experimental pressure, and the overall minimum and maximum experimental pressure curves. No convergence of the calculated cycles towards a mean cycle is observed and most cycles are within the experimental envelope. Cyclic variability is reflected by LES simulations, with qualitatively accurate amplitude of the cyclic pressure variations.

Vermorel et al. also demonstrated in their case (no fuel injections) that a strong cyclic variation in the tumble motion was the source of the turbulence intensity fluctuations, which caused the engine combustion cyclic variations [82]. They argued that the coherent structures (tumble around y-axis being the strongest), induced by the inlet valves, fluctuate from cycle to cycle. The cyclic variations of the gas motion are illustrated in Figure 4.15, which shows the velocity magnitude in a vertical cutting plane during intake for the first four cycles. Although the global flow structure is comparable, important local variations of velocity magnitude can be observed from one cycle to another.

FIGURE 4.14 Evolution of LES predicted cylinder pressure for the nine consecutive cycles and the measured pressure boundaries and mean.

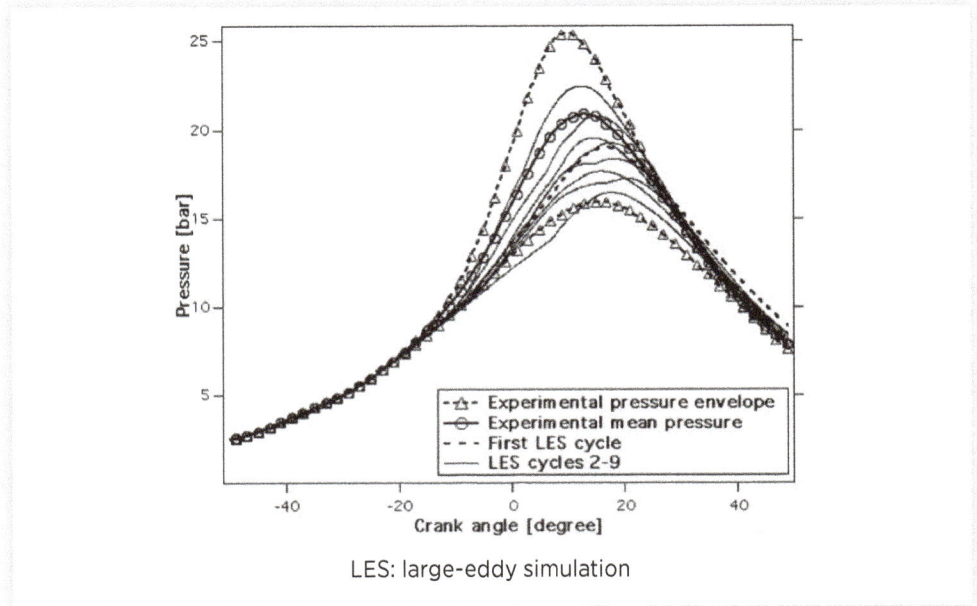

LES: large-eddy simulation

FIGURE 4.15 Computed velocity magnitude in a plane crossing the center line of the valves at 235° BTDC during intake stroke for the first four cycles showing the cyclic variations in gas motion.

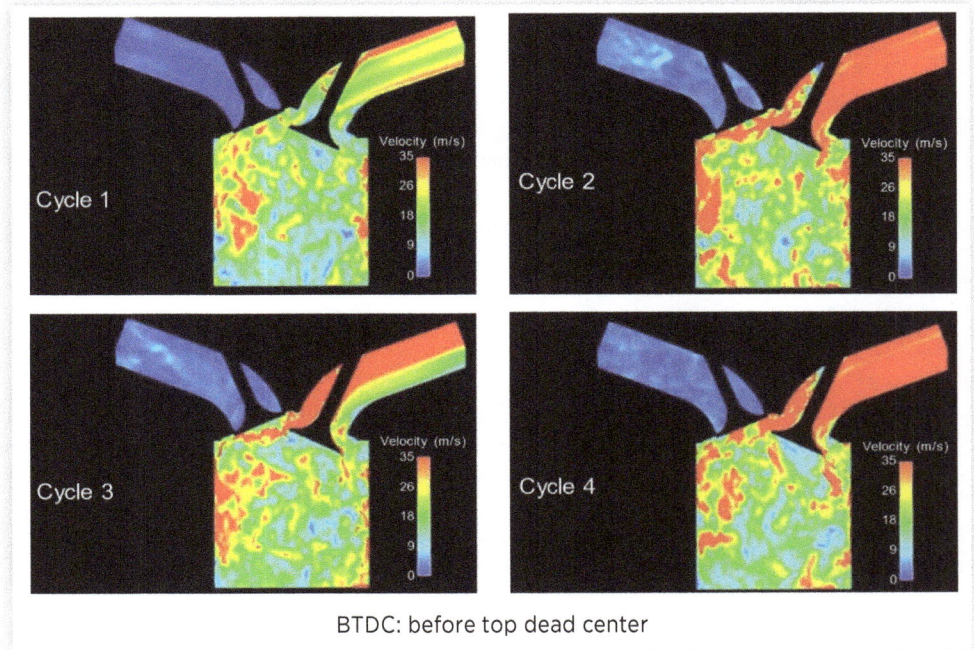

BTDC: before top dead center

4.4.2.3 Low-Temperature Spray Combustion

Hu et al. studied LTC in a diesel engine [79]. A RANS-based RNG k-ε model and a dynamic structure LES model in KIVA3V together with the CHEMKIN code were used to simulate the spray, ignition, and combustion in the Cummins N-14 single-cylinder engine. To save computing time, a 45° sector mesh with the largest mesh size of 1.2×1.2×1.5mm was used for both two-simulation approaches, since there are eight fuel sprays evenly distributed in the symmetric combustion chamber. Only one simulation cycle was performed for both modeling approaches. The researchers of this study viewed this type of LES simulation as a type of "engineering" LES and they justified its validity because the primary objective was to demonstrate that LES could predict more detailed flow structures.

Figure 4.16 shows a set of images that compare the model predicted gas temperature with the experimentally observed chemiluminescence. In this case, *n*-heptane flue is injected under an injection pressure of 160 MPa at the time of –22° ATDC, and the injection duration is 7 CA. The engine is operated at 1200 rpm and a low engine load.

FIGURE 4.16   Experimentally imaged ignition chemiluminescence and model-predicted in-cylinder gas temperature for the early injection LTC.

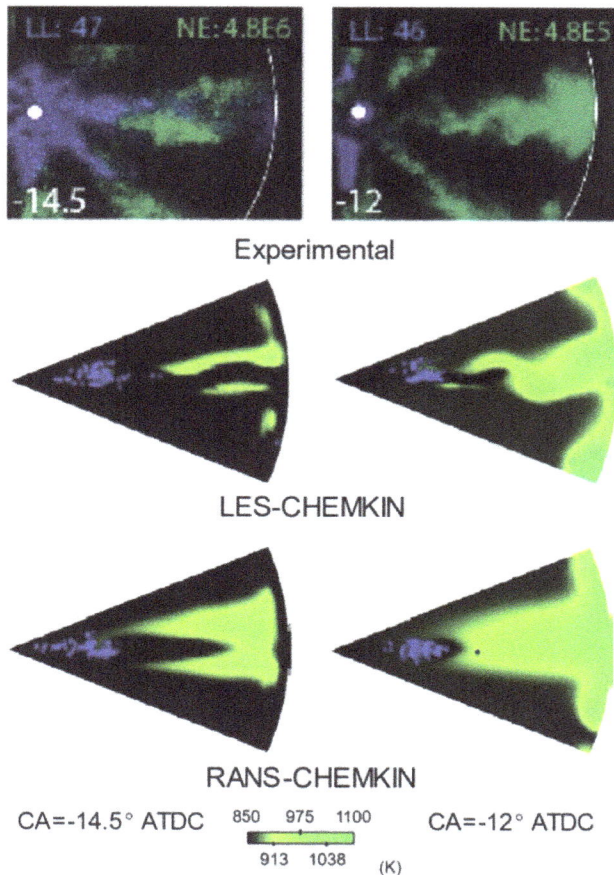

LTC: low-temperature combustion

The first row in Figure 4.16 are the simultaneous images of laser Mie scattering from the liquid spray droplets (blue) and the ignition chemiluminescence (green). The second and third rows are the LES–CHEMKIN and RANS–CHEMKIN model predicted gas temperature (green) in a plane along the nominal jet axis. Liquid droplets are also shown in blue color. It should be noted that the simulation results contain spray images only from the result of one nozzle. All the model images use the same temperature scale, as indicated on the bottom line in Figure 4.16.

It is seen that the RANS images are smeared by their nature of ensemble-averaging whereas the LES images show the more detailed structure and have features that are seen in the instantaneous experimental images. Particularly for the time of –12° ATDC, LES shows the same oscillatory structures on the jet axis that are seen in the experimental images. It is interesting to note that although both RANS and LES approach can predict the combustion pressure and heat release at the same level of fidelity as reported by Hu et al. [79], such a coarse numerical treatment in the LES modeling can reflect the details of the flow structures as supported by the experiments. Thus, LES models should be used for simulation of combustion regimes in which auto ignition is more random in space and time, such as knocking, HCCI, and PCCI.

4.4.2.4 Ignition Effects on DI Gasoline Combustion

In the work to study the effects of spark configurations (location and orientation of the electrodes), Fontanesi et al. [84] carried out LES combustion simulations for a production 3L V8 turbocharged gasoline direct injection engine of Ferrari S.p.A. under 7000 rpm and full load condition. The computations were performed with Star-CD® and a Smagorinsky stress model was adopted. The computational domain consists of 236,000 cells at TDC and 1,500,000 cells at BDC (including the ports) and the return time for a single cycle is about 24 CPU hours on a 32-core Linux cluster (only for the compression and combustion). Twenty cycles were performed for each spark configuration case. However, precomputed intake flow results were mapped to each cycle at each case.

Strong cyclic variations in the cylinder pressure as well as in other parameters were resolved by the LES modeling. Figure 4.17 shows that the correlation between the crank angle at which 50% of the fuel mass is consumed (MFB50 or $\Delta\varphi_{50}$) and peak pressure is very strong for all the configurations (Figure 4.17a). It is seen that a wide range of the peak pressure exists, which is affected by the fuel burn rate in some ways. Also, specific spark configurations (S0, S1, and S2) leads to a different scatter of MFB50s and peak pressures. The MFB50 is closely related to the peak pressure value for each individual cycle. The relationship between the crank angle at which 10% of fuel mass is burned (MFB10 or $\Delta\varphi_{10}$) and maximum pressure is given in Figure 4.17b, in which a scattered distribution is seen again, and a weak correlation between the MFB10 and peak pressure is seen as well. This work demonstrated the complicated influence of design details on engine performance and the aid that LES engine simulations can offer to understand the inherent reasons.

Fontanesi et al. concluded that the simple re-orientation of the spark plug electrode and, therefore, the aerodynamic interaction of the electrodes with the surrounding flow patterns, does not introduce statistically significant combustion variations, which confirms the experimental practice in which there is no preferred orientation for the spark plug [84]. The modification of the spark plug position (shifted 2 mm toward the intake side in the study) unexpectedly resulted in reductions in both peak pressure levels and combustion stability. This suggests that the spark plug was moved into a non-favorable location for ignition. Thus, modification of the spark plug position should be accompanied by targeted

FIGURE 4.17 Correlations of the main combustion phasing (MFB50), and flame development time (MFB10) with combustion peak pressure. S0 and S1 denote different orientations of the electrodes, and S2 denotes a different location of the electrodes.

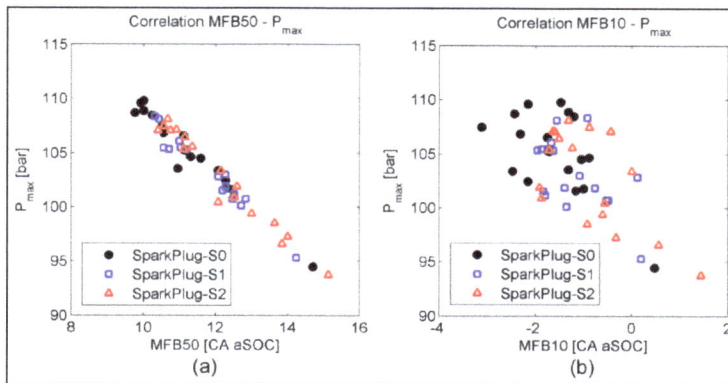

changes in the injection phasing and orientation, especially for the analyzed combustion chamber architecture.

4.4.2.5 Stratified-Charge DI Gasoline Combustion

Multiple-cycle LES have been recently performed for a single-cylinder, four-stroke cycle, spray-guided DI spark-ignition engine operating in a stratified globally fuel-lean mode by Kazmouz et al. [85]. One of the objectives of the study was to identify the causes that lead to misfire or robust combustion. The engine was operated at 1300 rpm and the overall equivalence ratio was 0.2. The fuel (iso-octane) is injected under 12 MPa from an eight-hole injector. The SOI is –25° ATDC and the start of ignition –22° ATDC. The experimental engine exhibits a high level of cyclic combustion variations under these conditions.

The LES were performed using STAR-CD with the use of a dynamic Smagorinsky model for sub-filter-scale turbulence. The computational domain includes the in-cylinder region, intake and exhaust ports, and a portion of the intake and exhaust runners. The maximum number of computational cells is approximately 4.7 million at BDC. The average cell size is approximately 1 mm, with finer meshes (approximately 0.25 mm) in the vicinity of the fuel injector and the spark plug. The simulations are run through 35 consecutive engine cycles, after discarding seven initial cycles to remove the influence of the arbitrary initial conditions. Instantaneous planar distributions of vapor equivalence ratio were measured in the engine in the vicinity of the fuel injector and spark plug. The measurement plane is offset by 3 mm from the spray-plug gap, and measurements were made at several instants during fuel injection and early combustion.

Comparisons between three experimental cycles and three LES cycles at –15° ATDC are shown in Figure 4.18. At this instant, the flame has not yet reached the measurement plane. The general locations and shapes of the individual spray plumes, the ranges of equivalence ratio values, and the degree of combustion variations that are observed experimentally are captured reasonably well in the simulations. The main difference between simulations and experiments is in the finer spatial structure that is seen in the experimental images.

FIGURE 4.18 Comparisons of simulated and measured instantaneous vapor equivalence ratios of a SCDI SI engine. The images are on a plane offset 3 mm from the spark-plug gap at –15° ATDC.

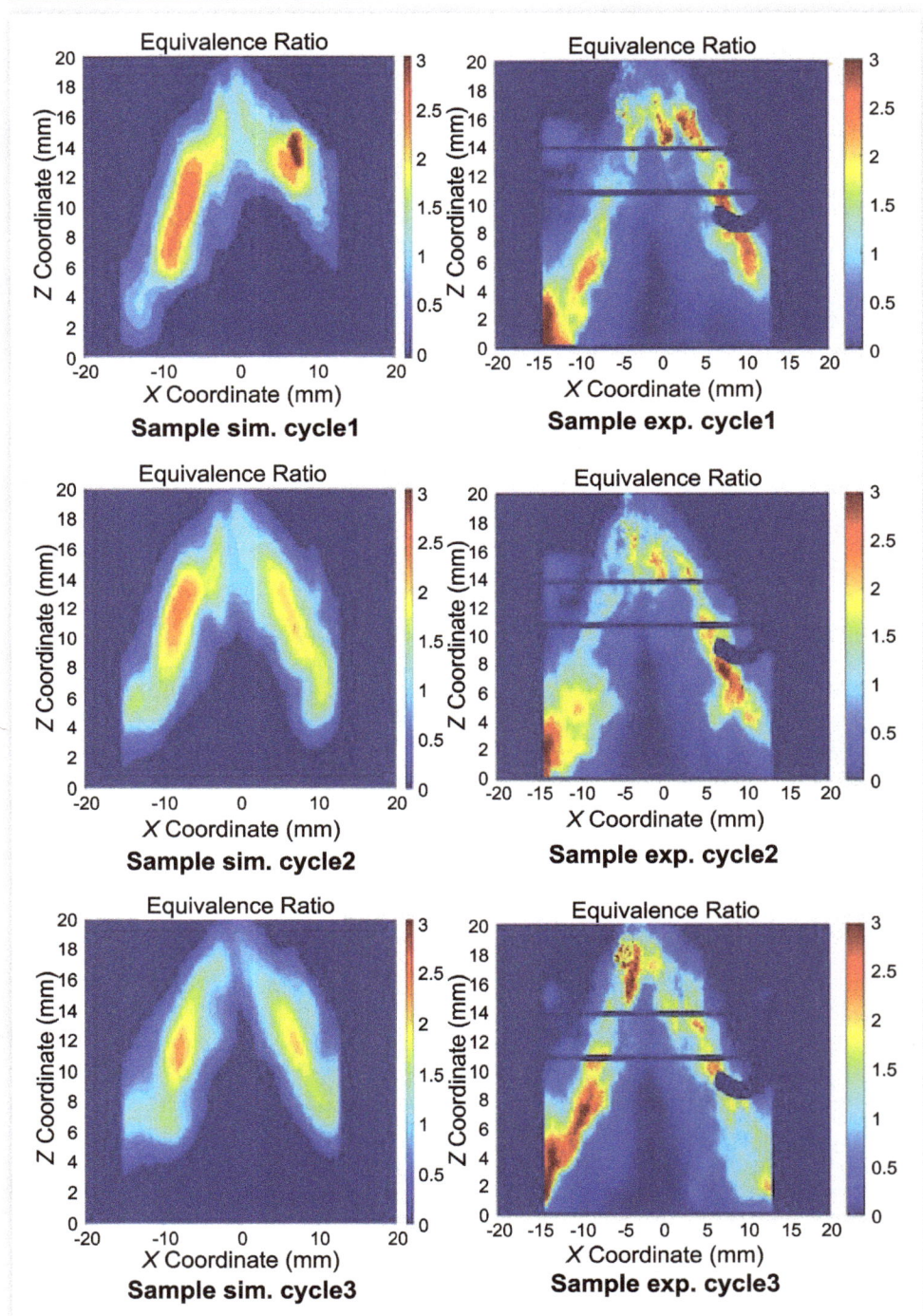

By analyzing the misfired LES cycle, it was seen that the influence of the early burn on subsequent flame development was subtler for stratified combustion. The local conditions in the vicinity of the spark gap at the time of ignition largely determine the subsequent flame development. It was also concluded that the velocity field imposed by the injection event, especially the leading edge of the spray jet, could create unfavorable conditions, which impede flame development and lead to slow burn or misfire cycles [85].

References

1. Tabaczynski, R.J. 1976. "Turbulence And Turbulent Combustion in Spark-Ignition Engines." *Progress in Energy And Combustion Science* 2, no. 3: 143–165.

2. Heywood, J.B. 2018. *Internal Combustion Engine Fundamentals*. 2nd ed. New York, USA: McGraw-Hill Education.

3. Han, Z., Fan, L., and Reitz, R.D. 1997. "Multidimensional Modeling of Spray Atomization And Air-Fuel Mixing in a Direct-Injection Spark-Ignition Engine." *SAE Transactions* 106, no. 3 (February): 1423–1441. DOI: 10.4271/970884

4. Hakariya, M., Toda, T., and Sakai, M. 2017. "The New Toyota Inline 4-Cylinder 2.5L Gasoline Engine." SAE Technical Paper No. 2017-01-1021. DOI: 10.4271/2017-01-1021

5. Wu, Z., Han, Z., Shi, Y., Liu, W., Zhang, J., Huang, Y., and Meng, S. 2021. "Combustion Optimization for Fuel Economy Improvement of A Dedicated Range-Extender Engine." *Proceedings of the Institution of Mechanical Engineers, Part D: Journal of Automobile Engineering* 235, no. 9: 2525–2539. DOI: 10.1177/0954407021993620

6. Kakuhou, A., Urushihara, T., Itoh, T., and Takagi, Y. 1999. "Characteristics of Mixture Formation in A Direct Injection SI Engine With Optimized In-Cylinder Swirl Air Motion." *SAE Transactions* 108, no. 3 (March): 550–558. DOI: 10.4271/1999-01-0505

7. Han, Z., Weaver, C., Wooldridge, S., Alger, T., Hilditch, J., McGee, J., Westrate, B., et al. 2004. "Development of A New Light Stratified-Charge DISI Combustion System for A Family Of Engines With Upfront CFD Coupling With Thermal And Optical Engine Experiments." *SAE Transactions* 113, no. 3 (March): 269–293. DOI: 10.4271/2004-01-0545

8. Arcoumanis, C., Bicen, A.F., and Whitelaw, J.H. 1983. "Squish And Swirl-Squish Interaction in Motored Model Engines." *Journal of Fluids Engineering* 105, no. 1 (March): 105–112. DOI: 10.1115/1.3240925

9. Miles, P.C. 2009. "Turbulent Flow Structure In Direct-Injection, Swirl-Supported Diesel Engines." In *Flow And Combustion In Reciprocating Engines*, edited by C.D. Arcoumanis and T. Kamimoto, 173–256. Berlin, Germany: Springer-Verlag.

10. Han, Z., Reitz, R.D., Yang, J., and Anderson, R.W. 1997. "Effects of Injection Timing on Air-Fuel Mixing in A Direct-Injection Spark-Ignition Engine." *SAE Transactions* 106, no. 3 (February): 848–860. DOI: 10.4271/970625

11. Liou, T.M., Hall, M., Santavicca, D.A., and Bracco, F.V. 1984. "Laser Doppler Velocimetry Measurements in Valved And Ported Engines." *SAE Transactions* 93, no. 2 (February): 935–948. DOI: 10.4271/840375

12. Tennekes, H., and Lumley, J.L. 1972. *A First Course in Turbulence*. Cambridge, Massachusetts: MIT Press.

13. Reynolds, W.C. 1980. "Modeling of Fluid Motions in Engines-An Introductory Overview." In *Combustion Modeling in Reciprocating Engines*, edited by J.N. Mattavi & C.A. Amman, 41–68. New York, NY: Plenum Press.

14. Hinze, J.O. 1975. *Turbulence*. 2nd ed. New York, USA: McGraw-Hill Education.

15. Corcione, F.E., and Valentino, G. 1994. "Analysis of In-Cylinder Flow Processes by LDA." *Combustion and Flame* 99, no. 2 (November): 387–394. DOI: 10.1016/0010-2180(94)90145-7

16. Collings, N., Roughton, A.W., and Ma, T. 1987. "Turbulence Length Scale Measurements in A Motored Internal Combustion Engine." SAE Technical Paper No. 871692. DOI: 10.4271/871692

17. Dinsdale, S., Roughton, A., and Collings, N. 1988. "Length Scale and Turbulence Intensity Measurements in A Motored Internal Combustion Engine." SAE Technical Paper No. 880380. DOI: 10.4271/880380

18. Ikegami, M., Shioji, M., and Nishimoto, K. 1987. "Turbulence Intensity And Spatial Integral Scale During Compression And Expansion Strokes in A Four-Cycle Reciprocating Engine." *SAE Transactions* 96, no. 4 (February): 399–413. DOI: 10.4271/870372

19. Fraser, R.A., and Bracco, F.V. 1988. "Cycle-Resolved LDV Integral Length Scale Measurements in An IC Engine." *SAE Transactions* 97, no. 3 (February): 222–241. DOI: 10.4271/880381

20. Miles, P.C., Megerle, M., Nagel, Z., Reitz, R.D., Lai, M.-C.D., and Sick, V. 2003. "An Experimental Assessment of Turbulence Production, Reynolds Stress And Length Scale (Dissipation) Modeling in A Swirl-Supported DI Diesel Engine." *SAE Transactions* 112, no. 3 (March), 1470–1499. DOI: 10.4271/2003-01-1072

21. Han, Z., Reitz, R.D., Corcione, F.E., and Valentino, G. 1996. "Interpretation Of K-E Computed Turbulence Length-Scale Predictions for Engine Flows." *Symposium (International) on Combustion* 26, no. 2: 2717–2723. DOI: 10.1016/S0082-0784(96)80108-1

22. Wu, C.T., Ferziger, J.H., and Chapman, D.R. 1985. *Department Of Mechanical Engineering Report (No. TF-21)*. Stanford, California: Stanford University.

23. Coleman, G.N., and Mansour, N.N. 1991. "Modeling the Rapid Spherical Compression of Isotropic Turbulence." *Physics of Fluids A: Fluid Dynamics* 3, no. 9: 2255–2259. DOI: 10.1063/1.857906

24. Tanner, F.X., Zhu, G., and Reitz, R.D. 2000. "Non-Equilibrium Turbulence Considerations for Combustion Processes in the Simulation of DI Diesel Engines." SAE Technical Paper No. 2000-01-0586. DOI: 10.4271/2000-01-0586

25. Bianchi, G.M., Pelloni, P., Zhu, G.S., and Reitz, R. 2001. "On Non-Equilibrium Turbulence Corrections in Multidimensional HSDI Diesel Engine Computations." SAE Technical Paper No. 2001-01-0997. DOI: 10.4271/2001-01-0997

26. Kuo, K.K. 1986. *Principles of Combustion*. Hoboken, New Jerset: John Wiley & Sons.

27. Warsi, Z.U.A. 2005. *Fluid Dynamics: Theoretical And Computational Approaches*. 3rd ed. Boca Raton, Florida: CRC Press.

28. Pope, S.B. 2000. *Turbulent Flows*. Cambridge: Cambridge University Press.

29. Celik, I., Yavuz, I., and Smirnov, A. 2001. "Large Eddy Simulations of In-Cylinder Turbulence For Internal Combustion Engines: A Review." *International Journal of Engine Research* 2, no. 2: 119–148. DOI: 10.1243/1468087011545389

30. Rutland, C.J. 2011. "Large-Eddy Simulations For Internal Combustion Engines—A Review." *International Journal of Engine Research* 12, no. 5: 421–451. DOI: 10.1177/1468087411407248

31. Pope, S.B. 1985. "PDF Methods For Turbulent Reactive Flows." *Progress in Energy and Combustion Science* 11, no. 2: 119–192. DOI: 10.1016/0360-1285(85)90002-4

32. Haworth, D.C. 2010. "Progress In Probability Density Function Methods For Turbulent Reacting Flows." *Progress in Energy and Combustion Science* 36, no. 2 (April): 168–259. DOI: 10.1016/j.pecs.2009.09.003

33. Reitz, R.D., and Rutland, C.J. 2014. "Multidimensional Simulation." In *Encyclopedia of Automotive Engineering*, edited by D. Crolla, D.E. Foster, T. Kobayashi, and N. Vaughan, 1–19). Hoboken, New Jersey: John Wiley & Sons.

34. Launder, B.E., and Spalding, D.B. 1972. *Mathematical Models of Turbulence*. Cambridge, Massachusetts: Academic Press.

35. Han, Z., and Reitz, R.D. 1995. "Turbulence Modeling of Internal Combustion Engines Using Rng K-E Models." *Combustion Science and Technology* 106, no. 4–6: 267–295. DOI: 10.1080/00102209508907782

36. Jones, W.P., and Launder, B.E. 1972. "The Prediction Of Laminarization With a Two-Equation Model Of Turbulence." *International Journal of Heat and Mass Transfer* 15, no. 2 (February), 301–314. DOI: 10.1016/0017-9310(72)90076-2

37. Gosman, A., and Watkins, A. 1977. "A Computer Prediction Method for Turbulent Flow And Heat Transfer In Piston/Cylinder Assemblies." In *Proceedings of a Symposium on Turbulent Shear Flows.* State College, PA: Pennsylvania State University.

38. Amsden, A.A., O'Rourke, P.J., and Butler, T.D. 1989. *KIVA-II: A computer program for chemically reactive flows with sprays* (LA-11560-MS). Los Alamos, New Mexico: Los Alamos National Laboratory.

39. Ramos, J.I., and Sirignano, W.A. 1980. "Axisymmetric Flow Model With And Without Swirl in a Piston-Cylinder Arrangement With Idealized Valve Operation." SAE Technical Paper No. 800284. DOI: 10.4271/800284

40. Morel, T., and Mansour, N.N. 1982. "Modeling of Turbulence in Internal Combustion Engines." SAE Technical Paper No. 820040. DOI: 10.4271/820040

41. El Tahry, S.H. 1983. "K-Epsilon Equation for Compressible Reciprocating Engine Flows." *Journal of Energy* 7: 345–353. DOI: 10.2514/3.48086

42. Grasso, F., and Bracco, F.V. 1983. "Computed And Measured Turbulence In Axisymmetric Reciprocating Engines." *AIAA Journal* 21, no. 4: 601–607. DOI: 10.2514/3.8119

43. Yakhot, V., and Orszag, S.A. 1986. "Renormalization Group Analysis of Turbulence. I. Basic Theory." *Journal of Scientific Computing* 1, no.1: 3–51. DOI: 10.1007/BF01061452

44. Yakhot, V., and Smith, L.M. 1992. "The Renormalization Group, the Ɛ-Expansion and Derivation of Turbulence Models." *Journal of Scientific Computing* 7, no. 1: 35–61. DOI: 10.1007/BF01060210

45. Yakhot, V., Orszag, S.A., Thangam, S., Gatski, T.B., and Speziale, C.G. 1992. "Development of Turbulence Models For Shear Flows by a Double Expansion Technique." *Physics of Fluids A: Fluid Dynamics* 4, no. 7: 1510–1520. DOI: 10.1063/1.858424

46. Speziale, C.G., Gatski, T.B., and Fitzmaurice, N. 1991. "An Analysis of RNG-Based Turbulence Models For Homogeneous Shear Flow." *Physics of Fluids A: Fluid Dynamics* 3, no. 9: 2278–2281. DOI: 10.1063/1.857963

47. Choudhury, D., Kim, S.E., and Flannery, W.S. 1993. "Calculation of Turbulent Separated Flows Using a Renormalization Group Based K-E Turbulence Model." *ASME Fluids Engineering Division* 149: 177–187.

48. Forster, D., Nelson, D.R., and Stephen, M.J. 1977. "Large-Distance And Long-Time Properties of a Randomly Stirred Fluid." *Physical Review A* 16, no. 2 (August): 732–749. DOI: 10.1103/PhysRevA.16.732

49. Fournier, J.D., and Frisch, U. 1983. "Remarks on the Renormalization Group in Statistical Fluid Dynamics." *Physical Review A* 28, no. 2 (August): 1000–1002. DOI: 10.1103/PhysRevA.28.1000

50. Smith, L.M., and Reynolds, W.C. 1992. "On the Yakhot–Orszag Renormalization Group Method for Deriving Turbulence Statistics and Models." *Physics of Fluids A: Fluid Dynamics* 4, no. 2: 364–390. DOI: 10.1063/1.858310

51. Lam, S.H. 1992. "On the RNG Theory of Turbulence." *Physics of Fluids A: Fluid Dynamics* 4, no. 5: 1007–1017. DOI: 10.1063/1.858517

52. Eyink, G.L. 1994. "The Renormalization Group Method in Statistical Hydrodynamics." *Physics of Fluids* 6, no. 9: 3063–3078. DOI: 10.1063/1.868131

53. Durbin, P.A., and Zeman, O. 1992. "Rapid Distortion Theory for Homogeneous Compressed Turbulence With Application to Modelling." *Journal of Fluid Mechanics* 242: 349–370. DOI: 10.1017/S0022112092002404

54. Lele, S.K. 1994. "Compressibility Effects on Turbulence." *Annual Review of Fluid Mechanics* 26: 211–254. DOI: 10.1146/annurev.fl.26.010194.001235

55. Bird, R.B., Stewart, W.E., and Lightfoot, E.N. 1960. *Transport Phenomena.* New York, USA: John Wiley & Sons.

56. Kong, S.C., Han, Z., and Reitz, R.D. 1995. "The Development And Application of a Diesel Ignition And Combustion Model for Multidimensional Engine Simulation." *SAE Transactions* 104, no. 3(February): 502–518. DOI: 10.4271/950278

57. Han, Z. 1996. Numerical study of air-fuel mixing in direct-injection spark-ignition and diesel engines (Publication No. 9634937). Madison, Wisconsin: University of Wisconsin-Madison. ProQuest Dissertations and Theses.

58. Patterson, M.A., Kong, S.C., Hampson, G.J., and Reitz, R.D. 1994. "Modeling the Effects of Fuel Injection Characteristics on Diesel Engine Soot And No$_x$ Emissions." *SAE Transactions* 103, no. 3 (March), 836–852. DOI: 10.4271/940523

59. Yang, X., Gupta, S., Kuo, T., and Gopalakrishnan, V. 2014. "RANS And Large Eddy Simulation Of Internal Combustion Engine Flows—A Comparative Study." *Journal of Engineering for Gas Turbines and Power* 136, no. 5 (May): article 051507. DOI: 10.1115/1.4026165

60. Perini, F., Zha, K., Busch, S., and Reitz, R. 2017. "Comparison of Linear, Non-Linear And Generalized RNG-Based K-Epsilon Models for Turbulent Diesel Engine Flows." SAE Technical Paper No. 2017-01-0561. DOI: 10.4271/2017-01-0561

61. Xie, M., and Jia, M. 2016. *Computational Combustion of Internal Combustion Engines* [In Chinese]. 3rd ed. Beijing: Science Press.

62. Wang, B.L., Miles, P.C., Reitz, R.D., Han, Z., and Petersen, B. 2011. "Assessment of RNG Turbulence Modeling And the Development of a Generalized RNG Closure Model." SAE Technical Paper No. 2011-01-0829. DOI: 10.4271/2011-01-0829

63. Kolmogorov, A.N. 1991. "The Local Structure of Turbulence in Incompressible Viscous Fluid for Very Large Reynolds Numbers." *Proceedings of the Royal Society A: Mathematical, Physical and Engineering Sciences* 434, no. 1890 (July): 9–13. DOI: 10.1098/rspa.1991.0075

64. Smagorinsky, J. 1963. "General Circulation Experiments With the Primitive Equations: I. The Basic Experiment." *Monthly Weather Review* 91, no. 3: 99–164. DOI: 10.1175/1520-0493(1963)091 <0099:Gcewtp>2.3.Co;2

65. Haworth, D.C. 1999. "Large-Eddy Simulation of In-Cylinder Flows." *Oil & Gas Science and Technology* 54, no. 2 (March–April): 175–185. DOI: 10.2516/ogst:1999012

66. Celik, I., Yavuz, I., Smirnov, A., Smith, J., Amin, E., and Gel, A. 2000. "Prediction of In-Cylinder Turbulence for IC Engines." *Combustion Science and Technology* 153, no. 1: 339–368. DOI: 10.1080/00102200008947269

67. Pomraning, E., and Rutland, C.J. 2002. "Dynamic One-Equation Nonviscosity Large-Eddy Simulation Model." *AIAA Journal* 40, no. 4: 689–701. DOI: 10.2514/2.1701

68. Deardorff, J.W. 1970. "A Numerical Study of Three-Dimensional Turbulent Channel Flow at Large Reynolds Numbers." *Journal of Fluid Mechanics* 41, no. 2 (March): 453–480. DOI: 10.1017/S0022112070000691

69. Lilly, D.K. 1967. "The Representation of Small-Scale Turbulence in Numerical Simulation Experiments." In *Proceedings of IBM Scientific Computing Symposium on Environmental Sciences, Yorktown Heights, November 14–16*, 195–210. White Plains, N.Y.: IBM Data Processing Division.

70. Speziale, C.G. 1998. "Turbulence Modeling for Time-Dependent RANS and VLES: A review." *AIAA Journal* 36, no. 2: 173–184. DOI: 10.2514/2.7499

71. Germano, M., Piomelli, U., Moin, P., and Cabot, W.H. 1991. "A Dynamic Subgrid-Scale Eddy Viscosity Model." *Physics of Fluids A: Fluid Dynamics* 3, no. 7: 1760–1765. DOI: 10.1063/1.857955

72. Lilly, D.K. 1992. "A Proposed Modification of the Germano Subgrid-Scale Closure Method." *Physics of Fluids A: Fluid Dynamics* 4, no. 3: 633–635. DOI: 10.1063/1.858280

73. Yoshizawa, A., and Horiuti, K. 1985. "A Statistically-Derived Subgrid-Scale Kinetic Energy Model for the Large-Eddy Simulation of Turbulent Flows." *Journal of the Physical Society of Japan* 54, no. 8: 2834–2839. DOI: 10.1143/JPSJ.54.2834

74. Kim, W.W., and Menon, S. 1995. "A New Dynamic One-Equation Subgrid-Scale Model for Large Eddy Simulations." Paper presented at the: *33rd Aerospace Sciences Meeting and Exhibit, 09–12 January 1995, Reno, NV, U.S.A.* DOI: 10.2514/6.1995–356

75. Sone, K., and Menon, S. 2003. "Effect of Subgrid Modeling on the In-Cylinder Unsteady Mixing Process in a Direct Injection Engine." *Journal of Engineering for Gas Turbines and Power* 125, no. 2 (April): 435–443. DOI: 10.1115/1.1501918

76. Brusiani, F., and Bianchi, G.M. (2008). "LES Simulation of ICE Non-Reactive Flows in Fixed Grids." SAE Technical Paper No. 2008-01-0959. DOI: 10.4271/2008-01-0959

77. Chumakov, S.G., and Rutland, C.J. 2005. "Dynamic Structure Subgrid-Scale Models for Large Eddy Simulation." *International Journal for Numerical Methods in Fluids* 47, no. 8-9: 911–923. DOI: 10.1002/fld.907

78. Jhavar, R., and Rutland, C.J. 2006. "Using Large Eddy Simulations to Study Mixing Effects in Early Injection Diesel Engine Combustion." SAE Technical Paper No. 2006-01-0871. DOI: 10.4271/2006-01-0871

79. Hu, B., Jhavar, R., Singh, S., Reitz, R.D., and Rutland, C.J. 2007. "Combustion Modeling of Diesel Combustion With Partially Premixed Conditions." SAE Technical Paper No. 2007-01-0163. DOI: 10.4271/2007-01-0163

80. Banerjee, S., Liang, T., Rutland, C., and Hu, B. 2010. "Validation of an LES Multi Mode Combustion Model for Diesel Combustion." SAE Technical Paper No. 2010-01-0361. DOI: 10.4271/2010-01-0361

81. Yu, R., Bai, X.S., Hildingsson, L., Hultqvist, A., and Miles, P.C. 2006. Numerical And Experimental Investigation of Turbulent Flows in a Diesel Engine." SAE Technical Paper No. 2006-01-3436. DOI: 10.4271/2006-01-3436

82. Vermorel, O., Richard, S., Colin, O., Angelberger, C., Benkenida, A., and Veynante, D. (2007). "Multi-Cycle LES Simulations of Flow And Combustion in a PFI SI 4-Valve Production Engine." *SAE Transactions* 116, no. 3 (April): 152–164. DOI: 10.4271/2007-01-0151

83. Schönfeld, T., and Rudgyard, M.J. 1999. "Steady And Unsteady Flow Simulations Using the Hybrid Flow Solver AVBP." *AIAA Journal* 37, no. 11:1378–1385. DOI: 10.2514/2.636

84. Fontanesi, S., d'Adamo, A., and Rutland, C.J. 2015. "Large-Eddy Simulation Analysis of Spark Configuration Effect on Cycle-To-Cycle Variability of Combustion And Knock." *International Journal of Engine Research* 16, no. 3: 403–418. DOI: 10.1177/1468087414566253

85. Kazmouz, S.J., Haworth, D.C., Lillo, P., and Sick, V. 2021. "Large-Eddy Simulations of a Stratified-Charge Direct-Injection Spark-Ignition Engine: Comparison With Experiment And Analysis of Cycle-To-Cycle Variations." *Proceedings of the Combustion Institute* 38, no. 4: 5849–5857. DOI: 10.1016/j.proci.2020.08.035

5

Fuel Sprays

In modern IC engines, liquid fuels are injected into the engine cylinder or intake port in forms of sprays that contain numerous small droplets. The injection pressures in recent engine applications range from 0.4 MPa in a PFI gasoline engine to 35 MPa in a DI gasoline engine, and 240 MPa in an automotive diesel engine. The droplets travel in the gas environment and evaporate under high gas temperature, changing phase from liquid to gaseous vapor, and the fuel vapors mix with the surrounding gases to form a mixture ready to burn. In many cases, sprays impinge on the surfaces of the walls, e.g., sprays impinge on the back surfaces of the intake valves in a PFI engine as intended and also on the surfaces of the piston or cylinder liner in a DI gasoline engine, which is not preferred. When spray-wall impingement occurs, part of the impinged spray will splash back, the other part will spread on the surfaces to form thin liquid films. These processes are affected by the interactions between the drops and gas during which momentum, mass, and heat are exchanged between the liquid and gas phases. In this chapter, the computational and physics models to simulate the spray processes are discussed with an emphasis on the fundamentals and the most used models in engine simulations. First, general descriptions of multidimensional spray modeling and dimensional parameters of spray are given. Second, atomization models of liquid jets and sheets are introduced. After that, models to describe the dynamical processes including drop secondary breakup, collision and coalescence, deformation and drag as well as turbulent dispersion are discussed; evaporation models of both single-component and multi-component treatments are presented. Finally, hydrodynamics and heat transfer models of spray-wall impingement are described.

5.1 General Description

5.1.1 Multidimensional Spray Modeling

Sprays form once the fuel is injected into the gas environment by an injector with high injection pressure, and consist of numerous droplets with different sizes, velocities, and other physical parameters. The fundamental mechanisms of the formation of discrete drops from an intact liquid jet or a liquid sheet have been under extensive experimental and theoretical study for many years [1,2]. However, the mechanisms of atomization are still not well understood due to the complicity of the phenomena. Direct measurements at or near the injector nozzle tip are extremely difficult because the spray in this region is optically dense and optical and laser diagnostics are generally ineffective [3]. This region is believed to be the most important area where liquid jet or sheet atomization (or primary break) takes place.

Recent progress has been made to observe the dynamics in the near nozzle region by using time-resolved x-radiography direct-imaging of hollow-cone sprays used in DI gasoline engines [4,5]. Complex density waves and unexpected axially asymmetric flows were found in the region. The completed primary breakup of the liquid sheet upon exiting the nozzle was also observed experimentally. These experiments provide additional new pieces of knowledge on spray atomization [6].

An important area of previous research on diesel sprays was to develop empirical correlations to relate the spray structure parameters with injection pressure, environment gas density, and fuel properties. The parameters usually include spray tip penetration, spray spreading angle, drop mean diameters. The correlations formed by Hiroyasu and co-workers at Hiroshima University are well recognized and still used in quasi-dimensional engine performance simulations [7,8].

Experimental observations have revealed that unstable waves grow up on the surface of a high-speed liquid jet or sheet [1]; the growth of the waves due to small disturbances owing to the interaction between the liquid and ambient gas is believed to initiate the liquid breakup process [2].

Reitz has suggested to divide the liquid jet atomization breakup phenomena into four breakup regimes to reflect the appearance of jets as the forces acting on the jet change [2,9,10]. These regimes are corresponding to different combinations of liquid inertia, surface tension, and aerodynamic forces acting on the jet, as shown in Figure 5.1. These include the Rayleigh, first wind-induced, second wind-induced, and atomization regimes. Observation and identification of surface wave instability for spray breakups have leaded to successful mathematical models to predict the resulting behaviors of sprays and drops [10,11].

The characteristics of the regimes are summarized as:

1. In the Rayleigh regime (Figure 5.1a), the low-velocity jet breakup is due to the unstable growth of surface waves caused by surface tension, and the sizes of the resulting droplets are larger than the jet diameter.

2. In the first wind-induced breakup regime (Figure 5.1b), as jet velocity is increased, forces due to the relative motion of the jet and the surrounding air are greater than the surface tension force, breakup occurs many nozzle diameters downstream of the nozzle and the sizes of the droplet are of the order of the jet diameter.

3. In the second wind-induced regime (Figure 5.1c), a further increase in jet velocity results in a breakup starting some distance downstream of the nozzle. The unstable growth of short-wavelength waves induced by the relative motion

between the liquid and surrounding air produces droplets whose average size is less than the jet diameter.

4. In the atomization regime (Figure 5.1d), as the jet velocity is further increased, jet breakup occurs at the nozzle exit and the sizes of the droplets are much smaller than the nozzle diameter. Aerodynamic interactions at the liquid/gas interface are believed to be one major component of the atomization mechanism in this regime.

Multidimensional models to compute fuel spray dynamics have been under development since the 1970s. However, simulation of diesel sprays in a 3D space was not plausible and feasible until the stochastic particle model by Dukowicz [12], instability-based spray atomization and breakup models by Reitz [9,10], drop dynamic models for collisions and coalescences by O'Rourke [13], and numerical methods by O'Rourke and co-workers [14] had been established in the 1980s. Their models have provided the framework of modeling sprays in IC engines, are of good computational efficiency and adequate accuracy, and widely used in engine combustion simulations to date.

The stochastic particle method solves the governing equations for the gas phase and the interactions between the droplets and the gas phase with a fully-interacting combination of Eulerian fluid and Lagrangian particle calculations. To compute spray dynamics, the idea of the Monte Carlo method is used. Discrete drops are introduced into the computation stochastically by a relatively small number of computational parcels of particles, and each computational parcel represents several droplets of identical size, velocity, and temperature. This method provides a basis on which some important new physical effects in spray calculations can be incorporated. In particular, much progress has been made in discovering the mechanisms that determine spray drop sizes with the use of this method.

FIGURE 5.1 Four breakup regimes of jet atomization breakup.

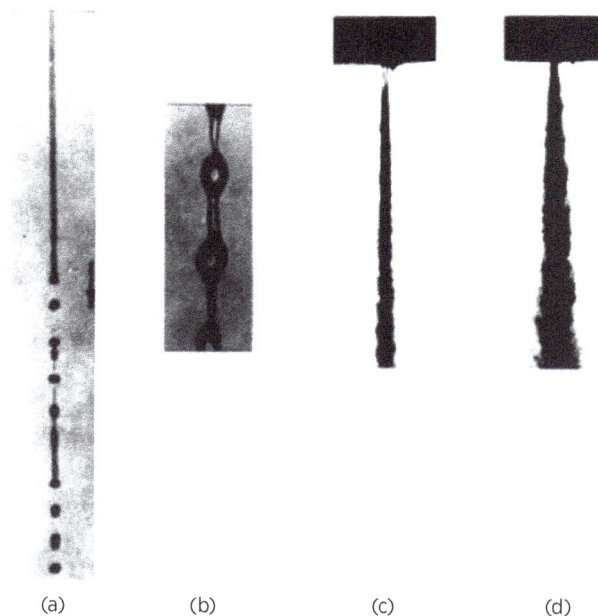

(a)   (b)   (c)   (d)

Reitz [10] describes fuel spray atomization by introducing the Blob method whereby "blobs" are injected (with sizes equal to the nozzle exit diameter), and the breakup of the blobs and the resulting drops is modeled using stability analysis for liquid jets. This method provides a way with theoretical analysis to initialize spray computation, and it also predicts various regimes of breakup as shown in Figure 5.1. The product drops are distinguished from the parent drop by having different drop sizes. In addition, Reitz and co-workers have also originated and improved several spray physical models for engine simulations. These models include but not limited to jet atomization [10], spray wall impingement [15], sheet atomization [16,17], drop secondary breakup [11], multi-component drop vaporization model [18,19], and film heat transfer [20,21].

O'Rourke [13] studied the consequent physics of the droplets after atomization and proposed the models to simulate collision and coalescence of drops. O'Rourke and co-workers also proposed a drop breakup model called the Taylor Analogy Breakup (TAB) model [22]. The model is based on the Taylor analogy between an oscillating and distorting droplet and a spring-mass system. The restoring force of the spring is analogous to the surface tension forces. The external force on the mass is analogous to the gas aerodynamic force. The damping forces due to liquid viscosity are added to the analogy. The most important contribution of O'Rourke and Amsden (and other members in the KIVA team at the NANL) was to integrate and implement the existing spray models into the KIVA family codes [14,23–25] with publicly accessible open source-codes. Therefore, the models can not only be used by others but can also be replaced easily with improvements or new models by researchers worldwide.

In the stochastic discrete particle modeling approach, physical phenomena of sprays or drops are described by the processes, which are represented by separated physical sub-models. These processes are conceptually illustrated in Figure 5.2. As can be seen, several sub-models are needed to compute spray atomization, drop dynamics including breakup, collision and coalescence, deformation and drag, turbulent dispersion, and drop evaporation. When spray wall impingement occurs, additional sub-models are required to simulate spray splash and the dynamics, heat transfer, and evaporation of fuel film. Although it is somewhat arbitrary to isolate these processes in this approach, it has been proven to be effective and accurate in engine engineering predictions. The sub-models provide mathematical formulations that are calculated and the results are fed back to the spray source terms in the governing equations. Models to describe the dynamic and thermal processes illustrated in Figure 5.2 are described later in this chapter.

FIGURE 5.2   Conceptual diagram showing spray processes.

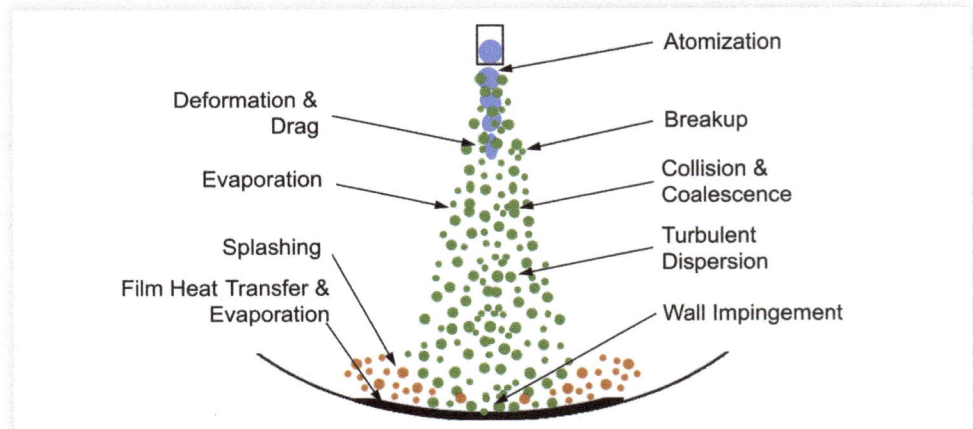

Courtesy of Zhiyu Han

It is pointed out that although the spray models were originally developed to simulate diesel injection and sprays in most cases, they have been successfully extended to sprays from pressure-swirl or multi-hole injectors in DI gasoline engines. Computation examples in this book are all carried out with the use of these models unless otherwise stated.

5.1.2 Structure Parameters of Sprays

Sprays are usually characterized by several parameters. Some macro dimensions are used to define the structure of a spray. The most important dimensional parameters are the spray tip penetration L and spray angle θ as illustrated in Figure 5.3, which shows a sketch of a jet spray. These parameters are obtained by measuring spray images taken in experiments. Spray experiments are usually carried out in an open or a closed vessel in which fuel is injected under certain pressure and temperature conditions.

Other important parameters are related to the sizes and size distributions of liquid droplets in a spray. Dependent on a definition, a weight-mean drop diameter (or radius) is given. In addition, certain types of sprays (e.g., splashing drops due to spray wall impingement) exhibit different forms of drop size distributions, represented by different mathematical functions. We will now discuss the commonly used mean spray sizes and size distributions in engine simulations.

In general, a mean diameter can be given as:

$$D_{ab} = \left(\frac{\sum N_i D_i^a}{\sum N_i D_i^b} \right)^{1/(a-b)} \tag{5.1}$$

FIGURE 5.3 A sketch diagram showing jet (diesel) spray structure.

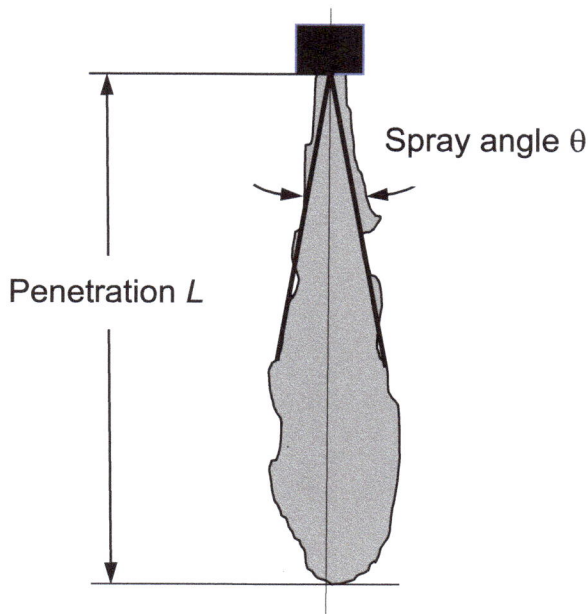

Spray angle θ

Penetration L

Courtesy of Zhiyu Han

where i denotes the size range considered, N_i is the number of droplets in size range i, and D_i is the middle diameter of size range i. Hence, when a is set to 3 and b is set to 2, the most used SMD in engine simulations is written as:

$$D_{32} = \frac{\sum N_i D_i^3}{\sum N_i D_i^2} \tag{5.2}$$

SMD represents the drop size at the mean of the drop surface area distribution. Therefore, consideration of the processes that depend on surface area (e.g., evaporation, combustion) with SMD makes more physical sense.

Another frequently used drop mean diameter in DI gasoline engine application is D_{v90} or $D_{0.9}$. It is the drop diameter such that 90% of the total spray volume is in drops with smaller diameters. D_{v90} measures the size of large drops in spray as well as the degree of uniformity of drop sizes, which is a smaller but important part of sprays with longer evaporation time and a likely cause of incomplete combustion and soot formation.

Sprays in engines are not homogeneous in size. A spray consists of numerous droplets of different sizes. However, the sizes of droplets fall into some sort of distribution from the smallest to the largest. Mathematics is applied to generalize drop size distributions, and the ones often used in engine simulations are discussed hereafter.

A χ^2 distribution is used in some models [14] based on diesel jet spray measurements. It is given as:

$$f(D) = \frac{1}{\bar{D}} e^{-D/\bar{D}} \tag{5.3}$$

and the cumulative distribution is then

$$V = 1 - exp\left(-\frac{D}{\bar{D}}\right)\left[1 + \frac{D}{\bar{D}} + \frac{1}{2}\frac{D^2}{\bar{D}} + \frac{1}{6}\frac{D^3}{\bar{D}}\right] \tag{5.4}$$

where $\bar{D}$ is a characteristic mean drop diameter. $\bar{D}$ relates to SMD (D_{32}) by

$$\bar{D} = \frac{1}{3}D_{32} \tag{5.5}$$

While the χ^2 distribution has been recommended for diesel sprays [22], it is found to overestimate the population of the large droplets when applied it to hollow-cone sprays in a DI gasoline engine, resulting in inaccuracy in predicting spray penetration and spray structure [16]. Therefore, the Rosin–Rammler distribution [33] has been proposed for sprays from sheet atomization.

The Rosin–Rammler cumulative distribution has the general form of

$$V = 1 - exp\left(-\frac{D^q}{\bar{D}}\right) \tag{5.6}$$

and the corresponding volume distribution is

$$\frac{dV}{dD} = \frac{qD^{q-1}}{\overline{D}^q} exp\left[-\frac{D^q}{\overline{D}} \right] \tag{5.7}$$

where q is the distribution parameter, which was set to be 3.5 in the work of Han et al. [16], and the characteristic mean diameter is given in a gamma function Γ as:

$$\overline{D} = D_{32}\Gamma(1 - q^{-1}) \tag{5.8}$$

A comparison of χ^2 and Rosin–Rammler volume distributions is shown in Figure 5.4. It is clearly seen that for a given SMD, the χ^2 distribution gives more large drops and less medium drops as compared to the Rosin–Rammler distribution.

FIGURE 5.4 Comparison of χ^2 and Rosin–Rammler distribution functions for two SMD cases. The SMD is indicated by the number by the curves.

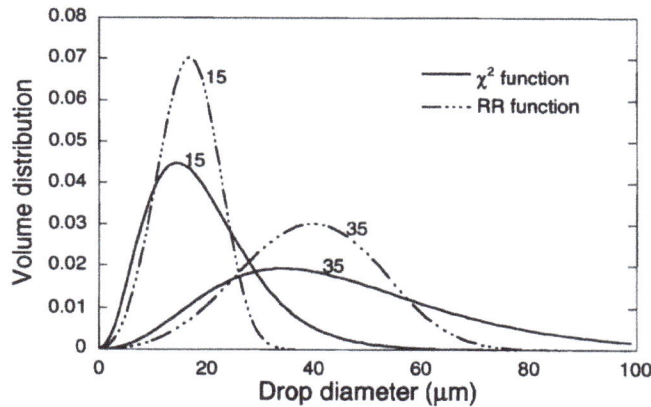

SMD: Sauter mean diameter

The Nukiyama–Tanasawa distribution functions [27] have been suggested in modeling the size distribution of secondary droplets from splash when a spray impinges on a wall. One form of the Nukiyama–Tanasawa distribution function that has been proposed by O'Rourke and Amsden [28] is given as:

$$f(D) = \frac{8}{\sqrt{\pi}} \frac{D^2}{D_{max}^3} exp\left[-\left(\frac{D}{D_{max}} \right)^{3/2} \right] \tag{5.9}$$

where D_{max} is the maximum drop diameter in the distribution. Another type of Nukiyama–Tanasawa function has been suggested by Han et al. [29], formulated as:

$$f(D) = \frac{2}{3}\frac{D^2}{\overline{D}^3} exp\left[-\left(\frac{D}{\overline{D}}\right)^{3/2}\right]$$

(5.10)

and the corresponding volume distribution and SMD are:

$$\frac{dV}{dD} = \frac{1}{4}\frac{D^5}{\overline{D}^6} \exp\left[-\left(\frac{D}{\overline{D}}\right)^{3/2}\right]$$

(5.11)

$$D_{32} = \frac{\Gamma(4)}{\Gamma(4/3)}\overline{D} = 2.16\overline{D}$$

(5.12)

It is worthwhile to summarize that despite the importance of drop size distribution, there are no models that can predict it directly so far. This is not surprising since a size distribution contains many dimensional scales that are beyond the current computer capability to resolve. Rather, a model usually predicts mean drop sizes, and a distribution function is then assigned to the mean sizes. The distribution function is often problem-dependent and can be obtained by curve-fitting experimental data. This method has been proven to be successful in spray simulations of engines.

5.2 Spray Atomization

5.2.1 Numerical Treatment of Fuel Injection

In simulating IC engines, fuels need to be introduced into the computational domains, which represent either the intake ports in PFI engines or cylinders in DI gasoline engines or diesel engines. This fuel induction process simulates fuel injection. Fuel injection simulation is made easy by using the stochastic particle model in the Lagrangian–Eulerian numerical approach as discussed in earlier chapters.

In the fuel injection model of the stochastic particle method by Dukowicz [12], each particle injected into or entering the computing mesh is assigned a velocity $\boldsymbol{u}_{pk}$, a radius r_k, and the number of particles in the group N_{pk}. The number of computational particles injected per cell per time step is denoted as K. The radius of each particle is then chosen from a uniform random distribution in the range $0 < r_k < r_{max}$ for $1 < k < K$.

If the particle mass flow into the cell is Q, the following equations can be written defining N_{pk}:

$$\sum_{k=1}^{K} N_{pk} m_k = Q\Delta t$$

(5.13)

$$f(r_k) = \frac{\alpha N_{pk}}{\sum_{l=1}^{K} N_{pl}}$$

(5.14)

where α is a proportionality constant and $f(r)$ is the initial particle size distribution function. These relationships in Equations (5.13) and (5.14) are sufficient to determine N_{pk}. More than one particle per cell must be injected ($K > 1$) in order to develop a distribution of particles that approximates the specified size distribution function $f(r_k)$.

It is assumed that the particle velocity distribution is independent of the size distribution and is very much case dependent. For a simple single-hole spray injector, the following procedure can be employed. If the mass flow Q is known, the magnitude of the injection velocity is

$$U_0 = \frac{4}{\pi} \frac{Q}{\rho_l d^2} \tag{5.15}$$

where ρ_l is the fuel density and d is the diameter of the injector orifice. Alternatively, if the pressure drop across the nozzle Δp is known, then

$$U_0 = C_D \sqrt{2\Delta p / \rho_l} \tag{5.16}$$

where C_D is the discharge coefficient of the nozzle. The transverse velocity is derived in terms of the initial spray angle using the relationship

$$\text{Max}(u_{pk}) = U_0 \tan\left(\frac{\theta}{2}\right) \tag{5.17}$$

where θ is the angle of the initial portion of the spray cone. Transverse velocities are then assigned to individual particles from a uniform random distribution in the range $0 < u_{pk} < Max(u_{pk})$. Since the magnitude of the drop velocity is V, this determines both components of velocity.

The initial particle size distribution can be defined by using some assumptions. In the simplest assumption, the drop size distribution can be presumed as a mono-disperse distribution (or uniform distribution). In this case, the initial radius can be set equal to the average size, e.g., Sauter mean radius (SMR), of the drops or most simply as equal to the radius of the internal hole of an injector. In KIVA, a χ^2 distribution is used as described earlier.

Note that by using the above fuel injection treatments, atomization of a liquid jet is implied to have completed at the exit of the injector nozzle and the resultant droplets are then modeled accordingly. This underlying assumption ignores the physical details of fuel atomization and hence may cause inaccuracy in computations. Atomization models will be discussed next.

5.2.2 Jet Atomization

Atomization (also called primary breakup) is a term to describe the breakup of an intact liquid jet or sheet at or near the exit of an injector. There are still some unknowns regarding atomization mechanisms as discussed earlier. Atomization models, nevertheless, are needed to supply the initial conditions at the injector nozzle exit for spray simulations, which include drop sizes, velocities, temperatures, etc. In Sec. 5.2.1, we described some basic, but simple, treatments (models) to provide these conditions. Although they are mathematically efficient, physics justifications are still needed.

Reprinted with permission from Ref. [10]. © Elsevier

FIGURE 5.5 Schematic diagram showing surface waves and breakup on a liquid jet.

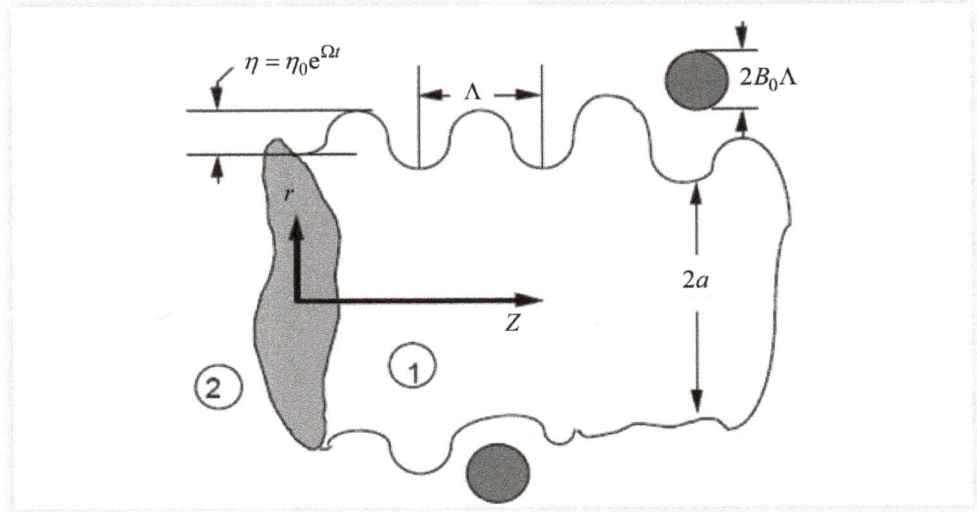

Many atomization models have been proposed for jet breakup in engine simulations [10,11,30–32]. Among these, the one most used is the Wave model by Reitz [10]. To develop the Wave model, a stability analysis is performed to consider the unstable growth of Kelvin–Helmholtz (KH) waves on the liquid surface, which results in a dispersion equation that relates the growth of an initial perturbation on a liquid surface of infinitesimal amplitude to its wavelength and to other physical and dynamic parameters of both the injected liquid and the ambient gas. The dispersion equation is solved for the maximum growth rate and the corresponding wavelength, which are then related to the sizes of the drops resulting from the breakup.

An infinitesimal axisymmetric displacement is imposed on the surface of a jet, illustrated in Figure 5.5, with the form as:

$$\eta = \mathcal{R}(\eta_0 e^{ikz + \omega t})$$

This leads to Equation (5.18), derived from the linearized hydrodynamical equations for the liquid and gas.

$$\omega^2 + 2v_1 k^2 \omega \left[\frac{I_1'(ka)}{I_0(ka)} - \frac{2kL}{k^2 + L^2} \frac{I_1(ka)}{I_0(ka)} \frac{I_1'(La)}{I_1(La)} \right]$$

$$= \left[\frac{\sigma k}{\rho_1 a^2} (1 - k^2 a^2) + \frac{\rho_2}{\rho_1} (U - i\omega / k)^2 k^2 \left(\frac{L^2 - k^2}{L^2 + k^2} \right) \frac{K_0(ka)}{K_1(ka)} \right] \left[\left(\frac{L^2 - k^2}{L^2 + k^2} \right) \frac{I_1(ka)}{I_0(ka)} \right] \tag{5.18}$$

Equation (5.18) relates the growth rate ω of an initial perturbation of infinitesimal amplitude η_0 to its wavelength λ (wavenumber $k = 2\pi / \lambda$). The relationship also includes the physical and dynamic parameters of the liquid jet and the surrounding gas. Subscript 1 refers to liquid properties while subscript 2 identifies gas properties. The liquid equations are solved with wave solutions of the form $\phi_1 = C_1 I_0(kr)\exp(ikz + \omega t)$ and $\psi_1 = C_2 r I_1(Lr)\exp(ikz + \omega t)$. Here, ψ_1 and ϕ_1 are the stream function and velocity potential, respectively, C_1 and C_2 are integration constants, $L_2 = k^2 + \omega/v_1$, v_1 is the liquid dynamic viscosity, I_0 and I_1 are modified Bessel functions of the first kind, K_0 and K_1 are modified Bessel functions of the second kind, U is the relative velocity between the liquid and gas, and the prime denotes differentiation.

Numerical solutions of Equation (5.18) have been generated for the maximum growth rate ($\omega = \Omega$) and the corresponding wavelength ($\lambda = \Lambda$), which results in

$$\Omega = \frac{(0.34 + 0.38We_2^{1.5})}{(1+Z)(1+1.4T^{0.6})}\left(\frac{\sigma}{\rho_1 a^3}\right)^{0.5} \tag{5.19}$$

$$\Lambda = 9.02a\frac{(1+0.45Z^{0.5})(1+0.4T^{0.7})}{(1+0.87We_2^{1.67})^{0.6}} \tag{5.20}$$

The drop Weber number We_1, gas Weber number We_2, drop Ohnesorge number Z, drop Reynolds number Re_1 and gas Taylor number T are defined as:

$$\begin{cases} We_1 = \dfrac{\rho_1 U^2 a}{\sigma} \\ We_2 = \dfrac{\rho_2 U^2 a}{\sigma} \\ Z = We_1^{0.5} / Re_1 \\ Re_1 = \dfrac{Ua}{v_1} \\ T = ZWe_2^{0.5} \end{cases} \tag{5.21}$$

The above analysis leads to the Reitz Wave breakup model by postulating that new drops are formed (with drop radius r) from a parent drop (with radius a) with

$$r = \begin{cases} B_0\Lambda & \left(B_0\Lambda \le a\right) \\[2ex] min\begin{cases} \left(\dfrac{3\pi a^2 U}{2\Omega}\right)^{0.33} \\[2ex] \left(\dfrac{3a^2\Lambda}{4}\right)^{0.33} \end{cases} & (B_0\Lambda > a,\ \text{one time only}) \end{cases} \tag{5.22}$$

It is assumed that (small) drops are formed with drop size proportional to the wavelength of the fastest growing or most probable unstable surface wave, and that the jet disturbance has frequency $\Omega/2\pi$ (a drop is formed each period) or that drop size is determined from the volume of liquid contained under one surface wave.

The rate of change of drop radius in a parent droplet is given as:

$$\frac{da}{dt} = -\frac{a-r}{\tau}\left(r \leqslant a\right) \tag{5.23}$$

where

$$\tau = \frac{3.726 B_1 a}{\Lambda \Omega} \tag{5.24}$$

B_0 and B_1 in Equations (5.22) and (5.24) are constants and were originally recommended to be 0.61 and 10.0, respectively, for diesel sprays. However, due to the uncertainty of the physics of the flow inside the injector nozzle orifice and the effects of these flows on spray jet atomization, the breakup time constant B_1 (which is related to the initial disturbance level on the liquid jet) cannot be determined on a fundamental basis. It has been found that B_1 vary from one injector to another in diesel combustion modeling [33], which is reasonable since the level of the initial disturbance is expected to vary with different injector designs.

Following each breakup event, the product drop parcel is given the same temperature and physical location as the parent. They were given the same velocity (magnitude) as the parent in the direction of the parent drop velocity vector V. The other velocity components v and w normal to V are given by

$$v = |V| \tan\left(\frac{\theta}{2}\right) sin\phi \tag{5.25}$$

$$w = |V| \tan\left(\frac{\theta}{2}\right) cos\phi \tag{5.26}$$

where θ is given to be uniformly distributed between 0 and Θ by

$$\tan\left(\frac{\Theta}{2}\right) = A_1 \frac{\Lambda \Omega}{U_0} \tag{5.27}$$

and ϕ is chosen at random on the interval $(0, 2\pi)$. U_0 is the injection velocity. A_1 is a model constant which depends on nozzle design [34].

This model has been validated with the use of the experimental data by Hiroyasu and Kadota [8]. A comparison of spray drop size is shown in Figure 5.6. The SMD data are the averages over the spray cross-section 65 mm downstream of the nozzle in the experiments. Good agreements are achieved. Reitz gives a more detailed analysis of the model performance [10].

Experiments on diesel sprays show that there is an intact core region of unbroken liquid within the spray near the nozzle exit. Reitz and Diwakar [30] used the Blob model [10] to introduce parcels of liquid into the computational cell, instead of assuming an intact liquid at the nozzle exit. The fuel "blobs" have a characteristic size equal to the nozzle hole diameter. They argue that the atomization of the injected liquid and the subsequent breakup of drops

FIGURE 5.6 SMD variation with distance from the nozzle for the sprays.

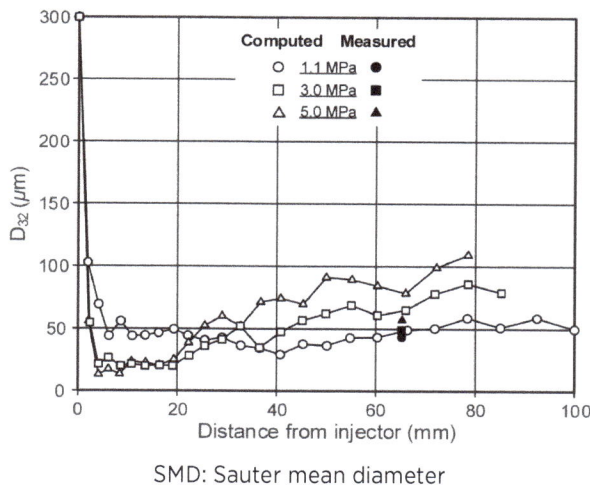

SMD: Sauter mean diameter

are indistinguishable processes within a dense spray. A core region is predicted to exist near the nozzle with the Blob model because, although the injected liquid breaks up due to its interaction with the surrounding gas as it penetrates the gas, there is a region of large discrete liquid particles near the nozzle, which is conceptually representing the intact core of a spray.

In the original Reitz Wave atomization model, it is assumed that the injected drop blobs or parcels have the same sizes as the nozzle exit diameter. The advantage of this assumption is that uncertainties related to the effects of the internal nozzle-orifice flow and nozzle geometry on the initial disturbances and the atomization process can be incorporated into just one model constant (B_1 in Equation (5.24)). However, this makes it difficult to model injectors with different nozzle geometries. Experiments have confirmed that processes such as super-cavitation take place at the nozzle exit under the normal injection conditions in a diesel engine and the measured flow velocity is close to the velocity calculated from the pressure drop assuming inviscid incompressible flow [35]. This indicates that an effective nozzle section area and a corresponding effective flow diameter should be used in the computation instead of the geometric nozzle exit area and diameter [36]. In this way, the contraction of the fuel jet can be included in the atomization model. Therefore, spray characteristics affected by different nozzle contraction effects (i.e., different discharge coefficients) can be modeled by introducing the discharge coefficient of an injector nozzle orifice C_D. The discharge coefficient relates the effective nozzle orifice radius r_e with the geometric radius r_0 as $r_e = \sqrt{C_D} r_0$ and the effective exit velocity is evaluated as $U_e = C_D \sqrt{2(p_1 - p_2)/\rho_l}$, and p_1 and p_2 are the nozzle orifice inlet and outlet pressure, respectively.

A more complete nozzle model was proposed by Sarre et al. [37]. In this model, if there is no cavitation, the exit velocity U_{mean} is used as the model injection velocity, which is evaluated as:

$$U_{mean} = C_D \sqrt{\frac{2(p_1 - p_2)}{\rho_l}}$$
(5.28)

The discharge coefficient C_D is estimated by

$$C_D = \frac{1}{\sqrt{K_{inlet} + f \cdot (L/D) + 1}} \quad (5.29)$$

where $f = \max(0.316Re^{-0.25}, 64/Re)$, which accounts for the effect of the Reynolds number Re of the fuel flow. K_{inlet} is inlet loss coefficients, and L/D is the orifice length/diameter ratio. If full cavitation is developed, the exit velocity and nozzle orifice radius are calculated by their effective values as:

$$U_e = U_{vena} - \frac{p_2 - p_{vapor}}{\rho_l U_{mean}} \quad (5.30)$$

$$r_e = r_0 \sqrt{U_{mean} / U_e} \quad (5.31)$$

where p_{vapor} is the vapor pressure of the fuel and U_{vena} is the velocity at the vena contracta, which is modeled by

$$U_{vena} = \frac{U_{mean}}{C_C} \quad (5.32)$$

and the contraction coefficient C_C is given as:

$$C_C = \left[\left(\frac{1}{C_{C0}} \right)^2 - 11.4 \frac{r_c}{D} \right]^{-0.5} \quad (5.33)$$

where r_c is the curvature of the orifice entrance and $C_{C0} = 0.62$. To determine occurrence of cavitation, the pressure at the smallest flow area p_{vena} is calculated as:

$$p_{vena} = p_1 - \frac{\rho_l}{2} U_{vena}^2 \quad (5.34)$$

If p_{vena} is lower than p_{vapor}, it is assumed that the flow is fully cavitating and the inlet pressure and discharge coefficient are calculated by

$$p_1 = p_{vapor} + \frac{\rho_l}{2} U_{vena}^2 \quad (5.35)$$

$$C_D = C_C \sqrt{\frac{p_1 - p_{vapor}}{p_1 - p_2}} \quad (5.36)$$

5.2.3 Sheet Atomization

Hollow-cone sprays resulting from pressure-swirl injectors have been used in DI gasoline engines [38–40]. Models for liquid sheet atomization have been proposed. Reitz and Diwakar [41] used an atomization model to give the initial drop size at the nozzle exit in which the SMD was proportional to the sheet breakup length, estimated from a stability analysis of liquid sheets performed by Clark and Dombrowski [42].

Miyamoto et al. [43] used another sheet atomization model for air-assisted hollow-cone injectors based on a sheet stability argument. They specified the initial drop size as a function of the thickness of the liquid sheet, liquid surface tension coefficient and density, and the gas-sheet relative velocity. The sheet thickness was computed by solving the coupled liquid and gas governing equations inside the injector and the subsequent drop breakup was not considered. Instead of assuming sheet atomization at the nozzle exit, Lee and Bracco [44] employed Lagrangian equations to solve for the motion of an intact sheet outside the injector. Both stripping breakup and atomization at the tip of the sheet were considered. However, a fine cone-shaped grid system was required for resolving the very thin and deformed liquid sheet, which is not practical in engine simulations.

Han et al. have proposed a sheet atomization model [16] and a modified version [45] for hollow-cone sprays from a pressure-swirl injector. Atomization is described using a method whereby "blobs" that represent the liquid sheet outside the injector nozzle are injected with sizes equal to the sheet thickness, following the approach used in the jet atomization model [10] as discussed earlier. The breakup of the blobs and the subsequent drops is modeled using the TAB model [22] in which the originally used χ^2 size distribution for the breakup drops is replaced by a Rosin–Rammler distribution for the hollow-cone sprays considered. Schmidt et al. [17] applied linearized instability analysis of the breakup of a viscous, liquid sheet and proposed a linearized instability sheet atomization (LISA) model.

In the following discussions, the models proposed by Han and co-workers [16] and the LISA model [17] are described in detail. In a pressure-swirl atomizer, angular momentum is imposed on the liquid to form a swirling motion. Under the action of centrifugal force, the liquid spreads out in the form of a conical sheet as soon as it leaves the orifice, and a hollow cone spray is formed due to the breakup of the sheet.

The basic idea of the sheet spray atomization and breakup model is illustrated in Figure 5.7. It is assumed that a conical liquid sheet with a length, L, and a thickness, h, is formed at the exit of the nozzle. Instead of assuming an intact sheet, discrete blobs are injected that have a characteristic size equal to the thickness of the sheet. Since blobs or parcels are used to represent the liquid sheet, they are assumed to experience negligible dynamic drag force and are not affected by turbulence dispersion. However, these parcels are subject to break up once certain breakup criteria are satisfied according to the breakup model. The subsequent droplets formed due to a parent blob breakup will be treated as ordinary drops that experience drag forces and are affected by gas turbulence. A blob will be also treated as an ordinary drop once its traveling distance is greater than the sheet breakup length L.

The sheet velocity V is defined as:

$$V = K_v \left[\frac{2(p_1 - p_2)}{\rho_l} \right]^{0.5}$$

(5.37)

FIGURE 5.7 Schematic diagram showing the conceptual liquid flow structure at the nozzle exit and the sheet breakup process.

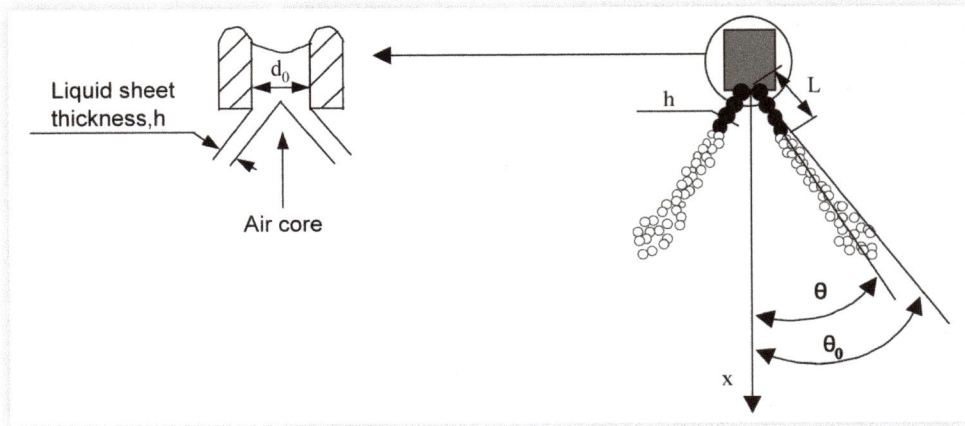

where p_1 is the fuel pressure within the injector, p_2 is the environmental pressure, ρ_l is the liquid density, and K_v is the velocity coefficient.

Due to the swirling motion of the liquid fuel, the liquid does not occupy the whole cross-sectional area of the nozzle hole; instead, air is sucked into the middle of the liquid to form an air core, which effectively blocks off the central portion of the nozzle orifice, as illustrated in Figure 5.7. A parameter X, which is defined as the ratio of the orifice area to the air core area, is used in the analysis. That is

$$X = \left(1 - \frac{2h}{d_0}\right)^2 \tag{5.38}$$

where d_0 is the nozzle orifice diameter. The velocity coefficient K_v can be derived based on inviscid analysis as:

$$K_v = C\left(\frac{1-X}{1+X}\right)^{0.5} \frac{1}{\cos\theta} \tag{5.39}$$

where $C = 1.1$ is used to account for discrepancy between the theory and experiments.

The thickness of the liquid sheet is estimated as:

$$h = \left[A\frac{12}{\pi}\frac{\mu_l \dot{m}_l}{d_0 \rho_l (p_1 - p_2)}\frac{1+X}{(1-X)^2}\right]^{0.5} \tag{5.40}$$

where $\dot{m}_l$ is the liquid mass flow rate and μ_l is the dynamic viscosity of the liquid [46]. The constant A is related to the nozzle geometry and is set to be 40.

Stability analyses have been conducted for describing the breakup of liquid sheets. Squire [47] studied an inviscid planar sheet and showed that the growth of sinuous waves eventually causes the sheet to break down. Clark and Dombrowski [42] extended that work by considering nonlinear effects and suggested the sheet breakup length to be

$$L = A \left[\frac{\rho_l \sigma K \ln(\eta/\eta_0)}{\rho^2 U^2} \right]^{1/3} \tag{5.41}$$

where A is a constant, ρ is the environmental gas density, σ is the liquid surface tension coefficient, U is the sheet-gas relative velocity, η is the wave amplitude when the sheet breaks up, and the parameter $ln(\eta/\eta_0)$ is determined experimentally to be equal to 12. K is used by Clark and Dombrowski to relate the sheet thickness h and its distance from the origin x, that is, $K = hx = $ const. It can be shown that $K = hL \cos \theta$ for a conical sheet. Hence, Equation (5.41) can be recast as:

$$L = B \left[\frac{\rho_l \sigma \ln(\eta/\eta_0) h \cos \theta}{\rho^2 U^2} \right]^{0.5} \tag{5.42}$$

where $B = A^{1.5}$ and is set to be 3.

Equations (5.37)–(5.42) provide the atomization or the initial blob information. A model is needed to compute the subsequent breakup processes of the blobs and the resulting drops, which will be discussed in Sec. 5.2.1.

The above sheet atomization model, together with the TAB model for drop breakup are validated with experiments [16]. A pressure-swirl injector was used and the sprays were injected into a vessel under ambient conditions. In the experiments, the charge injection devices (CID) camera imaging technique and a diffraction-based particle sizing system like a Malvern system were used [48]. Size measurements were performed 39 mm downstream of the nozzle on the centerline of the injector.

The computed spray images are compared to the experimental ones in Figure 5.8, where fuel is injected into a vessel with room conditions by a pressure-swirl injector with an injection pressure of 4.86 MPa. The computations capture the evolutions of the spray structures overall as well as the details showing the vortex-shape near the spray periphery, which agree well with the experiments. A hollow-cone structure presents inside the spray as shown by the computed two-dimensional slices across through the centerline of the spray. Note that the main spray exhibits a large spray cone angle, and there is an initial spray slug leading the main spray. It is believed that the fully developed swirl motion inside the injector leads to the large cone angle of the main spray while the initial spray is formed during the initial stages of the injection, at which time the angular momentum of the liquid within the injector has not yet been fully built up. The transit phenomena from the initial spray slug to the main spray could not be predicted by the atomization breakup models and two parts of the sprays were computed with different initial conditions.

Spray tip penetrations under the same conditions are shown in Figure 5.9, and the local SMDs of the spray droplets passing through the measurement beam region are shown in Figure 5.10. Good agreements between the computations and experiments are seen except that the predicted SMD is somewhat larger than the measured. Figure 5.11 compares the experimental and computational local drop size distributions. The experimental result is averaged over time. For the computed results, drop sizes at different delay times and those averaged over time are both shown. It is seen that while the computed temporal distributions present some scatters, the computed averaged distribution agrees with the measured one reasonably well. Han et al. provides more data comparisons [16].

Reprinted with permission from Ref. [16]. © Begell House Inc

FIGURE 5.8 Computed images of sprays from a pressure-swirl injector with comparison to experimental ones. Top row: experimental; Middle row: computed; Bottom row: computed two-dimensional slice through the center of the sprays showing the internal structure of the spray. The delay time is 0.7, 1.7, 2.7, 3.7, and 4.7 msec, respectively, from left to right.

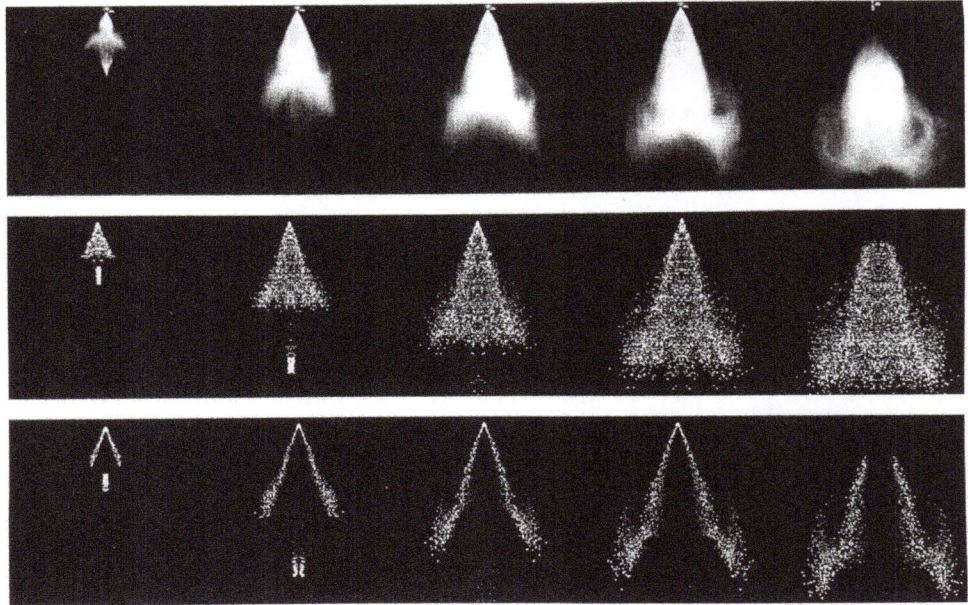

FIGURE 5.9 Spray tip penetration versus delay time.

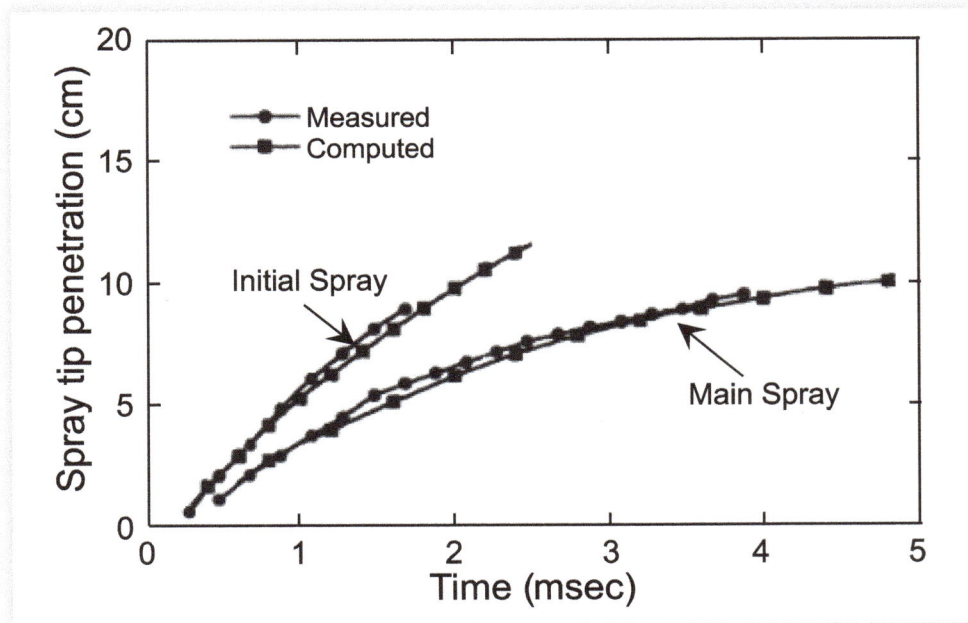

Reprinted with permission from Ref. [16]. © Begell House Inc

FIGURE 5.10 Local SMD versus delay time.

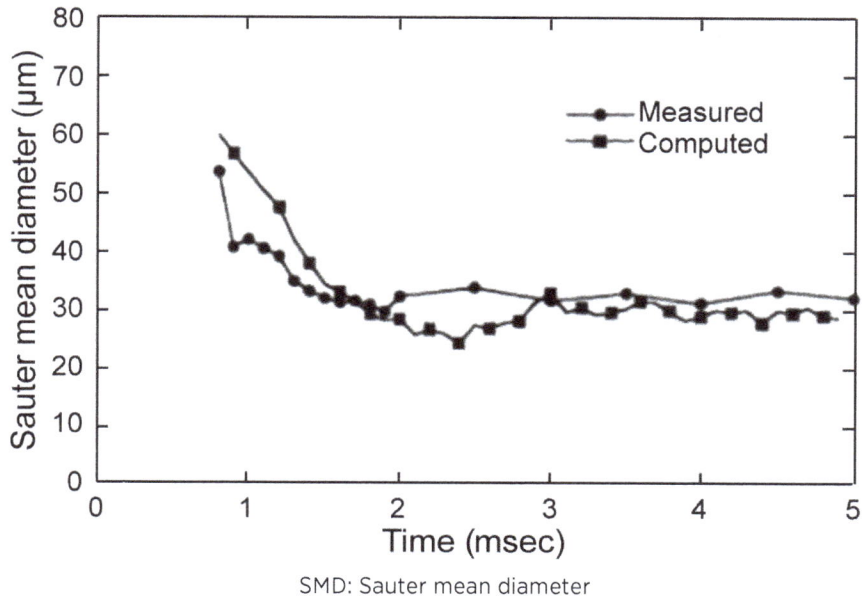

SMD: Sauter mean diameter

FIGURE 5.11 Comparison of measured and computed local drop size distribution.

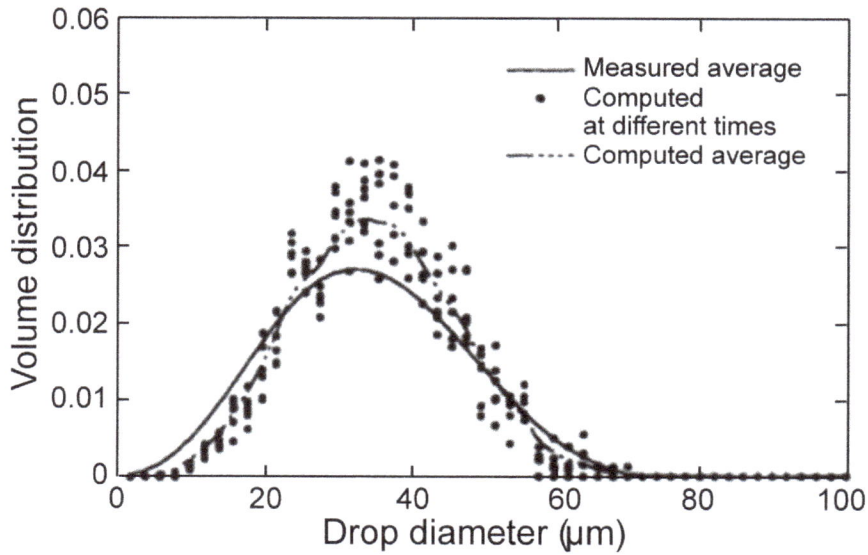

The LISA model was proposed to model liquid sheet breakup for a pressure-swirl injector [17,49], discussed earlier in the chapter. The sheet breakup process is modeled based on wave stability theory. The model assumes that a two-dimensional, viscous, incompressible liquid sheet moves with relative velocity to the gas U through a quiescent, inviscid, incompressible gas medium. A spectrum of infinitesimal disturbances,

$$\eta = \eta_0 \exp(ikx + \omega t) \tag{5.43}$$

is imposed on the initially steady motion that produces fluctuating velocities and pressures for both the liquid and the gas, where η_0 is the initial wave amplitude, $k = 2\pi/\lambda$ is the wavenumber, and $\omega = \omega_r + i\omega_i$ is the growth rate of the surface disturbances. A dispersion relation with its simplified form is given in Equation (5.44) with the assumption of short waves as:

$$\omega_r = -2v_l k^2 + \sqrt{4v_l^2 k^4 + \frac{\rho}{\rho_l}U^2 k^2 - \frac{\sigma k^3}{\rho_l}} \tag{5.44}$$

where v_l is the liquid kinematic viscosity, A critical Weber number of $We_g = \rho U^2 h/\sigma = 27/16$ is derived [49], below which the breakup process is dominated by long waves and above which it is dominated by short waves. As the Weber number is typically well above 27/16 for sheet breakup applications of pressure-swirl hollow-cone sprays, it is reasonable to assume that short waves are responsible for breakup.

The physical mechanism of sheet disintegration proposed by Dombrowski and Johns [50] is adopted to predict the drop sizes produced from the primary breakup process. Ligaments are assumed to form from the sheet breakup process once the unstable waves reach a critical amplitude. Since the growth rate for short waves is independent of the sheet thickness, the onset of ligament formation, or breakup length, can be formulated based on an analogy with the breakup length of cylindrical liquid jets [2]. If the surface disturbance has reached a value of $\eta_b = \eta_0 \exp(\Omega_s \tau)$ at breakup, the breakup time τ_b can be evaluated as:

$$\tau_b = \frac{1}{\Omega_s} \ln\left(\frac{\eta_b}{\eta_0}\right) \tag{5.45}$$

where Ω_s is the maximum growth rate obtained from Equation (5.44). Thus, the sheet will break up at the length L_b given by

$$L_b = V\tau_b = \frac{V}{\Omega_s} \ln\left(\frac{\eta_b}{\eta_0}\right) \tag{5.46}$$

where the quantity $\ln(\eta_b/\eta_0)$ is typically set to 12 based on the work of Dombrowski and Hooper [51]. Here V is the absolute velocity of the liquid sheet. The diameter of the ligaments formed at the point of breakup can be obtained from a mass balance. If it is assumed that the ligaments are formed from tears in the sheet once per wavelength, the resulting diameter of the ligament d_L is given by

$$d_L = \sqrt{\frac{16h}{K_s}} \tag{5.47}$$

where K_s is the wave number corresponding to the maximum growth rate Ω_s.

To calculate the film thickness, it is assumed that the sheet is in the form of a cone with the vertex at a point behind the injector orifice. The film half-thickness h at the breakup position L_b is approximately

$$h = \frac{d_f \cos(\theta)(d_0 - d_f)}{4L_b \sin(\theta) + d_0 - d_f} \tag{5.48}$$

where d_0 is the injector hole diameter, d_f is the thickness of the liquid film related to the liquid mass flow rate given later, and θ is the half-angle of the sheet cone.

If it is assumed that breakup occurs when the amplitude of the unstable waves is equal to the radius of the ligament, one drop will be formed per wavelength. A mass balance gives the drop size d_D as:

$$d_D = \left(\frac{3\pi d_L^2}{K_L}\right)^{1/3} \tag{5.49}$$

where the most unstable wavelength K_L is given by

$$K_L = \frac{1}{d_D}\left[\frac{1}{2} + \frac{3\mu_l}{2(\rho_l \sigma d_L)^{0.5}}\right]^{-1/2} \tag{5.50}$$

To calculate the sheet atomization, the initial conditions of the fuel flow are needed. A liquid film surrounding an air cone is formed due to the swirling motion of the liquid within the pressure-swirl injector. The mass flow rate is given by

$$\dot{m} = \pi \rho_l V \cos(\theta) d_f (d_0 - d_f) \tag{5.51}$$

where the injection velocity V is computed from the pressure drop across the injector exit, which is given as:

$$V = C_d \sqrt{\frac{2\Delta p}{\rho_l}} \tag{5.52}$$

and the effective discharge coefficient C_d is given as:

$$C_d = max\left(0.7, \ \frac{4\dot{m}}{\pi d_0^2 \rho_l \cos\theta}\sqrt{\frac{\rho_l}{2\Delta p}}\right) \tag{5.53}$$

5.3 Drop Dynamics

5.3.1 Secondary Breakup

We have discussed models for atomization or primary breakup of liquid jet or sheet in Sec. 5.1. Furthermore, the droplets resulting from the atomization processes will further break up into smaller droplets as evidenced by experiments [52,53]. These secondary breakup processes are simulated by the Wave model and the TAB model, which are based on completely different theories but give very similar results. In addition, these two models have been widely used in engine multidimensional simulations.

The model of Reitz [10] for primary breakup based on KH instability analysis has been assumed to be valid for the drop secondary breakup. Equations (5.22) to (5.24) are applicable in the KH drop breakup model and they are given here with the notation of KH as:

$$r_d = B_0 \Lambda_{KH} \tag{5.54}$$

$$\frac{dr}{dt} = -\frac{r - r_d}{\tau_{KH}} \tag{5.55}$$

where r is the radius of the parent parcel (which contains identical fuel droplets), and r_d is the radius of the new droplets. The breakup time is given by

$$\tau_{KH} = \frac{3.726 B_1 r}{\Lambda_{KH} \Omega_{KH}} \tag{5.56}$$

where Λ_{KH} and Ω_{KH} are calculated by Equations (5.19) and (5.20). B_0 and B_1 are model constants as stated in Sec. 5.2.1

Rayleigh–Taylor (RT) instability is believed to be responsible for additional droplet breakup due to rapid decelerations. The RT model is usually used in conjunction with the KH model to predict instabilities on the surface of the drop that grow until a certain characteristic breakup time when the drop finally breaks up. A hybrid KH-RT model has been proposed by Beale and Reitz [11], in which both models are used together to predict the secondary breakup of the droplets. The RT model is also based on wave instability theory. The maximum growth rate Ω_{RT} and the corresponding wavelength Λ_{RT} are given by

$$\Omega_{RT} = \sqrt{\frac{2}{3\sqrt{3\sigma}} \frac{[-a_t(\rho_l - \rho)]^{\frac{3}{2}}}{\rho_l + \rho}} \tag{5.57}$$

$$\Lambda_{RT} = 2\pi \Omega_{RT} \sqrt{\frac{3\sigma}{-a_t(\rho_l - \rho)}} \tag{5.58}$$

where a_t is the deceleration in the direction of travel. When the wavelength is smaller than the droplet diameter, the RT waves are assumed to be growing on the surface of the droplet. The wave growth time is then tracked. When it reaches its RT breakup timescale τ_{RT}, the drop is assumed to break up. τ_{RT} and the radius r_d of the new droplets are given as:

$$\tau_{RT} = \frac{C_\tau}{\Omega_{RT}}$$

(5.59)

$$r_d = 2C_{RT}\Lambda_{RT}$$

(5.60)

where C_τ and C_{RT} are model constants, 1.0 and 0.1, respectively.

The Reitz Wave atomization and breakup model is summarized in Figure 5.12. First, the Blob model is used to induce "blobs" of fuel liquid into the computational cell. The KH-RT model (Equations (5.22)–(5.24)) is then applied for calculating atomization of the "blobs." The KH and the RT models are then used together (Equations (5.54)–(5.60)) to compute the secondary breakup of the droplets. The activity of the RT model is the instability conditions in Equation (5.59).

The TAB model [22] is based on Taylor's analogy between an oscillating and distorting drop and a spring-mass system. The external forces acting on the mass, the restoring force of the spring, and the damping force are analogous to the gas aerodynamic force, the liquid surface tension force, and the liquid viscosity force, respectively. The parameters and constants in the TAB model equations have been determined from theoretical and experimental results. The force balance gives

$$\frac{d^2 y}{dt^2} + \frac{5\mu_l}{\rho_l r^2}\frac{dy}{dt} + \frac{8\sigma}{\rho_l r^3}y - \frac{2}{3}\frac{\rho U^2}{\rho_l r^2} = 0$$

(5.61)

where y is the normalized (by the drop radius) drop distortion parameter. U is the relative velocity between the gas and droplet, r is the droplet radius, σ is the liquid surface tension coefficient, and μ_l is the liquid viscosity. The exact solution of Equation (5.61) can be obtained by assuming constant coefficients, thus

FIGURE 5.12 Depiction of the Reitz Wave atomization and breakup model.

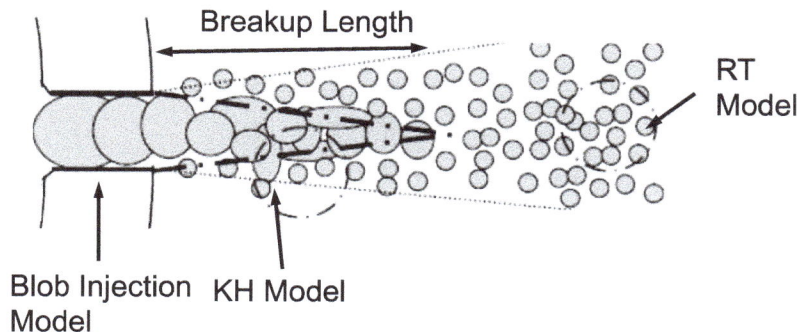

Courtesy of Zhiyu Han

$$y(t) = \frac{We}{12} + e^{-t/t_d}\left[\left(y(0) - \frac{We}{12}\right)\cos\omega t + \frac{1}{\omega}\left(\dot{y}(0) + \frac{y_0 - We/12}{t_d}\right)\sin\omega t\right] \tag{5.62}$$

where We is the Weber number. t_d is the viscous damping time and ω is the oscillation frequency, and they are given as $t_d = 2\rho_d r^2/5\mu_l$ and $\omega^2 = 8\sigma/\rho_d r^3 - 1/t_d^2$.

It is assumed that breakup occurs if and only if $y > 1$. When this condition is reached, the droplet breaks up into smaller droplets with sizes determined by an energy balance taken before and after the breakup as:

$$\frac{r}{r_{32}} = 1 + \frac{8K}{20} + \left(\frac{6K-5}{120}\right)\frac{\rho_l r^3}{\sigma}\left(\frac{dy}{dt}\right)^2 \tag{5.63}$$

where r_{32} is the SMR of the product drops. K is a model constant that varies with nozzle designs and should be determined experimentally. A value of 10/3 was suggested with the support of shock experiments.

The TAB model uses the χ^2 distribution for the distribution of sizes of the product drops to account for the resulted drop size variations due to the complicated interactions among the forces acting on the droplets. In addition, during the breakup process, there will be collisions and coalescences of the product drops, resulting in collisional broadening of the size distribution.

The velocity of the product drops normal to the path of the parent drop V_n is calculated by the follow equation as:

$$\tan\frac{\theta}{2} = \frac{V_n}{U} = C_v\frac{\sqrt{3}}{3}\sqrt{\rho/\rho_l} \tag{5.64}$$

where C_v is a model constant varying with nozzle designs. For hole nozzles with the ratio of inlet length and diameter of 12, $C_v = 1$ is suggested.

It is unclear which is superior, the Wave or TAB model. Fundamentally, surface wave instability and the resulting droplet detachment from bulk liquid have been observed experimentally in many cases for both high-speed liquid jets and sheets. Although there is some doubt about the validity of the analogy between the drop aerodynamics and spring-mass dynamics, some successful applications of the TAB model to predict diesel sprays [22] and gasoline sprays [16] have been reported. Another difficulty to assess the accuracy of these models arises from uncertainties in the model constants. However, it is interesting to note that although the Wave model and TAB model are derived from two very different approaches, the two models reach very similar results under extreme conditions [54]

5.3.2 Collision and Coalescence

Collision phenomena of drops have been under investigation for a long time. A comprehensive summary of this subject is given by Yarin et al. [55]. Interest in spray combustion has motivated investigations on the stability of colliding drops and the outcome from the drop collisions [56–61].

FIGURE 5.13 Depiction of the outcome of binary drop collisions.

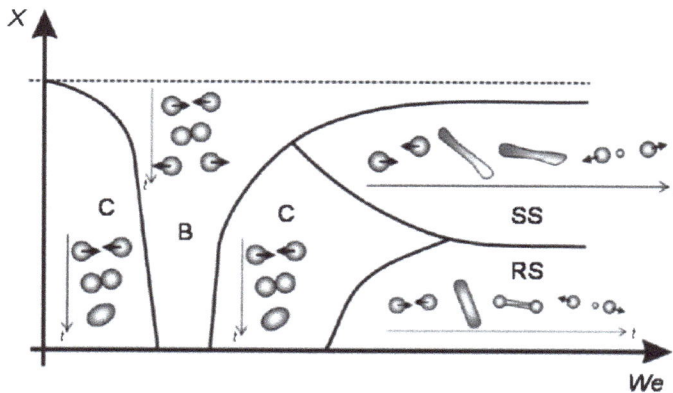

C: coalescence; B: bouncing; SS: stretching separation; RS: reflexive separation

For colliding drops of the same liquid, the collision outcome is determined by the sizes d_1 and d_2 and the relative velocity magnitude of the two colliding drops, the impact parameter $X(X = 2x/(d_1 + d_2))$, where x is the offset distance of the two colliding drops) at the instant of collision, and the physical properties of the drop liquid in contact with the ambient air. For colliding drops of equal size with a given Ohnesorge number, the outcome from a collision is presented in the form of nomograms (X, We), as in Figure 5.13. Four possible regimes (coalescence, bouncing, stretching separation, and reflexive separation) are identified as a function of the dimensionless impact parameter X and the drop Weber number We. As can be seen, coalescence occurs at low collision Weber number. Bouncing (separation without changes of drop sizes) occurs as the impact parameter X increases, but at a low Weber number. As the collision Weber number increases to a certain value, the collision results in small drops (collision followed by breakup).

Modeling of drop collision and coalescence in fuel sprays has been inspired by experimental observations of binary drop collisions. O'Rourke [13] developed a method for calculation of droplet collisions and coalescences, having made probably the first major extension of the stochastic particle method for engine spray simulations. Consistent with the stochastic particle method, collisions are calculated by a statistical, rather than a deterministic, approach. The probability distributions governing the number and nature of the collisions between two drops are sampled stochastically. The method was initially applied to diesel sprays, where it was found that coalescences caused a seven-fold increase in the mean drop size [13]. Many subsequent studies [33,41] have corroborated the importance of drop collisions in diesel-type sprays.

The drop collision and coalescence model of O'Rourke has been implemented in the KIVA family codes. In this model, a collision frequency is defined as the probability per unit time that the two drops will collide. That is, a collision frequency f_{12} between drops in parcels 1 and 2 (parcel 1 contains the smaller drops) is calculated for all parcels in each computational cell with a volume of V_{cell}. That is

$$f_{12} = \frac{\pi N_2 (r_1 + r_2)^2 E_{12} |v_1 - v_2|}{V_{cell}} \qquad (5.65)$$

where N_2 is the number of drops in parcel 2, v is the drop velocity vector, and r is the radius of the drops. The number of collisions n is assumed to follow the Poisson distribution as:

$$P(n) = \frac{(f_{12}\Delta t)^n}{n!} \exp(-f_{12}\Delta t) \qquad (5.66)$$

where Δt is the computational time step and $P(n)$ is chosen stochastically from the uniform distribution in the interval [0,1]. The collision efficiency factor E_{12} accounts for alternations of the collision frequency due to interactions of the flow fields near the drops with the similar reasoning to that used in the kinetic theory of gases. Most computer modeling studies have assumed that the value of E_{12} is approximately 1.0.

Two outcome regimes from drop collision are considered in the O'Rourke model [13]. They are either coalescence or grazing collision depending on whether the collision impact parameter b is greater or less than a critical value b_{cr}. In the coalescence regime, surface tension forces dominate over inertia forces of the drops at low relative velocities, and the two drops will combine to form a new bigger drop whose size, temperature, and velocity satisfy the conservations of mass, momentum, and energy before and after the collision. While in the grazing collision regime, the collision Weber number increases, the drop inertia forces dominate, and the drops separate. The colliding drops maintain their sizes and temperatures but undergo velocity changes.

The collision impact parameter b is defined as:

$$b^2 = (r_1 + r_2)^2 Y \qquad (5.67)$$

where Y is a random number in the interval [0,1]. If $b < b_{cr}$, coalescence occurs, and if $b \geq b_{cr}$, each collision is a grazing collision. The critical value b_{cr} of the collision impact parameter is given as:

$$b_{cr}^2 = (r_1 + r_2)^2 \, min\left(1.0, \frac{2.4(\gamma^3 - 2.4\gamma^2 + 2.7\gamma)}{We_l}\right) \qquad (5.68)$$

Where $\gamma = \frac{r_2}{r_1} \geq 1.0$ and We_l is the liquid Weber number defined by the wrelative velocity if the colliding drops.

In a grazing collision, the velocities of the drops after the collision are given as:

$$v_1' = \frac{r_1^3 v_1 + r_2^3 v_2 + r_2^3 (v_1 - v_2)\dfrac{b - b_{cr}}{r_1 + r_2 - b_{cr}}}{r_1^3 + r_2^3} \qquad (5.69)$$

$$v_2' = \frac{r_1^3 v_1 + r_2^3 v_2 + r_1^3 (v_2 - v_1)\dfrac{b - b_{cr}}{r_1 + r_2 - b_{cr}}}{r_1^3 + r_2^3} \qquad (5.70)$$

where v_1 and v_2 are the drop velocities before collision, and v_1' and v_2' are the velocities after collision.

It is noted that the possible formation of satellite droplets in the collision is not included in the O'Rourke model. In addition, shattering collisions that will result in a breakup of the colliding drops are not considered in the model. Another shortcoming of the O'Rourke model is that it considers collision events in the same computational cell and the outcome of the model, therefore, depends on mesh size. To overcome this problem, Schmidt and Rutland [62] used a secondary collision mesh that attempts to maintain a predetermined average number of parcels per collision cell, which improved the mesh-dependency issue. However, this collision mesh approach has the disadvantage that two parcels that are spatially very close to each other will be prevented from colliding if they belong to different collision cells.

Munnannur and Reitz [63] proposed a Radius-of-Influence (ROI) collision model. The ROI model considers potential collision between every pair of drop parcels whose distance $D_{1,2}$ is smaller than the maximum of their influence radii R_1 and R_2 that is $D_{1,2} \le max(R_1, R_2)$. The collision frequency f_{12} is then computed as:

$$f_{12} = \pi N_2 (r_1 + r_2)^2 \left| \mathbf{v}_1 - \mathbf{v}_2 \right| / V_{col} \tag{5.71}$$

The collision volume V_{col} is based on the radii of influence:

$$V_{col} = \frac{4}{3} \pi (R_1 + R_2)^3 \tag{5.72}$$

Thus, the influence of mesh topology is removed and the collision events only depend on the droplet parcel distribution in space. This method makes physical senesce and it is simple to implement. However, the choice of value of R seems arbitrary. A value of 2 mm was used in the simulation of non-evaporating diesel sprays [63].

Models to account for reflexive separation and stretching separation have been proposed [64,65]. In the model for binary droplet collisions applicable for moderate-to-high Weber numbers (typically 40 or above), bounce, coalescence, reflexive separation, and stretching separation were considered as possible collision outcomes [65]. Fragmentations in stretching and reflexive separations were modeled by assuming that the interacting droplets form an elongating ligament that either breaks up by capillary wave instability or retracts to form a single droplet. Predictions of collision outcomes, number of satellites from separation processes, and post-collision characteristics (namely, drop size and velocity) were compared with available experimental data for deterministic collisions of mono- and poly-disperse streams of fuel droplets, and the agreement was found to be reasonable

5.3.3 Drag, Deformation, and Turbulent Dispersion

Spray drops bear dynamic drag forces when they are traveling in gas. To account for the drag forces, we now define the droplet acceleration force $\mathbf{F}$ that determines the trajectories of individual droplets. A droplet moves according to its velocity, which is:

$$\mathbf{v} = \frac{d\mathbf{x}_d}{dt}$$

If only drag force and gravity force are considered, the acceleration of a droplet is written as:

$$\frac{d\boldsymbol{v}}{dt} = \boldsymbol{F}_D + \boldsymbol{g}$$

Hence, $\boldsymbol{F}$ has contributions due to aerodynamic drag $\boldsymbol{F}_D$ and gravity force $\boldsymbol{g}$:

$$\boldsymbol{F} = \boldsymbol{F}_D + \boldsymbol{g} = \frac{3}{8}\frac{\rho}{\rho_d}\frac{|\boldsymbol{u}+\boldsymbol{u}'-\boldsymbol{v}|}{r}(\boldsymbol{u}+\boldsymbol{u}'-\boldsymbol{v})C_{D,s} + \boldsymbol{g} \qquad (5.73)$$

The drag coefficient $C_{D,s}$ is given by

$$C_{D,s} = \begin{cases} \dfrac{24}{Re_d}\left(1+\dfrac{1}{6Re_d^{\frac{2}{3}}}\right) & Re_d < 1000 \\[3mm] 0.424 & Re_d \geq 1000 \end{cases} \qquad (5.74)$$

where

$$Re_d = \frac{2\rho|\boldsymbol{u}+\boldsymbol{u}'-\boldsymbol{v}|r}{\mu} \qquad (5.75)$$

Here, μ is the gas viscosity. $\boldsymbol{u}$ and $\boldsymbol{v}$ are the gas and drop velocity, respectively.

The gas turbulence velocity $\boldsymbol{u}'$ is added to the local mean gas velocity $\boldsymbol{u}$ when calculating a droplet's drag. It is assumed that each component of $\boldsymbol{u}'$ follows a Gaussian distribution with $\sqrt{2/3k}$ and k is the specific turbulent kinetic energy of the gas. Thus, it is assumed

$$G(\boldsymbol{u}') = (4/3\pi k)^{-3/2}exp(-3|\boldsymbol{u}'|^2/4k) \qquad (5.76)$$

The value of $\boldsymbol{u}'$ is chosen once every turbulence correlation time t_{turb} and is otherwise held constant. The correlation time is given by

$$t_{turb} = min\left(\frac{k}{\varepsilon}, c_{ps}\frac{k^{3/2}}{\varepsilon}\frac{1}{|\boldsymbol{u}+\boldsymbol{u}'-\boldsymbol{v}|}\right) \qquad (5.77)$$

This means that t_{turb} is the minimum of an eddy breakup time and a time for the droplet to traverse an eddy. In this equation, c_{ps} is an empirical constant with value 0.16432 and ε is the turbulence energy dissipation rate.

The droplets are assumed to be spherical in Equation (5.74), which is only true for very small droplets. When the relative velocity between the droplet and the surrounding gas is high, the drop deforms, which will change its drag coefficient. Its deformation depends on the Reynolds number and oscillation amplitude of the drop. The drop drag coefficient has been empirically related to the magnitude of the drop deformation by Liu et al. [66]. The relation is given as:

$$C_D = C_{D,s}(1 + 2.632\,y) \tag{5.78}$$

where C_D is the drag coefficient accounted for drop deformation, $C_{D,s}$ is the drag coefficient of the spherical droplets given by Equation (5.74), y is a dimensionless parameter describing drop distortion and is proportional to the displacement of the drop's surface from its equilibrium position divided by the drop radius (see Equation (5.62))

It has been found that vapors surrounding the evaporating liquid drops impact the drag forces of the drops [67–69]. To consider the effects of drop vaporization on drop drag forces, the following equation is suggested following Eisenklam et al. [67]:

$$C_D{}^* = \frac{C_D}{1 + B} \tag{5.79}$$

where B is the Spalding mass transfer number given by Equation (5.83) and the properties needed given in Equation (5.74) are evaluated at the one-third reference condition using the mixture rules as will be discussed in Sec. 5.4.

The dispersion (or diffusion) of particles by turbulence is important in many industrial and energy-related processes [70]. The gas phase turbulence is modulated (reduced) since a portion of the turbulence energy is used to do work in dispersing the spray droplets, as reviewed by Faeth [71].

Turbulence effects on the spray droplets are currently modeled with a Monte Carlo method [14]. A fluctuating velocity $\boldsymbol{u}'$ is added to the gas velocity u, where each component of $\boldsymbol{u}'$ is randomly chosen from a Gaussian distribution with standard deviation $\sqrt{2/3k}$ in the computational cell in which droplet is located. The fluctuating velocity $\boldsymbol{u}'$ is treated as a piecewise constant function of time, changing discontinuously after passage of turbulence correlation time t_{turb}, determined by Equation (5.77). The work done by the fluctuating velocity components in dispersing the spray droplets is then subtracted from the turbulent kinetic energy. The sum $\boldsymbol{u} + \boldsymbol{u}'$ becomes the gas velocity that the droplet "sees". It is not only used in calculating the drag as discussed but also used in computing the drop's heat and mass exchange with the gas and its breakup to account for the effects of the gas turbulence.

5.4 Evaporation

Fuel spray evaporation is important in an IC engine. It becomes even more crucial to DI gasoline engines since the time available for spray evaporation and air–fuel mixing at high speeds is limited relative to engine crank angles. Spray evaporation also plays a dominant role at cold starts of diesel and gasoline engines when temperatures of the gas and cylinder walls are low, which is not favorable for fuel evaporation. In engine modeling, the results of evaporation directly influence air–fuel mixing temporally and, therefore, the subsequent chemical reactions. Hence, accurate models are needed to calculate spray evaporation and related mass and heat transfer.

Despite the fact that fuel comprises many components with different volatilities, only a single component is usually used to represent a particular fuel in evaporation modeling. For example, iso-octane is often selected to represent gasoline and tetradecane for diesel. While single-component evaporation models have been satisfactory under engine warm-up conditions, they, however, cause large discrepancies under engine cold-start conditions [72]. Models to account for the effects of multiple components in fuels have been developed. In this section, evaporation models for both single-component fuel and multi-component fuel are discussed.

5.4.1 Single-Component Evaporation

A variation of the low-pressure droplet vaporization model by Spalding [73] has been adopted in the KIVA codes. A modified version of the KIVA model was suggested and used in simulation of DI gasoline engines by Han et al. [74]. The rate of fuel drop mass change due to evaporation is given as:

$$\frac{dm_d}{dt} = -2\pi r \rho_g DBSh \tag{5.80}$$

where m_d is the drop mass, ρ_g is the environment gas density, D is the molecular mass diffusivity of the fuel vapor to the gas, B is the Spalding mass transfer number, and Sh is the Sherwood number for mass transfer. The subscripts d and g refer to the fuel drop and the immediate environment gas surrounding the drop, respectively. With the assumption of a Lewis number of unity (i.e., thermal diffusivity equals mass diffusivity), Equation (5.80) can be rewritten as:

$$\frac{dm_d}{dt} = -2\pi r \left[\frac{k}{c_p}\right]_g BSh \tag{5.81}$$

where k is the thermal conductivity and c_p is the specific heat. In order to solve Equations (5.80) or (5.81), the drop temperature must be described and it can be determined from an energy balance involving the heat conduction from the gas, that is

$$m_d c_{pF} \frac{dT_d}{dt} - L\frac{dm_d}{dt} = 4\pi r Q_d \tag{5.82}$$

where c_{pF} is the fuel specific heat, L is the latent heat, and Q_d is the heat conduction rate to the fuel drop.

Equations (5.81) and (5.82) constitute the two basic governing equations for droplet evaporation. The Spalding mass transfer number is defined as:

$$B = \frac{Y_{F,s} - Y_{F,\infty}}{1 - Y_{F,s}} \tag{5.83}$$

where Y_F is the mass fraction of fuel vapor. The subscripts s and ∞ refer to the drop surface and free stream conditions, respectively. The heat conduction rate to the droplets is given as:

$$Q_d = \frac{k_g (T_\infty - T_d)}{2r} Nu \tag{5.84}$$

Extensive research has been devoted to the determination of Nu, the Nusselt number, for an evaporating drop under forced convection. The following correlation is suggested [71,75]:

$$Nu = (2.0 + 0.6 Re_d^{1/2} Pr_g^{1/3}) \frac{\ln(1+B)}{B} \tag{5.85}$$

Similarly, with the assumption of a Lewis number of unity, the Sherwood number is given as:

$$Sh = (2.0 + 0.6 Re_d^{1/2} Sc_g^{1/3}) \frac{\ln(1+B)}{B} \tag{5.86}$$

where Re_d is the Reynolds number evaluated at the relative velocity between the drop and the gas, and Pr_g and Sc_g are the molecular Prandtl and Schmidt numbers, respectively.

However, the accuracy of Equations (5.80)–(5.86) is very much dependent on how the thermophysical properties are evaluated. It has been recommended by Lefebvre [76] to use the one-third rule where the properties are evaluated at a reference temperature and composition given by:

$$T_r = T_s + \frac{1}{3}(T_\infty - T_s) \tag{5.87}$$

$$Y_{F,r} = Y_{F,s} + \frac{1}{3}(Y_{F,\infty} - Y_{F,s}) \tag{5.88}$$

$$Y_{A,r} = 1 - Y_{F,S} \tag{5.89}$$

Here the subscript r refers to the reference condition and A is the free stream gas (air). Equations (5.87)–(5.89) are used to calculate the relevant thermophysical properties of the vapor–air mixture immediately surrounding the evaporating fuel drop. The mixture rule is given as:

$$\Psi_g = Y_{A,r} \cdot \Psi_A(T_r) + Y_{F,r} \cdot \Psi_v(T_r) \tag{5.90}$$

where the subscript v refers to the fuel vapor and Ψ is a symbol that stands for c_p, k, or μ (molecular viscosity) of the environmental gas. The reference gas density is evaluated as:

$$\rho_{g,s} = \left(\frac{Y_{A,r}}{\rho_A} + \frac{Y_{F,r}}{\rho_v} \right)^{-1} \tag{5.91}$$

The physical properties embodied in Re_d, Pr_g, and Sc_g in Equations (5.85) and (5.86) are evaluated using the mixture rules given by Equations (5.90) and (5.91), and at the reference temperature and composition.

Formulations of Equations (5.87), (5.88), and (5.91) consider the effects of both temperature and composition. This treatment seems more physically meaningful. Physically, the mass flux from evaporation dynamically affects the flow field surrounding the droplet, and evaporation causes large temperature gradients and changes the chemical composition. Both effects directly affect the fluid properties adjacent to the droplet. By comparing various treatments of the reference conditions, Hubbard et al. [77] found that the one-third rules of Equations (5.87) and (5.88) gave the best results.

To determine the mass transfer number given by Equation (5.83), the surface mass fraction $Y_{F,s}$ is obtained from

$$Y_{F,s} = \left[1 + \frac{M_A}{M_F} \left(\frac{p}{p_v} - 1 \right) \right]^{-1} \tag{5.92}$$

where p_v is the fuel vapor pressure at the drop surface, p is the ambient pressure, and M_A and M_F are the molecular weights of air and fuel, respectively. The vapor pressure can be estimated from the Clausius–Clapeyron equation as:

$$p_v = \exp\left(a - \frac{b}{T_s - c} \right) \tag{5.93}$$

where a, b, and c are fuel-dependent constants. The drop interior temperature is assumed to be uniform and equals the surface temperature T_s. Lefebvre [76] and Vargraftik [78] gives the constants in Equation (5.93) and other properties of liquid fuel, fuel vapor, and air.

The detailed treatments of the physical properties can make significant differences in calculation of the rate of fuel drop mass change [74]. For example, the mass diffusivity is calculated by $(k/c_p)_g$ in Equation (5.81), which has been shown to be valid under low pressure conditions [79]. An empirical correlation for the mass diffusivity $(\rho D)_{air}$ of the free stream gas (air) is used in KIVA. The exclusion of the effect of the fuel vapor on the surrounding gas leads to a much lower mass diffusivity. Furthermore, the effects of fuel evaporation on the physical properties of the environment gas are assumed to be minimal. The environmental gas properties (c_p, k, μ, and ρD) are evaluated using the free stream gas (air) at the one-third reference temperature, but the environmental gas density is calculated using the air at the free stream temperature.

A computational comparison of a single drop evaporation shows that the lower mass diffusivity results in a higher equilibrium drop temperature which gives a greater mass transfer number B. A net result of the higher drop temperature and the lower diffusivity leads to a slower vaporization rate and, hence, a longer drop lifetime with the use of the KIVA evaporation model. Figure 5.14 shows the computed evaporation of an iso-octane droplet in

FIGURE 5.14 Evaporation of a 25°C iso-octane droplet in quiescent air at 0.3 MPa and 300°C.

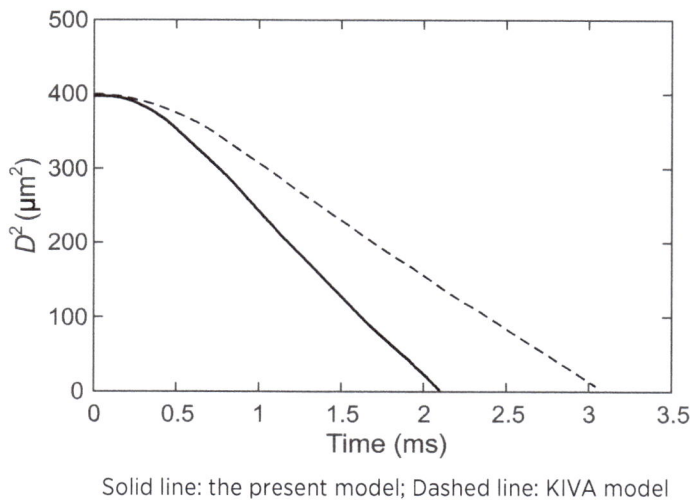

Solid line: the present model; Dashed line: KIVA model

quiescent air, in which the "present model" is the one being discussed in this section. Very different drop lifetimes are seen with different treatments of the physical properties by the present model and KIVA model [24]. The present model predicts more accurate droplet lifetimes. It is noted that the drop liquid density changes with temperature so that the drop size increases initially due to expansion. This expansion phenomenon cannot be captured if a constant fuel density is assumed.

The effect of drop evaporation on drop drag coefficient has been discussed in Sec. 5.3.3. Figure 5.15 demonstrates this effect by comparing the penetrations of a single vaporizing iso-octane droplet. Three different treatments of the drop drag coefficient are compared. The present model [74] considers the effect of drop evaporation as described in Equation (5.79), the KIVA model [24] uses Equation (5.74) and results in a somewhat shorter penetration and a much longer droplet lifetime, and the Yuen's model [68] also uses Equation (5.74), and the free stream density and viscosity to calculate the Reynold number. Yuen's Model leads to a somewhat longer penetration and about the same droplet lifetime of the present model.

In Equation (5.82), the drop temperature is assumed to be uniform and the transient heat transfer inside the droplet is neglected. This can lead to over-prediction or under-prediction of the evaporation mass flux, depending on the ambient temperature conditions. To account for heat transfer inside of a droplet [80], the energy balance at the interface can be given as:

$$4\rho_d \pi r_d^2 \dot{r}_d L(T_{d,s}) = 4\pi r_d^2 (Q_i + Q_d) \tag{5.94}$$

where Q_i is the heat flux from inside the droplet to the surface. It is modeled as a convective heat transfer process and is given as:

FIGURE 5.15 Drop penetration and lifetime (length of lines) computed with different models.

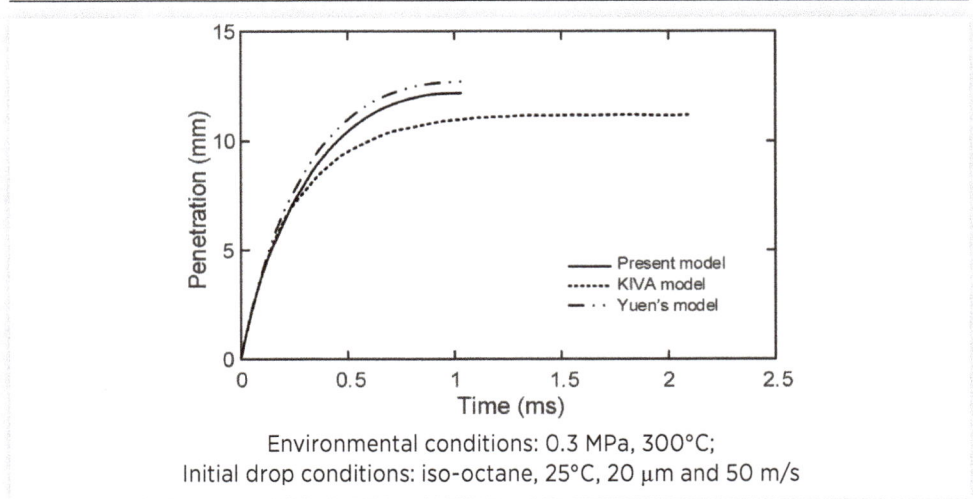

Environmental conditions: 0.3 MPa, 300°C;
Initial drop conditions: iso-octane, 25°C, 20 µm and 50 m/s

$$Q_i = \frac{k_l}{\delta_e}(T_d - T_{d,s}) \tag{5.95}$$

where k_l is the liquid thermal conductivity and δ_e is the unsteady equivalent thickness of the thermal boundary layer and calculated from the thermal diffusivity α_l as:

$$\delta_e = \sqrt{\pi \chi \alpha_l t} \tag{5.96}$$

where $\chi = 1.86 + 0.86 tanh[2.225 \log(Pe_l/30)]$ and Pe_l is the Peclet number of the droplet. The heat conduction rate to the drop, Q_d, is also computed at the drop surface temperature. The average temperature of the droplets T_d is replaced by the surface temperature of the droplets $T_{d,s}$.

Note that since the effective heat transfer coefficient in Equation (5.95) is coupled with the vaporization rate, the surface temperature of the droplet is, therefore, determined by solving two balance equations iteratively, assuming a quasi-steady heat transfer process [80].

5.4.2 Multi-Component Evaporation

Most practical fuels have multiple components and the boiling points of the components range from 300 to 500 K. This makes practical fuels significantly different from pure substance, which has only one boiling point, for vaporization process. Therefore, multicomponent representations for fuels are needed to simulate the air–fuel preparation process in engines more accurately, particularly under cold engine conditions.

Two types of multicomponent evaporation models have been developed, namely, the continuous multicomponent (CMC) models [18,80,81] and discrete multicomponent (DMC) models [19,82]. In a CMC model, the fuel composition is represented as a continuous distribution function with respect to an appropriate parameter such as molecular weight. This treatment enables a reduction of computational load while maintaining the predictability of the complex behavior of the vaporization of multicomponent fuels. However, when this model is applied to combustion simulations, especially with detailed reaction chemistry, describing the multicomponent features of the fuel is inevitably limited, making it difficult to model the consumption of individual components appropriately. In addition, tracking the behaviors of an individual component in the evaporation process, which is of interest in some cases, is difficult in a CMC model.

On the contrary, a DMC model tracks the individual components of the fuel during the evaporation process and allows coupling with the reaction kinetics of the individual fuel components. Although the computational cost of a DMC is high due to the increased additional transport equations for fuels with many components, it is becoming more affordable as computational capacity has improved substantially. This issue can be mitigated by using only few components that represent the low volatile and high volatile components of the fuel in engine simulations.

We will first briefly introduce the basic idea of the CMC model [18,80]. The DMC model of Zeng and Lee [84], which has been implemented in the Ford in-house CFD code MESIM [83], will be then discussed.

In the CMC model, it is assumed that each species in the mixture can be characterized by the value of one variable (e.g., molecular weight) I and the amount of substance can be expressed via the distribution functions $f_p(I)$ and $f_v(I)$ for the liquid and vapor phases, respectively. The general molar distribution function for the composition of the systems is then described as:

$$G_p(I) = x_F^p f_p(I) + \sum_{s=1}^{N} x_s^p \delta(I - I_s) \tag{5.97}$$

where the p represents v or l, denoting the properties of the vapor or liquid, respectively. x is the mole fraction, N is the total number of discrete species, and δ the Dirac delta function. Subscripts s and F denote the properties of the discrete species and fuel, respectively. The distribution has the property

$$\int_0^\infty G_p(I)dI = 1, \int_0^\infty f_p(I)dI = 1, \sum_{s=1}^{N} x_s^p = 1 - x_F^p \tag{5.98}$$

For the continuous system of the liquid fuel, x_s^p is zero and x_F^p is unity so that $G_p(I)$ becomes equal to $f_p(I)$. The same distribution function for liquid phase and vapor phase is chosen as:

$$f(I) = \frac{(1-\gamma)^{\alpha-1}}{\beta^\alpha \Gamma(\alpha)} exp\left(-\frac{I-\gamma}{\beta}\right) \tag{5.99}$$

where a and b are parameters that determine the shape of the distribution, Γ is the Gamma function, and γ is the origin of the distribution function. The mole fraction of a species I in the fuel is given as $x_i = G_p(I)_i \Delta I_i$ with ΔI_i being the interval in I.

The governing equations for the liquid phase, fuel vapor phase, and the surrounding gas, as well as vapor–liquid equilibrium and properties of the fuel species are developed on the base of the distribution functions, which will not be discussed here.

In the DMC model [82], the drop vaporization consists of a gas phase part determining the drop vaporization rate and the heat flux to the droplet, and a liquid phase part determining the surface parameters of liquid phase. For the gas phase, the quasi-steady-state model is applied. From film theory, the evaporation rate for component *i* is given by

$$\omega_i = \frac{D_{im} Sh_i \xi_i B_i}{d} \qquad (5.100)$$

where D_{im} is the mass diffusivity coefficient of component i, d is the drop diameter, and Sh_i is the Sherwood number of component *i* and given by

$$Sh_i = 2 + 0.6 Re^{1/2} Sc_{gi}^{1/3} \qquad (5.101)$$

ξ_i accounts for the effects of vaporization on mass transfer and is defined as:

$$\xi_i = \frac{z_i}{e^{z_i} - 1} \qquad (5.102)$$

where

$$z_i = \sum \omega_i d / D_{im} Sh_i \qquad (5.103)$$

and B_i is the mass transfer number of component *i*. It is given by

$$B_i = \frac{Y_{gi,\,s} - Y_{gi,\infty}}{1 - Y_{gi,s} \sum \omega_i / \omega_i} \qquad (5.104)$$

In Equation (5.104), $Y_{gi,\infty}$ is the gas phase mass fraction of component *i* in the surroundings, which is evaluated as the value of the cell where the droplet is located. $Y_{gi,s}$ is the gas phase mass fraction of component *i* at the drop surface and it is determined by the thermodynamic phase equilibrium at the drop surface. The ratio of mole fractions between the gas phase and liquid phase is given by

$$x_{gi,s} = \frac{x_{li,s} P_{sat}(T_s)}{P} \qquad (5.105)$$

where $x_{gi,s}$ and $x_{li,s}$ are the gas phase and liquid phase mole fractions of component *i*, respectively. P and P_{sat} are the ambient pressure and the saturated pressure at the surface temperature, respectively.

The heat flux to the droplet is expressed as:

$$q = \frac{k_g Nu \zeta_T (T_\infty - T_s)}{d} - \sum \omega_i h_{vi}$$ (5.106)

where Nu is given by

$$Nu = 2 + 0.6 Re^{1/2} Pr_g^{1/3}$$ (5.107)

h_{vi} is the vaporization enthalpy of component i. ζ_T is defined as:

$$\zeta_T = \frac{z_T}{e^{z_T} - 1}$$ (5.108)

where

$$z_T = \sum \omega_i Cp_{gi} / (k_i Nu_i / d)$$ (5.109)

and Cp_{gi} is the heat capability of component i.

Due to the diffusion resistance inside the droplet and preferential vaporization, the property distributions are non-uniform in general. Thus, an ordinary differential equation with respect to time t is used to describe the evolution of this non-uniformity [82]:

$$\frac{d\Phi_d}{dt} = \frac{0.2\phi R / D - \Phi_d}{R^2 / \varsigma_1^2 D}$$ (5.110)

where Φ can be the temperature T or mass fraction Y_i, $\varsigma_1 = 4.4934$. Subscript d represents the difference between the surface and average values. For temperature, D is the effective thermal diffusivity coefficient and

$$\phi = qD / k_l^e$$ (5.111)

where q is the heat flux to a droplet and k_l^e is the effective thermal conductivity. For mass fraction, D is the effective mass diffusivity coefficient and

$$\phi = [\omega_i - (\sum \omega_i) Y_{lim}] / \rho_l$$ (5.112)

where ω_i) is the vaporization rate of component i, Y_{lim} is the mass fraction of component i, and ρ_l is the liquid density.

In Equation (5.110), the effective diffusivities rather than the physical diffusivities are used to account for the effect of internal circulation. Equation 5.113 gives the ratio of effective diffusivity to physical diffusivity as:

$$\begin{cases} \dfrac{k_l^e}{k_l} = 1.86 + 0.86 \tanh\left[2.245 \log\left(\dfrac{\mathrm{Re}\,Pr_{li}}{30}\right)\right] \\[4mm] \dfrac{\Gamma_{li}^e}{\Gamma_{li}} = 1.86 + 0.86 \tanh\left[2.245 \log\left(\dfrac{\mathrm{Re}\,Sc_{li}}{30}\right)\right] \end{cases}$$

(5.113)

where Γ is mass diffusion coefficient, Re the drop Reynolds number, Pr_{li} and Sc_{li} liquid phase Prantdl and Schmidt numbers, respectively.

The integration of Equation (5.110) with respect to time gives the difference between surface and average values of temperature and mass fractions. By using the overall species mass conservation and energy conservation equations,

$$\frac{dT_m}{dt} = \frac{6q}{\rho_l Cp_l d}$$

(5.114)

$$\frac{d(\rho_l \pi d^3 Y_{lim}/6)}{dt} = \pi d^2 w_i$$

(5.115)

the average values of temperature and mass fractions can be obtained. Thus, the surface values can be found by simple algebraic operations.

The CMC model discussed accounted for the effects of finite diffusion within the droplets by a set of ordinary differential equations, without resolving spatial governing equations of the mass and energy transport inside the droplets. Validation computations of vaporization of bi-component single droplets show that the mass fraction distributions depend greatly on the modeling details for the liquid phase [82]. The model could predict the mass fraction evolution accurately, while the infinite diffusion model gives poor predictions compared with experimental data.

5.5 Spray Wall Impingement

Fuel spray wall impingement phenomena occur in IC engines including PFI engines, DI gasoline engines, and automotive diesel engines. The spray wall impingement is deliberately utilized to aid optimal stratified-charge formation in a specially designed bowl in some DI gasoline engines. In a PFI engine, sprays are injected in the intake ports with the targets of the back surfaces of the intake valves, resulting in clouds of droplets and air–vapor mixture, and films near the valve-seat region. However, spray wall impingement also causes problems. For example, it causes decreased engine response of transient operations and hydrocarbon emissions at cold starts in PFI engines [84,85]. In DI gasoline engines, the spray wall impingement results in wall-wetting, which is evidently believed to be the source of soot and HC emissions [86,87]. While in DI diesel engines, the spray wall impingement may lead to unacceptable hydrocarbon, CO, and soot emissions as well as deteriorated fuel consumption [85]. Hence, a better understanding of fuel spray wall impingement will be helpful in engine combustion development. Furthermore, as multidimensional modeling is becoming a more indispensable tool for engine design, there is a crucial need for accurate impingement models to improve engine simulation predictability. Recent effects to further reduce soot emissions and to control knock combustion in turbocharged DI gasoline engines increase the demand to understand the details of spray-wall interactions both experimentally and computationally [88].

The task of spray/wall impingement modeling is to predict the outcomes after the impingement including the changes of the size, velocity, and temperature of the droplets through the mass, heat, and momentum transport processes occurring near the wall. The hydrodynamical and thermal behaviors of the liquid film on the surface also needed to be resolved. Several models have been proposed and applied to study engine spray wall impingement processes. Naber and Reitz [15] have made perhaps the first attempt to develop a model for multidimensional engine simulation. In their model, the outcome of a drop impingement was categorized in three regimes, i.e., sticking, reflecting, and sliding, depending upon the incident drop Weber number. Although this model has been used in the simulation of diesel engines, there are issues associated with it. Their model neglects the splash phenomenon whereby impinged drops produce many smaller droplets (called secondary droplets here-after) flying outward away from the impingement location. Splash can be important since it affects near wall spray dispersion and vaporization, and in turn influences fuel-air mixing. Fuel films on the piston surface were not considered in their model either.

Recognizing film formation in an engine, particularly in a diesel and a PFI gasoline engine, some researchers included film hydrodynamics in their impingement models [29,89–94]. These models solve similar mass, momentum, and energy equations using a thin-film assumption. However, two different numerical methods are adopted in tracking the film. One method treats the film as continuous fluid [91–92] and another tracks the film using Lagrangian numerical particles [29,89,94]. The particle tracking method in principle can be applied to a relatively coarse surface grid (required to limit computer resources in engineering applications) without serious deterioration in simulation accuracy.

In modeling of spray/wall (or film) interactions, different regimes and corresponding transition criteria have been suggested [15,28,29,89,90,92]. In most cases, splash models for the size and velocity of the secondary droplets were formulated by either directly curve-fitting impingement experimental data or using some information of existing single-drop impingement experiments and various arguments of mass, momentum, and energy redistribution among incoming drops, outgoing droplets, and liquid film. Due to the complexity of the problem and the lack of detailed experimental data, many empirical correlations and very broad assumptions had to be introduced in these models.

O'Rourke and Amsden [94] proposed the particle-tracking film method. They also developed wall-function models for the transport of vapor mass, momentum, and energy in the turbulent boundary layers above the film in KIVA3V [25,95]. Trujillo et al. [96] found that neglecting the inertial term of the film momentum equation in their work led to significant under spreading of the film away from the impingement site. This inertial term and a pressure term due to impingement have recently been included in an improved version of the models by O'Rourke and Amsden [28].

In the following sections, spray impingement regimes in SI engines are first discussed. A splash breakup model based on instability analysis and formulation of the velocity of the secondary droplets using jet impingement theory are presented. Models of wall film dynamics, heat transfer and evaporation are then described.

5.5.1 **Spray Impingement Regimes**

Spray impingement in an SI engine is complicated and, in general, the details associated with it are very difficult to measure experimentally under real engine conditions due to transient dense spray and high ambient temperature and pressure conditions. In a PFI engine, splashing and film formation phenomena can be observed in the intake-port/valve regions and on the surfaces of the head and piston [97,98]. For a DI gasoline engine, although there might be evidence indicating liquid film on a quartz piston surface that is used for optical access [99], the observation may not be conclusive for a real engine. It has been shown that under atmospheric conditions, the Leidenfrost temperature is about 80°C higher when an iso-octane drop impinges on a quartz plate than on an aluminum plate [100] due to their very different thermal conductivities.

For a drop levitated on a hot surface, researchers have concluded that the drop experiences different physical processes depending on the surface temperature T_s, resulting in wetting or non-wetting (no film) of the surface and varied drop lifetimes. As shown in Figure 5.16, when the surface temperature T_s, is below the boiling point of the liquid (point b), the drop rests on the surface in the shape of a lens and gradually evaporates. As T_s reaches point c (the so-called "Nukiyama point"), the lens-shaped drop flattens into a film, violent vaporization occurs and the drop disappears at the highest rate due to the maximum heat transfer. In the wetting region (a–c), liquid drops always adhere to the wall surface. Above the Nukiyama point, the flattened drop disintegrates and some satellite droplets form that bounce above the surface. The drop lifetime is increased until the so-called "Leidenfrost temperature" point d is reached. In this transition region (c–d), unstable drop adhesion occurs. At and above the Leidenfrost temperature, a thin layer of vapor forms between the surface and the drop; spheroidal vaporization then takes place. Drops do not directly adhere to the wall surface in this non-wetting region (>d).

FIGURE 5.16   Schematic diagram showing drop lifetime on a hot surface.

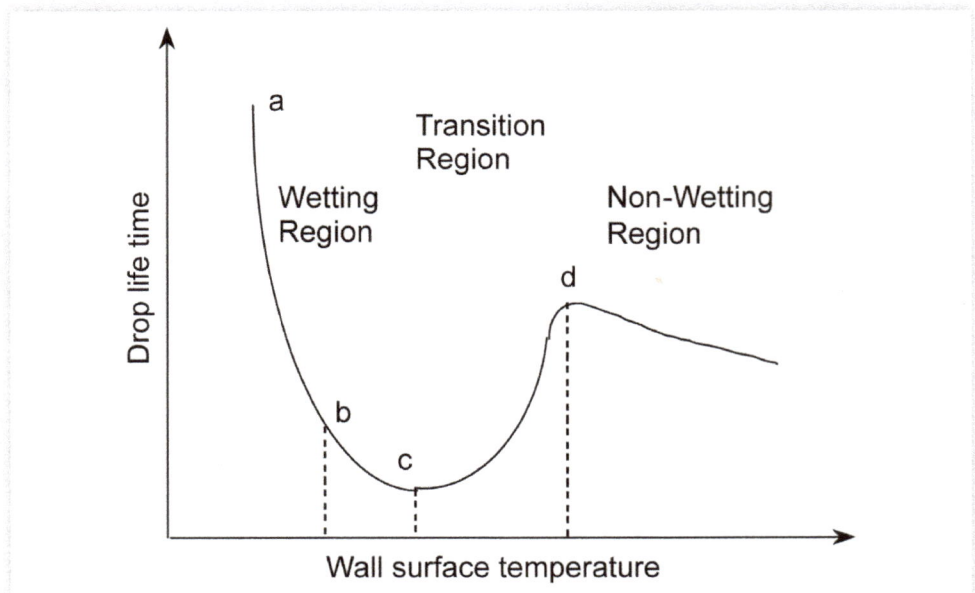

Reprinted with permission from Ref. [29]. © SAGE Publications

It has been found that the specific values of these characteristic temperature points change with liquid type. However, the Leidenfrost temperature is mostly independent of drop sizes for hydrocarbon fuels [101]. It has been also found experimentally that as the ambient pressure is increased, the lifetime curve in Figure 5.16 moves to the higher temperature range, and the Nukiyama and Leidenfrost temperatures become higher. Interestingly, when the pressure reaches the fuel's critical pressure value, the Leidenfrost phenomenon disappears [100]. The Nukiyama and Leidenfrost temperatures define the boundaries for the wetting and non-wetting regions. It is of particular interest to plot available data [100–103] for hydrocarbon drops on a metal plate together, as shown in Figure 5.17. Both Nukiyama and Leidenfrost temperatures become higher as the ambient pressure is increased. Also, gasoline exhibits higher values of these temperature points due to the heavy-end hydrocarbon components. Typical ranges of the wall temperature and ambient pressure, which fuel droplets experience in an SI engine, are also plotted in Figure 5.17.

As clearly indicated, films exist in a PFI engine as fuels are injected in the intake ports at or under atmospheric pressure conditions. A DI gasoline engine usually operates under two distinct modes. At low engine loads, fuels are injected into the cylinder during the late compression stroke to form a stratified-charge (SCDI); thus, the impinged drops experience a higher ambient pressure. It is seen that under the SCDI conditions the impinged sprays result in wall-wetting, even when heptane fuels are used. The higher ambient pressure in this case further ensures the existence of a wall film. On the other hand, at high loads, fuels are injected during the intake stroke to form a homogeneous-charge (HCDI). The relatively high wall temperature and low ambient pressure put the impinged drops into the transition region if heptane fuels are used, in which wetting may not occur. However, for gasoline fuels, wall-wetting is most likely to occur in this case, as shown.

FIGURE 5.17 Spray impingement regime in an SI engine.

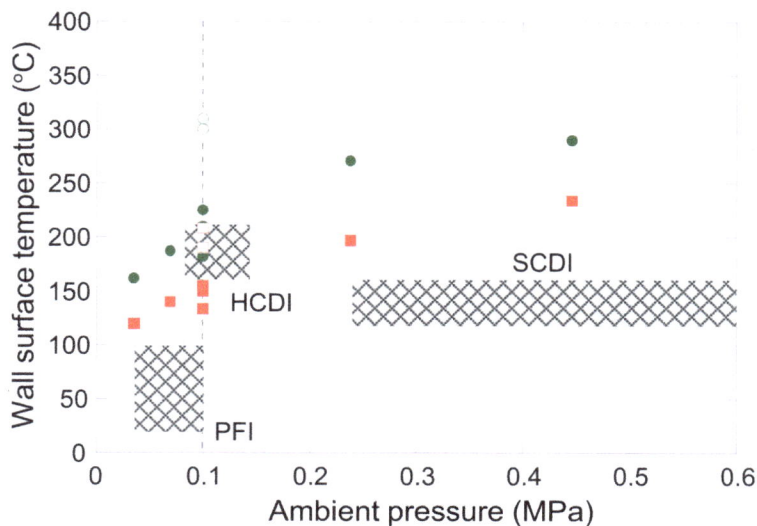

Experimental data: squares, Nukiyama point; circles, Leidenfrost point. Solid symbol, heptane; open symbol, gasoline
SI: spark ignition

In summary, the Leidenfrost phenomena most likely do not occur in a PFI and DI gasoline engine under normal operating conditions. The impinged sprays result in wall-wetting so that the liquid fuel directly contacts the wall. However, whether the liquid fuel forms a continuous film or spotted small wetting areas in an SCDI engine due to the short lifetime of the drops remains to be investigated. Since the wetting regime is present, impingement models for both PFI and DI engines should include the treatment of a liquid film.

Experimental works have been carried out to characterize the drop-wall collision phenomena. Various hydrodynamic impact modes on hot surfaces can be identified, as shown in Figure 5.18. At the deposition mode (A) there is no rebound observed, while at the partial rebound (B) only a part of the drop rebounds. The total rebound (C) can be subdivided into three different categories C1, C2, and C3. During the receding of the drop in

FIGURE 5.18 Drop impact onto a hot surface at different temperatures. Hydrodynamic impact modes: deposition (A), partial rebound (B), total rebound (C), and atomization (D).

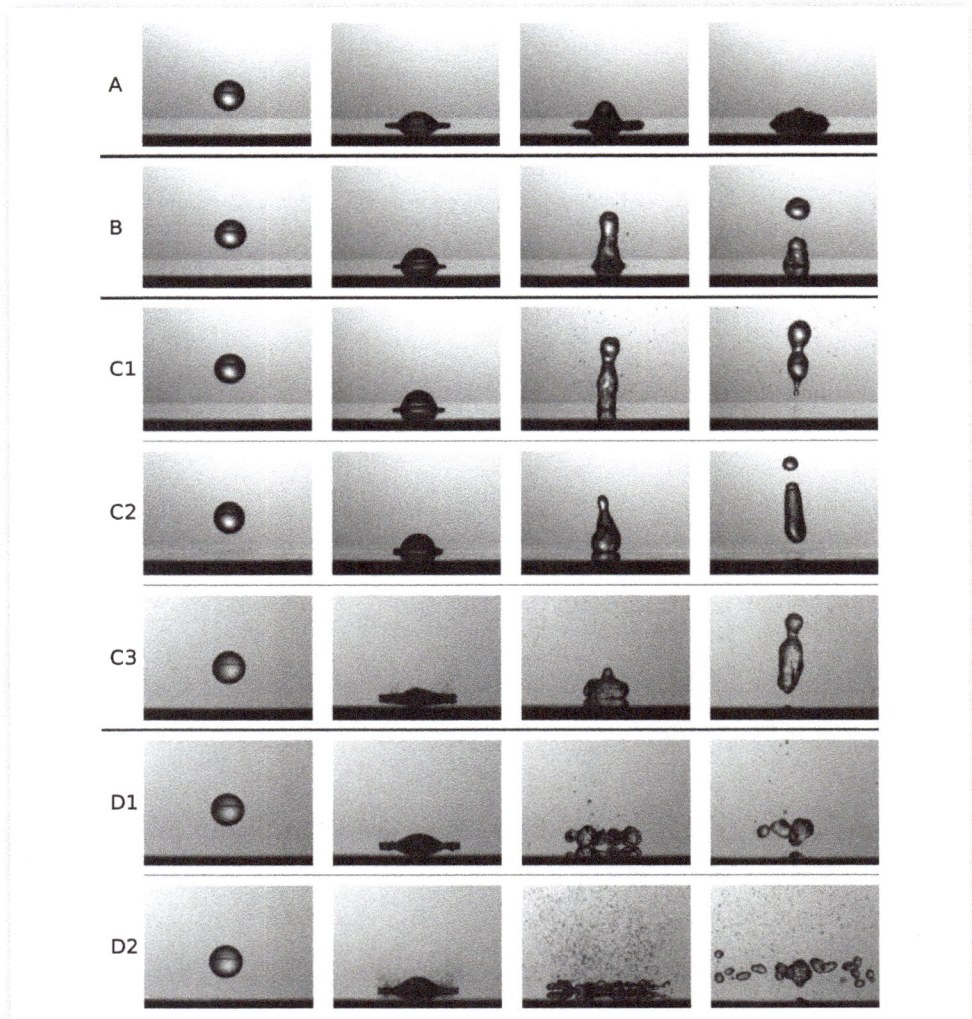

Mode C1, secondary droplets will be formed at the contact line, while in Mode C3 secondary droplets will appear during the drop advancing motion in the case of very fast rising bubbles through the thin lamella of the drop. This is very close to the atomization stage. Mode C2 does not result in any secondary droplets at the rebound event. At the atomization stage (D) the drop breaks up in the case of lamella instabilities and multiple secondary droplets are formed. Fast rising and growing bubbles in the lamella produce holes and, hence, instabilities. The lamella breaks up and the drop disintegrates. This phenomenon can be very weak as the primary drop breaks up into a few parts (D1), or very strong (D2) as the drop disintegrates into multiple secondary droplets.

In modeling drop-wall interactions, typical post-impingement scenarios as observed in experiments are usually identified. For spray impingement in an engine, Bai and Gosman [90] assumed four post-impingement regimes, namely stick, rebound, spread, and splash. Stanton and Rutland [92] also adopted this assumption. For a wetted surface, the stick regime occurs as the impact energy is very low and the drop adheres to the film in nearly spherical form. As the impact energy increases, the air layer trapped between the drop and the surface causes low energy loss and the drop rebounds. When the impact energy is further increased to a certain level, the impinging drop spreads and merges with the film. Finally, splash occurs at high impact energy.

Among these impingement regimes, splash is the most important and difficult one to treat. One of the limitations in most of the previous studies was that the effects of surface roughness and the film on the splash threshold were neglected.

Mundo et al. [104] measured drop impingement on a dry surface and found that for splash to occur, the parameters of the impinging drop collapsed to a dimensionless parameter K, which must be greater than 57.7 as:

$$K = We_n^{0.5} Re_n^{0.25} > 57.7 \qquad (5.116)$$

Here, We_n and Re_n are the Weber number and Reynolds number, respectively, of the impinging drops based on the normal velocity components. They also found this threshold independent of drop incident angle with the normal velocity being used in K for non-dimensional surface roughness β in the range of 0.03–0.86. This threshold parameter is consistent with an earlier study of Stow and Hadfield [105].

Han et al. [29] proposed a formation considering the effects of surface roughness on splash threshold for a dry surface by curve-fitting experimental data. The correlation is given as:

$$H_{cr,dry} = 1500 + \frac{650}{\beta^{0.42}} \qquad (5.117)$$

Splash occurs if and only if

$$H = We_n Re_n^{0.5} \geq H_{cr,dry} \qquad (5.118)$$

The equation indicates that for the surface roughness parameter $\beta \ll 10^{-3}$, the parameter $H_{cr,dry}$ increases sharply and splash is suppressed. A smoother surface suppresses the occurrence of splash because the outflowing fluid at the contact line between the impinging drop and the surface cannot be redirected in a direction normal to the wall [106].

For a wetted surface, a splash criterion was proposed by Yarin and Weiss [107], who studied a train of drops impinging on a plate. They find that splash occurred if a dimensionless impact velocity u is greater than 18, which is evaluated as:

$$u = w_0 (\rho_l / \sigma)^{0.25} \upsilon^{-0.125} f^{-0.375} \tag{5.119}$$

This criterion is essentially equivalent to that proposed by Mundo et al. [106] if the impact frequency f is replaced with w_0/D. Cossali et al. [108] reported an experimental study of splash by impinging a single drop on a smooth surface covered by a thin layer of the same liquid film, varying the height of the film and the impact drop parameters. They found that an increase of the film height (for $\delta = h/D > 0.1$) leads to an increase of the splash onset Weber number suppressing the occurrence of splash due to the increased kinetic energy dissipation in the drop deformation process. Also, the surface morphology does not influence the onset of splash significantly if $\delta \gg \beta$.

By including the effects of existing film on the onset of splash with the support of experimental data [108], a new splash threshold that takes account of the wall-surface roughness and pre-existed film height has been proposed by Han et al. [29]. The threshold is

$$H_{cr} = \left(1500 + \frac{650}{\beta^{0.42}} \right) [1 + 0.1 Re_n^{0.5} \min(\delta, 0.5)] \tag{5.120}$$

where β is the dimensionless roughness parameter, δ is the dimensionless film thickness. This threshold is valid in the shallow film ($\sim \delta < 16$) case on relatively smooth surfaces ($\delta \gg \beta$), and engine walls usually satisfy these conditions. A good summery of other empirical splash criteria can be found in the work of Moreira et al. [109].

For other regimes, the transition criteria for a wetted wall are proposed as [29,92]:
Stick:

$$We_n \leq 5 \tag{5.121}$$

Rebound:

$$5 < We_n \leq 10 \tag{5.122}$$

Spread:

$$We_n > 10 \text{ and } We_n Re_n^{0.5} < H_{cr} \tag{5.123}$$

Splash:

$$We_n Re_n^{0.5} \geq H_{cr} \tag{5.124}$$

5.5.2 **Post Impingement Outcomes**

Four regimes are identified in the model of Han et al. [29]. Let's first discuss the secondary droplet size from the splash. Experimental observations indicate that upon impingement, a crown-alike liquid sheet forms and propagates outwards as shown in Figure 5.19. Capillary waves present on the crown lead to the formation of a free rim. Cusps and liquid jets form and break up into small droplets due to Raleigh–Taylor instability. The crown continues to grow during the late splash phase and finally collapses onto the wall. Based on the experimental observations, it is assumed that the process of disintegration of the crown sheet is like that of a conical sheet issued from an atomizer to a first approximation. This analogy can be supported by the experimental evidence that both longitudinal and azimuthal waves exist on the crown [108] and similar wave structures on the crown and on a liquid cone from a swirl atomizer have been found by Levin and Hobbs [110].

Atomization due to sheet disintegration has been subjected to study for a long time. The postulation of a wave instability induced atomization mechanism by many researchers has led to success in predicting atomized drop size. A widely used model was proposed by Dombrowski and Johns [50] in which they correlated drop sizes with the wavelengths that grow on the surface of the sheet. They derived a dispersion equation for the growth rate of long waves with infinitesimal amplitude for a viscous sheet and assumed that the wavelength with the largest growth rate is responsible for the breakup of the sheet. The sheet is broken up at a one-half wavelength interval into fragments. The fragments contract by surface tension into unstable ligaments that subsequently break down into droplets.

FIGURE 5.19 Photographs showing the splash evolution (Delay time: (a): 0.4 msec, (b): 2.4 msec, (c): 6.4 msec, (d): 12.4 msec).

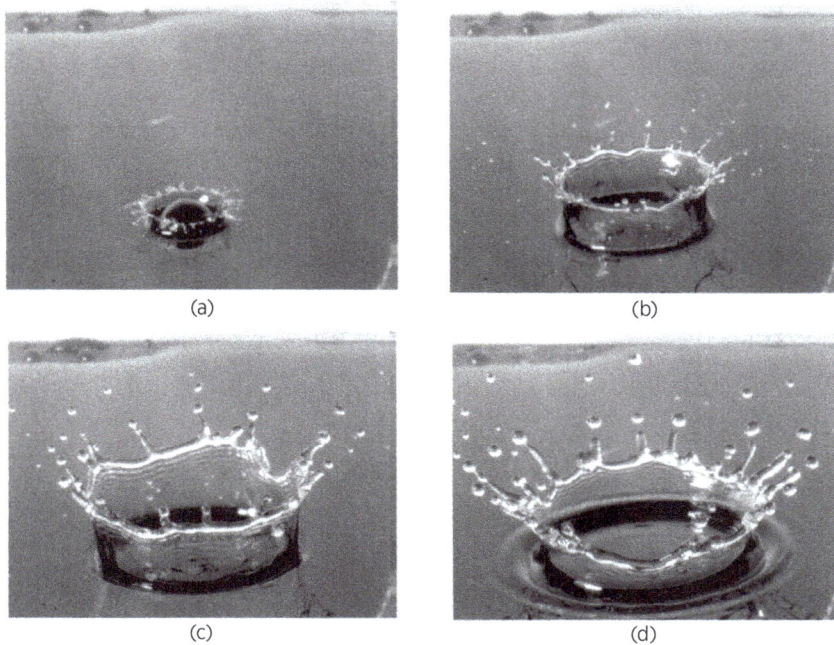

(a)

(b)

(c)

(d)

Test conditions: water droplets, atmospheric environment. Impingement parameters: $\delta = h/D = 0.5$, $Oh = \mu/\sqrt{d\sigma\rho} = 0.0022$

Reprinted with permission from Ref. [29]. © SAGE Publications

FIGURE 5.20 Schematic diagram showing splash processes.

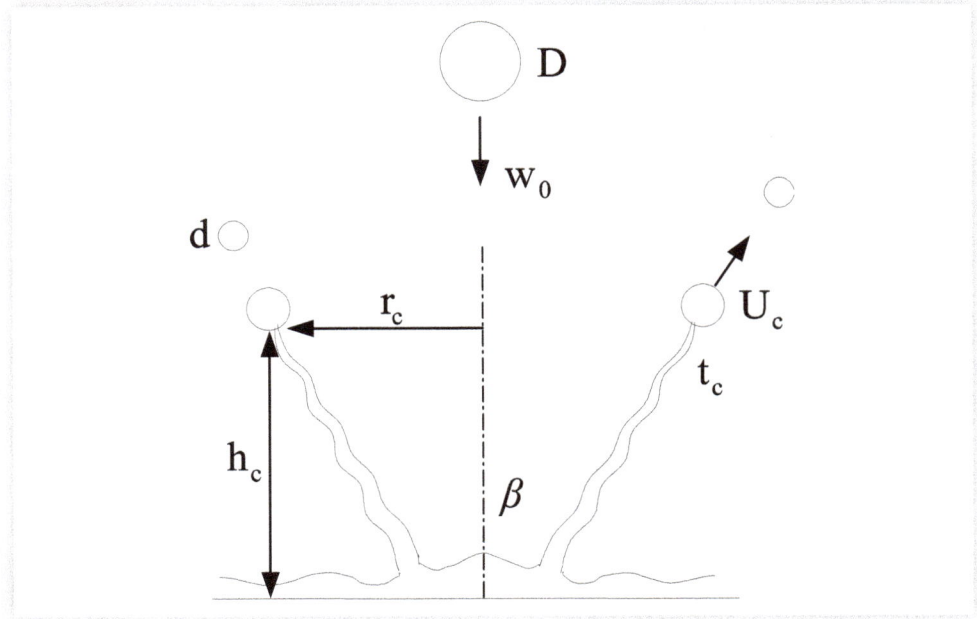

It should be noted that the disintegration mechanism of the crown could be more complicated than that of a sheet from an atomizer due to the formation and breakup of the jets at the top of the crown. Yarin and Weiss [107] have provided a simple explanation of the formation of the cusps (jets); however, a complete theory about the relation between the crown perturbations and the raising of the jets has not yet been developed. To simplify the problem, the details of the jets are ignored and the jets and the rim are together treated as the "ligaments" since the sizes of the jets are similar to those of the rim, as seen in the experiments. These ligaments break down into droplets. Instability analysis for an incompressible viscous sheet [50,111] can be applied to the crown sheet schematically shown in Figure 5.20.

Since the jets and the rim are treated together as ligaments, the jets are not shown separately in the schematic. From the instability analysis, the equation for the size of the secondary droplets is given as:

$$d = \left(\frac{3\pi}{\sqrt{2}}\right)^{1/3} d_r \left[1 + \frac{3\mu}{(\rho_1 \sigma d_r)^{1/2}}\right]^{1/6} \tag{5.125}$$

where d_r is the diameter of the free rim. For long wave growth it is assumed that the free rim is formed twice per wavelength and

$$d_r = \sqrt{\frac{8\sigma t_c}{\rho_g U_c^2}} \tag{5.126}$$

Now the task is to find the thickness t_c and relative velocity U_c of the crown. In a theoretical study of drop impact processes, Yarin and Weiss [107] derived an equation describing the radial position of the crown during its propagation as a function of time. The equation is given as:

$$\frac{r_c}{D} = \frac{w_0^{0.5}}{6^{0.25}\,\pi^{0.5}\,\nu^{0.125}\,D^{0.25}\,f^{0.375}}\,\tau^{0.5} \tag{5.127}$$

where $\tau = 2\pi f t$ is non-dimensional time (t being the dimensional time). Notice that Equation (5.127) gives the square-root dependence on time. Replacing f by w_0/D gives

$$\frac{r_c}{D} = \frac{1}{6^{0.25}\,\pi^{0.5}}\,Re^{0.125}\,\tau^{0.5} \tag{5.128}$$

The crown propagation speed can be cast by taking the derivative of Equation (1.128) as:

$$U_c = \frac{1}{\sin\beta}\frac{dr_c}{dt} \tag{5.129}$$

which gives

$$U_c = \frac{\pi^{0.5}}{6^{0.25}\sin\beta}\,Re^{0.125}\,\frac{w_0}{\tau^{0.5}} \tag{5.130}$$

To obtain the crown thickness, the remaining volume of the crown excluding the secondary droplets must first be assumed to be proportional to that of the impact drop as:

$$V_{crown} = C_0\frac{\pi}{6}D^3 \tag{5.131}$$

where C_0 is a proportionality constant, which will be given later. On the other hand, it can be clearly shown that

$$V_{crown} = \pi t_c h_c^2\left(2\frac{r_c}{h_c} - \tan\beta\right) \tag{5.132}$$

By approximating $\tan\beta = r_c/h_c$, equating Equations (5.131) and (5.132), and using Equation (5.128), we get

$$t_c = \frac{C_0\pi\tan\beta}{6^{0.5}}\frac{1}{Re^{0.25}\tau}D \tag{5.133}$$

Substituting U_c and t_c into Equations (5.125) and (5.126) gives

$$\frac{d}{D} = 1.88 \times \sqrt{8} \sin\beta \tan^{0.5}\beta \sqrt{C_0} \frac{1}{We^{0.5} Re^{0.25}} \sqrt{\frac{\rho_l}{\rho_g}} \tag{5.134}$$

Equation (5.134) states that the size of the secondary droplets is a function of the Weber and Reynolds numbers of the impact drop. It is also affected by the ratio of the liquid density and the ambient gas density, indicating an important aerodynamic aspect of splashing processes. Note that C_0 is a proportionality constant and is given a value of 0.25 [29]. Thus, the mean size of the secondary droplets is finally obtained as:

$$\frac{d_m}{D} = \frac{3}{We^{0.5} Re^{0.25}} \sqrt{\frac{\rho_l}{\rho_g}} \tag{5.135}$$

Interestingly, by equating the secondary droplet size with the impact drop size, Equation (5.135) gives a splash threshold of $We^{0.5}Re^{0.25} \approx 85$ for a liquid with a density of $1.0 \, \text{g/cm}^3$ under atmospheric conditions, which is in good agreement with the experimental observation (see Equation (5.116)). Hence, it can be concluded that the ambient gas density also plays an important role in the onset of a splash. This is of particular significance to DI gasoline and diesel engines. The higher in-cylinder gas density as the injected sprays are encountered would help splash to occur at a lower impact velocity.

Equation (5.135) provides a mean size of the secondary droplets. A size distribution of one type of Nukiyama–Tanasawa function is assumed [29]:

$$f(d) = \frac{2}{3} \frac{d^2}{d_m^3} exp\left(-\left(\frac{d}{d_m}\right)^{3/2}\right) \tag{5.136}$$

The corresponding volume distribution is

$$\frac{dV}{dd} = \frac{1}{4} \frac{d^5}{d_m^6} exp - \left(\frac{d}{d_m}\right)^{3/2} \tag{5.137}$$

Knowing the distribution, it can be shown that the SMD, d_{32} is related to the distribution mean d_m as:

$$d_{32} = \frac{\Gamma(4)}{\Gamma(4/3)} d_m = 2.16 d_m \tag{5.138}$$

where Γ is the gamma function.

Comparisons of Equation (5.136) and the experimental data are shown in Figure 5.21. The equation can reproduce the experimental data well.

The velocity of the secondary droplets u is given as:

$$u = wn + v\zeta(\cos\varphi e_t + \sin\varphi e_p) \tag{5.139}$$

FIGURE 5.21 Comparison of measured size distribution and model fitting.

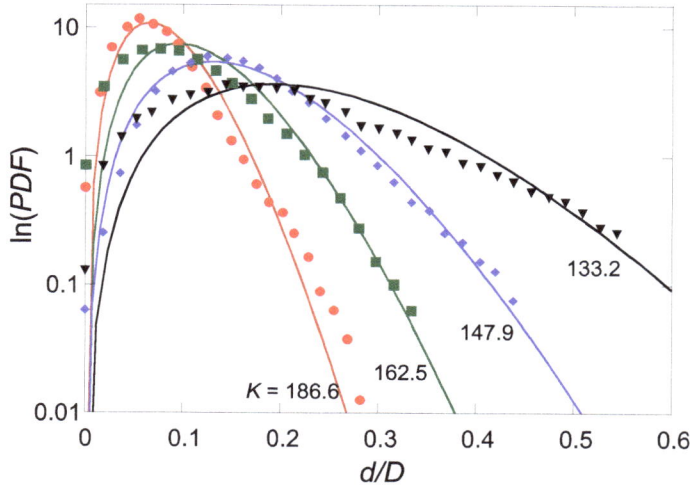

where n is the unit normal to the wall surface, e_t is the unit vector tangent to the surface and in the plane made of n and the incident drop velocity, and $e_p = n \times e_t$. The quantity w is the normal velocity component and is chosen from the following Nukiyama–Tanasawa function:

$$P(w) = \frac{4}{\sqrt{\pi}} \frac{w^2}{w_m^3} exp\left[-\left(\frac{w}{w_m} \right)^2 \right]$$

(5.140)

where the distribution mean w_m (at which the probability of w achieves the maximum value) changes with incident angle α (measured from the normal), and azimuthal angle φ as:

$$w_m = (0.1 + \xi \cos\varphi) w_0$$

(5.141)

where the function ξ is given as:

$$\xi = \frac{1}{900} \alpha$$

(5.142)

The azimuthal angle φ is defined as the angle that the tangential velocity makes with the vector e_t in the plane of the wall and lies in the interval $[-\pi, \pi]$. It is statistically chosen according to the distribution suggested by Naber and Reitz [15]:

$$\varphi = -\frac{\pi}{\gamma} \ln[1 - P(1 - e^{-\gamma})]$$

(5.143)

where P is a random number in the interval $[0,1]$ and y is a parameter related to angle α by

$$\sin\alpha = \frac{e^{y}+1}{e^{y}-1}\frac{y^2}{y^2+\pi^2} \tag{5.144}$$

The physical meaning of Equation (5.141) is that the most possible normal velocity of secondary droplets reaches the highest value in front of the impact point ($\varphi = 0°$) and decreases from that location with the azimuthal angle for an oblique impingement. It obtains the lowest value right behind the impact point ($\varphi = 180°$). For a normal impact, α becomes zero, and w is then uniformly distributed along the azimuthal direction.

The tangential velocity v of a secondary droplet is described by a normal distribution as:

$$N(v) = \frac{1}{\sqrt{2\pi}\delta}\exp\left[-\frac{(v-\overline{v})^2}{2\delta^2}\right] \tag{5.145}$$

where

$$\delta = c_1 v_0 + c_2 w_0 \tag{5.146}$$

$$\overline{v} = \sqrt{A v_0^2 + B w_0^2} \tag{5.147}$$

The constants in Equations (5.146) and (5.147) are set to be $A=0.7$, $B=0.03$, $c_1=0.1$, and $c_2=0.02306$. Justifications for these values are given by Han et al. [29].

Equations (5.145) to (5.147) give a uniform distribution of the magnitude of tangential velocity along the azimuthal direction that is determined randomly by using Equation (5.143). This profile is not true for an oblique impingement in which droplets travel slower in the directions with $\varphi > 0$ than those with $\varphi = 0$ according to experimental sprays images [89]. Elliptical spray shapes can be observed in experiments for an oblique impingement due to the non-uniform tangential velocity distribution. The function ζ in Equation (5.139) is then introduced to reflect this trend.

In deriving ζ, the jet analogy continues to be used, assuming that the spreading of secondary droplets follows that of a liquid sheet resulting from jet impingement. In a study of atomization of the sheet formed by the oblique impact of two equal jets, Ibrahim and Przekwas [112] derived an analytical expression to describe the shape of the sheet. Because of symmetry about the plane along which the liquid sheet produced by the impinging jet flows, their results can be applied to spray impinging on a solid wall. Thus, the radial distance between the edge of the sheet and the point of impact at any azimuthal angle φ (normalized by that at $\varphi = 0$) can be cast. It is assumed to be proportional to the radial velocity and ζ is obtained as:

$$\zeta = \frac{\sin^2\left[(\pi/2)(1-2\alpha/\pi)^{\left(1-\frac{|\varphi|}{\pi}\right)}\right]}{\cos^2\alpha}\exp\left(-\frac{y}{\pi}|\varphi|\right) \tag{5.148}$$

Splash results in the rebounding of many smaller secondary droplets from the impingement location. Experiments and theoretical analysis show that a fraction of the impinged drop will remain on the surface. Although it has been found that a splashing drop may entrain liquid from the wall film [113], a qualitative description remains unavailable at this time.

Detailed knowledge of the total mass fraction of secondary droplets is very limited although it is believed to be influenced by surface conditions. Experimental data of Yarin and Weiss [107] suggest that the ratio of the total secondary droplet's mass m and the incident drop mass M increases rapidly from zero to a value of about 0.75 and then remains constant. A correlation that fits the data is given as [29]:

$$\frac{m}{M} = 0.75\{1 - \exp[-10^{-7}(H - H_{cr})^{1.5}]\} \tag{5.149}$$

Equations (5.135) and (5.149) are derived based on a single-drop impingement scenario and this is valid for this case, as discussed above. However, computations indicated a considerable underprediction of the secondary droplet size and mass deposition rate for PFI sprays (which can be up to 50%) using these equations directly. To account for the effects of multiple-drop interactions, as a first approximation, an effective impingement hypothesis is proposed. The justification will be given later. The assumption is that the post-impingement properties of a group of impinging drops could be statistically represented by those of a single drop (called effective drop) having a mean size characterizing that group of drops. Numerically, the size of the effective drop is calculated first and a so-called impingement effectiveness factor is then defined. In each computational cell, an effective drop diameter D_{eff} is calculated from all the already impinging parcels at each time step as:

$$D_{eff} = min\left(\sqrt{\sum D_i^2}, 3D_{32}\right) \tag{5.150}$$

The first term on the right-hand size of Equation (5.150) represents surface area conservation and the second one comes from the experimental observations, which indicate the size of the crown base is about three times that of the incident drop [110]. The impingement effectiveness factor $f_{eff,i} = D_{eff}/D_i$ is applied to each numerical parcel to give

$$\frac{d}{D} = \frac{3}{We^{0.5}Re^{0.25}}\sqrt{\frac{\rho_1}{\rho_g}}f_{eff,i} \tag{5.151}$$

$$\frac{m}{M} = 0.75\{1 - \exp[-10^{-7}(H - H_{cr})^{1.5}\}\frac{1}{f_{eff,i}} \tag{5.152}$$

Equations (5.151) and (5.152) return to Equations (5.135) and (5.149), respectively, if only one parcel collides ($f_{eff,i} = 1$), recovering the single-drop impact case.

In the stick and spread regimes, incident drops are treated as part of the liquid film. Their energy and momentum are completely transferred to the film. For the rebound regime, the drop is reflected away from the wall with its size unchanged. Bai and Gosman [90] suggested calculating the velocity of the rebounding droplets as:

$$w = -ew_0 \tag{5.153}$$

$$v = \frac{5}{7}v_0 \tag{5.154}$$

The quantity e is the so-called "restitution coefficient" and is given as:

$$e = 0.993 - 1.76\theta + 1.56\theta^2 - 0.49\theta^3 \tag{5.155}$$

where θ is the impingement angle measured from the impingement plane.

Computations with the use of the above models are assessed with experiments [29]. The test cases are listed in Table 5.1. One is reported by Nagaoka et al. [89] (named NKN case), and the other by Trujillo et al. [96] (named TMLP case). In both experiments, sprays impinge on a flat plate under room conditions. The nozzle is a single-hole pintle type used in PFI engines. In the TMLP case, iso-octane fuel is injected upon a flat plexiglass flat plate. A phase Doppler particle analyzer (PDPA) was used to measure the secondary droplet sizes at several locations 5 mm above the plate surface.

TABLE 5.1 Experimental cases of spray impinging on a flat wall surface [29]

PARAMETER	NKN	TMLP
Fuel	n-Heptane	Iso-octane
Injection pressure (kPa)	384	370
Injection pulse width (msec)	4	8.45
Spray cone angle (°)	18	7
Initial drop SMD (mm)	230	380
Distance to wall (mm)	70	100
Angle to wall (°)	90, 60, 30	60, 45, 30

Reprinted with permission from Ref. [29]. © SAGE Publications

Computations were performed using KIVA3V implemented with the above impingement models. Figure 5.22 shows the computed spray structure in the NKN case from the perpendicular direction. The symbol X represents the experimentally measured boundary. Both experiment and computation reflect the interesting change in spray shape with the impingement angle. The secondary droplets form a symmetric circular structure for the normal impingement. However, as the impinging spray is inclined, the outer boundary of the spray becomes ellipse-like, with a larger radius in front of the impingement location. The spray shape becomes more elliptic as the impingement angle is decreased from 60° to 30°. Also, both experiment and computation indicate that some secondary droplets fly opposite to the incident direction behind the impingement location, even for the 30° impingement case. This is due to the combination of effects due to splash and gas entrainment that is induced by the impinging spray.

Figure 5.23 shows the computed spray structure in the TMLP case 12 msec ASI for 60°, 45°, and 30° impingement angles as seen from the side and from the top, respectively. In this case, the surface roughness height was set to be 0.1 μm to represent the very smooth plexiglas surface. Again, the spray shape is seen to become more and more elliptic as the

FIGURE 5.22 Spray structure of a PFI-engine spray impacting on a wall, seeing from the surface normal direction.

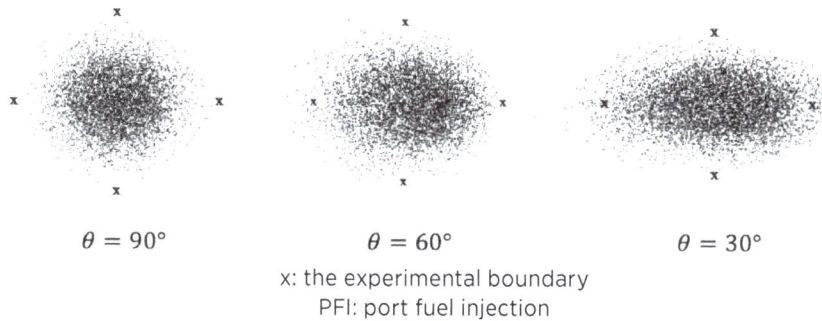

$\theta = 90°$ $\theta = 60°$ $\theta = 30°$

x: the experimental boundary
PFI: port fuel injection

Reprinted with permission from Ref. [29]. © SAGE Publications

FIGURE 5.23 Computed spray structure of a PFI-engine spray impacting on a wall. Impingement angle is 60°, 45°, and 30° from left to right.

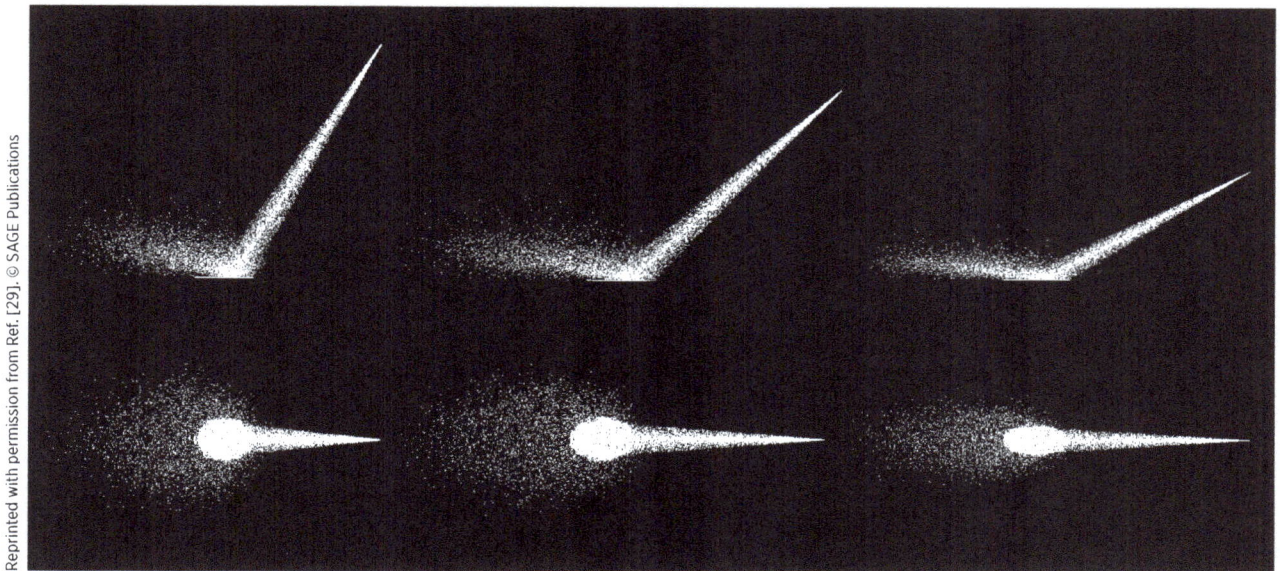

Reprinted with permission from Ref. [29]. © SAGE Publications

impingement angle is reduced, as discussed before. The computed time-averaged droplet sizes are compared with the measurements as shown in Figure 5.24 for the three impingement angle cases. The monitoring locations are at four fixed locations ($\varphi = 0$, front distance from the center of the impingement location = 20, 40, 60, and 80 mm) and 5 mm above the surface. It is seen that both experiment and computation indicate that the spray drop sizes reduce significantly by the impingement, from about 300 µm before the impingement to about 120 µm after the impingement. The post-impingement droplet sizes are relatively insensitive to the impingement angle. The averaged droplet size becomes larger when the droplets are farther away from the impingement location.

FIGURE 5.24 Computed and measured secondary droplet size.

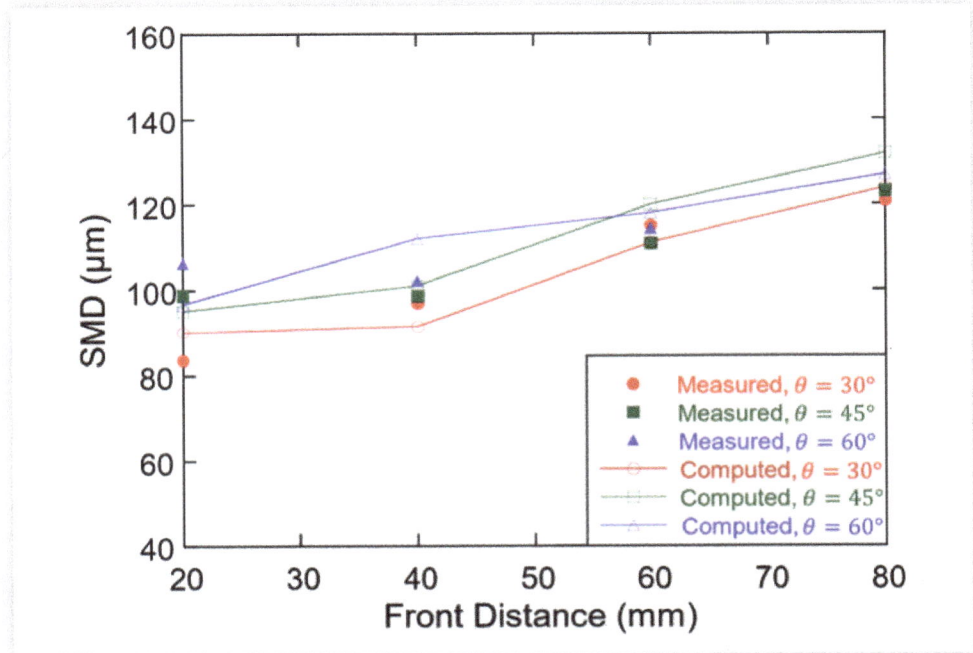

While models of spray/wall impingement have obtained some successes, more research works are needed in this area. The models are derived based on a single drop impingement scenario. However, for a more complicated scenario case of spray impingement, multiple drops impinge on a surface successively and/or simultaneously. Complicated interactions among the formed crowns are most likely to occur and may result in overlapping, edge destruction and intersection, and collapse of the crowns. These effects would impact the outcomes of the impingement (e.g., secondary droplet sizes and liquid deposition) due to irregular and early disintegration of the crowns. Thus, there is a critical need of detailed experimental studies on spray-wall interactions under engine-like conditions to reveal the physics for numerical development.

Spray impingement experiments by Sivakumar and Tropea [114] have revealed some differences between the single-drop collision on a wall and the spray impingement on a wall. The latter is shown in Figure 5.25, in which the irregularity of the crown formation and drop-crown interactions are clearly seen. Importantly, all experimental data of nondimensional crown radius (r_c/D) lie close to a growth rate proportional to $\tau^{0.2}$ in the experiments of Sivakumar and Tropea [114]. This is considerably different from the $\tau^{0.5}$ growth rate predicted by Yarin and Weiss [107], which is used in Equation (5.127). This new proportionality implies the rim radial size and, therefore, the secondary drop size would be larger than those predicted by the proportionality of Yarin and Weiss [117] based on mass conservation or Equation (5.134). These experimental evidences justify, in part, the underestimation of the secondary drop size given by Equation (5.134).

FIGURE 5.25 Photographic sequence showing the impinging drop crown expansion and splash.

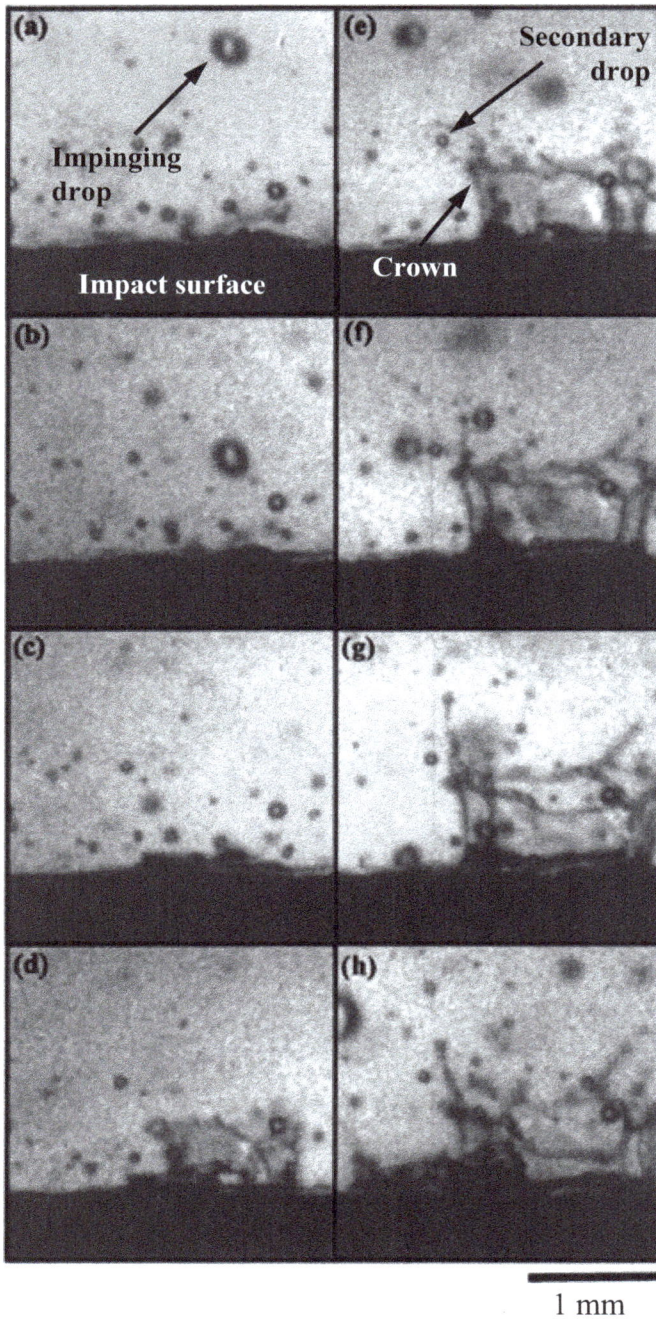

1 mm

Number concentration of impinging spray: 2.78 drops/mm³.
For the impinging drop: d=241 μm, U=14.3 m/s, h_0/d=0.37, and We=676

5.5.3 Wall Film Hydrodynamics and Heat Transfer

Spray impinging on the walls of an engine generally leads to liquid film formation. Liquid fuel initially accumulates on the wall after the impingement, and subsequently spreads on it as a thin film. In this section, the film models of hydrodynamics, heat transfer, and evaporation will be described following the framework of O'Rourke and Amsden [28,94] with additions by Deng et al. [21,116] and Han and Xu [115]. In these models, hydrodynamics of fuel films has been computed with the use of the Lagrangian numerical particles to represent the films. Some basic assumptions are applied to simplify the problem, which are:

1. Film is very thin (usually in the order of 10^{-5} m).

2. Film thicknesses are much smaller than radii of curvature of the walls; the effects of rotational motion are ignored.

3. Liquid flow in the films is laminar and the liquid velocities in the film are tangent to the wall and vary linearly with height above the wall.

4. Temperatures on walls (intake-valve, cylinder liner, cylinder head, and piston) are less than the boiling temperatures; the liquid film is in direct contact with the wall and ignore other heat transfer regimes.

With these assumptions, applying the conservation laws gives the governing equations for the film. The mass conservation equation is

$$\frac{\partial \rho_l h}{\partial t} + \nabla_s \cdot \left[\rho_l (\boldsymbol{u}_f - \boldsymbol{u}_w) h \right] = \dot{M} \tag{5.156}$$

where ρ_l is the liquid film density, h is the film thickness, $\boldsymbol{u}_f$ is the mean film velocity, and $\boldsymbol{u}_w$ is the wall velocity, ∇_s is the surface gradient operator. $\dot{M}$ is the mass source per unit wall area due to impingement, or film vaporization:

$$\dot{M} = \dot{M}_{imp} + \dot{M}_{vap} \tag{5.157}$$

The impingement source $\dot{M}_{imp}$ is

$$\dot{M}_{imp} = \iiint \frac{4}{3} \pi r^3 \rho_l \boldsymbol{v} \cdot \boldsymbol{n} f(\boldsymbol{x}_s, \boldsymbol{v}, r, T_d, t) d\boldsymbol{v} dr dT_d \tag{5.158}$$

where f is the spray droplet distribution function, n is the unit normal to the wall pointing into the gas, and $\mathbf{x}_s$ is a point on the wall surface. The vaporization source $\dot{M}_{vap}$ will be given later.

Liquid film spreads on the wall that the spray impacting on under the influence of several forces: shear stress force exerted on the top of the film by the gas flow, viscous forces in the film, body force, film inertial, film pressure gradient, and the force exerted by the impinging spray. This last force was found to be especially important in determining the film spreading process [115]. Under these forces, the film momentum equation is given as:

$$\rho_l h \left\{ \frac{\partial \boldsymbol{u}_f}{\partial t} + [(\boldsymbol{u}_f - \boldsymbol{u}_w) \cdot \nabla_s] \boldsymbol{u}_f \right\} + h \nabla_s p_f = \tau_w \boldsymbol{t}$$

$$- \mu_l \frac{\boldsymbol{u}_f - \boldsymbol{u}_w}{h/2} + \dot{\boldsymbol{P}}_{imp} - (\dot{\boldsymbol{P}}_{imp} \cdot \boldsymbol{n}) \boldsymbol{n} + \quad (5.159)$$

$$\dot{M}_{imp}[(\boldsymbol{u}_w \cdot \boldsymbol{n}) \boldsymbol{n} - \boldsymbol{u}_f] + \delta p_f \boldsymbol{n} + \rho_l p h \boldsymbol{g}$$

In Equation (5.159), τ_w is the shear stress on the gas-side of the wall film, $\boldsymbol{t}$ is the unit tangent to the wall in the direction $\boldsymbol{u}_f - \boldsymbol{u}_w$, μ_l is the film viscosity (film temperature-dependent), T_f is the mean film temperature, $\dot{\boldsymbol{P}}_{imp}$ is the momentum source/area due to impingement, p_f is the film (impingement) pressure, δp_f is the pressure difference across film to bring about $\boldsymbol{u}_f - \boldsymbol{u}_w) \cdot \boldsymbol{n} = 0$, and $\boldsymbol{g}$ the is acceleration due to gravity. The momentum source due to impingement is given by

$$\dot{\boldsymbol{P}}_{imp} = -\iiint \frac{4}{3} \pi r^3 \rho \boldsymbol{v} \boldsymbol{v} \cdot \boldsymbol{n} f(\boldsymbol{x}_s, \boldsymbol{v}, r, T_d, t) d\boldsymbol{v} dr dT_d \quad (5.160)$$

To obtain the energy conservation of the film, we assume the temperature profile of the film to be piecewise linear, varying from the wall temperature T_w to T_f in the lower half of the film and from T_f to a gas surface temperature T_s in the upper half of the film. The film energy equation is then given as:

$$\rho_l h C_{vl} \left\{ \frac{\partial T_f}{\partial t} + [(\boldsymbol{u}_f - \boldsymbol{u}_w) \cdot \nabla_s] T_f \right\} = \lambda_l \left[\frac{T_s - T_f}{h/2} - \frac{T_f - T_w}{h/2} \right] + \dot{Q}_{imp} - I_l \dot{M}_{imp} \quad (5.161)$$

where C_{vl} is the film liquid specific heat, λ_l is the liquid heat conductivity, I_l is the liquid internal energy at temperature T_f, and T_w is the wall temperature. The source term is

$$\dot{Q}_{imp} = -\iiint \frac{4}{3} \pi r^3 \rho_l I_l \boldsymbol{v} \cdot \boldsymbol{n} f(\boldsymbol{x}_s, \boldsymbol{v}, r, T_d, t) d\boldsymbol{v} dr dT_d \quad (5.162)$$

The term in braces in Equation (5.161) is the time rate-of-change of temperature following a liquid film element moving along the wall surface. In addition to this unsteady term and the heat conduction terms, the mean film temperature can change due to spray wall impingement. In the work of O'Rourke and Amsden [28,94], Equation (5.161) ignores the small changes in mean film temperatures due to fuel vaporization and the changes in film temperatures due to fluctuations of temperature and velocity within the film. These assumptions were relaxed in an effort of Deng et al. [21] to develop an alternative heat transfer model, which will be discussed later.

In order to calculate the film surface temperature, we need the interface conservation condition relating the gas-side heat transport to the film, the energy used to vaporize the fuel, and the liquid-side heat transport due to conduction:

$$\dot{Q}_w = \dot{M}_{vap} L(T_s) + \lambda_l \frac{T_s - T_f}{h/2} \tag{5.163}$$

where L is the latent heat of vaporization. $\dot{Q}_w$ is given in Equation (5.168).

O'Rourke and Amsden [94] derived the wall functions considering the effects of film evaporation. Two major assumptions are made in their derivation in the fully turbulent region of the boundary layer. First, it is assumed that total transport is independent of the normal coordinate to the wall and is the sum of transport due to turbulent diffusion and due to convection by the vaporization velocities. Second, as in non-vaporizing boundary layers, there is a linear variation of the turbulent diffusivity with distance from the wall. Thus, the mass vaporization rate is given by

$$\dot{M}_{vap} = H_Y ln\left(\frac{1-Y_v}{1-Y_{vs}}\right) \tag{5.164}$$

where

$$H_Y = \begin{cases} \dfrac{\rho c_\mu^{1/4} k^{1/2}}{y_c^+ Sc_l + \dfrac{Sc_t}{\kappa} \ln\left(\dfrac{y^+}{y_c^+}\right)} & y^+ > y_c^+ \\[4ex] \dfrac{\rho c_\mu^{1/4} k^{1/2}}{y^+ Sc_l} & y^+ \le y_c^+ \end{cases} \tag{5.165}$$

and Y_v is the fuel vapor mass fraction at y^+, $Y_{vs} = Y_{veq}(T_s)$ is the equilibrium vapor mass fraction at film surface temperature, k is the turbulent kinetic energy, c_μ is the k-ε turbulence model constant of 0.09, Sc_l and Sc_t are the laminar and turbulent Schmidt numbers, respectively, and κ is the Karmann constant of 0.433.

The dimensionless normal coordinate y^+ is given by

$$y^+ = \frac{yc_\mu^{1/4} k^{1/2}}{v_0} \tag{5.166}$$

where v_0 is the laminar kinematic viscosity. It is assumed that the transition between the fully turbulent region and the laminar profile near the wall occurs at a value y_c^+ of 11.05, independent of the mass vaporization rate. The boundary layer shear stress τ_w and heat flux $\dot{Q}_w$ are given as:

$$\frac{\tau_w}{\rho|\mathbf{u}-\mathbf{u}_w|c_\mu^{1/4} k^{1/2}} = \begin{cases} \dfrac{M^*}{e^{BM^*}(y^+)^{M^*/\kappa} -1} & y^+ > y_c^+ \\[3ex] \dfrac{M^*}{e^{M^* y+} -1} & y^+ \le y_c^+ \end{cases} \tag{5.167}$$

$$\frac{\dot{Q}_w}{\rho c_p c_\mu^{1/4} k^{1/2}(T-T_s)} = \begin{cases} \dfrac{M^*}{e^{Pr_l M^* y_c^+}\left(\dfrac{y^+}{y_c^+}\right)^{M^* Pr_t/\kappa} - 1} & y^+ > y_c^+ \\[3ex] \dfrac{M^*}{e^{Pr_l M^* y^+} - 1} & y^+ \le y_c^+ \end{cases} \tag{5.168}$$

where B = 5.5, Pr_l and Pr_t are the laminar and turbulent Prandtl numbers, respectively, and u is the gas velocity. The dimensionless vaporization rate M* is defined as:

$$M^* = \frac{\dot{M}_{vap}}{\rho c_\mu^{1/4} k^{1/2}} \tag{5.169}$$

In the limit of small M^*, Equations (5.167) and (5.168) reduce to the standard wall functions for the turbulent boundary layers above non-vaporizing surfaces as discussed in Sec. 3.3. In this case, the shear speed u^* and wall shear stress τ_w are related as:

$$u^* = \sqrt{\tau_w / \rho} \tag{5.170}$$

If k-ε models are used for the gas turbulence, then $u^* = k^{1/2} C_\mu^{1/4}$.

Noticing the importance of the force exerted by the impinging spray, accounted by Equation (5.160), Han and Xu [115] modified the original treatment of O'Rourke and Amsden [28,94] and this has improved the prediction considerably. The impingement momentum source term $\dot{P}_{imp}$ includes the contributions from both sticking and splashing spray particles that impinge on the face a where particle p resides at one time step.

$$\dot{P}_{imp,a} = \left[\sum_p \rho V_p(u_p^{n+1} - u_p^n)\right](|A_a|\Delta t)^{-1} \tag{5.171}$$

where V_p is the liquid volume associated with particle p and A_a is the area projection vector of the face a. In Equation (5.171), if a spray particle splashes, the secondary droplet velocity u_p^{n+1} is calculated by Equation (5.139), and the droplet size is given by Equations (5.135) and (5.136). If a particle becomes a film particle, it will stick on the wall and the momentum of it is deposited into the film.

In the meantime, the incident droplet velocity u_p^n is treated as:

$$u_p^n = w_{p,0}^n n + v_{p,0}^n(\cos\varphi e_t + \sin\varphi e_p) \tag{5.172}$$

where $w_{p,0}$ and $v_{p,0}$ are the normal and tangential velocity of the incident droplet. φ, e_t, e_p and n have been defined earlier.

Note that in the original model of O'Rourke and Amsden [28], the incident droplet velocity u_p^n is treated as $u_p^n = w_{p,0}^n n + v_{p,0}^n e_t$. Since all the incident particles are coming in along the impingement direction e_t, the source term $\dot{P}_{imp,a}$ of the film particles is dominant along the impingement directions which causes the film front to move excessively in those directions.

FIGURE 5.26   Experimental and computed film images: (a) 10 msec ASOI; (b) 15 msec ASOI; (c) 20 msec ASOI; (d) 25 msec ASOI.

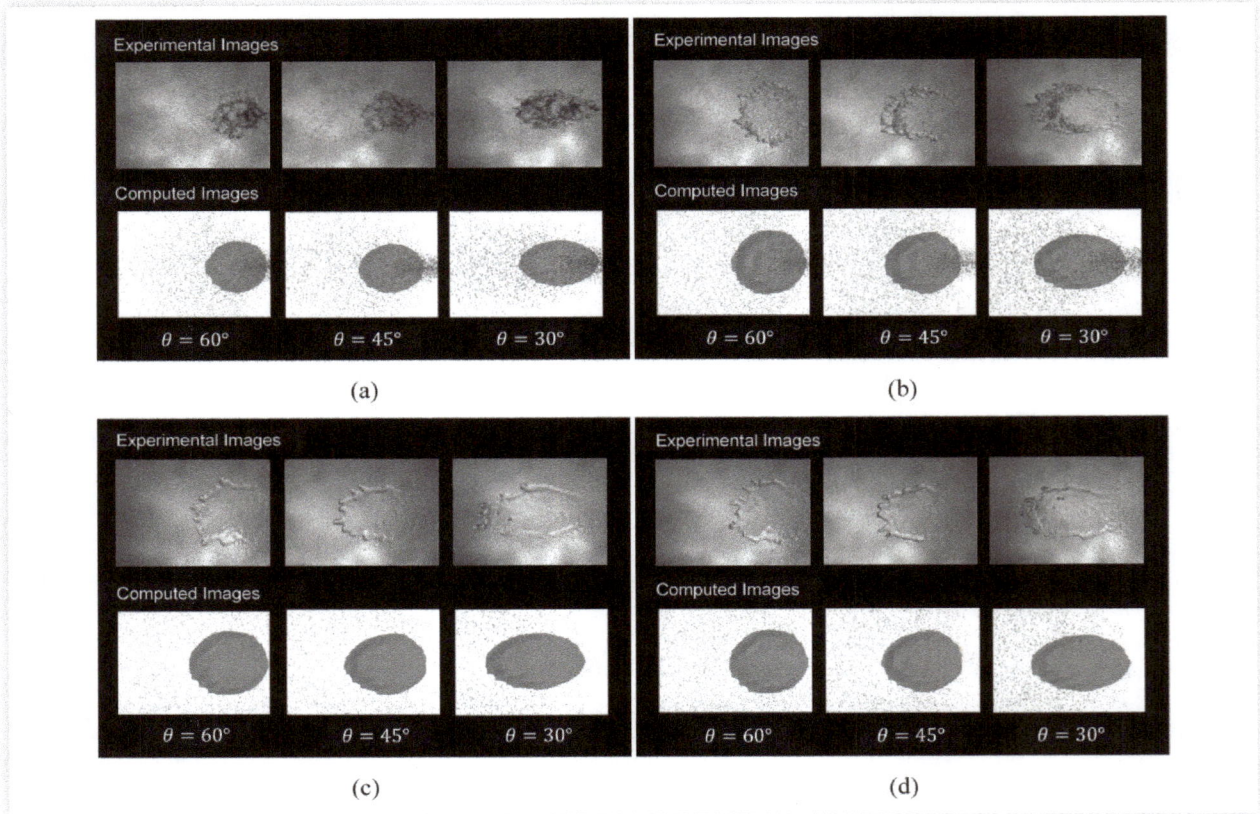

The models of O'Rourke and Amsden with the addition of improvements by Han and Xu were assessed with the use of the measurements of Mathews et al. [117]. In their experiments, an unintensified charged-coupled device (CCD) camera was used to take images of the fuel film. An optical, non-intrusive technique was used to determine the thickness of the deposited fuel film. Detailed experiment conditions and spray characteristics are the same as those in Figure 5.23 and as listed in Table 5.1.

Comparison of the experimental and computed film images as a time sequence are given in Figure 5.26 for three different impingement angles: 60°, 45°, and 30°. At 10 msec ASI, the film is at the early stage of its development. The impingement of the incoming droplets is displayed by the particles coming in from the right side of the images. At this time, the film surfaces are very rough as the experiments show, indicating the impingement of incoming droplets is still taking place and the film mass is accumulating. The liquid on the surface is pushed away from the impingement point in the impingement and other directions. The film forms in elliptic shape spreading outwards. In comparison, the extension and shape of the film as a function of the impingement angle are well captured by the model.

At 15 msec ASOI, there are still some incoming droplets, while the liquid film continues to develop its shape. The film surfaces still show some roughness, although they are smoother than the ones at 10 msec ASOI. A film front is developed and moves outwards. At 20 msec ASOI, there is no more impingement taking place. The liquid films are close to the end of

FIGURE 5.27 Evolution of the film front length.

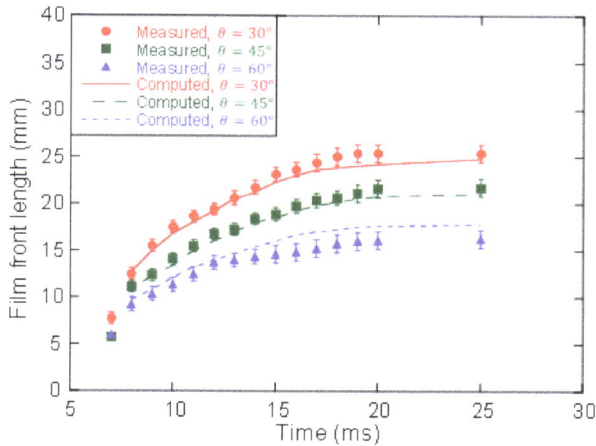

their development and become very smooth, and the film front is clearly established. At 25 msec ASOI, the images of the liquid film are very similar to the ones at 20 msec ASOI; the front is not moving, indicating the films reach a steady state. Overall, the model can predict the film shape and spreading evolution in good agreement with the experimental images. However, the details of the experimental images such as the "fingers" at the film edges and the unsteadiness of the films are not resolved by the computations.

The data for film front displacement were also provided by the experiments of Mathews et al. [117]. Measurements were taken from 30 averaged images. The film front displacement is measured as a distance between the spray impingement point and the front of the film along the impingement direction. Values of the error represent the largest deviation from the averages. In Figure 5.27, the computed film front versus time is compared with the measurements. The computed film front is defined as the distance between the position where the film front thickness reaches 5 μm and the impingement point along the impingement direction. The film front results clearly show that as the impingement angle reduces, the film slides further away from the impingement point. For all three impingement angles, the computed film front that indicates the development of the films agrees very well with the experiment results. This confirms the good agreement of the film images.

The thicknesses of the fuel films deposited on the plexiglass surface were also measured. Measurements were taken along the centerline of injection as well as a single transverse cross section. An averaged value of the film thickness along the centerline of injection and transverse direction are shown in Figures 5.28 and 5.29, respectively. Also shown in these figures are the corresponding computed results. These film thickness curves indicate that there is an accumulation of liquid along the periphery of the film as have already been seen in the film images. The computed film thickness values can reflect this characteristic of the film: the peak value at the edge of the film indicating more accumulation of the liquid there. And simulations capture the steep slope of the film front. However, the model overpredicts the peak values of the film thickness at the film front edges, and underpredicts the thickness in the central region along the transverse direction. The unsymmetric features of the films indicated by the thickness distributions along the transverse direction are not captured by the modeling.

FIGURE 5.28 Film thickness distributions along the impingement direction.

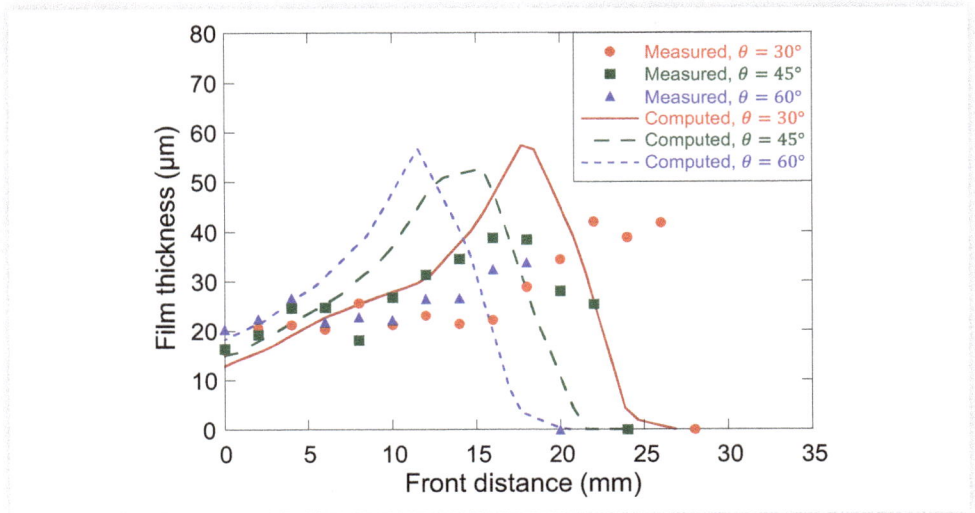

FIGURE 5.29 Film thickness distributions along the transverse direction of the impingement.

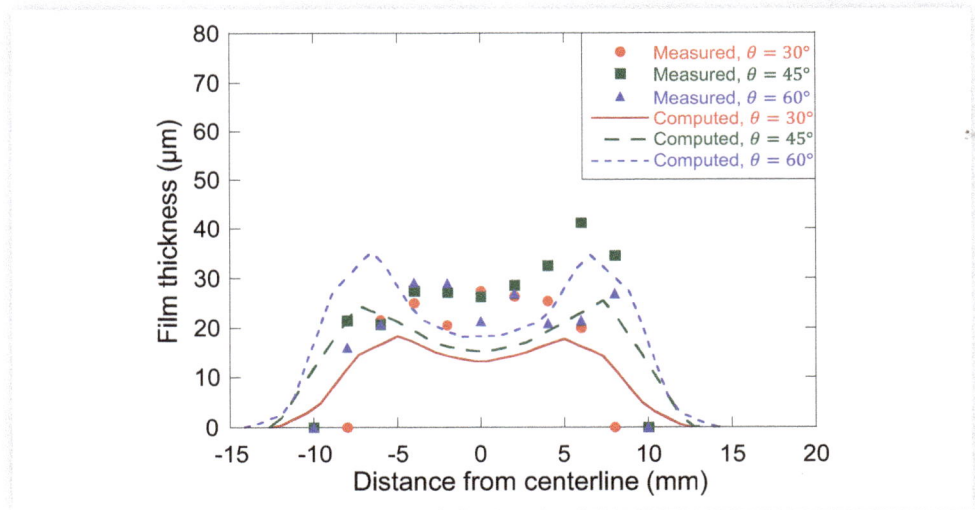

The above models have been applied in DI gasoline engine simulations for combustion system development (see Sec. 7.5). However, the situation may be different for a diesel spray impingement since diesel sprays are injected under a much higher injection pressure (in the order of 200 MPa, one order higher than DI gasoline spray injection pressure). The impact momentum of diesel sprays acting on the walls is much higher than that of gasoline sprays. This raises new challenges for modeling diesel spray wall impingement.

Katsura et al. [118] conducted diesel spray impingement experiments and they found that after impingement, a large quantity of the droplets was located close to the wall. At the peripheral region of the impinging spray, a vortex was observed where the droplet density was large and turbulent mixing occurred between the spray and the surrounding gas. This observation motivated Deng et al. [116] to suspect that the missing lift force term may have effects on the simulation accuracy for diesel spray impingement. Thus, Deng et al. proposed a modified model [116] to add the Saffman lift force [119,120] and droplet-air relative velocity modulation based on the theoretical gas jet velocity profile.

The magnitude of the lift force is given by Saffman as:

$$F_{Saff} = 1.61 \mu_f d \, |u - v| \sqrt{Re_G} \tag{5.173}$$

where μ_f is the viscosity of the surrounding gas, d is the drop diameter, u is the fluid velocity, v is the drop velocity, and $Re_G = (d^2 \rho_f / \mu_f)|du/dy|$ is the shear Reynolds number. However, F_{Saff} is derived based on the assumption that the Reynolds number $Re_r = (\mu_f \rho_f d)|u - v|$ based on the velocity difference is much less than Re_G, which is not suitable for diesel-like spray impingement. The correlation of Mei [121] was adopted, which is:

$$F_L / F_{Saff} = \begin{cases} (1 - 0.3314 \gamma^{1/2}) \exp\left(-\dfrac{Re_r}{10}\right) + 0.3314 \gamma^{1/2} & Re_r \leq 40 \\[2ex] 0.0524(\gamma Re_r)^{1/2} & Re_r > 40 \end{cases} \tag{5.174}$$

where

$$\gamma = \frac{d}{2\,|u - \gamma|} \frac{du}{dy} \tag{5.175}$$

Thus, the lift force F_L of the sliding droplets that are not in the stick regime is calculated in the direction normal to the wall.

Now, let us discuss wall film heat transfer modeling furthermore. Although research on this topic has been progressive, challenges to accurately model the local heat transfer at the spray wall impingement region remain. Part of the reason is attributed to the multiple physical processes that the fuel droplets undergo depending on the wall surface temperature, as shown in Figure 5.16. Other parts come from the complexity of spray impingements under wide ranges of temperature and pressure at different engine operating conditions in different engine combustion systems, which raise doubts about the generality of any single model.

A comprehensive review on heat transfer upon spray wall impingement is given by Moreira et al. [109]. They summarized the droplet–wall heat transfer phenomena into four regimes:

1. Single-phase film-evaporation ($T_w < T_{sat}$): Heat transfer occurs mainly by conduction and free convection as the wall surface temperature T_w is lower than the droplet's saturation temperature T_{sat}, without phase change.

2. Nucleate boiling ($T_{sat} < T_w < T_{CHF}$): Vapor bubbles form close to the wall and move by buoyancy up to the liquid–air interface. The heat is removed by vaporization and increases with the surface temperature up to a maximum at the critical heat flux temperature (T_{CHF}).

3. Transition ($T_{CHF} < T_w < T_{LF}$): As the vaporization rate increases, an insulating vapor layer forms at the liquid–solid interface and the heat flux decreases down to a local minimum at the Leidenfrost temperature (T_{LF}).

4. Film boiling/Leidenfrost regime ($T_w > T_{LF}$): A stable vapor layer forms, which precludes contact between the droplet and the surface and through which heat is transferred by conduction. Radiation starts to play a non-negligible role only at higher temperatures, and in the case of fuel droplets, ignition may also occur, after which a slight decrease in droplet lifetime curve occurs.

The complexity of the heat transfer in the spray-wall impingement regions is already recognized with these mechanisms plus the local gas flow fluctuations and the interactions among the drops and between them and the walls induced by the spray impingement. There is perhaps no single model or correlation that can cover the wide ranges of physics for heat transfer of engine spray impingement. Thus, the law of wall analogy to model heat transfer in spray-wall interactions may not be accurate in some situations since it involves many assumptions that are invalid in real engines. Significant underpredictions of the wall heat flux in the diesel spray impingement [122] were found by Deng et al. [21] with the use of the law-of-the-wall model of O'Rourke and Amsden [94]. Fortunately, the current research interest of spray-wall impingements is mainly focused on cold starts of engines in which the wall temperatures are lower than the fuel saturation temperature. The assumption of single-phase film evaporation is valid.

Considering the complexity of the problem, research works have been carried out to evaluate the local heat transfer of spray wall impingement at macroscales to formulate correlations for the local heat transfer coefficient at spray impact. Experiments have been performed to provide measurement data for both diesel and gasoline spray cases under engine-like conditions. A summary of the correlations is given by Moreira et al. [109].

Eckhause and Reitz [20] formulated a spray/wall heat transfer model based on the experimental observation [123] that for diesel engines, where the normal in-cylinder surface temperature ranges from 400–600 K, the appropriate hydrodynamic regime is the wetting regime. The model does not account for Leidenfrost effects, but it does consider flooded and non-flooded regimes depending on whether a liquid film is determined to be present on the surface, respectively. In the flooded case, heat transfer is modeled based on boundary layer correlations. In the non-flooded regime, heat transfer is modeled by considering correlations for individual drops impinging on a surface.

Along with this approach, Deng et al. [21] formulated a model that is based on the correlations obtained from experiments. The model balances the energies of the film region from these four terms: convection between the gas and the film ($\dot{Q}_{g\text{-}to\text{-}f}$), heat transfer between the film and the wall ($\dot{Q}_{w\text{-}to\text{-}f}$), vaporization of the film ($\dot{Q}_{vap}$), and energy exchanges among the rebounding/sliding droplets and the film ($\dot{Q}_{imp}$) due to spray impingement. The energy balance at each computational timestep Δt for the spray impacted area A_f is given as:

$$\frac{\Delta(m_f c_{p,f} T_f)}{A_f \Delta t} = \dot{Q}_{w\text{-}to\text{-}f} + \dot{Q}_{g\text{-}to\text{-}f} + \dot{Q}_{vap} + \dot{Q}_{imp} \tag{5.176}$$

where, the mass, specific heat capacity, temperature, and area of the film are denoted by m_f, $c_{p,f}$, T_f, and A_f, respectively. For the splashed mass ratio $\dot{m}$, Deng et al. [21] also proposed a correlation as:

$$\dot{m} = \left[0.1 + 0.4 \cdot \min\left(\frac{h}{d}, 1\right)\right]\left(\frac{C}{We}\right)^{0.25} \tag{5.177}$$

where

$$C = \sqrt{We_{inj}} = \sqrt{\frac{\rho_l U_{inj}^2 d_{noz}}{\sigma_l}} \tag{5.178}$$

and We_{inj} is the Weber number of the droplet calculated from the normal component of the impact velocity, where ρ_l and σ_l are the density and surface tension of the liquid, respectively, U_{inj} is the liquid injection velocity and d_{noz} is the nozzle diameter.

In the Equation (5.176), the four terms of the energy exchange rates are calculated in each computational cell and their correlations are given by Deng et al. [21], which are summarized here without citing the details.

For heat transfer between the film and the wall:

$$\dot{Q}_{w-to-f} = H_{Tfw}(T_w - T_f), \qquad Nu = 0.34\frac{We^{0.94}}{Re^{0.53}Pr^{0.33}} = \frac{H_{Tfw}h}{\lambda_l} \tag{5.179}$$

For convection between the gas and the film:

$$\dot{Q}_{g-to-f} = H_{Tgf}(T_g - T_f), \qquad Nu = 0.03Re_{inj}^a = \frac{H_{Tgf}h}{\lambda_g} \tag{5.180}$$

$$a = 0.82 - \frac{0.32\left[1 - 1.95\left(\dfrac{x}{D_{noz}}\right)^{1.8} + 2.23\left(\dfrac{x}{D_{noz}}\right)^2\right]^{-1}}{\left[1 - 0.21\left(\dfrac{z}{D_{noz}}\right)^{1.25} + 0.21\left(\dfrac{z}{D_{noz}}\right)^{1.5}\right]} \tag{5.181}$$

$$Re_{inj} = \frac{\rho_g U_{inj}}{\mu_g}D_{noz}\sqrt{\frac{\rho_l}{\rho_g}} \tag{5.182}$$

For heat exchanges due to film vaporization:

$$\dot{Q}_{vap} = \dot{M}_{vap} \cdot L(T_f), \qquad \dot{M}_{vap} = \rho H_Y ln\left(\frac{1-Y_\infty}{1-Y_s}\right), \qquad \frac{H_Y}{H_{Tgf}} = \frac{1}{\rho c_p}\left(\frac{Pr}{Sc}\right)^{2/3} \tag{5.183}$$

For heat exchanges due to impinging droplets on film:

$$\dot{Q}_{imp} = H_{T_{imp}}(T_d - T_f)\frac{\pi}{4}\sqrt{\frac{\rho_l d^3}{\sigma_l}}, \qquad Nu = 1.4 Re^{1/2} = \frac{H_{T_{imp}} d}{\lambda_l} \qquad (5.184)$$

In Equations (5.179) to (5.184), *We, Re,* and *Pr* are the Weber, Reynolds, and Prandtl numbers, respectively, of the impingement droplet calculated based on its axial velocity, *Nu* is the average value for the impinging droplets, *h* is the film thickness, λ_l and λ_g are the liquid and gas thermal conductivity, respectively, Re_{inj} is the Reynolds number of the injection calculated based on injection velocity and nozzle distance to the impact wall, U_{inj} is the liquid injection velocity, D_{noz} is the nozzle diameter, H_Y is the mass transfer coefficient, Y_∞ is the mass fraction above the film, and Y_s is the equilibrium vapor mass fraction at film temperature.

Assessment of the model with several sets of experimental measurements of diesel sprays impinging on a wall impingement under engine-like conditions is also reported by Deng et al. [21]. In general, satisfactory reproductions of the measured data have been achieved. However, application of these measurement-based correlations should be within the ranges of the test conditions under which the experiments were performed.

An example case is cited in which the measured heat flux data of Senda et al. [124] were used to compare the computed one. A transient diesel spray was impinged onto a cool plate (343 K) in a vessel at several different pressure/temperature combinations and different impingement distances. The ambient air temperature was varied from 500 to 1000 K with air pressures of 1.5–3.0 MPa. The heat flux at the impingement-point and other various radial locations was measured. The comparisons of the measured and computed heat flux at two locations (center of the impingement and 2.5 mm away from the center) and under two different ambient air temperatures (500 and 1000 K) are shown in **Figure 5.30**.

FIGURE 5.30 Heat flux of a diesel spray impingement.

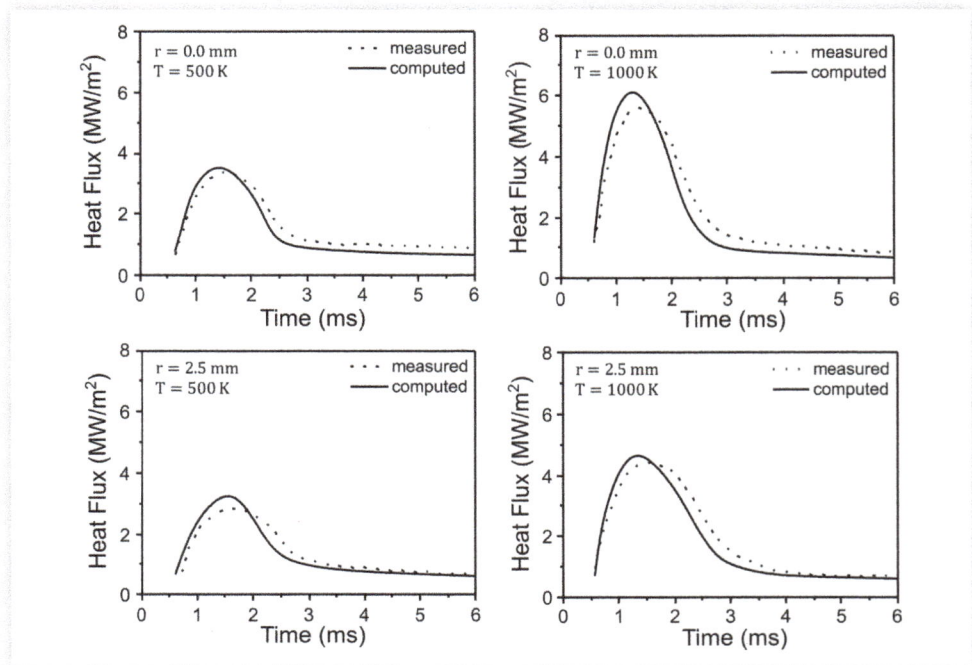

Good agreements are seen between the measured data and the computed ones. Note that a higher ambient temperature leads to a higher heat flux through the wall due to the higher film temperature. On the other hand, large spatial variations are seen in the impingement region, which indicate the complexity of the impingement heat transfer, and suggest that the computational mesh sizes should be fine enough to resolve these variations.

References

1. Lefebvre, A.H., and McDonell, V.G. 2017. *Atomization and Sprays*. 2nd ed. Boca Raton, Florida: CRC press.

2. Reitz, R.D., and Bracco, F.V. 1982. "Mechanism of Atomization of a Liquid Jet." *The Physics of Fluids* 25, no. 10: 1730–1742. DOI: 10.1063/1.863650

3. Chigier, N., and Reitz, R.D. 1996. "Regimes of Jet Breakup And Breakup Mechanisms (Physical Aspects)." *Progress in Astronautics and Aeronautics* 166: 109–134.

4. Cai, W., Powell, C.F., Yue, Y., Narayanan, S., Wang, J., Tate, M.W., Renzi, M.J., Ercan, A., Fontes, E., and Gruner, S.M. 2003. "Quantitative Analysis of Highly Transient Fuel Sprays by Time-Resolved X-Radiography." *Applied Physics Letters* 83, no. 8: 1671–1673. DOI: 10.1063/1.1604161

5. Liu, X., Im, K.-S., Wang, Y., Wang, J., Tate, M.W., Ercan, A., Schuette, D.R., and Gruner, S.M. 2009. "Four Dimensional Visualization of Highly Transient Fuel Sprays by Microsecond Quantitative X-Ray Tomography." *Applied Physics Letters* 94, no. 8; article 084101. DOI: 10.1063/1.3048563

6. Heindel, T. 2018. "X-Ray Imaging Techniques to Quantify Spray Characteristics in the Near Field." *Atomization and Sprays* 28, no. 11: 1029–1059. DOI: 10.1615/AtomizSpr.2019028797

7. Hiroyasu, H. 1985. "Diesel Engine Combustion and Its Modeling." In *The Proceedings of the International Symposium on Diagnostics And Modeling of Combustion in Internal Combustion Engines (Comodia), September 4–6, 1985, Tokyo, Japan*, 53–75.

8. Hiroyasu, H., and Kadota, T. 1974. "Fuel Droplet Size Distribution in Diesel Combustion Chamber." *SAE Transactions* 83, no. 3: 2615–2624. DOI: 10.4271/740715

9. Reitz, R.D. 1978. Atomization And Other Breakup Regimes Of A Liquid Jet (Publication No. 7907964). Princeton, New Jersey: Princeton University. ProQuest Dissertations and Theses.

10. Reitz, R.D. 1987. "Modeling Atomization Processes in High-Pressure Vaporizing Sprays." *Atomisation Spray Technology* 3, no. 4: 309–337.

11. Beale, J.C., and Reitz, R.D. 1999. "Modeling Spray Atomization With the Kelvin-Helmholtz/Rayleigh-Taylor Hybrid Model." *Atomization and Sprays* 9, no. 6: 623–650. DOI: 10.1615/AtomizSpr.v9.i6.40

12. Dukowicz, J.K. 1980. "A Particle-Fluid Numerical Model for Liquid Sprays." *Journal of Computational Physics* 35, no. 2 (April): 229–253. DOI: 10.1016/0021-9991(80)90087-X

13. O'Rourke, P.J. 1981. Collective Drop Effects on Vaporizing Liquid Sprays (Publication No. 8203229). Princeton, New Jersey: Princeton University. ProQuest Dissertations and Theses.

14. Amsden, A.A., O'Rourke, P.J., and Butler, T.D. 1989. *KIVA-II: A Computer Program For Chemically Reactive Flows With Sprays (LA-11560-MS)*. Los Alamos, New Mexico: Los Alamos National Laboratory.

15. Naber, J.D., and Reitz, R.D. 1988. "Modeling Engine Spray/Wall Impingement." SAE Technical Paper No. 880107. DOI: 10.4271/880107

16. Han, Z., Parrish, S.E., Farrell, P.V., and Reitz, R.D. 1997. "Modeling Atomization Processes of Pressure-Swirl Hollow-Cone Fuel Sprays." *Atomization and Sprays* 7, no. 6: 663–684. DOI: 10.1615/AtomizSpr.v7.i6.70

17. Schmidt, D.P., Nouar, I., Senecal, P., Rutland, J., Martin, J., Reitz, R.D., and Hoffman, J.A. 1999. "Pressure-Swirl Atomization in the Near Field." *SAE Transactions* 108, no. 3 (March): 471–484. DOI: 10.4271/1999-01-0496

18. Lippert, A.M., and Reitz, R.D. 1990. "Modeling of Multicomponent Fuels Using Continuous Distributions With Application to Droplet Evaporation And Sprays." SAE Technical Paper No. 972882. DOI: 10.4271/972882

19. Ra, Y., and Reitz, R.D. 2009. "A Vaporization Model for Discrete Multi-Component Fuel Sprays." *International Journal of Multiphase Flow* 35, no. 2 (February): 101–117. DOI: 10.1016/j.ijmultiphaseflow.2008.10.006

20. Eckhause, J.E., and Reitz, R.D. 1995. "Modeling Heat Transfer to Impinging Fuel Sprays in Direct-Injection Engines." *Atomization and Sprays* 5, no. 2: 213–242. DOI: 10.1615/AtomizSpr.v5.i2.60

21. Deng, P., Han, Z., and Reitz, R.D. 2016. "Modeling Heat Transfer in Spray Impingement Under Direct-Injection Engine Conditions." *Proceedings of the Institution of Mechanical Engineers, Part D: Journal of Automobile Engineering* 230, no. 7: 885–898. DOI: 10.1177/0954407015596284

22. O'Rourke, P.J., and Amsden, A.A. 1987. "The TAB Method for Numerical Calculation of Spray Droplet Breakup." SAE Technical Paper No. 872089. DOI: 10.4271/872089

23. Amsden, A.A., Ramshaw, J.D., O'Rourke, P.J., and Dukowicz, J.K. 1985. *KIVA: A Computer Program for Two- And Three-Dimensional Fluid Flows With Chemical Reactions And Fuel Sprays* (LA-10245-MS). Los Alamos, New Mexico: Los Alamos National Laboratory.

24. Amsden, A.A. 1993. *KIVA-3: A KIVA Program With Block-Structured Mesh for Complex Geometries* (LA-12503-MS). Los Alamos, New Mexico: Los Alamos National Laboratory.

25. Amsden, A.A. 1997. *KIVA-3V: A Block-Structured KIVA Program for Engines With Vertical or Canted Valves* (LA-13313-MS). Los Alamos, New Mexico: Los Alamos National Laboratory.

26. Rosin, P., and Rammler, E. 1933. "Laws Governing the Fineness of Powdered Coal." *Journal of Institute of Fuel* 7: 29–36.

27. Nukiyama, S., and Tanasawa, Y. 1939. "Experiments on the Atomization Of Liquids in an Air Stream, Report 3, on the Droplet-Size Distribution in a Atomized Jet." *Transactions of JSME* 5: 62–67.

28. O'Rourke, P.J., and Amsden, A.A. 2000. "A Spray/Wall Interaction Submodel for the KIVA-3 Wall Film Model." *SAE Transactions* 109, no. 3 (March): 281–298. DOI: 10.4271/2000-01-0271

29. Han, Z., Xu, Z., and Trigui, N. 2000. "Spray/Wall Interaction Models for Multidimensional Engine Simulation." *International Journal of Engine Research* 1, no. 1: 127–146. DOI: 10.1243/1468087001545308

30. Reitz, R.D., and Diwakar, R. 1987. "Structure of High-Pressure Fuel Sprays." *SAE Transactions* 96, no. 5 (February): 492–509. DOI: 10.4271/870598

31. Yi, Y., and Reitz, R.D. 2004. "Modeling the Primary Breakup of High-Speed Jets." *Atomization and Sprays* 14, no. 1: 53–80. DOI: 10.1615/AtomizSpr.v14.i1.40

32. Shinjo, J. 2018. "Recent Advances in Computational Modeling of Primary Atomization of Liquid Fuel Sprays." *Energies* 11, no. 11: 2971. DOI: 10.3390/en11112971

33. Kong, S.C., Han, Z., and Reitz, R.D. 1995. "The Development And Application of a Diesel Ignition And Combustion Model for Multidimensional Engine Simulation." *SAE Transactions* 104, no. 3 (February): 502–518. DOI: 10.4271/950278

34. Reitz, R.D., and Bracco, F.B. 1979. "On the Dependence of Spray Angle And Other Spray Parameters on Nozzle Design And Operating Conditions." SAE Technical Paper No. 790494. DOI: 10.4271/790494

35. Chaves, H., Knapp, M., Kubitzek, A., Obermeier, F., and Schneider, T. 1995. "Experimental Study of Cavitation in the Nozzle Hole of Diesel Injectors Using Transparent Nozzles." SAE Technical Paper No. 950290. DOI: 10.4271/950290

36. Han, Z., Uludogan, A., Hampson, G.J., and Reitz, R.D. 1996. "Mechanism of Soot And No$_x$ Emission Reduction Using Multiple-Injection in a Diesel Engine." *SAE Transactions* 105, no. 3 (February), 837–852. DOI: 10.4271/960633

37. Sarre, C.V.K., Kong, S.C., and Reitz, R.D. 1999. "Modeling the Effects of Injector Nozzle Geometry on Diesel Sprays." *SAE Transactions* 108, no. 3 (January): 1375–1388. DOI: 10.4271/1999-01-0912

38. Zhao, F., Lai, M.C., and Harrington, D.L. 1999. "Automotive Spark-Ignited Direct-Injection Gasoline Engines." *Progress in Energy and Combustion Science* 25, no. 5: 437–562. DOI: 10.1016/S0360-1285(99)00004-0

39. VanDerWege, B.A., Han, Z., Iyer, C.O., Muñoz, R.H., and Yi, J. 2003. "Development And Analysis of A Spray-Guided DISI Combustion System Concept." *SAE Transactions* 112, no. 4 (October): 2135–2153. DOI: 10.4271/2003-01-3105

40. Han, Z., Weaver, C., Wooldridge, S., Alger, T., Hilditch, J., McGee, J., Westrate, B., et al. 2004. "Development of a New Light Stratified-Charge DISI Combustion System for a Family Of Engines With Upfront CFD Coupling With Thermal And Optical Engine Experiments." *SAE Transactions* 113, no. 3 (March): 269–293. DOI: 10.4271/2004-01-0545

41. Reitz, R.D., and Diwakar, R. 1986. "Effect of Drop Breakup on Fuel Sprays." *SAE Transactions* 95, no. 3 (February): 218–227. DOI: 10.4271/860469

42. Clark, C., and Dombrowski, N. 1972. "Aerodynamic Instability And Disintegration of Inviscid Liquid Sheets." *Proceedings of the Royal Society of London. A. Mathematical Physical Sciences* 329, no. 1579 (September): 467–478. DOI: 10.1098/rspa.1972.0124

43. Miyamoto, T., Kobayashi, T., and Matsumoto, Y. 1996. "Structure of Sprays From an Air-Assist Hollow-Cone Injector." *SAE Transactions* 105, no. 3 (February): 1058–1070. DOI: 10.4271/960771

44. Lee, C.F., and Bracco, F.V. 1995. "Comparisons of Computed And Measured Hollow-Cone Sprays in an Engine." *SAE Transactions* 104, no. 3 (February): 569–594. DOI: 10.4271/950284

45. Han, Z., Xu, Z., Wooldridge, S.T., Yi, J., and Lavoie, G. 2001. "Modeling of DISI Engine Sprays With Comparison to Experimental In-Cylinder Spray Images." *SAE Transactions* 110, no. 3 (September): 2376–2386. DOI: 10.4271/2001-01-3667

46. Rizk, N., and Lefebvre, A.H. 1985. "Internal Flow Characteristics of Simplex Swirl Atomizers." *Journal of Propulsion and Power* 1, no. 3: 193–199. DOI: 10.2514/3.22780

47. Squire, H.B. 1953. "Investigation of the Instability of a Moving Liquid Film." *British Journal of Applied Physics* 4, no. 6: 167–169. DOI: 10.1088/0508-3443/4/6/302

48. Parrish, S.E., and Farrell, P.V. 1997. Transient Spray Characteristics of a Direct-Injection Spark-Ignited Fuel Injector. SAE Technical Paper No. 970629. DOI: 10.4271/970629

49. Senecal, P., Schmidt, D.P., Nouar, I., Rutland, C.J., Reitz, R.D., and Corradini, M. 1999. "Modeling High-Speed Viscous Liquid Sheet Atomization." *International Journal of Multiphase Flow* 25, no. 6–7 (September–November): 1073–1097. DOI: 10.1016/S0301-9322(99)00057-9

50. Dombrowski, N., and Johns, W.R. 1963. "The Aerodynamic Instability And Disintegration of Viscous Liquid Sheets." *Chemical Engineering Science* 18, no. 3 (March): 203–214. DOI: 10.1016/0009-2509(63)85005-8

51. Dombrowski, N., and Hooper, P.C. 1962. "The Effect of Ambient Density on Drop Formation in Sprays." *Chemical Engineering Science* 17, no. 4 (April): 291–305. DOI: 10.1016/0009-2509(62)85008-8

52. Kennedy, J., and Roberts, J. 1990. "Rain Ingestion in a Gas Turbine Engine." In *Proceedings of 4th ILASS Americas Meeting, May 21–23, Hartford, CT,* pp. 154.

53. Wu, P., Hsiang, L., and Faeth, G. 1993. "Aerodynamic Effects on Primary And Secondary Spray Breakup." Paper presented at the *First International Symposium on Liquid Rocket Combustion Instability, Pennsylvania State University, University Park, PA.*

54. Reitz, R.D. 1996. "Computer Modeling of Sprays." Spray Technology Short Course Note. Pittsburgh, PA.

55. Yarin, A.L., Roisman, I.V., and Tropea, C. 2017. *Collision Phenomena in Liquids And Solids.* Cambridge: Cambridge University Press.

56. Ashgriz, N., and Poo, J. 1990. "Coalescence and separation in binary collisions of liquid drops." *Journal of Fluid Mechanics* 221 (April): 183–204. DOI: 10.1017/S0022112090003536

57. Jiang, Y., Umemura, A., and Law, C. 1992. "An Experimental Investigation on the Collision Behaviour of Hydrocarbon Droplets." *Journal of Fluid Mechanics* 234 (April): 171–190. DOI: 10.1017/S0022112092000740

58. Orme, M. 1997. "Experiments on Droplet Collisions, Bounce, Coalescence And Disruption." *Progress in Energy and Combustion Science* 23, no. 1: 65–79. DOI: 10.1016/S0360-1285(97)00005-1

59. Qian, J., and Law, C.K. 1997. "Regimes of Coalescence And Separation In Droplet Collision." *Journal of Fluid Mechanics* 331 (May): 59–80. DOI: 10.1017/S0022112096003722

60. Brenn, G., Valkovska, D., and Danov, K.D. 2001. "The Formation of Satellite Droplets by Unstable Binary Drop Collisions." *Physics of Fluids* 13, no. 9: 2463–2477. DOI: 10.1063/1.1384892

61. Gotaas, C., Havelka, P., Jakobsen, H.A., Svendsen, H.F., Hase, M., Roth, N., and Weigand, B. 2007. "Effect of Viscosity on Droplet-Droplet Collision Outcome: Experimental Study And Numerical Comparison." *Physics of Fluids* 19, no. 10: article 102106. DOI: 10.1063/1.2781603

62. Schmidt, D.P., and Rutland, C. 2000. "A New Droplet Collision Algorithm." *Journal of Computational Physics* 164, no. 1 (October): 62–80. DOI: 10.1006/jcph.2000.6568

63. Munnannur, A., and Reitz, R.D. 2009. "A Comprehensive Collision Model for Multi-Dimensional Engine Spray Computations." *Atomization and Sprays* 19, no.7: 597–619. DOI: 10.1615/AtomizSpr.v19.i7.10

64. Post, S.L., and Abraham, J. 2002. "Modeling the Outcome of Drop–Drop Collisions in Diesel Sprays." *International Journal of Multiphase Flow* 28, no. 6 (June): 997–1019. DOI: 10.1016/S0301-9322(02)00007-1

65. Munnannur, A., and Reitz, R.D. 2007. "A New Predictive Model for Fragmenting And Non-Fragmenting Binary Droplet Collisions." *International Journal of Multiphase Flow* 33, no. 8 (August): 873–896. DOI: 10.1016/j.ijmultiphaseflow.2007.03.003

66. Liu, A.B., Mather, D., and Reitz, R.D. 1993. "Modeling the Effects of Drop Drag And Breakup on Fuel Sprays." *SAE Transactions* 102, no. 3 (March): 83–95. DOI: 10.4271/930072

67. Eisenklam, P., Arunachalam, S.A., and Weston, J.A. 1967. "Evaporation Rates And Drag Resistance of Burning Drops." *Symposium (International) on Combustion* 11, no. 1: 715–728. DOI: 10.1016/S0082-0784(67)80197-8

68. Yuen, M.C., and Chen, L.W. 1976. "On Drag of Evaporating Liquid Droplets." *Combustion Science and Technology* 14, no. 4–6: 147–154. DOI: 10.1080/00102207608547524

69. Renksizbulut, M., and Yuen, M.C. 1983. "Numerical Study of Droplet Evaporation in a High-Temperature Stream." *Journal of Heat Transfer* 1052 (May): 389–397. DOI: 10.1115/1.3245591

70. Lázaro, B.J., and Lasheras, J.C. 1989. "Particle Dispersion in a Turbulent, Plane, Free Shear Layer." *Physics of Fluids A: Fluid Dynamics* 1, no. 6: 1035–1044. DOI: 10.1063/1.857394

71. Faeth, G.M. 1983. "Evaporation And Combustion of Sprays." *Progress in Energy and Combustion Science* 9, no. 1: 1–76. DOI: 10.1016/0360-1285(83)90005-9

72. Xu, Z., Yi, J., Curtis, E., and Wooldridge, S. 2009. "Applications of CFD Modeling in GDI Engine Piston Optimization." *SAE International Journal of Engines* 2, no. 1 (June): 1749–1763. DOI: 10.4271/2009-01-1936

73. Spalding, D.B. 1953. "The Combustion of Liquid Fuels." *Symposium (International) on Combustion* 4, no. 1: 847–864. DOI: 10.1016/S0082-0784(53)80110-4

74. Han, Z., Reitz, R.D., Claybaker, P.J., Rutland, C.J., Yang, J., and Anderson, R.W. 1996. "Modeling the Effects of Intake Flow Structures on Fuel/Air Mixing in a Direct-Injected Spark-Ignition Engine." *SAE Transactions* 105, no. 4 (May): 960–977. DOI: 10.4271/961192

75. Ranz, W., and Marshall, W.R. 1952. "Evaporation From Drops, Parts I & II." *Chemical Engineering Progress* 48: 141–146.

76. Lefebvre, A.H. 1989. *Atomization and Sprays.* Bristol: Taylor & Francis.

77. Hubbard, G., Denny, V., and Mills, A. 1975. "Droplet Evaporation: Effects of Transients And Variable Properties." *International Journal of Heat Mass Transfer* 18, no. 9 (September): 1003–1008. DOI: 10.1016/0017-9310(75)90217-3

78. Vargraftik, N. 1975. *Tables on the Thermophysical Properties of Liquids And Gases in Normal And Dissociated States.* New York: Hemisphere Publishing Corp.

79. Curtis, E.W. 1991. A Numerical Study Of Spherical Droplet Vaporization in a High Pressure Environment (Publication No. 9205538). Madison, Wisconsim: University of Wisconsin-Madison. ProQuest Dissertations and Theses.

80. Ra, Y., and Reitz, R.D. 2004. "A Model for Droplet Vaporization for Use in Gasoline And HCCI Engine Applications." *Journal of Engineering for Gas Turbines and Power* 126, no. 2 (April): 422–428. DOI: 10.1115/1.1688367

81. Zhu, G.S., and Reitz, R.D. 2002. "A Model for High Pressure Vaporization of Droplets of Complex Liquid Mixtures Using Continuous Thermodynamics." *International Journal of Heat and Mass Transfer* 45 (January): 495–507. DOI: 10.1016/S0017-9310(01)00173-9

82. Zeng, Y., and Lee, C.F. 2002. "A preferential vaporization model for multicomponent droplets and sprays." *Atomization and Sprays* 12, no. 1–3: 163–186. DOI: 10.1615/AtomizSpr.v12.i123.90

83. Zeng, Y., Han, Z. 2001. *Implementation Of Multicomponent Droplet And Film Vaporization Models Into The KIVA-3V Code* (SRR-2001-0165). Dearborn, Michigan: Ford Research Laboratory.

84. Curtis, E.W., Aquino, C.F., Trumpy, D.K., and Davis, G.C. 1996. "A New Port And Cylinder Wall Wetting Model To Predict Transient Air/Fuel Excursions in a Port Fuel Injected Engine." SAE Technical Paper No. 961186. DOI: 10.4271/961186

85. Heywood, J.B. 2018. *Internal Combustion Engine Fundamentals*. 2nd ed. New York: McGraw-Hill Education.

86. Han, Z., Yi, J., and Trigui, N. 2002. "Stratified Mixture Formation And Piston Surface Wetting in a DISI Engine." SAE Technical Paper No. 2002-01-2655. DOI: 10.4271/2002-01-2655

87. Hilditch, J., Han, Z., and Chea, T. 2003. "Unburned Hydrocarbon Emissions From Stratified Charge Direct Injection Engines." SAE Technical Paper No. 2003-01-3099. DOI: 10.4271/2003-01-3099

88. Fansler, T.D., Trujillo, M.F., and Curtis, E.W. 2020. "Spray-Wall Interactions in Direct-Injection Engines: An Introductory Overview." *International Journal of Engine Research* 21, no. 2: 241–247. DOI: 10.1177/1468087419897994

89. Nagaoka, M., Kawazoe, H., and Nomura, N. 1994. "Modeling Fuel Spray Impingement on a Hot Wall for Gasoline Engines." *SAE Transactions* 103, no. 3 (March): 878–896. DOI: 10.4271/940525

90. Bai, C., and Gosman, A.D. (1995). "Development of Methodology for Spray Impingement Simulation." *SAE Transactions* 104, no. 3 (February): 550–568. DOI: 10.4271/950283

91. Bai, C., and Gosman, A.D. 1996. "Mathematical Modelling of Wall Films Formed by Impinging Sprays." *SAE Transactions* 105, no. 3 (February): 782–796. DOI: 10.4271/960626

92. Stanton, D.W., and Rutland, C.J. 1996. "Modeling Fuel Film Formation And Wall Interaction in Diesel Engines." SAE Technical Paper No. 960628. DOI: 10.4271/960628

93. Foucart, H., Habchi, C., Le Coz, J.F., and Baritaud, T. 1998. "Development of a Three Dimensional Model of Wall Fuel Liquid Film for Internal Combustion Engines." SAE Technical Paper No. 980133. DOI: 10.4271/980133

94. O'Rourke, P.J., and Amsden, A.A. 1996. "A Particle Numerical Model for Wall Film Dynamics in Port-Injected Engines." *SAE Transactions* 105, no. 3 (October): 2000–2013. DOI: 10.4271/961961

95. Amsden, A.A. 1999. *KIVA-3V, release 2: Improvements to KIVA-3V* (LA-13608-MS). Los Alamos, New Mexico: Los Alamos National Laboratory.

96. Trujillo, M., Mathews, W., Lee, C.F., and Peters, J. 1998. "A Computational And Experimental Investigation of Spray/Wall Impingement." In *Proceedings of the 11th Annual Conference on Liquid Atomization and Spray System, ILASS-Americas, May 17–20, Sacramento, CA*, 111–114.

97. Meyer, R., and Heywood, J.B. 1999. "Effect of Engine And Fuel Variables on Liquid Fuel Transport Into the Cylinder in Port-Injected SI Engines." SAE Technical Paper No. 1999-01-0563. DOI: 10.4271/1999-01-0563

98. Witze, P.O. 1999. "Diagnostics for The Study of Cold Start Mixture Preparation in a Port Fuel-Injected Engine." SAE Technical Paper No. 0148-7191. DOI: 10.4271/1999-01-1108

99. Salters, D., Williams, P., Greig, A., and Brehob, D. 1996. "Fuel Spray Characterisation Within an Optically Accessed Gasoline Direct Injection Engine Using a CCD Imaging System." *SAE Technical Paper No. 0148-7191.* DOI: 10.4271/961149

100. Temple-Pediani, R. 1969. "Fuel Drop Vaporization Under Pressure on a Hot Surface." *Proceedings of the Institution of Mechanical Engineers* 184, no. 1: 677–696. DOI: 10.1243/PIME_PROC_1969_184_053_02

101. Xiong, T., and Yuen, M. 1991. "Evaporation of a Liquid Droplet on a Hot Plate." *International Journal of Heat Mass Transfer* 34, no. 7 (July): 1881–1894. DOI: 10.1016/0017-9310(91)90162-8

102. Abu-Zaid, M. 1994. "An Experimental Study of the Evaporation of Gasoline And Diesel Droplets on Hot Surfaces." *International Communications in Heat and Mass Transfer* 21, no. 2 (March–April): 315–322. DOI: 10.1016/0735-1933(94)90029-9

103. Tamura, Z., and Tanasawa, Y. 1958. "Evaporation And Combustion of a Drop Contacting With a Hot Surface." *Symposium (International) on Combustion* 7, no. 1: 509–522. DOI: 10.1016/S0082-0784(58)80086-7

104. Mundo, C., Sommerfeld, M., and Tropea, C. 1995. "Droplet-Wall Collisions: Experimental Studies of the Deformation And Breakup Process." *International Journal of Multiphase Flow* 21, no. 2 (April): 151–173. DOI: 10.1016/0301-9322(94)00069-V

105. Stow, C.D., and Hadfield, M.G. 1981. "An Experimental Investigation of Fluid Flow Resulting From the Impact of a Water Drop With an Unyielding Dry Surface." *Proceedings of the Royal Society of London. A. Mathematical Physical Sciences* 373, no. 1755 (January): 419–441. DOI: 10.1098/rspa.1981.0002

106. Mundo, C., Sommerfeld, M., and Tropea, C. 1998. "On the Modeling of Liquid Sprays Impinging on Surfaces." *Atomization and Sprays* 8, no. 6: 625–652. DOI: 10.1615/AtomizSpr.v8.i6.20

107. Yarin, A.L., and Weiss, D.A. 1995. "Impact of Drops on Solid Surfaces: Self-Similar Capillary Waves, And Splashing as a New Type Of Kinematic Discontinuity." *Journal of Fluid Mechanics* 283 (April): 141–173. DOI: 10.1017/S0022112095002266

108. Cossali, G.E., Coghe, A., and Marengo, M. 1997. "The Impact of a Single Drop on a Wetted Solid Surface." *Experiments in Fluids* 22, no. 6 (April): 463–472. DOI: 10.1007/s003480050073

109. Moreira, A.L.N., Moita, A.S., and Panão, M.R. 2010. "Advances And Challenges in Explaining Fuel Spray Impingement: How Much of Single Droplet Impact Research is Useful?" *Progress in Energy And Combustion Science* 36, no. 5 (October), 554–580. DOI: 10.1016/j.pecs.2010.01.002

110. Levin, Z., and Hobbs, P.V. 1971. "Splashing of Water Drops on Solid And Wetted Surfaces: Hydrodynamics And Charge Separation." *Philosophical Transactions of the Royal Society of London. Series A, Mathematical And Physical Sciences* 269, no. 1200 (May): 555–585. DOI: 10.1098/rsta.1971.0052

111. Li, X., and Tankin, R. 1991. "On the Temporal Instability of a Two-Dimensional Viscous Liquid Sheet." *Journal of Fluid Mechanics* 226: 425–443. DOI: 10.1017/S0022112091002458

112. Ibrahim, E., and Przekwas, A. 1991. "Impinging Jets Atomization." *Physics of Fluids A: Fluid Dynamics* 3, no. 12: 2981–2987. DOI: 10.1063/1.857840

113. Weiss, D.A., and Yarin, A.L. 1999. "Single Drop Impact Onto Liquid Films: Neck Distortion, Jetting, Tiny Bubble Entrainment, And Crown Formation." *Journal of Fluid Mechanics* 385 (April): 229–254. DOI: 10.1017/S002211209800411X

114. Sivakumar, D., and Tropea, C. 2002. "Splashing Impact of a Spray Onto a Liquid Film." *Physics of Fluids* 14, no. 12: L85–L88. DOI: 10.1063/1.1521418

115. Han, Z., and Xu, Z. 2004. "Wall Film Dynamics Modeling for Impinging Sprays in Engines." *SAE Technical Paper No. 2004-01-0099.* DOI: 10.4271/2004-01-0099

116. Deng, P., Jiao, Q., Reitz, R.D., and Han, Z. 2015. "Development of an Improved Spray/Wall Interaction Model for Diesel-Like Spray Impingement Simulations." *Atomization and Sprays* 25, no. 7: 587–615. DOI: 10.1615/AtomizSpr.2015011000

117. Mathews, W., Lee, C.-F., & Peters, J.E. 2003. "Experimental Investigations of Spray/Wall Impingement." *Atomization and Sprays* 13, no. 2–3: 223–242. DOI: 10.1615/AtomizSpr.v13.i23.40

118. Katsura, N., Saito, M., Senda, J., and Fujimoto, H. 1989. "Characteristics of a Diesel Spray Impinging an a Flat Wall." SAE Technical Paper No. 890264.

119. Saffman, P.G.T. 1965. "The Lift on a Small Sphere in a Slow Shear Flow." *Journal of Fluid Mechanics* 22, no. 2 (March): 385–400. DOI: 10.1017/S0022112065000824

120. Saffman, P.G.T. 1968. "Corrigendum to the Lift on a Small Sphere in a Slow Shear Flow." *Journal of Fluid Mechanics* 31, no. 3: 624.

121. Mei, R. 1992. "An Approximate Expression for the Shear Lift Force on a Spherical Particle at Finite Reynolds Number." *International Journal of Multiphase Flow* 18, no. 1 (January): 145–147. DOI: 10.1016/0301-9322(92)90012-6

122. Wolf, R.S., and Cheng, W.K. 1989. "Heat Transfer Characteristics of Impinging Diesel Sprays." SAE Technical Paper No. 890439. DOI: 10.4271/890439

123. Naber, J.D., and Farrell, P.V. 1993. "Hydrodynamics of Droplet Impingement on a Heated Surface." *SAE Transactions* 102, no. 3 (March): 1346–1361. DOI: 10.4271/930919

124. Senda, J., Fukimoto, H., and Yamamoto, K. 1991. "Heat Flux Between Impinged Diesel Spray And Flat Wall." SAE Technical Paper No. 912460.

6

Combustion and Pollutant Emissions

6.1 Overview

Chemical reactions of the fuel and air mixture occur during the combustion processes in IC engines. The desired result of combustion is the released chemical energy (heat), which is converted to mechanical work through the piston motion in a reciprocating engine. There are also undesired products from combustion including harmful pollutants such as PM, NO_x, CO, and unburned hydrocarbons. In an SI engine, fuel and air are mixed during the induction and compression stroke to usually form a homogeneous mixture before the electrical spark ignites the mixture. The deposition of electrical energy from the spark ionizes the gas and heats it to several thousand degrees Kelvin. At temperatures above 1000 K, chemical reactions are initiated. These generate a flame kernel that grows at first by laminar, then by turbulent flame propagation. In a diesel engine, several liquid fuel sprays are injected into the combustion chamber with hot compressed air and the fuel evaporates and mixes partially with the air before autoignition occurs. The autoignition process happens quite randomly and independently at some locations near the peripheries of the sprays. These ignition kernels then initiate rapid combustion of the premixed fraction of the mixture followed by diffusive combustion under partially premixed and/or non-premixed conditions.

Accurate modeling of the combustion process and pollutant formations is an important goal of engine simulation. Models of combustion and emissions will be discussed in this chapter. In the governing equations of reactive flows in Chapter 3, combustion models are needed to represent the chemical source terms $\dot{\rho}^c_k$ and $\dot{Q}^c$. Various models have been developed for SI gasoline engines, compression ignition diesel engines, and LTC engine concepts to simulate a variety of engine combustion phenomena including SI, fuel autoignition, laminar and turbulent combustion, premixed, non-premixed, and partially premixed combustion, kinetics-controlled and turbulence-mixing-controlled combustion, and flame propagation combustion. Reitz and Rutland [1,2], Haworth [3], and Peters [4] have given good summaries on engine combustion models.

The simplest combustion model, at the early stage of engine multidimensional simulation, is the mean-reaction-rate model that adopts highly simplified chemical kinetic schemes, one or at most a few global or quasi-global kinetic reaction equation(s) for conversion of reactants to products based on a temperature-dependent Arrhenius approximation [5,6]. These models essentially neglect the effects of turbulent fluctuations in species concentrations and temperature on the mean chemical production rates and generally exhibit a far strong dependence of overall burn rate on chemical kinetics. The reaction rate given by these models is usually very high, that is, energy is released in a non-realistically short time period. However, the Arrhenius parameters can be viewed as adjustable model coefficients that are tuned to yield good agreement with experimental measurements, and this type of model can be used in a narrow range of operating conditions for parametric studies [7].

To reflect the physics of turbulent combustion in engine modeling, turbulence mixing-controlled flame propagation models are developed. A simpler, but more computationally efficient, approach is the laminar and turbulent CTC model [8]. This model follows the ideas of the eddy breakup (EBU) model [9] and eddy dissipation model (EDM) [10] but includes the effects of chemical kinetics when they become important. The EBU model assumes that the burn rate of the fuel-oxidizer mixture is determined by the turbulent mixing rate, instead of by the chemical rate. The fuel and oxidizer are contained in two separated eddies, the rate of the reaction is comparable to the rate of the dissipation of the turbulent eddies (EBU time). The mean chemical conversion rate is then taken to be inversely proportional to a turbulence eddy turnover time $\tau = k/\varepsilon$ in the case of a k-ε turbulence model.

The EBU model was formulated primarily for premixed combustion. The main idea is to replace the chemical timescale of an assumed one-step reaction with the turbulent timescale. The model eliminates the influence of chemical kinetics. In case chemical kinetics is also important, the CTC models allow either chemical kinetics or turbulence to be rate controlling. The CTC models have been formulated for SI combustion [11] and for diesel combustion [12], which have been widely used in engine simulations. The CTC model is discussed in detail in Sec. 6.2.

Flamelet models are another group of widely used methods to simulate engine combustion. These models track turbulent flame propagations with transportation equations of representative progress variable(s). For premixed combustion such as that in HCSI engines, the Bray–Moss–Libby (BML) model [13,14] takes a normalized temperature or a normalized product mass fraction as the progress variable c; and the coherent flame model (CFM) [15–17] extends the progress variable c to flame surface density Σ and uses it as the progress variable. Parallelly, the non-reacting scalar G is introduced in the level set method [18] as the progress variable. Thus, G-equation models are proposed to simulate SI engine combustion [19–21]. Flamelet models are discussed in detail in Sec. 6.3.

For non-premixed (or partially premixed) combustion, which occurs in compression ignition diesel engines and SCSI engines, the mixture fraction Z is used to play a role similar to the scalar G in premixed combustion in determining the flame surface [22]. Based on this, a representative interactive flamelet (RIF) model has been proposed by Pitsch et al. [23,24]. The model assumes that there exists a thin flame sheet, such that turbulent time and length scales are large compared to the chemical ones; the reaction zone cannot be intruded by turbulent eddies. The flame sheet can only be stretched by the turbulent movement and the chemical structure of the flame remains, since chemical reactions are fast enough to compensate for disturbances. Combustion essentially takes place in the vicinity of the surface of the stoichiometric mixture. Thus, the fluid dynamics and the chemical processes can be numerically separated (but still be coupled via the mixture fraction equations), and a model describing non-premixed turbulent combustion using detailed chemistry can be derived. The chemical state is described by the conserved scalar Z and its fluctuation. The mean values of the species mass fractions are determined from the laminar flamelet calculation by a presumed β-function PDF for Z. Similarly, a transport equation for the number density of soot particles is derived based on Z [23].

The G-equation flamelet models using the non-reacting scalar G rather than the progress variables are based on the level set approach [25], in which an iso-scalar surface $G(x, t) = G_0$ (G_0 normally set equal to zero) divides the flow field into two regions, unburned ($G < G_0$) and burned region ($G > G_0$). A transport equation of G was introduced by Williams [26]. In a G-equation model, flame propagation is driven by the bulk fluid velocity of the unburned mixture ahead of the flame front and the laminar flame speed normal to the flame. The turbulent G-equation models have been successfully applied to combustion simulations for SI engines by Dekena and Peters [20], Tan and Reitz [21,27], and Liang and Reitz [28]. More descriptions of these models will be given in Sec. 6.3.

An alternative approach that is apart from the RANS methodology is the PDF model. In a PDF model, the turbulent flow is viewed as a random medium and described using a probabilistic mathematical model in which the PDF is computed by solving its transport equation, which can be deduced from the Navier–Stokes equations. In the PDF transport equation, the terms of convection, mean pressure gradient, and chemical reaction source appear in a closed form [29–31]. Thus, in PDF combustion models, the emphasis shifts from modeling of the chemical source terms to modeling of molecular transport processes. In these models, joint velocity-composition PDF transport equations for velocity and reactive scalars are solved, usually by using the Lagrangian Monte Carlo particle methods. The gas-phase flow is represented by many particles, each of which contains information about its position, velocity, temperature, composition, and so on. The main drawback of the PDF models is that they suffer from a statistical error that decreases slowly with the number of particles N_p per computational cell. The error is proportional to $N_p^{-0.5}$. For an acceptable numerical accuracy, far more than a hundred particles must be present in each cell [4]. Thus, the PDF models are prevented from most engine simulations.

The models discussed are for mixing-controlled combustion. However, for the case in which chemical kinetics dominates, such as in HCCI engines, it is reasonable to assume that turbulence plays a lesser role in the combustion (chemical reactions) process. In this case, combustion is assumed to occur volumetrically in each computational cell, and is then modeled by using detailed reaction chemistry mechanisms. This will hereafter be called the sub-grid direct chemistry (SGDC) approach in this book. Thus, CFD codes are integrated with the CHEMKIN code [32] used for chemical kinetics solutions. Applications of this approach are given by Kong et al. [33] who integrated KIVA3V [34] and CHEMKIN to form the KIVA–CHEMKIN model, and by Senecal et al. [35] who developed the SAGE detailed chemical kinetics solver in the CONVERGE code [36]. The SGDC models will be further discussed in Sec. 6.5.

There are numerous model variants in the mentioned modeling approaches, and there are also other types of models for engine simulations, for example, the Conditional Moment Closure (CMC) [37–39]. However, each model has its own applicable range. When LES is used for engine turbulence modeling, the models discussed can be used as long as they are updated to be theoretically consistent with LES and adapted to work with LES variables [40].

Ignition occurs during the early stage of combustion in engines. Models to simulate SI and compression autoignition are included in Sec. 6.6. In addition, models for NO_x and soot formations are presented in the last section. It is worthwhile to note that some engine simulation application examples are given while the physics models are described. The example cases will provide a better understanding of the models and the essences of engine combustion modeling.

6.2 Characteristic-Time Combustion Model

6.2.1 Model Formulation

As mentioned, the CTC models consider turbulence effects on the combustion process. With the CTC models, the time rate of change of the partial density of species m due to conversion from one chemical species to another is given by

$$\frac{dY_m}{dt} = -\frac{Y_m - Y_m^*}{\tau_c} \tag{6.1}$$

where Y_m is the mass fraction of species m, Y_m^* is the local and instantaneous thermodynamic equilibrium value of the mass fraction, and τ_c is the characteristic time to achieve such equilibrium. The characteristic time is assumed to be the sum of a laminar (kinetic) timescale τ_l and a turbulent timescale τ_t, i.e.,

$$\tau_c = \tau_l + f\tau_t \tag{6.2}$$

where f is a delay coefficient that gradually introduces the controlling role of turbulent effects in a developing flame kernel. The laminar timescale is modeled based on Arrehenius kinetics and the turbulent timescale is assumed to be proportional to the eddy turnover time k/ε which is calculated from the turbulence model.

Different formulations of τ_l, τ_t and f have been proposed for modeling gasoline engine combustion and for diesel engine combustion. One of the differences among the formulations is in the delay coefficient f. In gasoline engines, the time of ignition is known and this is followed by a well-defined flame growth process. Hence, in the model of Abraham et al. [11], when the flame kernel grows to be comparable to the turbulence eddy size, it is assumed to become influenced by the turbulence. Thus, f is given as:

$$f = 1 - e^{-\frac{(t-t_s)}{\tau_d}} \tag{6.3}$$

where $\tau_d = C_{m1} l / S_L$ and $(t - t_s)$ is the time after spark, S_L is the laminar flame speed, l is the turbulence integral length scale, and C_{m1} is a model constant with a typical value of 7.4.

However, the combustion process is more complicated in diesel engines than in gasoline engines. The wide range of equivalence ratio and non-homogeneous spray droplet combustion makes the monitoring of flame kernel growth difficult. In the model of Kong et al. [12], it is assumed that laminar chemistry initiates combustion, which is then influenced gradually by turbulence. To account for the separate effects of laminar chemistry and turbulence, the presence of the products is used as an indicator of mixing effects following the initiation of combustion events, and is formulated as:

$$f = \frac{1 - e^{-r}}{0.632} \tag{6.4}$$

where r is the local ratio of the amount of products to that of total reactive species at each point in the combustion chamber, i.e.,

$$r = \frac{Y_{CO_2} + Y_{H_2O} + Y_{CO} + Y_{H_2}}{1 - Y_{N_2}} \tag{6.5}$$

The parameter r indicates the completeness of combustion in a specific region. Its value varies from 0 (no combustion yet) to 1 (complete consumption of fuel).

The assumption used in the model of Kong et al. that combustion occurs at the molecular scale as a result of collisions between reacting molecules is justified in a diesel engine. Subsequently, combustion is strongly influenced by turbulence since it has significant effects on the transport properties and the preparation (mixing) of the reactants. The delay coefficient f changes from 0 to 1 accordingly, depending on the local conditions. In other words, turbulence starts to have an influence only after combustion events have already been observed. Eventually, the combustion will be dominated by turbulent mixing effects in the regions of $\tau_l \ll \tau_t$. However, the laminar timescale is not negligible near the injector regions where the high injection velocity makes the turbulent timescale very small. In view of the overall diesel combustion process of diesel engines, the general picture, premixed first and then mixing control, coincides with the above model formulation.

The idea of naming the function f as a delay coefficient comes from modeling SI engines and the growth of a flame kernel. When the flame kernel grows to be comparable to the turbulence eddy size, it becomes influenced by the turbulence. However, the combustion process is more complicated in diesel engines than that in SI engines. The function f must be formulated differently. The way to account for the separate effects of laminar chemistry and turbulence is to use the appearance of products as an indicator of mixing effects following the initiation of combustion events. By using this combustion model, the chemical source term in the species continuity equation, and the chemical heat release in the energy equation are computed. Since the total chemical timescale also includes the turbulent timescale, the effects of turbulence on mean reaction rates are accounted for.

Seven species are considered necessary to predict thermodynamic equilibrium temperatures accurately: fuel, O_2, N_2, CO_2, CO, H_2, and H_2O in the model of Kong et al. [12]. Among these seven species, six reactive species (i.e., all except N_2) are accounted for in order to solve the local and instantaneous thermodynamic equilibrium values Y_m^*. The laminar timescale is derived from the correlated one-step reaction rate from a single droplet auto-ignition experiment as:

$$\tau_l = \frac{1}{A} \left[fuel \right]^{0.75} \left[O_2 \right]^{-1.5} e^{E/RT} \tag{6.6}$$

where $A = 1.54 \times 10^{10}$ and E=77.3 kJ/mol·K if tetradecane is used.

The turbulent timescale τ_t is proportional to the eddy turnover time

$$\tau_t = C_2 \frac{k}{\varepsilon} \tag{6.7}$$

where $C_2 = 0.1$ if the RNG k-ε model is used. The turbulence parameters k and ε are calculated from the k-ε turbulence model.

Keeping this in mind, the chemical source term in the species Equation (3.7) and the chemical heat release in the energy Equation (3.11), for given computational timestep Δt are found from

$$\dot{\rho}_m^c = -\frac{\rho \left(Y_m - Y_m^* \right)}{\Delta t} (1 - e^{-\Delta t / \tau_c}) \tag{6.8}$$

$$\dot{Q}^c = -\sum_m \frac{\dot{\rho}_m^c}{W_m} (\Delta h_f^0)_m \tag{6.9}$$

where $(\Delta h_f^0)_m$ is the heat of formation of species m.

Note that the above formulations are used for high-temperature reactions, that is, for the reactions after autoignition has taken place. They shall be combined with the ignition models, discussed in Sec. 6.4, to simulate the overall combustion processes in a diesel engine. A critical temperature of the reacting mixture is set to be the transition criterion from the low-temperature ignition to high-temperature reaction, which is suggested to be 1000 K. Thus, the ignition model applies whenever and wherever the mixture temperature is lower than 1000 K to simulate the low temperature chemistry. If the temperature is higher than 1000 K, the above CTC combustion model is activated for describing the high temperature chemistry.

The CTC model was also applied to SI engines [41–43]. In these cases, the decay coefficient f by Abraham et al. [11] was used. In the meantime, flamelet turbulent combustion models were developed and applied to SI engines.

6.2.2 Diesel Engine Combustion Simulation

Three medium- and heavy-duty diesel engines were simulated to evaluate the CTC model by Kong et al. [12]. The specifications of the three engines, namely the Caterpillar, Tacom, and Cummins engines, are listed in Table 6.1. Since the engine combustion chamber is symmetric about the cylinder axis with evenly distributed injector nozzle orifices, the simulations were performed on a sector mesh to save computational time, which is demonstrated schematically in Figure 6.1. The KIVA2 code [6] was implemented with the RNG k-ε turbulence model [44] and the Wave spray breakup model [45]. The Shell multistep ignition kinetics model was used to simulate the low temperature chemistry during the ignition delay period. The computations used tetradecane ($C_{14}H_{30}$) as the fuel due to the similar C/H ratio. The cylinder pressures were measured with a pressure transducer, and the so-called measured HRRs were calculated based on the measured pressures by using the equations in Chapter 2. Kong et al. provides more experiments and computations [12]. Note that ignition, soot, and NO_x emissions were also predicted and compared with the experiments for the same engines, the results will be discussed in Sec. 6.7.

FIGURE 6.1 Schematic diagram of the combustion chamber geometry and computational grid (the raised sector) of a Caterpillar engine at TDC. Sprays represented by the particles are shown for reference.

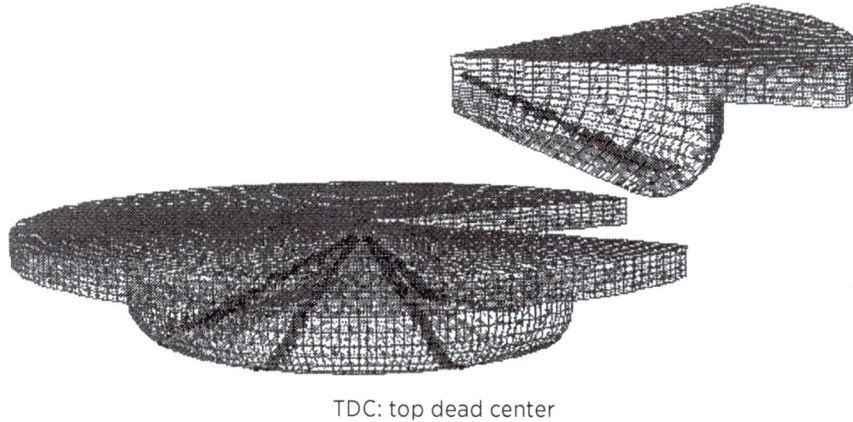

TDC: top dead center

TABLE 6.1 Engine specifications and operating conditions

	Caterpillar	Tacom	Cummins
Cylinder bore × stroke (mm)	137.6 × 165.1	114.3 × 114.3	140 × 152
Displacement (L/cylinder)	2.44	1.173	2.34
Compression ratio	15.1	16.0	10
Number of nozzle orifice × diameter (mm)	6 × 0.259	8 × 0.18	8 × 0.203
Spray angle (° from cylinder head)	27.5	7.5	14.0
Combustion chamber type	Quiescent	Quiescent	Quiescent
Piston crown	Mexican hat	Mexican hat	Mexican hat
Inlet air pressure (kPa)	184	180	166
Inlet air temperature (K)	310	334	423
Intake valve closure (° ATDC)	−147	−140	−150
Swirl ratio	1.0	1.0	1.0
Engine speed (rpm)	1600	1500	1200
Fuel (in experiment)	Amoco Premier #2	Amoco Premier #2	Blended (68% heptamethylnonane and 32% hexadecane)
Injection system	Common Rail	Jerk Pump	Cummins CELECT system
Injection pressure (MPa)	90	100 (peak)	84 (peak)
Fuel injected (g/cycle/cylinder)	0.1622	0.067	Case 1: 0.0553 Case 2: 0.1113
Overall equivalence ratio	0.46	0.3	Case 1: 0.25 Case 2: 0.5
Injection duration (CA)	21.5	25	16
Start of injection (° ATDC)	−15, −11, −5	−10.5	−10.0

Figure 6.2 shows the results of the Caterpillar engine. The computed cylinder pressures and HRRs are seen in excellent agreement with the measured data while the injection timing was varied. No case-by-case change of any model constant was made. HRR represents combustion evolution in which heat is released with time. The two peaks seen in a HRR curve correspond to the premixed combustion and diffusion combustion, respectively. The CTC model can accurately predict these two combustion regimes quantitively with the consideration of the effects of gas turbulence and laminar flame propagation, while a one-step global Arrhenius approximation totally failed to predict these phenomena [12].

FIGURE 6.2   The Caterpillar engine: Predicted (dashed lines) and measured (solid lines) engine cylinder pressure and HRR under different injection timings (from the top row to bottom row: −15°, −11°, and −5° ATDC).

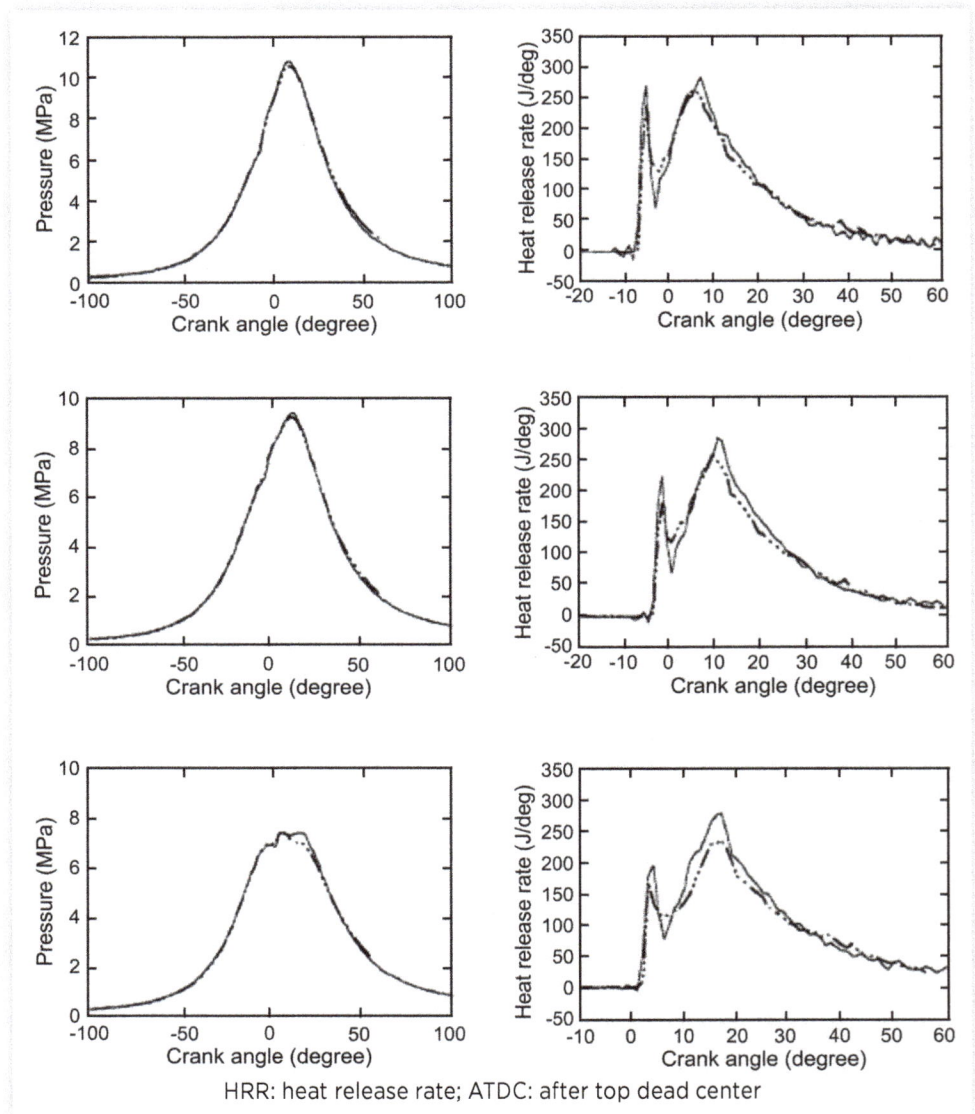

HRR: heat release rate; ATDC: after top dead center

Reprinted with permission from Ref. [12]. © SAE International

Notice that accurate prediction of the premixed burn, in terms of ignition timing and premixed burn amount, is critical to the overall success of combustion computations and to accurate emission computations. Good agreement in the cylinder pressures can be obtained only if the premixed burn is well reproduced, i.e., a good match of the HRR curves. The choice of the fuel for computation is based on the C/H ratio and the results are not particularly sensitive to the total fuel energy since they are somewhat similar for common hydrocarbon fuels. However, the total energy released during the combustion process should be examined carefully in numerical simulations. Integration of the HRR (J/deg) with respect to CA yields the total energy of the injected fuel. The total areas underneath the measured and predicted HRR curves must be consistent.

The modeling results of the Tacom engine are given in Figure 6.3. This engine has a shallower bowl and a considerably smaller displacement than the Caterpillar engine. Also, the spray angle is only 7.5° from the head, which confines the fuel spray inside the combustion chamber. In this case, the fuel spray tends to vaporize faster and no strong wall impingement is observed, except for when some droplets hit the far side of the piston surface. A pump-line fuel injection system was used in the Tacom engine. It was found necessary to change the spray break-up time constant B_1 from 60 (for the Caterpillar engine) to 40 (for the Tacom engine). The constant B_1 is related to injector nozzle structure and it is expected to vary with injector designs as explained in Sec. 5.2.2. All other model constants were kept the same as for the Caterpillar engine discussed. Again, a good agreement is obtained in the pressure prediction. However, prediction of the premixed combustion heat release peak was missed, possibly due to the combined inaccuracy of spray evaporation and autoignition prediction.

FIGURE 6.3   The Tacom engine: Predicted (dashed lines) and measured (solid lines) engine cylinder pressure and HRR.

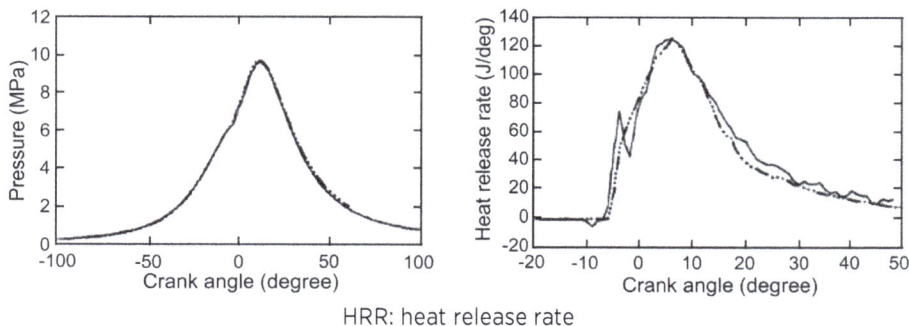

HRR: heat release rate

Reprinted with permission from Ref. [12]. © SAE International

The CTC model was also applied to model a Cummins N-series engine. The fuel was modeled using hexadecane ($C_{16}H_{34}$) and the ignition kinetic parameters were changed correspondingly. In this case, the fuel spray is confined in the combustion chamber and spray wall impingement is insignificant. The same value of the spray breakup time constant used for the Tacom engine cases (i.e., 40) was adopted for the Cummins engine cases and was found to give good modeling results. Figure 6.4 shows comparisons of computed and measured cylinder pressure and HRR cases in which the engine load is varied as listed in Table 6.1. The agreement is good as can be seen. Note that in these cases, the ignition model constant had to be adjusted empirically to account for the fuel effect (blended with 68% heptamethylnonane and 32% hexadecane) used in the experiment to best fit the ignition timing.

The Cummins engine: Predicted (dashed lines) and measured (solid lines) engine cylinder pressure and HRR at two engine loads.

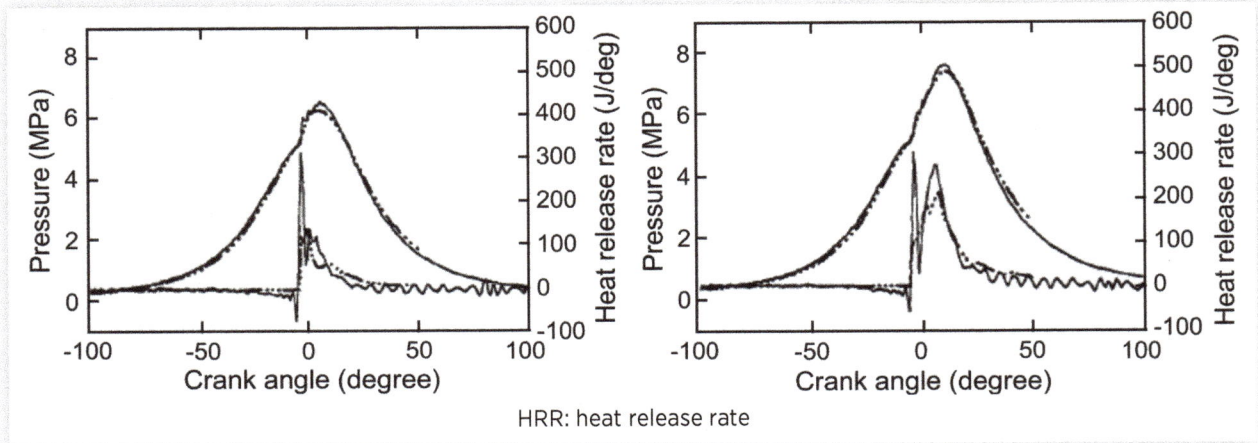

HRR: heat release rate

The model assessment of the three engines under different operating conditions verifies the good predictivity of the CTC model for diesel combustion in terms of cylinder pressure and HRR. These parameters are directly related to engine performance and thermal efficiency. In fact, the predicted soot and NO_x emissions are also in good agreements with experiments as will be discussed in Sec. 6.6, which indicate the in-cylinder temperature predictions are also good since the NO_x formation is very sensitive to local gas temperature. However, few model constants must be tuned from engine to engine for good results as discussed in Kong et al. [12]. This model-constant tuning requirement is not surprised for a simply formulated CTC model to predict very complicated combustion phenomena.

While the CTC model works well, some modifications to the model have been suggested. For example, Xin et al. [46] argued that the assumption that the modeled species approach their thermodynamic equilibrium at the same time is questionable. Instead, it is likely that these species approach their equilibrium partial densities at different rates or characteristic times, depending on the nature of the reactions and mixing processes involved. Thus, a faster rate is assigned to the conversion of CO and H_2, or $\tau_{c,CO} = \tau_{c,H2} = 0.2\tau_{c,fuel}$ for the fastest CO and H_2 conversion. Their study showed that this treatment does not alternate the computed global parameters much (e.g., cylinder pressure and HRR); however, it changes the local chemistry resulting in a higher temperature distribution. The resultant higher gas temperatures, therefore, impact the NO_x and soot prediction substantively. They also considered the effects of exhaust gas recirculation (EGR) or residual gas on the chemical kinetics and suggested a correlation of the model constant A in the laminar timescale (Equation (6.6)) as a function of the residual mass fraction (including EGR), and a modified calculation of the ratio r in Equation (6.5). With the modifications, Xin et al. [46] achieved success in prediction of a single-cylinder Caterpillar engine (see Table 6.1) under six sets of different operation conditions that simulate the six-mode US federal transient test cycle for heavy-duty diesel engines. The computed cylinder pressures and NO_x emissions agreed well with the measured data across all the six modes,. However, the prediction accuracy of soot emissions deteriorated at the light-load engine conditions, possibly due to the neglect of the contribution of soluble organic fraction (SOF) in their soot model.

6.3 Flamelet Methods

The concept of ensemble of laminar flamelets was first introduced by Williams [18], while the theoretical foundation was developed by Peters [4,22]. The turbulent flame can be viewed as ensemble of thin reaction-diffusion layers, called flamelets, embedding within a non-reacting turbulent flow. The flamelet concept considers the turbulent flame as an aggregate of thin, laminar, locally 1D flamelet structures present within the turbulent flow field.

The flamelet concept focuses on the location of the flame surface and not on the reactive scalars themselves. That location is defined as an iso-surface of a nonreacting scalar quantity, for which a suitable field equation is derived. For non-premixed combustion, the mixture fraction Z is that scalar quantity; for premixed combustion the scalar G is introduced. Once the equations that describe the statistical distributions of Z and G are solved, the profiles of the reactive scalars normal to the surface are calculated using flamelet equations. These profiles are assumed to be attached to the flame surface and are convected with it in the turbulent flow field.

Since the scalar quantities Z and G are nonreacting, their transport equations do not contain a chemical source term, and turbulent modeling assumption used for nonreacting scalars can be applied. Flamelet modeling thus permits a decoupling between detailed chemical kinetics and turbulent hydrodynamics, while maintaining tight local coupling between chemical kinetics and molecular transport. Flamelet models have been formulated for premixed, non-premixed, and partially premixed combustion regimes. For non-premixed flame, the model assumes that the terms involving transients and gradients parallel to the instantaneous surface of the constant mixture fraction to be small. By assuming equal diffusivity of all species, the species conservation equations can locally and instantaneously be transformed into a stationary laminar flamelet equation [22]. The flamelet structure is precalculated by solving the 1D flamelet equations. The results are stored in a structured table. The composition state space can be determined by looking up the table according to the mixture fraction and its dissipation rate. The mean values of the compositions are obtained usually through a presumed PDF approach. Applications of this method to model diesel engine combustion can be found in [47–49].

For turbulent premixed flame, a unity Lewis number and an infinitely thin flame structure are assumed and the species transport equation is transformed into a single balance equation, e.g., G-equation [20]. The reaction rate is computed from the laminar burning velocity, a correction factor representing turbulence stretch, and flame surface density. For turbulent partially premixed flame, which is common in SCDI gasoline engine, it is usually assumed that the partially premixed flame is a combination of a diffusion flame and a premixed flame, and it thus is modeled using a hybrid model [50,51].

6.3.1 Level Set G-Equation Model

The G-equation flamelet model tracks the propagating flame using a level set method. Williams [26] first suggested a transport equation of a nonreacting scalar G for laminar flame propagation. Peters [4] subsequently extended this approach to the turbulent flame regime. In the G-equation method, an iso-scalar surface $G(\boldsymbol{x}, t) = G_0$ (G_0 normally set equal to zero) defines the flame front and it divides the flow field into two regions, unburned ($G < G_0$) and burned region ($G > G_0$). Flame propagation is driven by the flow velocity $\boldsymbol{v}_f$ of the unburned mixture ahead of the flame front, and the laminar flame speed S_l normal to the flame. The rate of change of flame position $\boldsymbol{x}_f$ or the flame propagation velocity is

$$\frac{dx_f}{dt} = v_f + nS_l \tag{6.10}$$

The normal vector n is defined as:

$$n = -\frac{\nabla G}{|\nabla G|} \tag{6.11}$$

and the transport equation for G can be derived as:

$$\frac{\partial G}{\partial t} + v_f \cdot \nabla G = S_l |\nabla G| \tag{6.12}$$

Peters [4] extended the G-equation to model turbulent flames. With Favre averaging, the equations for the mean $\tilde{G}$ and the variance G'' of G are given as:

$$\bar{\rho}\frac{\partial \tilde{G}}{\partial t} + \bar{\rho}\tilde{v}_f \cdot \nabla \tilde{G} = \bar{\rho}_u S_t^0 |\nabla \tilde{G}| - \bar{\rho} D_t \tilde{\kappa} |\nabla \tilde{G}| \tag{6.13}$$

$$\bar{\rho}\frac{\partial \widetilde{G''^2}}{\partial t} + \bar{\rho}\tilde{v}_f \cdot \nabla \widetilde{G''^2} = \nabla_{\parallel} \cdot \left(\bar{\rho}_u D_t \nabla_{\parallel} \widetilde{G''^2} \right) + 2\bar{\rho} D_t (\nabla \tilde{G})^2 - c_s \bar{\rho}\frac{\tilde{\varepsilon}}{\tilde{\kappa}}\widetilde{G''^2} \tag{6.14}$$

where $\tilde{v}_f$ is the fluid velocity, $\bar{\rho}_u$ is the unburned gas density, and $\bar{\rho}$ is the gas density at the mean location of the turbulent flame defined by $G(x, t) = G_0 = 0$. D_t is the turbulent diffusivity. $\nabla_{\parallel}$ denotes the tangential gradient operator and $\tilde{k}$ and $\tilde{\varepsilon}$ are the TKE and its dissipation rate from the k-ε turbulence models, respectively. $\tilde{\kappa}$ is the mean flame front curvature and can be expressed in terms of the level set function $\tilde{G}$ as:

$$\tilde{\kappa} = \nabla \cdot \left(-\frac{\nabla \tilde{G}}{|\nabla \tilde{G}|} \right) \tag{6.15}$$

The turbulent burning velocity S_t^0 is given as:

$$\frac{S_t^0}{S_l^0} = 1 - \frac{a_4 b_3^2}{2b_1}\frac{l}{l_F} + \left[\left(\frac{a_4 b_3^2}{2b_1}\frac{l}{l_F} \right)^2 + a_4 b_3^2 \frac{u'l}{s_l^0 l_F} \right]^{1/2} \tag{6.16}$$

where S_l^0 is the reference laminar burning velocity (flame speed) of a planar unstretched flame. The model constants are $a_4 = 0.78$, $b_1 = 2.0$, $b_3 = 1.0$, and $c_s = 2.0$. The turbulence intensity is $u' = \sqrt{2\tilde{k}/3}$, l and l_F are the turbulence integral length scale and laminar flame thickness, respectively. The turbulence integral length scale can be calculated in the k-ε model. The flame thickness can be estimated by [52]:

$$l_F = \frac{(\lambda/c_p)_0}{\rho_u S_l^0} \tag{6.17}$$

In this equation, the heat conductivity λ and heat capacity c_p are evaluated at the inner temperature.

In the implementation of these G-equations to KIVA3V for engine simulations, considering the ALE numerical method used in the KIVA code, the flow velocity $\tilde{v}_f$ in the convection terms of the G-equations is replaced by $\tilde{v}_f - v_{vertex}$ to account for the velocity of the moving vertex [21] so that

$$\overline{\rho}\frac{\partial\tilde{G}}{\partial t} + \overline{\rho}\left(\tilde{v}_f - v_{vertex}\right)\cdot\nabla\tilde{G} = \overline{\rho}_u S_t^0\left|\nabla\tilde{G}\right| - \overline{\rho}D_t\tilde{\kappa}\left|\nabla\tilde{G}\right| \tag{6.18}$$

Also, as pointed by Liang and Reitz [28], Equation (6.16) is for the fully developed turbulent flame and a progress variable I_p is introduced to model the increasingly disturbing effect of the surrounding eddies on the flame front surface as the spark-ignition kernel flame grows from the laminar stage into the fully developed turbulent stage. Thus, for an unsteady turbulent flame, the burning velocity is modified as:

$$\frac{S_t^0}{S_t^0} = 1 + I_p\left\{-\frac{a_4 b_3^2}{2b_1}\frac{l}{l_F} + \left[\left(\frac{a_4 b_3^2}{2b_1}\frac{l}{l_F}\right)^2 + a_4 b_3^2\frac{u'l}{s_t^0 l_F}\right]^{1/2}\right\} \tag{6.19}$$

where

$$I_P = \left[1 - \exp\left(-C_{m2}\frac{t-t_0}{\tau}\right)\right]^{1/2} \tag{6.20}$$

where t_0 is the spark time. The model constant C_{m2} (order of 1.0) is regarded as tunable for different engines, and it should the same for a given simulated engine.

For the unstretched laminar flame speed S_l^0, many correlations have been proposed (e.g., Amirante et al. [53]). Liang and Reitz [28] suggested the correlation of Metgalchi et al. [54] as:

$$S_l^0 = S_{l,ref}^0\left(\frac{T_u}{T_{u,ref}}\right)^\alpha\left(\frac{P}{P_{ref}}\right)^\beta F_{dil} \tag{6.21}$$

where the subscript *ref* means the reference condition of 298 K and 1atm. F_{dil} is a factor accounting for the dilution effect. The fuel-type independent exponents α and β were correlated as functions of equivalence ratio ϕ as: $\alpha = 2.18 - 0.8(\phi - 1)$ and $\beta = -0.16 + 0.22(\phi - 1)$

The reference flame speed is

$$S_{l,ref}^0 = B_M + B_2\left(\phi - \phi_M\right)^2 \tag{6.22}$$

Values for B_M, B_2, and ϕ_M for propane and iso-octane are given by Liang and Reitz [28]. However, Equation (6.21) is only applicable for $0.6 < \phi < 1.7$, the range of which is too narrow for stratified-charge engine combustion. Hence, they modified the constants in the correlation of Gülder [55] with a better match of experimental data and suggested that

$$S_{l,ref}^{0} = \omega \phi^{\eta} exp\left(-\xi(\phi-\sigma)^2\right) \tag{6.23}$$

The constants are $\omega = 26.9$, $\eta = 2.2$, $\xi = 3.4$, $\sigma = 0.84$ for iso-octane.

The presence of diluent due to internal residual and/or EGR has a significant effect on the laminar flame speed. This effect is accounted for by F_{dil} in Equation (6.24). One expression is suggested by Metghalchi et al. [54] for all fuel types as:

$$F_{dil} = 1 - fY_{dil} \tag{6.24}$$

where Y_{dil} is the mass fraction of the diluent. f is an experimentally determined constant and is recommended by Liang and Reitz [28] as:

$$\begin{cases} f = 2.1 + 1.33 \cdot Y_{dil} & 0 < Y_{dil} < 0.2 \\ \\ f = 2.5 & 0.2 < Y_{dil} < 0.476 \end{cases} \tag{6.25}$$

After the flame front has passed, the mixture within the mean flame brush tends to the local and instantaneous thermodynamic equilibrium. The species conversion rate and the associated primary heat release at the flame front are calculated accordingly. To calculate the density changes of the species in the cells containing the mean flame front, Tan and Reitz [21] considered seven species including fuel, O_2, N_2, CO_2, H_2O, CO, H_2 in their study, and suggested that

$$\dot{\rho}_m^c = \rho\left(Y_{m,u} - Y_{m,b}\right)\frac{A_{f,i}}{V_i}S_t^0 \tag{6.26}$$

where ρ is the average density of the mixture in cell i, $Y_{m,u}$ and $Y_{m,b}$ are the mass fractions (i.e., fractions of the total mass in the cell) of species m in the unburned and burned mixtures, respectively, $A_{f,i}$ is the mean flame front area, and V_i is the cell volume.

Tan and Reitz gives the detailed numerical implementation scheme [21]. When detailed chemistry was adopted, Liang and Reitz [28] used a modified scheme for the numerical implementation to treat the issue raised due to a large number of intermediate species.

6.3.2 SI Engine Combustion Simulation

The G-equation model by Reitz's group at the University of Wisconsin-Madison (Madison, Wisconsin) have been applied to a number of engine simulations [21,27,28,57–59]. Typical results will be discussed hereafter.

A Caterpillar diesel-converted propane gas engine with a premixed homogeneous-charge was simulated. The engine specifications and operating conditions are given in Table 6.2. The effects of ignition timing and EGR are demonstrated in Figure 6.5 in which the measured and computed cylinder pressures at different spark timings are compared. Overall, the simulations capture the trends under varied conditions and the agreement between the prediction and measurement is reasonably good in quantity. As the spark is retarded, the peak pressure reduces. With a higher EGR rate, the laminar flame speed decreases, so does the turbulent flame speed, which leads to a reduced cylinder pressure as the EGR rate is changed from 0 to 5% and then to 10%, spark timing remaining the same.

FIGURE 6.5 Predicted and measured cylinder pressures: (a) at different ignition timings at 1600 rpm; (b) effects of EGR at spark time of −40° ATDC at 1600 rpm; (b) effects of engine speed at spark timing of −10° ATDC.

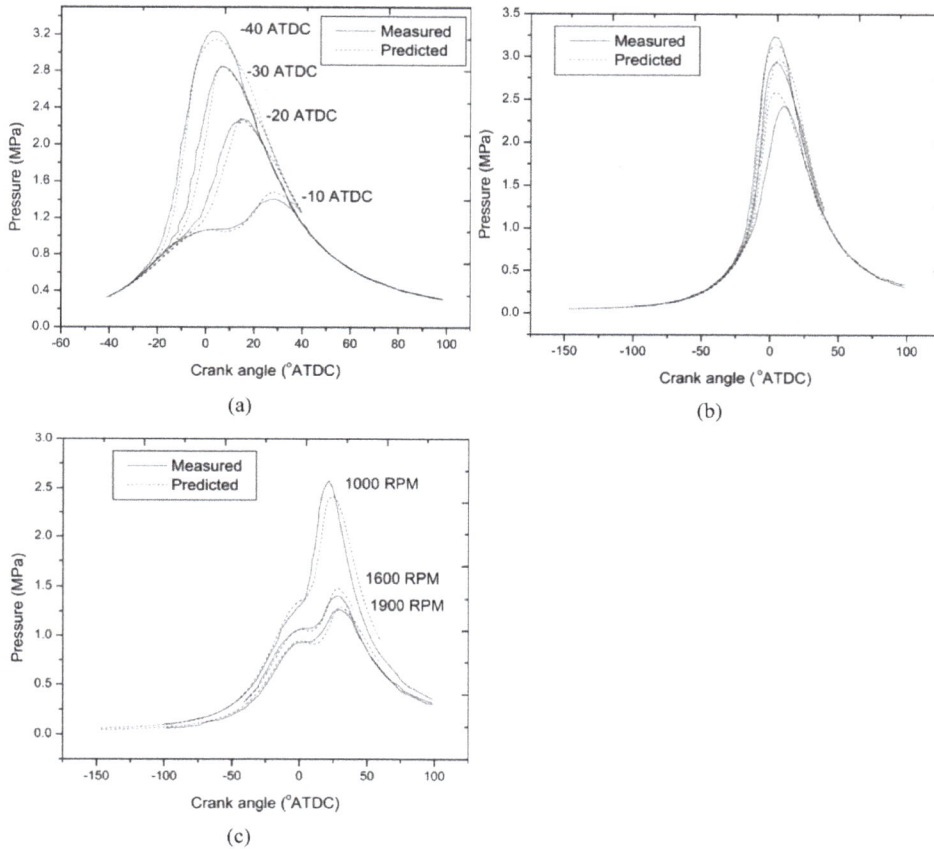

(a)

(b)

(c)

Adapted with permission from Ref. [21]. © SAE International

TABLE 6.2 Specifications and operating conditions of the simulated Caterpillar Engine

Engine	Caterpillar 3400 SI Gas Retrofit
Bore × Stroke (mm)	137.2 × 165.1
Compression ratio	10.1
Fuel	C_3H_8 (propane)
Equivalence ratio	Stoichiometric
Engine speed (rpm)	1000, 1600, 1900
Intake Valve Closure (° ATDC)	−147.0
Cylinder pressure at IVC (kPa)	51.0
Spark timing (° ATDC)	−10, −20, −30, −40
Spark duration (msec)	2.0
EGR (%)	0, 5, 10

Reprinted with permission from Ref. [21]. © SAE International

The predicted flame structure and temperature contours are shown in Figure 6.6. The diagrams show the propagation of the mean flame front from the spark plug, which is located at the center of the engine head. The black contour line denotes the locations of the mean flame front ($G_0 = 0$ iso-surface). For the chemistry, a 100-species, 539-reaction propane mechanism was used. As can be seen, the temperature of the mixture immediately behind the turbulent flame brush is above 2500 K, which is approximately equal to the local equilibrium temperature.

FIGURE 6.6   In-cylinder temperature contours at different crank angles.

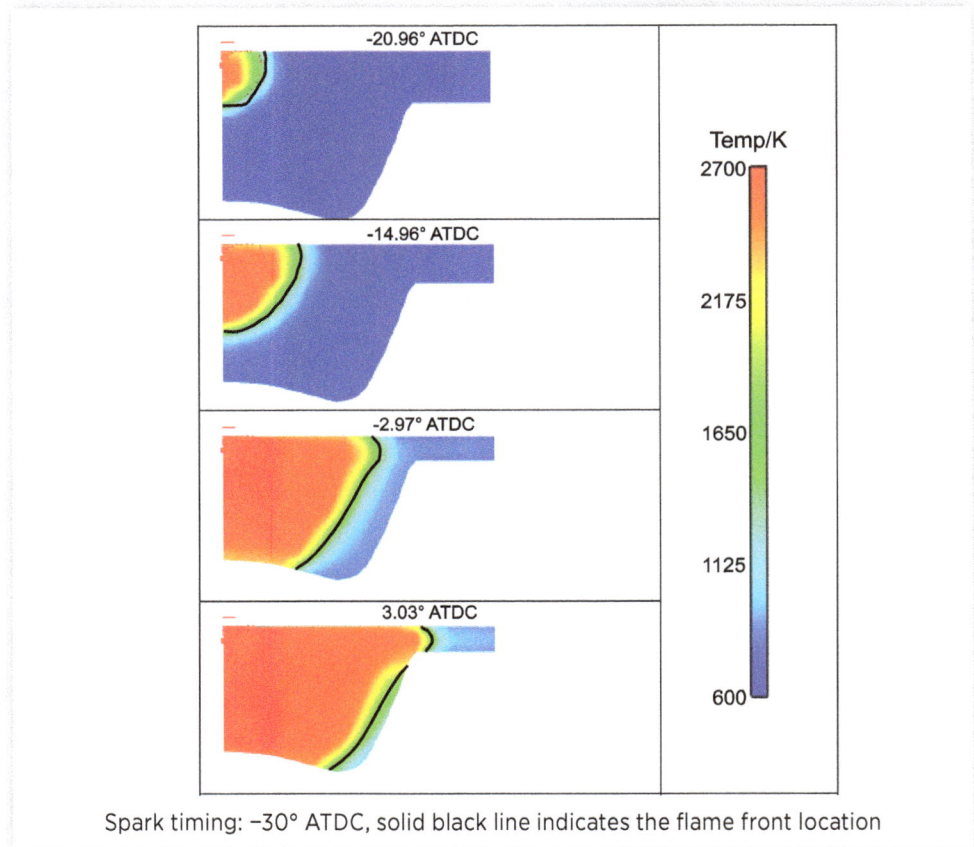

Spark timing: −30° ATDC, solid black line indicates the flame front location

The predicted in-cylinder species mass fractions of C_3H_8 (fuel), CO_2, CO, OH, NO, and NO_2 at −6 CA ATDC are shown in Figure 6.7. After the flame front passes, the fuel molecules are consumed, generating CO and CO_2. The subsequent chemistry process behind the flame brush is governed by the CO oxidation reactions, the H_2_O_2 system reactions, and the NO_x formation mechanism. The mass fraction of NO reaches its peak value in the highest temperature region, as expected. Most of the NO_2 is generated right ahead of the mean flame front. This is because NO_2 formation is favored under relatively low temperature combustion conditions. The peak mass fraction of NO is two orders of magnitude higher than that of NO_2, which is the main composition of the NO_x emissions in this case.

FIGURE 6.7   Predicted species mass fractions at −6° ATDC (Spark timing: −30° ATDC).

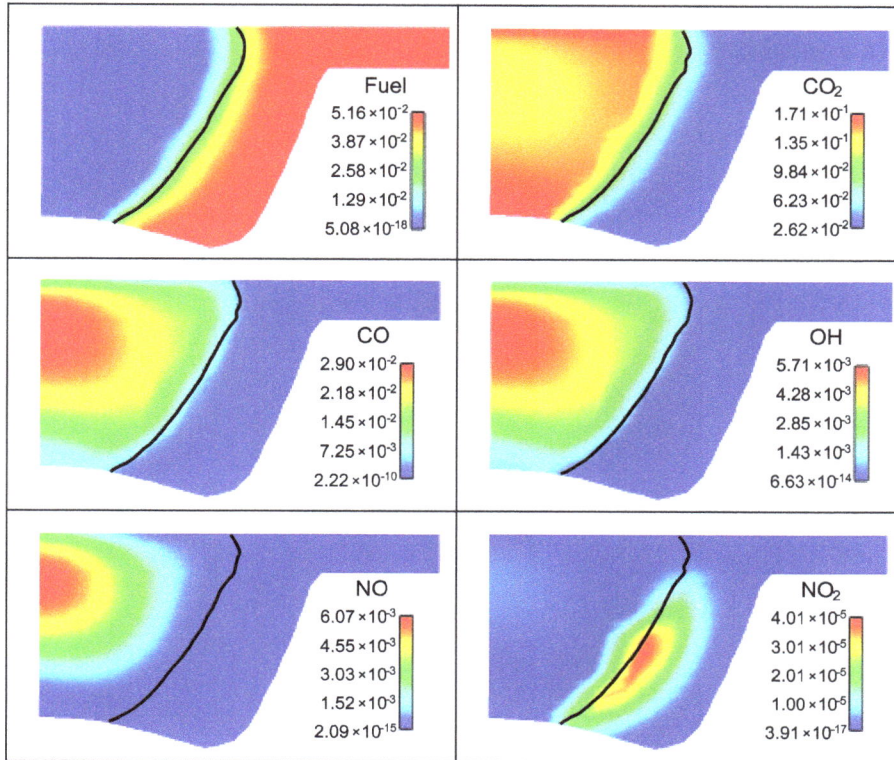

6.4 Sub-Grid Direct Chemistry Approach

6.4.1 Description of the Method

Recent research on gasoline HCCI combustion and diesel LTC has motivated scientists to resolve the combustion phenomena with detailed chemical mechanisms in engine simulations. For example, in HCCI combustion, the fuel and air are premixed and compression ignited. However, the mixture is very fuel lean. Although these mixtures are typically too lean to support flame-type combustion, they react and burn volumetrically as they are compressed to autoignition temperatures by the piston.

To simulate the HCCI combustion process in which chemical kinetics dominate, turbulence-mixing controlled combustion models are not applicable. Instead, volumetric heat release in each computational cell is assumed and a chemical kinetics mechanism is used for solutions of chemical species changes and heat release, which are integrated directly into the combustion source terms in the energy and species conservation equations. This is called the SGDC approach in this book.

Usually, a system of stiff ordinary differential equations that govern the rate of change of the chemical species involved in the reactions is solved in a chemistry solver. Considering a system consisting of N species and n elementary reactions, the elementary reactions are written in a general form as:

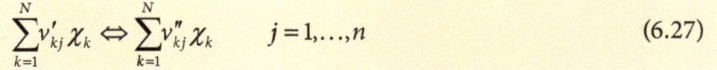

$$\sum_{k=1}^{N} v'_{kj} \chi_k \Leftrightarrow \sum_{k=1}^{N} v''_{kj} \chi_k \qquad j = 1, \dots, n \tag{6.27}$$

where χ_k is the symbol for the k-th species; v'_{kj} and v''_{kj} are the molar stoichiometric coefficients of the reactants and products, respectively. χ'_{kj} and v''_{kj} are integer numbers for elementary reactions and may be non-integers for non-elementary reactions. Each reaction fulfills element and mass conservation. So that

$$\sum_{k=1}^{N} \left(v'_{kj} - v''_{kj} \right) W_k = 0 \tag{6.28}$$

where W_k is the molecular weight of the k-th species. For kinetic reactions, the rate of progress of the j-th reaction $\dot{\omega}_j$ is written using the molar concentration $[X_k] = \rho Y_k / W_k$

$$\dot{\omega}_j = K_{fj} \prod_{k=1}^{N} [X_k]^{v'_{kj}} - K_{rj} \prod_{k=1}^{N} [X_k]^{v''_{kj}} \tag{6.29}$$

where K_{fj} and K_{rj} are the forward and reverse reaction rates of the j-th reaction, respectively. K_{fj} and K_{rj} are usually computed using the empirical Arrhenius expression:

$$K_{fj} = A_{fj} T^{\beta_{fj}} exp\left(-\frac{E_{fj}}{RT} \right) \tag{6.30}$$

$$K_{rj} = A_{rj} T^{\beta_{bj}} exp\left(-\frac{E_{bj}}{RT} \right) \tag{6.31}$$

The pre-exponential constant A_{fj} and A_{rj}, temperature exponent β_{fj} and β_{bj}, and activation energy E_{fj} and E_{bj} are given by the chemical kinetic scheme. The reverse reaction rate can be also related to the forward rate through the equilibrium constant by

$$K_{rj} = K_{fj} / K_{cj} \tag{6.32}$$

And the equilibrium constant K_{cj} is determined from

$$K_{cj} = \left(\frac{P_{atm}}{RT} \right)^{\sum_{k=1}^{N} v_{kj}} exp\left(\frac{\Delta S_j^0}{R} - \frac{\Delta H_j^0}{RT} \right) \tag{6.33}$$

where ΔS_j^0 and ΔH_j^0 are the change of specific entropy and enthalpy of j-th reaction, respectively:

$$\Delta S_j^0 = \sum_{k=1}^{N} v_{kj} S_k^0 \tag{6.34}$$

$$\Delta H_j^0 = \sum_{k=1}^{N} v_{kj} H_k^0 \tag{6.35}$$

With the reaction rates $\dot{\omega}_j$ determined, the chemical source term in the mass conservation equation of the k-th species (Equation (3.7)) is given by:

$$\dot{\rho}_k^c = W_k \sum_{j=1}^{n} v_{kj} \dot{\omega}_j \tag{6.36}$$

where the overall stoichiometric coefficient of the k-th species in the j-th reaction is

$$v_{kj} = v_{kj}'' - v_{kj}' \tag{6.37}$$

The chemical heat release term in the energy conservation equation (Equation (3.11)) is given by

$$\dot{Q}^c = \sum_{j=1}^{n} \dot{\omega}_j \Delta H_j^0 \tag{6.38}$$

Solving Equations (6.27)–(6.33) for elementary kinetic and equilibrium reactions can be complicated and is usually outside of the CFD hydrodynamic solver. Therefore, in the SGDC method, one often integrates the CFD codes like KIVA with CHEMKIN. CHEMKIN is a widely used software tool for solving complex chemical kinetics problems. It was originally developed at Sandia National Laboratories (Livermore, California) and is now part of ANSYS software toolset. This has been done by many researchers such as Kong et al. [33] who formed a KIVA3V–CHEMKIN code for HCCI simulation, Liang et al. [57] a KIVA3V–G-Equation–CHEMKIN code for SI engine knock prediction, and Li et al. [60] a KIVA3V–CHEMKIN code for biodiesel combustion.

When a CFD code, such as KIVA, is integrated with CHEMKIN, the computation procedure can be demonstrated in Figure 6.8. The CFD code solves for the flow field, and the pressure p_i, temperature T_i, and species mass fraction $Y_{i,k}$ of the flow at the computational cell i at every timestep are sent to CHEMKIN, which then calculates the new state of the species mass fraction $Y_{i,k}'$ and heat release $\dot{q}_i$, based on the chemical kinetics, and then returns them to the CFD cell i.

In the SGDC method, each computational cell is treated as a "well-stirred" reactor undergoing volumetric heat release. Thus, combustion modeling is assumed to be managed by two parts: fluid dynamics and chemistry. Fluid dynamics resolves mixing (momentum, energy, and species fractions) of the flow field, together with thermodynamics, to provide mean pressure, temperature, composition, and velocity at the grid level. Chemistry returns the change of species concentrations (function of temperature) and released heat averaged over the cell. Therefore, sub-grid turbulence chemistry interactions are not considered. It is apparent that the overall modeling accuracy relies on several factors:

FIGURE 6.8 Computation flow-chart for KIVA–CHEMKIN simulations.

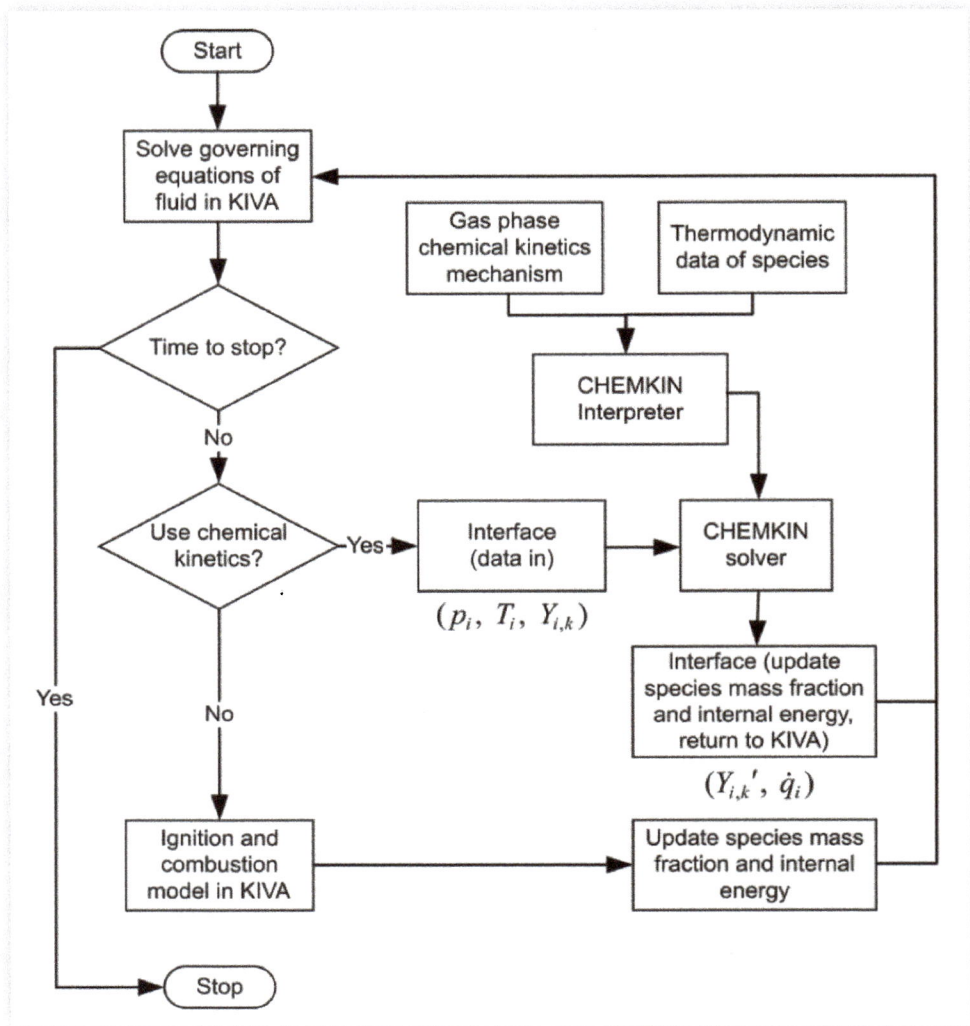

1. Turbulence modeling: This is still important at the resolved grid scale. While the RANS RNG k-ε model is acceptable for engineering modeling, turbulence modeling can be improved by LES models.

2. The level of details in a reaction mechanism of the chemistry: Although a detailed mechanism of a single hydrocarbon, e.g., 857 species and 3606 reactions for iso-octane [61], which is the typical surrogate fuel for gasoline, is complicated enough, research on engine cold-start emissions has shown the importance of multicomponent fuel. Balance between the accuracy and computing time forces one to use a reduced mechanism.

3. Computational grid size: This plays a significant role due to the nature of the cell "reactor". The smaller the grid size, the more details of turbulent flow field

can be resolved, and so the thermal field. The grid size sensitivity study by Pomraning et al. [62] indicated that a 10^{-4}-meter grid-size is needed in order to achieve converged RANS modeling, which means the computation time would increase by 1000 times to refine the currently-used 10^{-3}-meter grid-sizes. This may not be acceptable. However, one advantage of the SGDC is that it points the clear directions to improve modeling accuracy which relies on the improvement of computer power. In the meantime, there are "knobs" that can be tuned to achieve satisfactory results. The knobs may include LES with a coarser mesh, reduced chemical mechanism, and adaptive mesh refinement, and so on.

Strictly speaking, the mean-reaction-rate model as briefly discussed in Sec. 6.1 belongs to the SGDC method. The difference is that only one or a few reaction(s) and few species are used to simulate the chemistry in the mean-reaction-rate model, while up to several thousands of reactions and some hundreds of species for a complete mechanism are adopted in the current SGDC method. Therefore, the former can be viewed as the simplest extreme and the latter with a comprehensive mechanism as the most complicated extreme of the same model.

Although the SGDC method was originally motivated by HCCI combustion research, it has been widely used for other types of combustion (premixed and non-premixed) in SI engines, diesel engines and engines with alternative fuels [60,63,64]. The key issue is to adopt simplified (or reduced) chemical kinetics of the surrogate fuels based on detailed kinetics mechanisms and experimental data, since the detailed chemical kinetics is not practical for current engine simulations.

Various chemical mechanisms have been developed to describe the oxidation of different fuels. Gasoline and diesel are multicomponent fuels and mechanisms are still not available for all the component fuel species. To simplify the problem, iso-octane and n-heptane are often used as surrogates in reduced chemistry models because they have similar chemical characteristics as those of gasoline and diesel, respectively. In most practical engine simulations, the comprehensive reaction mechanisms are simplified to one order-less species and reactions using mechanism reduction methods. The reduction methods will be discussed in Sec. 6.5.

6.4.2 HCCI Combustion Simulation

Numerous publications on HCCI combustion modeling can be found in the literature as a result of intensive research in the past twenty years. Early modeling efforts have focused on the development of modeling methodology using detailed chemistry, i.e., the SGDC method. Autoignition of the mixture as well as the effects of fuel properties and thermal stratification was of interest. NO_x, CO, and hydrocarbon emissions were also modeled. As an example, the work of Kong [65] is presented here. Many other simulations can be found in the literature.

The studied engine is a Volvo TD100 truck engine that is modified to operate on one cylinder in the HCCI mode. Table 6.3 lists the engine specifications. The fuel is injected into the intake port approximately 30 cm upstream of the inlet valve. The intake air is heated by an electrical heater located upstream of the fuel injector. Two different pistons are used in the experiments, including a disc-shaped piston with a pancake-shaped combustion chamber, and a square-bowl piston to generate stronger turbulence.

TABLE 6.3 Engine specifications of the HCCI study

Displacement (cm³)	1600
Bore × Stroke (mm × mm)	120.65 × 140
Connecting rod length (mm)	260
Compression ratio	11.2
Intake valve open (ATDC)/close (ABDC)	5 /13
Exhaust valve open (BBDC)/close (BTDC)	39 /10
Engine speed (rpm)	1200
Fuel system	Iso-octane, port injection
Equivalence ratio	0.4
Inlet pressure (bar)	2.0 (supercharged)
Exhaust pressure (bar)	2.3
Estimated wall temperature (K)	450 (disc piston)
	460 (square piston)

Reprinted with permission from Ref. [65]. © Jon Wiley and Sons

A CHEMKIN–KIVA3V code is used. An iso-octane reaction mechanism consisting of 79 species and 398 reactions is used to simulate the complex chemistry of both low and high temperature regimes. The mechanism has been validated by simulating the ignition delay of iso-octane at constant pressure at various initial conditions. During the compression process, the mixture experiences a wide range of temperature and pressure histories, and, in most cases, autoignition occurs as a result of low temperature chemistry. The chemical reactions soon shift to the high temperature chemistry that characterizes the combustion phase, and the mixture is then burned within a very short timescale.

The computations use a 0.5-degree sector mesh for the disc-shaped piston cases and a 90-degree sector mesh for the square-bowl piston cases to shorten the simulation time. The computations were started at IVC assuming a uniform mixture distribution. The initial mixture temperatures at IVC are estimated by considering fuel vaporization, heat transfer in the inlet manifold and the cylinder, and mixing with the internal residual gas. Note that there were uncertainties associated with the initial temperature estimations. The model results showed that the combustion predictions are very sensitive to the initial temperature.

Table 6.4 lists the engine conditions for the experiments and simulations. Note that the experimental intake air temperatures need to be approximately 10 K higher in the disc-shaped cases to obtain the same ignition timing as those in the square bowl cases.

TABLE 6.4 Operating conditions for HCCI cases

	Tintake(K)	Fuel rate (g/s)	CAD of peak HRR
Case	Disc shape piston		
D1	430	0.6194	10.4
D3	451	0.6047	1.0
Case	Square bowl piston		
S1	418	0.6487	10.8
S3	440	0.6246	0.8

Reprinted with permission from Ref. [65]. © Jon Wiley and Sons

The model is also applied to simulate other cases in Table 6.4. Figure 6.9 shows the results for Cases S1 and D1. The ignition timing and combustion duration are well predicted for Case S1 using the estimated initial temperature of 442 K at IVC. Notably, the slightly higher predicted cylinder pressure before ignition is caused by the energy release due to the so-called low temperature or "cool-flame" chemistry. The experiments have shown that the combustion duration is prolonged in the square-bowl piston cases due to enhanced heat loss. The present model also predicted the same trend for the two comparable cases (Case D1 versus Case S1).

FIGURE 6.9   (a) Cylinder pressure and HRR for Case S1; (b) HRR and in-cylinder average gas temperatures for Cases S1 and D1.

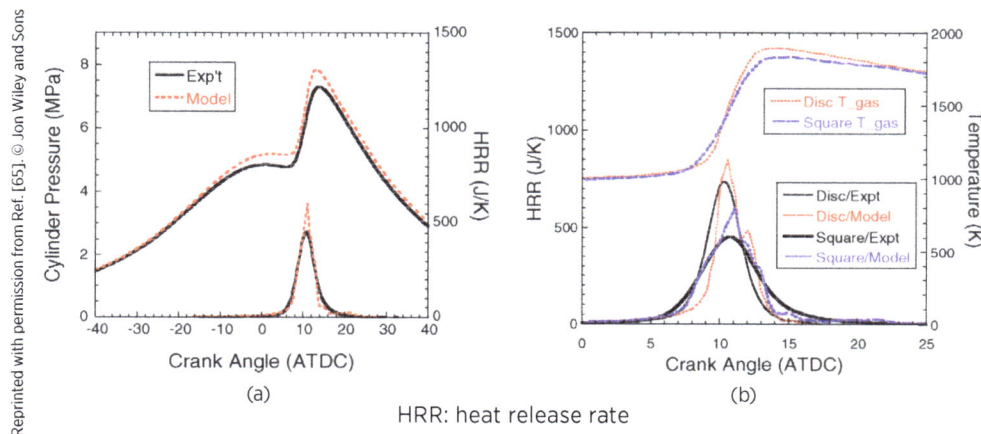

HRR: heat release rate

Ignition occurred at almost the same time in both geometries but the square-bowl piston case has a lower peak heat release and longer combustion duration. Figure 6.9 also shows that the peak in-cylinder average gas temperature in the square-bowl piston case is slightly lower than that of the disc piston case. Since the total amounts of energy release in both cases are the same, the difference is believed to be due to the different wall heat transfer rates in the different geometries.

Figure 6.10 demonstrates good levels of agreement in-cylinder pressure and heat release for Cases D3 and S3 as well. As can be seen, since the intake gas temperature is 20 K higher, the autoignition occurs earlier, which results in the peak pressure occurs 10 degrees earlier and a much higher pressure raise in both piston cases, when compared with the cases in Figure 6.9. The sensitivity of engine performance with the intake temperature is captured in both experiments and computations.

The predicted NO_x emissions are also compared with the engine-out measurements in Figure 6.11. The NO_x data reported by the model is the sum of NO and NO_2. The NO_x predictions are also very sensitive to the initial temperatures. The accuracy of the NO_x predictions is strongly related to that of the combustion predictions. Results showed that the accurate predictions of cylinder pressure (i.e., thermodynamics states) greatly help with the NO_x predictions using the current models. The model can predict the trend in combustion and emissions despite the uncertainties associated with the initial conditions and CFD models. The present model requires minimal empiricism and the simulations are carried out in the same way as regular engine combustion simulations. Thus, it is possible to use the model as a tool to explore premixed HCCI combustion under various engine geometries and operating conditions to help provide guidelines for engine design.

Reprinted with permission from Ref. [65]. © Jon Wiley and Sons

FIGURE 6.10   Cylinder pressure and HRR for: (a) Case S3 and (b) Case D3.

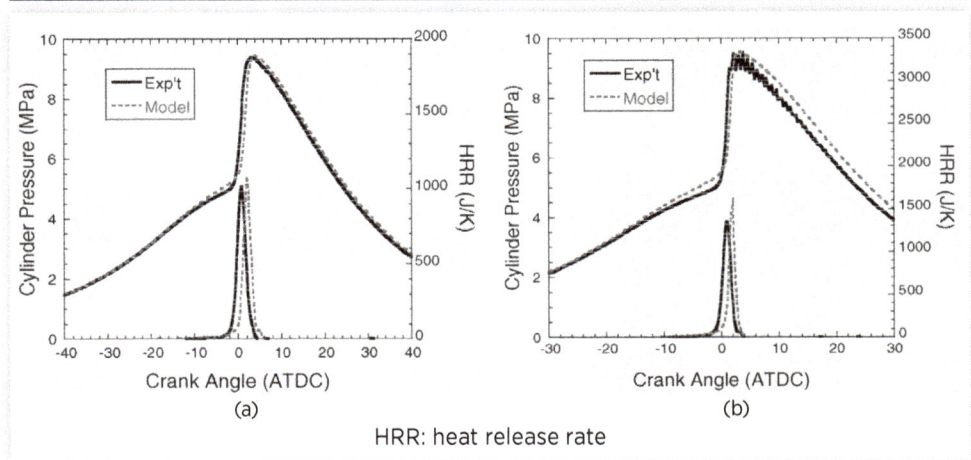

HRR: heat release rate

FIGURE 6.11   Measured and predicted NO_x emissions of HCCI combustion.

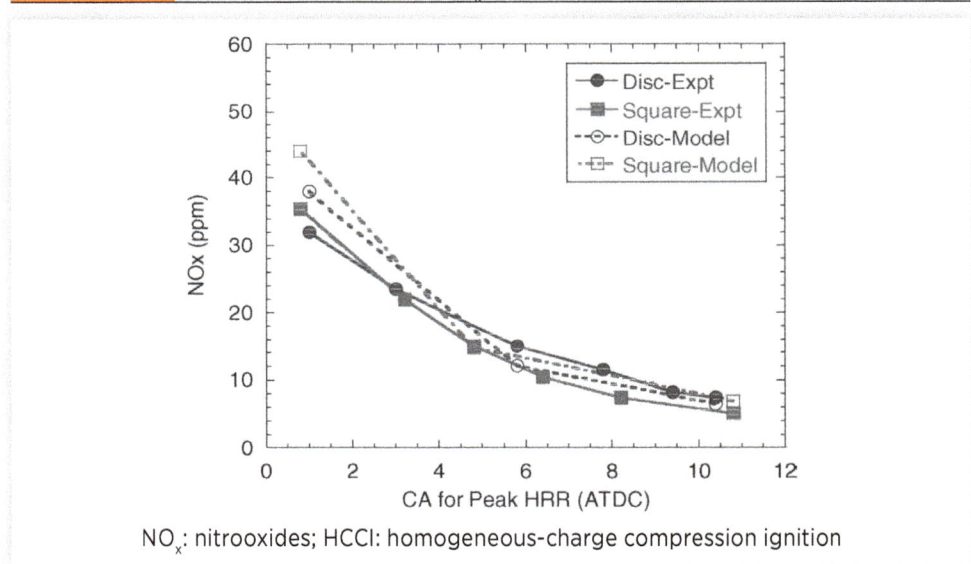

NO_x: nitrooxides; HCCI: homogeneous-charge compression ignition

Reprinted with permission from Ref. [65]. © Jon Wiley and Sons

6.5 Chemical Reaction Mechanism and Its Reduction

As discussed earlier, detailed chemical kinetics is needed in order to resolve the physics and chemistry occurring in an engine where chemistry dominates the reaction processes such as those in gasoline HCCI and diesel LTC. When a detailed reaction mechanism is used in combustion modeling, practical engine simulation is often prohibitive due to unaffordable computational time. Although surrogate fuels (i.e., n-heptane for diesel, iso-octane for gasoline) are used, the detailed mechanisms for themselves are huge. For example, the detailed n-heptane mechanism consists of 561 species and 2539 reactions [66], iso-octane of 857 species and 3606 reactions [61], and methyl decanoate (MD) for biodiesel of 2878 species and 8555 reactions [67]. Advances in chemical kinetics in terms of the number of the species and of reactions have largely paralleled the increase in computer capabilities. The current size of a surrogate fuel mechanism is over the order of 10^3 for species and 10^4 for reactions [68]. Thus, reduced mechanisms of smaller sizes, which can represent their corresponding comprehensive mechanisms over a wide range of conditions are necessary.

Many mechanism reduction methods have been proposed over the last few decades [69,70], employing different numerical approaches and emphasizing different physical and chemical aspects. The aim of each method is the same, that is to identify and remove redundant species and reactions, and hence produce simplified mechanisms with fewer numbers of species and reactions (usually, at least one order less) that are still able to reproduce the main features of their corresponding detailed mechanisms over the conditions of interest.

Skeletal reduction is typically the first step of mechanism reduction, which eliminates unimportant species and reactions from detailed mechanisms with the use of sensitivity analysis [71,72], principal component analysis [73,74], Jacobian analysis and computational singular perturbation [75,76], directed relation graph (DRG) [77], and other DRG-based methods [78,79]. After the skeletal reduction, other methods can be applied to further reduce the size of the mechanism, e.g., through lumping methods [80–82] that group the correlated species.

The DRG method uses a directed graph to map the coupling of species and consequently find unimportant species for removal based on selected target species and an acceptable error threshold. It has been shown to be a particularly efficient and reliable method to reduce large reaction mechanisms. In particular, the DRG method is based on reaction rate and species flux analyses, and does not involve Jacobian matrix evaluation and factorization. Therefore, it features a low reduction cost compared to most other reduction methods. Further development of the DRG method branched into two major directions: DRG-aided sensitivity analysis (DRGASA) [83,84], which performs sensitivity analysis on species not removed by DRG to further reduce the mechanism, and DRG with error propagation (DRGEP) [78], which considers the propagation of error due to species removal down graph pathways.

An approach that integrates the major aspects of DRGEP and DRGASA is DRG with error propagation and sensitivity analysis, DRGEPSA [85]. DRGEPSA first applies DRGEP to efficiently remove many unimportant species followed by DRGASA to further remove unimportant species, producing an optimally small skeletal mechanism for a given error limit. Thus, the DRGEPSA approach is able to overcome the weaknesses of each, specifically the inability of that DRGEP to identify all unimportant species and the fact that DRGASA shields unimportant species from removal.

In the study of biodiesel engine combustion, Li [86] proposed a so-called group species elimination (GSE) method to reduce the very comprehensive MD mechanism by Lawrence Livermore National Laboratory (LLNL). The GSE method is based on the DRGEPSA method in which a group of species that lead to either negative or positive error coefficient in the brute-force sensitivity analysis (BFSA) [83] is simultaneously removed from the initial mechanism. Compared to other methods of species elimination based on BFSA, the GSE does a better job of reducing the computational time of mechanism reduction because it does not need to delete the unimportant species one by one and assess the induced error every time.

The GSE reduction procedure begins with simulations of constant volume autoignition using the detailed reaction mechanism. Like other reduction methods, GSE uses ignition delay as the main parameter to assess the overall performance of the resulting reduced mechanism. In addition, GSE uses accumulated heat release as the second parameter, which prevents the elimination of species that could lead to incomplete heat release due to the change of the reaction path. GSE and DRGEPSA methods are similar in that both use DRGEP to perform the first phase of reduction. In the second phase, however, GSE uses BFSA. Li argues that the absolute value of the induced error δ_k in ignition delay used in DRGEPSA cannot reflect the error direction with respect to the reference mechanism. Therefore, although the error δ_k of some species may be small, removal of these species could result in ignition delay deviation over the allowed maximum limit. This may be the reason why DRGEPSA experiences premature termination of the reduction procedure as mentioned by its authors [85].

Instead, GSE defines a brute-force sensitivity coefficient as:

$$e_{k,i} = \frac{\tau_{k-el,i} - \tau_{b,i}}{\tau_{b,i}} \times 100\% \qquad (6.39)$$

where the subscript i represents the i-th test case, $\tau_{k-el,i}$ is the ignition delay of the mechanism with the removal of the species k at the i-th case, and $\tau_{b,i}$ is the ignition delay of the beginning mechanism. It is obvious that $e_{k,i}$ is the error with respect to the beginning mechanism. Similarly, the brute-force sensitivity coefficient of accumulated heat release is defined as:

$$eq_{k,i} = \frac{Q_{k-el,i} - Q_{b,i}}{Q_{b,i}} \times 100\% \qquad (6.40)$$

where $Q_{k-el,i}$ is the accumulated heat release of the mechanism with the removal of the species k at the i-th case.

Note that $e_{k,i}$ is used for the error analysis and $eq_{k,i}$ is used as the second parameter, which prevents the elimination of species that could lead to incomplete heat release due to the change of the reaction path. The ignition delay error between the beginning mechanism after DRGEP and a detailed mechanism is

$$E_{b,i} = \frac{\tau_{b,i} - \tau_{d,i}}{\tau_{d,i}} \times 100\% \qquad (6.41)$$

where $\tau_{d,i}$ is the ignition delay of the detailed mechanism at the i-th case.

Based on $e_{k,i}$, a group of species with positive and negative values of $e_{k,i}$ will be removed simultaneously in the error limit as their positive and negative effects will cancel each other. This is the main idea of the GSE method. A mechanism reduction procedure involving analysis of $e_{k,i}$, $eq_{k,i}$, and $E_{b,i}$ in the GSE method was also proposed. The GSE was implemented in the so-called Mechanism Reduction Code (MRC) to perform reduction computations. Li gives the reduction procedure and MRC instruction in detail [86], which are omitted here for brevity.

The GSE method has been assessed in a wide range of test cases [86]. Few examples are given here. A reduced mechanism (SA99) of iso-octane consisting of 99 species and 314 reactions is generated from the detailed mechanism of 857 species and 3606 reactions [61]. The comparison of the ignition delays from the computation of the SA99 reduced mechanism and the detailed mechanism and the shock tube measurement [87,88] is shown in Figure 6.12. It is seen that the SA99 agrees with the detailed mechanism well under the condition ranges of 1.520–4.154 MPa, equivalence ratio (ϕ) 0.5–2.0, and 650–1350 K, and the maximum error is −29.82%. When compared with the measured data, the SA99 mechanism carries over most of the deviations of the detailed mechanism. Both the mechanisms exhibit relatively larger errors in the negative temperature coefficient (NTC) region and good agreement in other regions.

FIGURE 6.12   Ignition delay of iso-octane.

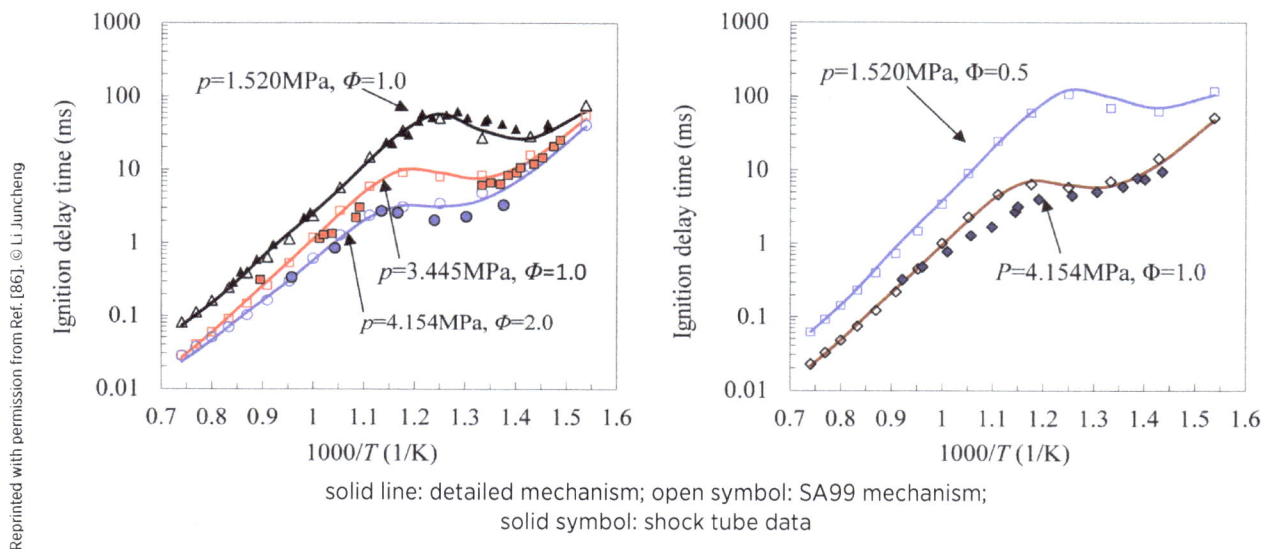

solid line: detailed mechanism; open symbol: SA99 mechanism;
solid symbol: shock tube data

To form the reduced mechanism of a surrogate fuel for biodiesel, the GSE method was applied to reduce the detailed mechanism of n-heptane and MD separately first and combine the two reduced mechanisms together. The reduced n-heptane mechanism consists of 60 species from 561 species of the detailed mechanism of LLNL [66], and the reduced MD mechanism consists of 87 species from 2878 species of the detailed mechanism of LLNL [67]. The final mechanism for biodiesel surrogate fuel with 111 species and 310 reactions (Bio111) has been proposed by Li [86]. A sample result from the extensive assessments is shown in Figure 6.13. The computed ignition delay of MD with Bio111 is compared with the shock tube measurement data [89], and the computed results with the LLNL MD mechanism are also shown in the graphs. As seen, Bio111 is consistent with the LLNL mechanism and can reproduce the measurement well.

Other methods have also been applied to generate reduced mechanisms of surrogate fuels for multidimensional engine combustion simulation of conventional SI and diesel engines, low-temperature HCCI and RCCI, and biodiesel. As examples, the detailed mechanism of iso-octane was reduced to 195 species and 647 reactions [90], 22 species and 42 reactions [57] for gasoline; that of n-heptane was reduced to 34 species and 77 reactions for diesel [91]; and of the blend of n-heptane and methyl butanoate was reduced to 56 species and 169 reactions [92]. The reduced mechanism size is usually dependent on the problem to be solved. A relatively larger size may be affordable for RANS modeling while a smaller size may be suitable only for LES modeling.

FIGURE 6.13 Predicted ignition delay of MD with comparison to the results from detailed mechanism and experiment.

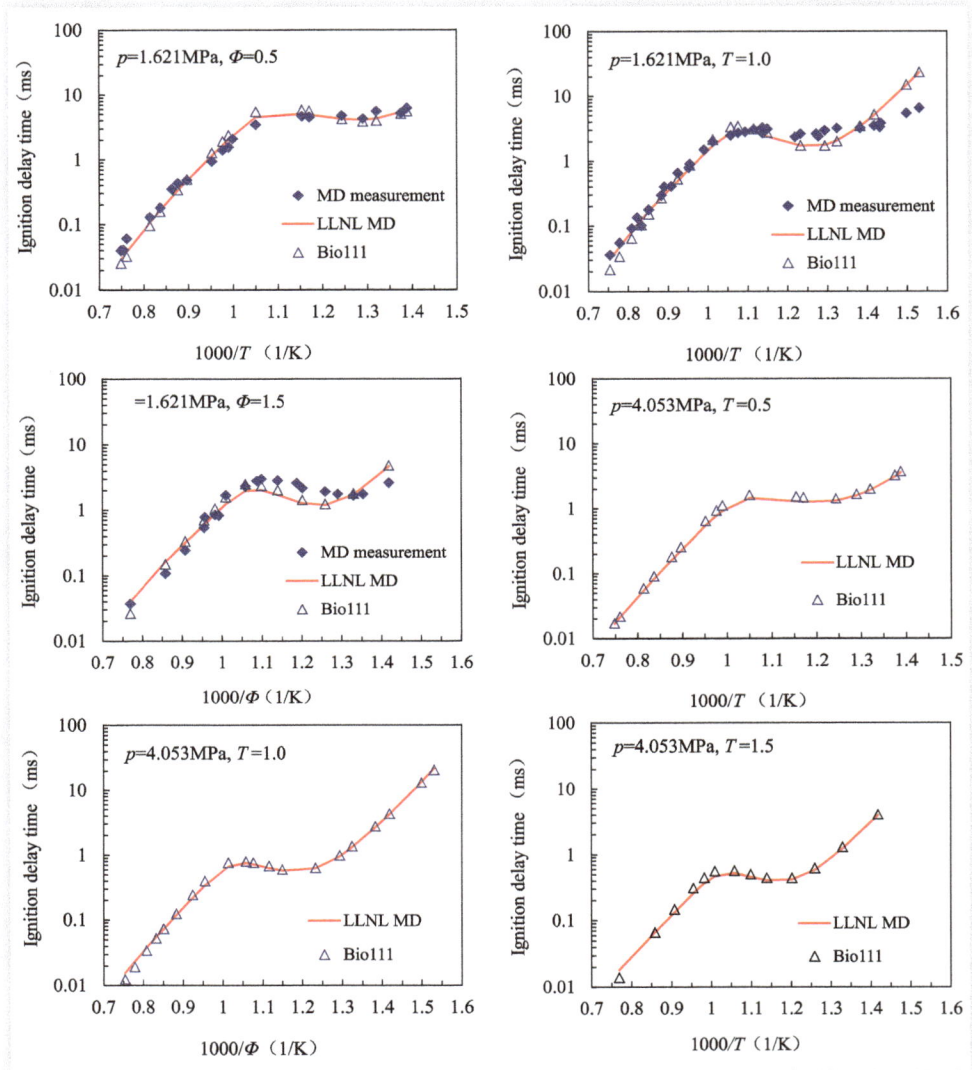

Research on improved numerical methods to accelerate chemistry solving process has been also carried out [90,93], one of which is called the adaptive multigrid chemistry (AMC) model [90]. The basic idea of the AMC model is to group thermodynamically similar (pressure, temperature, species composition) cells in engine combustion simulations to reduce the calling frequency to the chemistry solver. Hence, the number of cells that need to be computed by chemistry solvers is reduced significantly. Two key steps are involved in this process. The first one is to map (group) eligible cells together and solve the grouped cells together with the chemistry solver; and the second one to redistribute the group information back onto the individual cells so that the gradients can be preserved. Specific algorithms have been developed to accomplish these two steps. The AMC model has been proven to reduce the computation time by one order while maintaining the same level of prediction accuracy for both HCCI and diesel engines. More details can be found in Shi et al. [94].

6.6 Ignition Models

6.6.1 Spark Ignition

In SI engines, the flame is initiated by the electrical discharge of the spark plug. The ignition process includes electrical discharge, plasma breakdown, and shock wave propagation. All these phenomena occur in a very short time period (less than 10^{-6} sec), and in a relatively small region (~1 mm as the spark gap). It is not practical to resolve the process in engine simulations because the computational grid sizes and time steps are much larger than the scales at which these phenomena take place. Thus, a phenomenological model is often used to simulate the SI process that is important for the flame kernel growth.

SI can be simply simulated by adding energy to the ignition cells empirically over the spark duration [6]. However, this method is sensitive to the computational grid size, and the effects of the surrounding flow and mixture conditions are not considered. Other models to account for more ignition physics have been also proposed [43,95–97]. Pyszczek et al. [97] considered four sub-models to simulate the spark channel evolution, electrical circuit, temperature diffusion, and ignition delay. Fan et al. [43] focused on the development of the early kernel surface initialized between the spark electrodes and developed a discrete particle ignition kernel (DPIK) model, in which a set of Lagrangian particles are used to represent the flame surface. One advantage of this model is that the sensitivity of ignition predictions to the numerical grid size is much reduced. An improved model was proposed by Tan and Reitz [21], which accounts for the spark discharge energy and for the effects of the flow turbulence on the kernel growth.

The flame kernel radius r_k is given by Fan et al. [43] as:

$$r_k = \left(\frac{T_{ad}}{T} S_l + \sqrt{2k/3} \right)(t - t_0) + r_0 \qquad (6.42)$$

where T is the local gas temperature, T_{ad} is the adiabatic flame temperature, k is the turbulent kinetic energy, and S_l is the laminar flame speed. $t - t_0$ is the elapsed time from the start of ignition, and r_0 is the initial kernel size, which is about 0.5–1.0 mm.

In the derivation of the improved DPIK model, it is assumed that the ignition kernel mass burning rate dm_k/dt is related to the increase in the flame kernel mass as:

$$\frac{dm_k}{dt} = \rho_u A_k (S_t + S_{plasma}) \qquad (6.43)$$

where ρ_u is unburned gas density, A_k is the flame kernel surface area, S_t is the turbulent flame speed, and S_{plasma} is the so-called plasma velocity, which results from the spark discharge energy. The sum of S_t and S_{plasma} is viewed as the effective kernel growth speed. From mass conservation and neglecting the effects of convection and pressure raise within the kernel, the particle's distance to the spark plug r_k is computed as:

$$\frac{dr_k}{dt} = \frac{\rho_u}{\rho_b}(S_t + S_{plasma}) \tag{6.44}$$

where ρ_b is the density of the burned gas. Considering the energy conservation in the ignition kernel, the plasma velocity can be given as:

$$S_{plasma} = \frac{\dot{Q}_e \eta}{4\pi r_k^2 \rho_u (e_b - h_u + p/\rho_b)} \tag{6.45}$$

where $\dot{Q}_e$ is the spark electrical energy discharge rate, η is the electrical energy transfer efficiency with a value of about 30%, and e_b and h_u are the internal energy of the burned mixture and the specific enthalpy of the unburned mixture, respectively.

The turbulent flame speed can be calculated by Equation (6.19) or by other correlations. Turbulent strain and curvature effects on the kernel flame are considered. More details of the laminar and turbulent flame speed are given by Tan and Reitz [21]. Once the ignition kernel exceeds a critical radius that is related to the integral turbulent length scale of the flow field, the ignition model switches to the combustion model. The chemistry in the kernel growth stage is treated based on the combustion model used, which can be a one-step reaction in the CTC model [43], or in the same way as the flame propagation in the G-equation combustion model since the G field is constructed by the positions of the kernel particles, thus providing the necessary information for the chemical heat release calculations [98].

6.6.2 Compression Ignition

In a diesel engine, compression autoignition occurs once the local thermal and chemical conditions become suitable, which then initiates the high-temperature combustion. Numerically, autoignition is not affected by the gas turbulence at the sub-grid level. Thus, ignition is included in the chemical kinetics calculation when the SGDC method is used. However, the Shell ignition model [99] has been implemented in the KIVA codes for diesel ignition. This can be used together with the CTC combustion model for computational efficiency. The Shell model was originally developed for the prediction of knock combustion in SI gasoline engines. Kong et al. implemented it into KIVA2, applied it for diesel autoignition, and achieved satisfactory results [12].

The Shell model consists of five species and eight generic reactions based on the degenerating branching characteristics of hydrocarbon auto-ignition. The reactions and species are as follows:

$RH + O_2 \rightarrow 2R^*$	K_q	R1
$R^* \rightarrow R^* + P + Heat$	K_p	R2
$R^* \rightarrow R^* + B$	$f_1 K_p$	R3
$R^* \rightarrow R^* + Q$	$f_4 K_p$	R4
$R^* + Q \rightarrow R^* + B$	$f_2 K_p$	R5
$B \rightarrow 2R^*$	K_b	R6
$R^* \rightarrow termination$	$f_3 K_p$	R7
$2R^* \rightarrow termination$	K_t	R8

In these reactions, R1 is the initiation reaction, R2–R6 are the propagation reactions in a chain-propagation cycle, and R7 and R8 are the two termination reactions. *RH* is the hydrocarbon fuel (C_nH_{2m}), R^* is the radical formed from the fuel, *B* is the branching agent, *Q* is a labile intermediate species, and *P* is oxidized products consisting of CO, CO_2, and H_2O in specified proportions. The expressions for K_q, K_p, K_b, K_t, f_1, f_2, f_3, and f_4 are all in an Arrhenius expression, all of which along with other parameters for a number of fuels are given in Kong et al. [12].

As said earlier, when applied to engine simulation, the ignition model applies whenever and wherever the mixture temperature is lower than 1000 K to simulate the low temperature chemistry. If the temperature is higher than 1000 K, the combustion model (e.g., the CTC) is activated for describing the high temperature reactions.

The ignition delay time is calculated and compared with the experiments [100]. The computation uses dodecane ($C_{12}H_{26}$) to simulate the #2 Diesel fuel in the experiments (#2 or #1 represents the grade of the commercial diesel fuel. #2 Diesel is more commonly used than #1). The results are shown in Figure 6.14. The behavior of ignition delay time decreasing with the increase of gas temperature and pressure is well captured by the model. However, as the ambient pressure increases, the discrepancy of the model rises when the ambient temperature is lower than about 800 K. Nevertheless, the Shell model and its implementation in KIVA codes perform well for practical diesel ignition simulation as demonstrated in simulation cases in Sec. 6.2.2.

FIGURE 6.14 Delay time of diesel fuel ignition by measurement (solid line) and by prediction (dashed line with symbols).

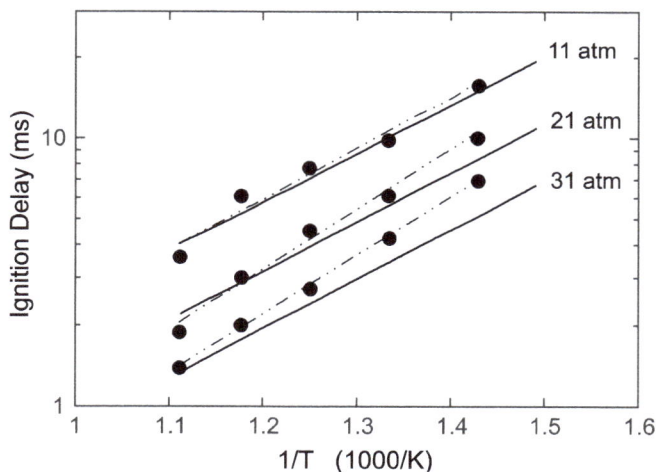

6.7 Models of NO_x and Soot Emissions

As the regulation of pollutant emissions becomes more and more stringent, expensive technologies (e.g., high-pressure injection, DPF, and SCR) have to be used to control pollutant formation during combustion and to reduce the pollutants in the tail-pipes through after-treatment devices. Thus, a major goal of engine combustion modeling is to predict pollutant emissions and to find ways to mitigate their formation during combustion. Of most concern are NO_x and PM or soot.

6.7.1 **NO$_x$ Emission Models**

The formation mechanism of NO$_x$ is well understood. While nitric oxide (NO) and nitrogen dioxide (NO$_2$) are usually grouped together as NO$_x$ emissions, NO is the predominant oxide of nitrogen produced inside the engine cylinder. For engine simulation, the extended Zel'dovich mechanism [101] has been widely used to predict NO emission. The mechanism consists of the following reactions:

$$N_2 + O \rightleftharpoons NO + N \tag{6.46}$$

$$N + O_2 \rightleftharpoons NO + O \tag{6.47}$$

$$N + OH \rightleftharpoons NO + H \tag{6.48}$$

The first two reactions, Equations (6.46) and (6.47), were first proposed by Zel'dovich for the formation of NO from atmospheric N$_2$. Both the reactions have large activation energies that result in the strong temperature dependence of NO formation rates. The third reaction, Equation (6.48) was added by Lavoie et al. [102] and it has a significant contribution.

With the partial equilibrium of the following reaction for the hydrogen radicals:

$$O + OH \rightleftharpoons O_2 + H \tag{6.49}$$

and a steady state assumption for N, which results from setting $d[N]/dt = 0$ in the rate equations resulting from Equations (6.46)–(6.48), the extended Zel'dovich mechanism can be written as a single rate equation for NO as:

$$\frac{d[NO]}{dt} = 2K_{1f}[O][N_2]\left\{\frac{1 - [NO]^2 / K_{12}[O_2][N_2]}{1 + K_{1b}[NO] / K_{2f}[O_2] + K_{3f}[OH]}\right\} \tag{6.50}$$

where $K_{12} = (K_{1f}/K_{1b})(K_{2f}/K_{2b})$ and the subscripts 1, 2, and 3 refer to Equations (6.46), (6.47), and (6.48), respectively. O, OH, O$_2$, and N$_2$ are assumed to be in local thermodynamic equilibrium. Heywood gives the reaction rate expressions [103].

The extended Zel'dovich mechanism only accounts for the NO formation. The formation reactions of NO and NO$_2$ can be calculated by the simplified mechanism based on the Gas Research Institute (GRI) NO$_x$ mechanism [104] where a detailed chemical mechanism is applied for the simulation, since more species are represented in the model. The simplified mechanism comprises the following reactions:

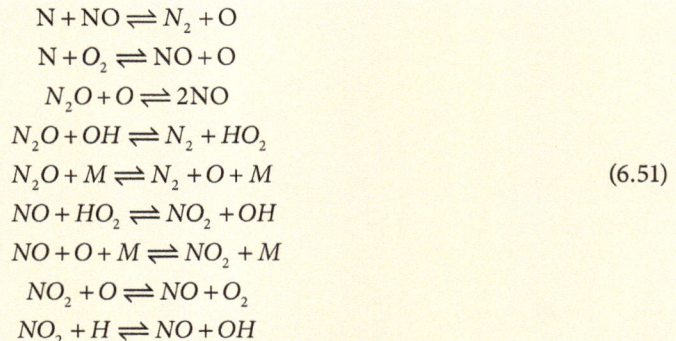

$$
\begin{aligned}
N + NO &\rightleftharpoons N_2 + O \\
N + O_2 &\rightleftharpoons NO + O \\
N_2O + O &\rightleftharpoons 2NO \\
N_2O + OH &\rightleftharpoons N_2 + HO_2 \\
N_2O + M &\rightleftharpoons N_2 + O + M \\
NO + HO_2 &\rightleftharpoons NO_2 + OH \\
NO + O + M &\rightleftharpoons NO_2 + M \\
NO_2 + O &\rightleftharpoons NO + O_2 \\
NO_2 + H &\rightleftharpoons NO + OH
\end{aligned}
\tag{6.51}
$$

This model has been extensively used in IC engine simulation, and an example has been given in Sec. 6.4.2 for the HCCI combustion modeling.

6.7.2 Soot Emission Models

The emissions of soot particles are very complicated. There is much about the soot formation process that is incompletely understood. A review of soot formation can be found in [103,105]. Conceptually, in a diesel engine, the soot particles form primarily from the carbon in the fuel. The formation process starts with a fuel molecule containing 12–22 carbon atoms with an H/C ratio of about 2, and ends up with particles typically a hundred nanometers or so in diameter, composed of spherules 10–25 nm in diameter. Soot formation takes place within the combusting fuel spray at temperatures of about 1600 K, and with much less air locally than is needed to fully burn the fuel. At temperatures higher than 1300 K, fast polymerization reactions are triggered leading to PAHs that are considered to be the building blocks for particulates in flames. In the overall soot formation process, oxidation of soot at the precursor, nuclei, and particle stages occurs as well.

In general, the soot particle formation process is considered to have three phases:

1. Particle formation, where the first condensed-phase material arises from the fuel molecules via their partial oxidation and/or pyrolysis products.

2. Particle growth, which includes both surface growth, coagulation, and aggregation.

3. Adsorption and condensation of hydrocarbons, which occurs after the cylinder gases have been exhausted from the engine, as these exhaust gases are diluted and cooled by mixing with air.

In a DI gasoline engine, Han et al. [106] found that wall-wetting is an important phenomenon affecting the engine-out smoke level. This was subsequently confirmed by optical engine observations [107,108] (see Sec. 7.5). Thus, they established a guideline to minimize the spray impingement on the surfaces of the combustion chamber including the piston top surface, cylinder linear, and valves in engine simulations for reduction of the engine-out soot emissions in DI gasoline engines [109]. Other sources of engine smoke emissions from wall-guided DI (WGDI) engines operating under stratified-charge conditions are probably formed from locally rich gaseous mixtures, incompletely volatilized liquid fuel drops.

While the understanding of the soot formation mechanism is accumulating, empirical soot models are still being widely used in multidimensional diesel engine simulations. In this type of model, only the total mass of soot particles is represented, in which soot is formed either directly from vaporized fuel or from an inception species. The most prominent examples of such models are the Hiroyasu model [110] and the Hiroyasu–Nagle and Strickland-Constable (NSC) model [111], which are widely used in diesel engine simulations. The Hiroyasu two-step soot model considers soot formation from a soot precursor and soot oxidation by oxygen, and the oxidation step is often replaced by the one from NSC [112]. Two empirical expressions for soot formation and soot oxidation are used. The net soot production rate is determined from

$$\frac{dM_s}{dt} = \frac{dM_{sf}}{dt} - \frac{dM_{so}}{dt} \qquad (6.52)$$

The soot formation rate is given by

$$\frac{dM_{sf}}{dt} = A_{sf} M_{fv} p^{0.5} \exp\left(-\frac{E_{sf}}{RT}\right) \tag{6.53}$$

where $A_{sf}=40$, M_{fv} is the fuel vapor mass which is treated as the soot precursor, p is the pressure in bar, and $E_{sf}=12{,}500$ cal/mol is the activation energy.

The NSC soot oxidation rate is given by

$$\frac{dM_{so}}{dt} = \frac{6M_c}{\rho_s D_s} M_s R_{ox} \tag{6.54}$$

where M_s is the soot mass, ρ_s is the soot density, D_s is the soot particle diameter, and M_c is the molecular weight of carbon. The reaction rate R_{ox} is given by

$$R_{ox} = \left(\frac{K_A p_{O_2}}{1+K_Z p_{O_2}}\right) x + K_B p_{O_2} (1-x) \tag{6.55}$$

and

$$x = \frac{p_{O_2}}{p_{O_2} + K_T / K_B} \tag{6.56}$$

where p_{O_2} is the partial pressure of oxygen. The model parameters are:

$$K_A = 20.0 \exp\left(-\frac{30000.0}{RT}\right) \tag{6.57}$$

$$K_B = 4.46 \times 10^{-3} \exp\left(-\frac{15200.0}{RT}\right) \tag{6.57}$$

$$K_T = 1.51 \times 10^5 \exp\left(-\frac{97000.0}{RT}\right) \tag{6.57}$$

$$K_Z = 21.3 \exp\left(-\frac{4100.0}{RT}\right) \tag{6.57}$$

where $R=1.98$ kcal/mol·K. The change of soot mass during one timestep Δt in a computational cell can be estimated as:

$$\Delta M_s = \left(\frac{\dot{M}_{sf}}{A_{sf}} - M_s \right) [1 - \exp(-A_{sf} \Delta t)] \qquad (6.58)$$

It is worthwhile to point out that when a simple reaction mechanism (e.g., CTC model) is employed, the soot precursor is usually set as the fuel vapor. Other species such as acetylene (C_2H_2) or benzene (C_6H_6) may be taken as soot precursors as long as they have been included in the reaction mechanism. Kong et al. [104] use this two-step soot model coupled with a detailed mechanism where fuel is depleted quickly to form intermediate hydrocarbon species once reactions start. Thus, acetylene is taken as the soot precursor in the soot model, i.e., M_{C2H2} is used in place of M_{fv} in the calculation. This is because acetylene is the most relevant species pertaining to soot formation in hydrocarbon fuels.

6.7.3 Model Predictions

Now let's discuss the model predictions. The predicted in-cylinder NO formation history in the Tacom engine (as discussed in Sec. 6.2.2, see Table 6.1) is shown in Figure 6.15. The Shell ignition and CTC combustion models are used. The computed evolution is compared with the total cylinder content dumping experiment data [113]. The extended Zel'dovich mechanism works very well to reproduce the experimental data. However, a scaling factor of 0.78 for the NO formation rate had to be used in order to match the experimental data quantitively [12]. The inconsistency is likely not due to the NO mechanism but due to the uncertainties in other physics models that may resolve inaccurate gas temperature and mixing, which may be explained by the effects of turbulence modeling as discussed in Sec. 4.3.2.

FIGURE 6.15   NO evolution in the Tacom engine.

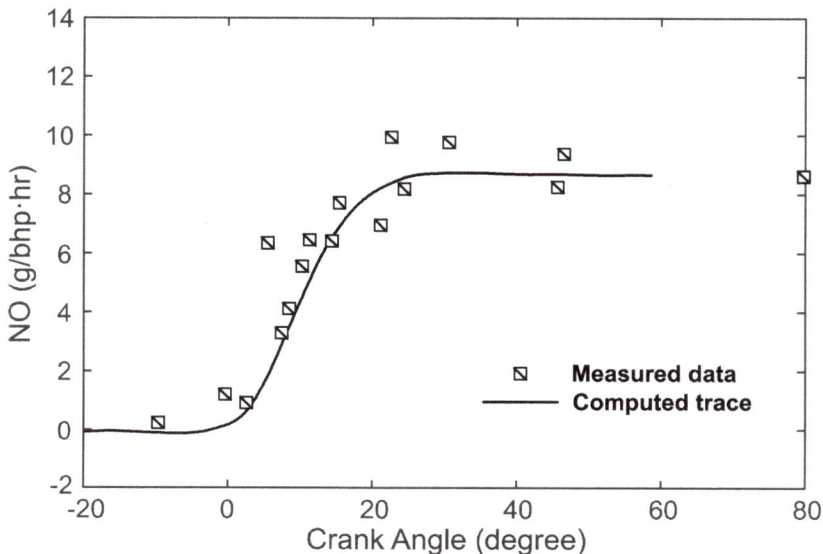

We also discussed the prediction of emissions under the six modes of engine operation conditions by Xin et al. [46] in Sec. 6.2.2 in which the Shell ignition and a modified CTC combustion model were used. The predicted NO_x and soot emissions are given in Figure 6.16. The information of the engine and operation conditions are, again, given in Table 6.1. Among all the six modes, the computed NO_x emissions agree well with the measured data. However, the soot prediction with the use of the Hiroyasu two-step model is not very good, in particular, at engine light loads. Part of the reason is possibly the neglect of the contribution of SOF in the soot model. Another reason is possibly the unsolved flow and thermal field.

FIGURE 6.16 Comparison of the measured and predicted soot and NO_x emissions under the engine operation conditions of the six modes.

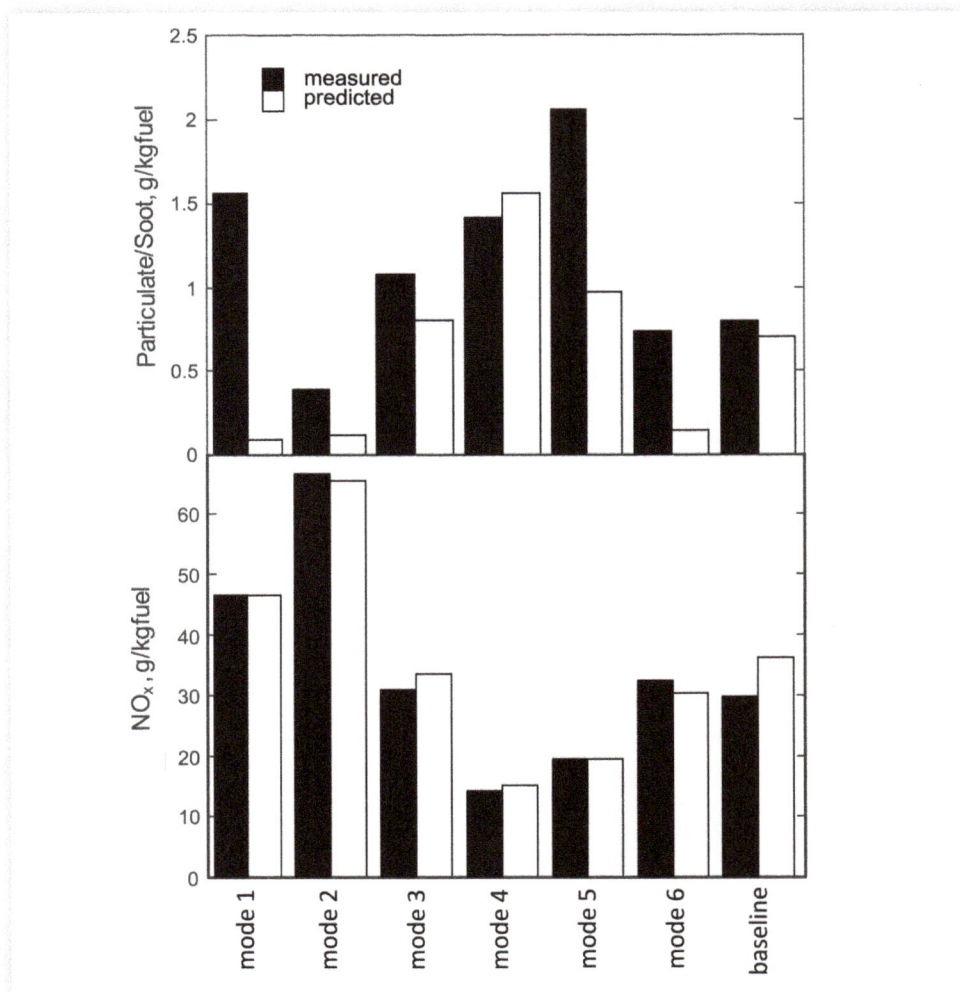

Kong et al. [104] modeled the same base Caterpillar engine by using more advanced computational models. In this case, the SGDC model in KIVA3V–CHEMKIN code with a skeletal reaction mechanism for n-heptane (29 species and 52 reactions) was used for both low-temperature (ignition) and high-temperature (diffusive flame) chemistry. The simplified GRI NO_x mechanism and the same Hiroyasu two-step soot model were adopted. Advances in spray and other models at the Engine Research Center at the University of Wisconsin-Madison were also included as well. The details of the model and engine conditions are given by Kong et al. [104].

The predicted soot and NO_x (i.e., sum of NO and NO_2) emissions were also compared with the measurements, as shown in Figure 6.17. It is seen from the figure that the overall trends of soot and NO_x are captured very well with respect to the SOI timing. It is of interest to note that engine-out soot emissions reach peak values when fuel is injected near TDC and then decrease when SOI is further retarded. The present model also predicts correctly the soot reduction seen at further retarded injection timing for all different EGR levels.

FIGURE 6.17 Effects of SOI and EGR on engine-out NO_x and soot emissions.

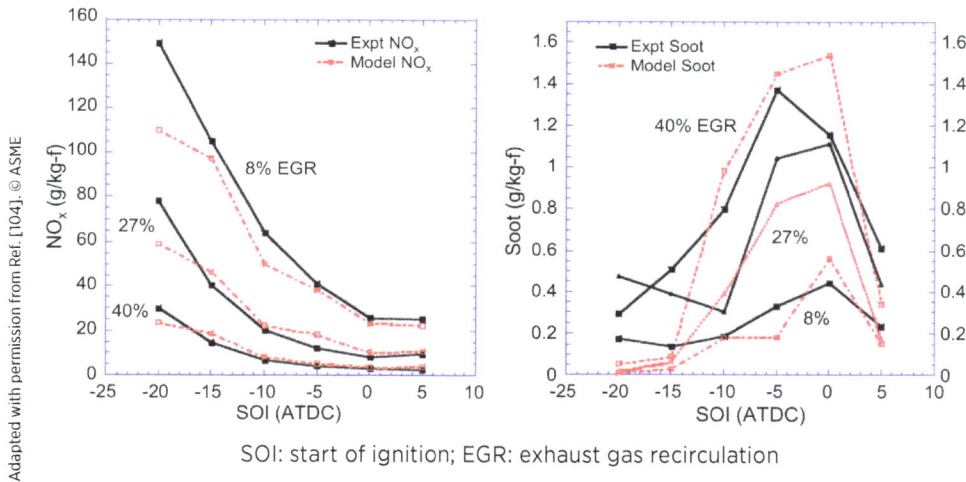

SOI: start of ignition; EGR: exhaust gas recirculation

More advanced soot models have been developed for IC engine simulation. The model by Kazakov and Foster [114] is also of empirical nature but represents soot by two numbers: soot mass fraction and particle number (PN). An alternative is to represent the particle size distributions by individual stochastic particles and use a Monte Carlo method to simulate their growth and oxidation [115]. Tao et al. [116] developed a multistep phenomenological soot model, which considers surface growth, soot inception and coagulation, soot oxidation by oxygen and OH, and soot precursor oxidation by oxygen.

Vishwanathan and Reitz [117] developed a new reduced n-heptane reaction mechanism that included PAHs, which are taken as the soot precursors. More features were added to the former model in modeling a WGDI stratified-charge gasoline engine [118]. Soot surface growth from both C_2H_2 and PAHs (up to four rings A1-A4 (pyrene)) are included once the soot particles are formed, thus the soot particle size is increased. Soot oxidation via both oxygen and the OH radical is also considered. The simulation showed that high soot levels are more likely to be formed in the rich regions near the wall films and successfully correlated soot formation with wall film fuel, as has also been observed in experiments.

References

1. Reitz, R.D., and Rutland, C.J. 1995. "Development And Testing of Diesel Engine CFD Models." *Progress in Energy and Combustion Science* 21, no. 2: 173–196. DOI: 10.1016/0360-1285(95)00003-Z

2. Reitz, R.D., and Rutland, C.J. 2014. "Multidimensional Simulation." In *Encyclopedia of Automotive Engineering, Part 1*, edited by D. Crolla, D.E. Foster, T. Kobayashi, and N. Vaughan, 1–19. Chichester: John Wiley & Sons.

3. Haworth, D.C. 2005. "A Review of Turbulent Combustion Modeling for Multidimensional In-Cylinder CFD." *SAE Transactions* 114, no. 3: 899–928.

4. Peters, N. 2000. *Turbulent Combustion*. Cambridge, UK: Cambridge University Press.

5. Diwakar, R. 1984. "Assessment of the Ability of a Multidimensional Computer Code to Model Combustion in a Homogeneous-Charge Engine." *SAE Transactions* 93, no. 2 (February): 85–108. DOI: 10.4271/840230

6. Amsden, A.A., O'Rourke, P.J., and Butler, T.D. 1989. *KIVA-II: A Computer Program for Chemically Reactive Flows With Sprays* (LA-11560-MS). Los Alamos, New Mexico: Los Alamos National Laboratory.

7. Tsao, K.C., and Han, Z. 1993. "An Exploratory Study on Combustion Modeling And Chamber Design of Natural Gas Engines." SAE Technical Paper No. 930312. DOI: 10.4271/930312

8. Reitz, R.D., and Bracco, F.V. 1983. "Global Kinetics Models And Lack of Thermodynamic Equilibrium." *Combustion and Flame* 53, no. 1–3 (November: 141–144. DOI: 10.1016/0010-2180(83)90013-5

9. Spalding, D.B. 1971. "Mixing And Chemical Reaction in Steady Confined Turbulent Flames." *Symposium (International) on Combustion* 13, no. 1: 649–657. DOI: 10.1016/S0082-0784(71)80067-X

10. Magnussen, B.F., and Hjertager, B.H. 1977. "On Mathematical Modeling of Turbulent Combustion With Special Emphasis on Soot Formation And Combustion." *Symposium (International) on Combustion* 16, no. 1: 719–729. DOI: 10.1016/S0082-0784(77)80366-4

11. Abraham, J., Bracco, F.V., and Reitz, R.D. 1985. "Comparisons of Computed And Measured Premixed Charge Engine Combustion." *Combustion and Flame* 60, no. 3 (June): 309–322. DOI: 10.1016/0010-2180(85)90036-7

12. Kong, S.C., Han, Z., and Reitz, R.D. 1995. "The Development And Application of a Diesel Ignition And Combustion Model for Multidimensional Engine Simulation." *SAE Transactions* 104, no. 3 (February): 502–518. DOI: 10.4271/950278

13. Bray, K.N.C., and Moss, J.B. 1977. "A Unified Statistical Model of the Premixed Turbulent Flame." *Acta Astronautica* 4, no.3–4 (February): 291–319. DOI: 10.1016/0094-5765(77)90053-4

14. Bray, K.N.C., and Libby, P.A. 1994. "Recent developments in BML model of premixed turbulent combustion." In *Turbulent Reacting Flows*, edited by P.A. Libby and F.A. Williams, 115–152. New York: Academic Press.

15. Marble, F.E., and Broadwell, J.E. 1977. *The Coherent Flame Model for Turbulent Chemical Reactions* (TRW-9-PU). Project SQUID. West Lafayette, Indiana: Purdue University.

16. Duclos, J.M., Veynante, D., and Poinsot, T. 1993. "A Comparison of Flamelet Models for Premixed Turbulent Combustion." *Combustion and Flame* 95, no. 1–2 (October): 101–117. DOI: 10.1016/0010-2180(93)90055-8

17. Musculus, M.P., and Rutland, C.J. 1995. "Coherent Flamelet Modeling of Diesel Engine Combustion." *Combustion Science and Technology* 104, no. 4–6: 295–337. DOI: 10.1080/00102209508907726

18. Williams, F.A. 1975. "Recent Advances in Theoretical Descriptions of Turbulent Diffusion Flames." In *Turbulent Mixing in Nonreactive and Reactive Flows*, edited by S.N.B. Murthy, 189–208. New York, NY: Springer.

19. Wirth, M., Keller, P., and Peters, N. 1993. "A Flamelet Model for Premixed Turbulent Combustion in SI-Engines." *SAE Transactions* 102, no. 3 (October), 2200–2213. DOI: 10.4271/932646

20. Dekena, M., and Peters, N. 1999. "Combustion Modeling With the G-equation." *Oil & Gas Science and Technology* 54, no. 2 (March–April: 265–270. DOI: 10.2516/ogst:1999024

21. Tan, Z., and Reitz, R.D. 2003. "Modeling Ignition And Combustion in Spark-Ignition Engines Using a Level Set Method." *SAE Transactions* 112, no. 3 (March): 1028–1040. DOI: 10.4271/2003-01-0722

22. Peters, N. 1984. "Laminar Diffusion Flamelet Models in Non-Premixed Turbulent Combustion." *Progress in Energy and Combustion Science* 10, no. 3: 319–339. DOI: 10.1016/0360-1285(84)90114-X

23. Pitsch, H., Wan, Y.P., and Peters, N. 1995. "Numerical Investigation of Soot Formation And Oxidation Under Diesel Engine Conditions." *SAE Transactions* 104, no. 4 (October): 938–949. DOI: 10.4271/952357

24. Pitsch, H., Barths, H., and Peters, N. 1996. "Three-Dimensional Modeling of No_x And Soot Formation in DI-Diesel Engines Using Detailed Chemistry Based on the Interactive Flamelet Approach." *SAE Transactions* 105, no. 4 (October): 2010–2024. DOI: 10.4271/962057

25. Sethian, J.A. 1999. *Level Set Methods And Fast Marching Methods*. Cambridge: Cambridge University Press.

26. Williams, F.A. 1985. "Turbulent Combustion." In *The Mathematics of Combustion*, edited by J. Buckmaster, 97–131. Philadelphia, PA: Society for Industrial and Applied Mathematics.

27. Tan, Z., and Reitz, R.D. 2004. "Development of a Universal Turbulent Combustion Model for Premixed And Direct Injection Spark/Compression Ignition Engines." SAE Technical Paper No. 2004-01-0102. DOI: 10.4271/2004-01-0102

28. Liang, L., and Reitz, R.D. 2006. "Spark Ignition Engine Combustion Modeling Using a Level Set Method With Detailed Chemistry." SAE Technical Paper No. 2006-01-0243. DOI: 10.4271/2006-01-0243

29. Pope, S.B. 1985. "PDF Methods for Turbulent Reactive Flows." *Progress in Energy and Combustion Science* 11, no. 2: 119–192. DOI: 10.1016/0360-1285(85)90002-4

30. Pope, S.B. 2000. *Turbulent Flows*. Cambridge: Cambridge University Press.

31. Haworth, D.C. 2010. "Progress in Probability Density Function Methods for Turbulent Reacting Flows." *Progress in Energy and Combustion Science* 36, no. 2 (April): 168–259. DOI: 10.1016/j.pecs.2009.09.003

32. Kee, R.J., Rupley, F.M., and Miller, J.A. 1989. *Chemkin-II: A Fortran Chemical Kinetics Package for the Analysis of Gas-Phase Chemical Kinetics* (SAND89-8009). Livermore, CA: Sandia National Laboratory. DOI: 10.2172/5681118

33. Kong, S.C., Marriott, C.D., Reitz, R.D., and Christensen, M. 2001. "Modeling And Experiments of HCCI Engine Combustion Using Detailed Chemical Kinetics With Multidimensional CFD." *SAE Transactions* 110, no. 3 (March): 1007–1018. DOI: 10.4271/2001-01-1026

34. Amsden, A.A. 1997. *KIVA-3V: A Block-Structured KIVA Program for Engines With Vertical or Canted Valves* (LA-13313-MS). Los Alamos, New Mexico: Los Alamos National Laboratory.

35. Senecal, P.K., Pomraning, E., Richards, K.J., Briggs, T.E., Choi, C.Y., McDavid, R.M., and Patterson, M.A. 2003. "Multi-Dimensional Modeling of Direct-Injection Diesel Spray Liquid Length And Flame Lift-Off Length Using CFD And Parallel Detailed Chemistry." *SAE Transactions* 112, no. 3 (March): 1331–1351. DOI: 10.4271/2003-01-1043

36. Richards, K.J., Senecal, P.K., and Pomraning, E. 2017. *CONVERGE Manual (v2.4)*. Madison, WI: Convergent Science Inc.

37. Klimenko, A.Y., and Bilger, R.W. 1999. "Conditional Moment Closure for Turbulent Combustion." *Progress in Energy and Combustion Science* 25, no. 6 (December): 595–687. DOI: 10.1016/S0360-1285(99)00006-4

38. Seo, J., Lee, Y., Han, I., Huh, K.Y., and Kim, H. 2008. "Extended CMC Model for Turbulent Spray Combustion in a Diesel Engine." *SAE Technical Paper* No. 2008-01-2411. DOI: 10.4271/2008-01-2411

39. Wright, Y.M., Boulouchos, K., De Paola, G., and Mastorakos, E. 2009. "Multi-Dimensional Conditional Moment Closure Modelling Applied to a Heavy-Duty Common-Rail Diesel Engine." *SAE International Journal of Engines* 2, no. 1 (April): 714–726. DOI: 10.4271/2009-01-0717

40. Rutland, C.J. (2011). "Large-Eddy Simulations for Internal Combustion Engines—A Review." *International Journal of Engine Research* 12, no. 5 (May): 421–451. DOI: 10.1177/1468087411407248

41. Reitz, R.D., and Kuo, T. 1989. "Modeling of HC Emissions Due to Crevice Flows in Premixed-Charge Engines." *SAE Transactions* 98, no. 4 (September): 922–939. DOI: 10.4271/892085

42. Hampson, G.J., Xin, J., Liu, Y., Han, Z., and Reitz, R.D. 1996. "Modeling of No_x Emissions With Comparison to Exhaust Measurements for a Gas Fuel Converted Heavy-Duty Diesel Engine." *SAE Transactions* 105, no. 4 (October): 1503–1517. DOI: 10.4271/961967

43. Fan, L., Li, G., Han, Z., and Reitz, R.D. 1999. "Modeling Fuel Preparation And Stratified Combustion in a Gasoline Direct Injection Engine." *SAE Transactions* 106, no. 3 (March): 105–119. DOI: 10.4271/1999-01-0175

44. Han, Z., and Reitz, R.D. 1995. "Turbulence Modeling of Internal Combustion Engines Using RNG K-E Models." *Combustion Science and Technology* 106, no. 4–6: 267–295. DOI: 10.1080/00102209508907782

45. Reitz, R.D. 1987. "Modeling Atomization Processes in High-Pressure Vaporizing Sprays." *Atomisation Spray Technology* 3, no. 4: 309–337.

46. Xin, J., Montgomery, D., Han, Z., and Reitz, R.D. 1997. "Multidimensional Modeling of Combustion for a Six-Mode Emissions Test Cycle on a DI Diesel Engine." *Journal of Engineering for Gas Turbines and Power* 119, no. 3 (July): 683–691. DOI: 10.1115/1.2817041

47. Lee, D., and Rutland, C.J. 2002. "Probability Density Function Combustion Modeling of Diesel Engines." *Combustion Science and Technology* 174, no. 10: 19–54. DOI: 10.1080/00102200290021489

48. Hu, B., and Rutland, C.J. 2006. "Flamelet Modeling With LES for Diesel Engine Simulations." *SAE Technical Paper* No. 2006-01-0058. DOI: 10.4271/2006-01-0058

49. Pauls, C., Grünefeld, G., Vogel, S., and Peters, N. 2007. "Combined Simulations And OH-Chemiluminescence Measurements of The Combustion Process Using Different Fuels Under Diesel-Engine Like Conditions." *SAE Transactions* 116, no. 4 (January): 1–17. DOI: 10.4271/2007-01-0020

50. Haworth, D.C. 2000. "A Probability Density Function/Flamelet Method for Partially Premixed Turbulent Combustion." In *Proceedings of the 2000 Summer Program* , 145–156. Stanford, CA: Center for Turbulence Research, NASA Ames/Stanford University.

51. Hu, B., Jhavar, R., Singh, S., Reitz, R., and Rutland, C. 2007. "Combustion Modeling of Diesel Combustion with Partially PremixedConditions." *SAE Technical Paper* No. 2007-01-0163. DOI: 10.4271/2007-01-0163

52. Göttgens, J., Mauss, F., and Peters, N. 1992. "Analytic Approximations of Burning Velocities And Flame Thicknesses of Lean Hydrogen, Methane, Ethylene, Ethane, Acetylene, And Propane Flames." *Symposium (International) on Combustion* 24, no. 1: 129–135. DOI: 10.1016/S0082-0784(06)80020-2

53. Amirante, R., Distaso, E., Tamburrano, P., and Reitz, R.D. 2017. "Laminar Flame Speed Correlations for Methane, Ethane, Propane And Their Mixtures, And Natural Gas And Gasoline for Spark-Ignition Engine Simulations." *International Journal of Engine Research* 18, no. 9 (July): 951–970. DOI: 10.1177/1468087417720018

54. Metghalchi, M., and Keck, J.C. 1982. "Burning Velocities of Mixtures of Air With Methanol, Isooctane, And Indolene at High Pressure And Temperature." *Combustion and Flame* 48: 191–210. DOI: 10.1016/0010-2180(82)90127-4

55. Gülder, Ö.L. 1984. "Correlations of Laminar Combustion Data for Alternative SI Engine Fuels." *SAE Technical Paper* No. 841000. DOI: 10.4271/841000

56. Metghalchi, M., and Keck, J. 1983. "Burning velocities of methanol, ethanol and iso-octane-air mixtures." In *19th Symposium (International) on Combustion*, 275. Pittsburgh, Pennsylvania: The Combustion Institute.

57. Liang, L., Reitz, R.D., Iyer, C.O., and Yi, J. 2007. "Modeling Knock in Spark-Ignition Engines Using a G-Equation Combustion Model Incorporating Detailed Chemical Kinetics." SAE Technical Paper No. 2007-01-0165. DOI: 10.4271/2007-01-0165

58. Singh, S., Reitz, R.D., Wickman, D., Stanton, D., and Tan, Z. 2007. "Development of a Hybrid, Auto-Lgnition/Flame-Propagation Model And Validation Against Engine Experiments And Flame Liftoff." *SAE Transactions* 116, no. 3 (April): 176–194. DOI: 10.4271/2007-01-0171

59. Yang, S., and Reitz, R.D. 2009. "Improved Combustion Submodels for Modelling Gasoline Engines With the Level Set G Equation And Detailed Chemical Kinetics." *Proceedings of the Institution of Mechanical Engineers, Part D: Journal of Automobile Engineering* 223, no. 5 (May): 703–726. DOI: 10.1243/09544070JAUTO1062

60. Li, J., Han, Z., Shen, C., and Lee, C.-f. 2014. "A Study on Biodiesel No$_x$ Emission Control With the Reduced Chemical Kinetics Model." *Journal of Engineering for Gas Turbines and Power* 136, no. 10: article 101505. DOI: 10.1115/1.4027358

61. Curran, H.J., Gaffuri, P., Pitz, W.J., and Westbrook, C.K. 2002. "A Comprehensive Modeling Study of Iso-Octane Oxidation." *Combustion and Flame* 129, no. 3 (May): 253–280. DOI: 10.1016/S0010-2180(01)00373-X

62. Pomraning, E., Richards, K., and Senecal, P.K. 2014. "Modeling Turbulent Combustion Using a RANS Model, Detailed Chemistry, And Adaptive Mesh Refinement." SAE Technical Paper No. 2014-01-1116. DOI: 10.4271/2014-01-1116

63. Singh, S., Reitz, R.D., and Musculus, M.P.B. 2006. "Comparison of The Characteristic Time (CTC), Representative Interactive Flamelet (RIF), And Direct Integration With Detailed Chemistry Combustion Models Against Optical Diagnostic Data For Multi-Mode Combustion in a Heavy-Duty DI Diesel Engine." *SAE Transactions* 115, no. 3 (April): 61–82. DOI: 10.4271/2006-01-0055

64. Ra, Y., Loeper, P., Reitz, R., Andrie, M., Krieger, R., Foster, D.E., Durrett, R., Gopalakrishnan, V., Plazas, A., Peterson, R., and Szymkowicz, P. 2011. "Study of High Speed Gasoline Direct Injection Compression Ignition (GDICI) Engine Operation in the LTC Regime." *SAE International Journal of Engines* 4, no. 1 (April): 1412–1430. DOI: 10.4271/2011-01-1182

65. Kong, S.C. 2010. "Simulation of Low Temperature Combustion in Engines." In *Handbook of Combustion, Vol 5,* 35–50. Weinheim: Wiley-VCH.

66. Curran, H.J., Gaffuri, P., Pitz, W.J., and Westbrook, C.K. 1998. "A Comprehensive Modeling Study of N-Heptane Oxidation." *Combustion and Flame* 114, no. 1(July): 149–177. DOI: 10.1016/S0010-2180(97)00282-4

67. Herbinet, O., Pitz, W.J., and Westbrook, C.K. 2008. "Detailed Chemical Kinetic Oxidation Mechanism for a Biodiesel Surrogate." *Combustion and Flame* 154, no. 3 (August): 507–528. DOI: 10.1016/j.combustflame.2008.03.003

68. Curran, H.J. 2019. "Developing Detailed Chemical Kinetic Mechanisms for Fuel Combustion." *Proceedings of the Combustion Institute* 37, no. 1: 57–81. DOI: 10.1016/j.proci.2018.06.054

69. Lu, T., and Law, C.K. 2009. "Toward Accommodating Realistic Fuel Chemistry in Large-Scale Computations." *Progress in Energy and Combustion Science* 35, no. 2 (April): 192–215. DOI: 10.1016/j.pecs.2008.10.002

70. Zhen, X., Wang, Y., and Liu, D. 2017. "An Overview of the Chemical Reaction Mechanisms for Gasoline Surrogate Fuels." *Applied Thermal Engineering* 124 (September): 1257–1268. DOI: 10.1016/j.applthermaleng.2017.06.101

71. Tomlin, A.S., Pilling, M.J., Turányi, T., Merkin, J.H., and Brindley, J. 1992. "Mechanism Reduction for the Oscillatory Oxidation Of Hydrogen: Sensitivity And Quasi-Steady-State Analyses." *Combustion and Flame* 91, no. 2 (November (September): 107–130. DOI: 10.1016/0010-2180(92)90094-6

72. Turanyi, T. 1990. "Sensitivity Analysis of Complex Kinetic Systems. Tools And Applications." *Journal of Mathematical Chemistry* 5, no. 3: 203–248. DOI: 10.1007/BF01166355

73. Vajda, S., Valko, P., and Turanyi, T. 1985. "Principal Component Analysis of Kinetic Models." *International Journal of Chemical Kinetics* 17, no. 1 (January): 55–81. DOI: 10.1002/kin.550170107

74. Vajda, S., and Turanyi, T. 1986. "Principal Component Analysis for Reducing the Edelson-Field-Noyes Model of the Belousov-Zhabotinskii Reaction." *The Journal of Physical Chemistry* 90, no. 8 (April): 1664–1670. DOI: 10.1021/j100399a042

75. Turanyi, T. 1990. "Reduction of Large Reaction Mechanisms." *New Journal of Chemistry* 14, no. 11: 795–803.

76. Valorani, M., Creta, F., Goussis, D.A., Lee, J.C., and Najm, H.N. 2006. "An Automatic Procedure for the Simplification of Chemical Kinetic Mechanisms Based on CSP." *Combustion and Flame* 146, no. 1–2: 29–51. DOI: 10.1016/j.combustflame.2006.03.011

77. Lu, T., and Law, C.K. 2005. "A Directed Relation Graph Method for Mechanism Reduction." *Proceedings of the Combustion Institute* 30, no. 1 (January): 1333–1341. DOI: 10.1016/j.proci.2004.08.145

78. Pepiot-Desjardins, P., and Pitsch, H. 2008. "An Efficient Error-Propagation-Based Reduction Method for Large Chemical Kinetic Mechanisms." *Combustion and Flame* 154, no. 1–2 (July): 67–81. DOI: 10.1016/j.combustflame.2007.10.020

79. Sun, W., Chen, Z., Gou, X., and Ju, Y. 2010. "A Path Flux Analysis Method for the Reduction of Detailed Chemical Kinetic Mechanisms." *Combustion and Flame* 157, no. 7 (July), 1298–1307. DOI: 10.1016/j.combustflame.2010.03.006

80. Ahmed, S.S., Mauß, F., Moréac, G., and Zeuch, T. 2007. "A Comprehensive And Compact N-Heptane Oxidation Model Derived Using Chemical Lumping." *Physical Chemistry Chemical Physics* 9, no. 9: 1107–1126. DOI: 10.1039/b614712g

81. Li, G., Rabitz, H., and Tóth, J. 1994. "A General Analysis of Exact Nonlinear Lumping in Chemical Kinetics." *Chemical Engineering Science* 49, no. 3: 343–361. DOI: 10.1016/0009-2509(94)87006-3

82. Ranzi, E., Dente, M., Goldaniga, A., Bozzano, G., and Faravelli, T. 2001. "Lumping Procedures in Detailed Kinetic Modeling of Gasification, Pyrolysis, Partial Oxidation And Combustion of Hydrocarbon Mixtures." *Progress in Energy and Combustion Science* 27, 1 (January): 99–139. DOI: 10.1016/S0360-1285(00)00013-7

83. Zheng, X.L., Lu, T.F., and Law, C.K. 2007. "Experimental Counterflow Ignition Temperatures And Reaction Mechanisms of 1,3-Butadiene." *Proceedings of the Combustion Institute* 31, no. 1(January): 367–375. DOI: 10.1016/j.proci.2006.07.182

84. Lu, T., and Law, C.K. 2008. "Strategies for mechanism reduction for large hydrocarbons: n-heptane." *Combustion and Flame* 154, no. 1–2 (July): 153–163. DOI: 10.1016/j.combustflame.2007.11.013

85. Niemeyer, K.E., Sung, C.J., and Raju, M.P. 2010. "Skeletal Mechanism Generation for Surrogate Fuels Using Directed Relation Graph With Error Propagation And Sensitivity Analysis." *Combustion and Flame* 157, no. 9 (September): 1760–1770. DOI: 10.1016/j.combustflame.2009.12.022

86. Li, J. 2014. Study on the Development of a Biodiesel Chemical Mechanism Model And its Application to Engine Combustion Analysis (Publication No. TK421.2). Changsha, China: Hunan University. China Doctoral Dissertations Full-text Database.

87. Fieweger, K., Blumenthal, R., and Adomeit, G. 1994. "Shock-Tube Investigations on the Self-Ignition of Hydrocarbon-Air Mixtures at High Pressures." *Symposium (International) on Combustion* 25, 1" 1579–1585. DOI: 10.1016/S0082-0784(06)80803-9

88. Fieweger, K., Blumenthal, R., and Adomeit, G. 1997. "Self-Ignition of SI Engine Model Fuels: A Shock Tube Investigation at High Pressure." *Combustion and Flame* 109, no. 4: 599–619. DOI: 10.1016/S0010-2180(97)00049-7

89. Wang, W., and Oehlschlaeger, M.A. 2012. "A Shock Tube Study of Methyl Decanoate Autoignition at Elevated Pressures." *Combustion and Flame* 159, no. 2 (February): 476–481. DOI: 10.1016/j.combustflame.2011.07.019

90. Shi, Y., Ge, H.W., Brakora, J.L., and Reitz, R.D. 2010. "Automatic Chemistry Mechanism Reduction of Hydrocarbon Fuels for HCCI Engines Based on DRGEP And PCA Methods With Error Control." *Energy & Fuels* 24, no. 3: 1646–1654. DOI: 10.1021/ef901469p

91. Patel, A., Kong, S., and Reitz, R.D. 2004. "Development And Validation of a Reduced Reaction Mechanism for HCCI Engine Simulations." SAE Technical Paper No. 2004-01-0558. DOI: 10.4271/2004-01-0558

92. Brakora, J.L., and Reitz, R.D. 2010. "Investigation Of No_x Predictions From Biodiesel-Fueled HCCI Engine Simulations Using a Reduced Kinetic Mechanism." SAE Technical Paper No. 2010-01-0577. DOI: 10.4271/2010-01-0577

93. Liang, L., Stevens, J.G., and Farrell, J.T. 2009. "A Dynamic Adaptive Chemistry Scheme For Reactive Flow Computations." *Proceedings of the Combustion Institute* 32, no. 1: 527–534. DOI: 10.1016/j.proci.2008.05.073

94. Shi, Y., Ge, H.W., and Reitz, R.D. 2011. *Computational Optimization of Internal Combustion Engines.* London: Springer London.

95. Duclos, J.M., and Colin, O. 2001. "Arc And Kernel Tracking Ignition Model For 3D Spark Ignition Engine Calculations." In *Proceedings of the Internaltional Symposium on Diagnostics and Modeling of Combustion in Internal Combustion Engines,* 343–350. Nagoya, Japan. DOI: 10.1299/jmsesdm.01.204.46

96. Dahms, R., Fansler, T.D., Drake, M.C., Kuo, T.W., Lippert, A.M., and Peters, N. 2009. "Modeling Ignition Phenomena in Spray-Guided Spark-Ignited Engines." *Proceedings of the Combustion Institute* 32, no. 2: 2743–2750. DOI: 10.1016/j.proci.2008.05.052

97. Pyszczek, R., Hahn, J., Priesching, P., and Teodorczyk, A. 2020. "Numerical Modeling of Spark Ignition in Internal Combustion Engines." *Journal of Energy Resources Technology* 142, no. 2 (February): article 022202. DOI: 10.1115/1.4044222

98. Tan, Z., and Reitz, R.D. 2006. "An Ignition And Combustion Model Based on the Level-Set Method for Spark Ignition Engine Multidimensional Modeling." *Combustion and Flame* 145, no. 1–2(April): 1–15. DOI: 10.1016/j.combustflame.2005.12.007

99. Halstead, M.P., Kirsch, L.J., and Quinn, C.P. 1977. "The Autoignition of Hydrocarbon Fuels at High Temperatures And Pressures—Fitting of a Mathematical Model." *Combustion and Flame* 30: 45–60. DOI: 10.1016/0010-2180(77)90050-5

100. Igura, S., Kadota, T., and Hiroyasu, H. 1975. "Spontaneous Ignition Delay of Fuel Sprays in High Pressure Gaseous Environment." *Transactions of the Japan Society of Mechanical Engineers* 41, no. 345: 1559–1566.

101. Heywood, J.B. 1976. "Pollutant Formation And Control in Spark-Ignition Engines." *Progress in Energy and Combustion Science* 1, no. 4: 135–164. DOI: 10.1016/0360-1285(76)90012-5

102. Lavoie, G.A., Heywood, J.B., and Keck, J.C. 1970. "Experimental And Theoretical Study of Nitric Oxide Formation in Internal Combustion Engines." *Combustion Science and Technology* 1, no. 4: 313–326. DOI: 10.1080/00102206908952211

103. Heywood, J.B. 2018. *Internal Combustion Engine Fundamentals.* 2nd ed. New York: McGraw-Hill Education.

104. Kong, S.C., Sun, Y., and Rietz, R.D. 2007. "Modeling Diesel Spray Flame Liftoff, Sooting Tendency, And No_x Emissions Using Detailed Chemistry With Phenomenological Soot Model." *Journal of Engineering for Gas Turbines and Power* 129, no. 1 (January): 245–251. DOI: 10.1115/1.2181596

105. Kittelson, D., and Kraft, M. 2014. "Particle Formation And Models." In *Encyclopedia of Automotive Engineering*, edited by D. Crolla, D.E. Foster, T. Kobayashi, and N. Vaughan, 1–23. Hoboken, New Jersey: John Wiley & Sons.

106. Han, Z., Yi, J., and Trigui, N. 2002. "Stratified Mixture Formation And Piston Surface Wetting in a DISI Engine." SAE Technical Paper No. 2002-01-2655. DOI: 10.4271/2002-01-2655

107. Wooldridge, S., Lavoie, G., and Weaver, C. 2003. "Convection Path for Soot And Hydrocarbon Emissions From the Piston Bowl of a Stratified Charge Direct Injection Engine." In *Proceedings of the Third Joint Meeting of the US Sections of the Combustion Institute*, March 16–19, 1–6. Chicago: University of Illinois.

108. Drake, M.C., Fansler, T.D., Solomon, A.S., and Szekely, G.A. 2003. "Piston Fuel Films as a Source of Smoke And Hydrocarbon Emissions From a Wall-Controlled Spark-Ignited Direct-Injection Engine." *SAE Transactions* 112, no. 3 (March): 762–783. DOI: 10.4271/2003-01-0547

109. Han, Z., Weaver, C., Wooldridge, S., Alger, T., Hilditch, J., McGee, J., Westrate, B., et al. 2004. "Development of a New Light Stratified-Charge DISI Combustion System for a Family of Engines With Upfront CFD Coupling With Thermal And Optical Engine Experiments." *SAE Transactions* 113, no. 3 (March): 269–293. DOI: 10.4271/2004-01-0545

110. Hiroyasu, H., and Kadota, T. 1976. "Models For Combustion And Formation of Nitric Oxide And Soot in Direct Injection Diesel Engines." *SAE Transactions* 85, no. 1 (February): 513–526. DOI: 10.4271/760129

111. Patterson, M.A., Kong, S.C., Hampson, G.J., and Reitz, R.D. 1994. "Modeling the Effects of Fuel Injection Characteristics on Diesel Engine Soot And No$_x$ Emissions." *SAE Transactions* 103, no. 3 (March): 836–852. DOI: 10.4271/940523

112. Nagle, J., and Strickland-Constable, R.F. 1962. "Oxidation of Carbon Between 1000–2000°C." In *Proceedings of the Fifth Carbon Conference, Pennsylvania State University, University Park, Pennsylvania, Vol. 1*, 154. New York: Pergammon Press.

113. Donahue, R.J., Borman, G.L., and Bower, G.R. 1994. "Cylinder-Averaged Histories of Nitrogen Oxide in a DI Diesel With Simulated Turbocharging." *SAE Transactions* 103, no. 4 (October): 1789–1801. DOI: 10.4271/942046

114. Kazakov, A., and Foster, D.E. 1998. "Modeling of Soot Formation During DI Diesel Combustion Using a Multi-Step Phenomenological Model." *SAE Transactions* 107, no. 4 (October): 1016–1028. DOI: 10.4271/982463

115. Balthasar, M., and Kraft, M. 2003. "A Stochastic Approach to Calculate the Particle Size Distribution Function of Soot Particles in Laminar Premixed Flames." *Combustion and Flame* 133, no. 3 (May): 289–298. DOI: 10.1016/S0010-2180(03)00003-8

116. Tao, F., Foster, D.E., and Reitz, R.D. 2006. "Soot Structure in a Conventional Non-Premixed Diesel Flame." *SAE Transactions* 115, no. 4 (April): 24–40. DOI: 10.4271/2006-01-0196

117. Vishwanathan, G., and Reitz, R.D. 2008. "Modeling Soot Formation Using Reduced PAH Chemistry in N-Heptane Lifted Flames With Application to Low Temperature Combustion." In *Proceedings of the ASME 2008 Internal Combustion Engine Division Spring Technical Conference, April 27–30, Chicago, Illinois, Vol. 131*, 29–36. New York, NY: ASME Digital Library.

118. Jiao, Q., and Reitz, R.D. 2015. "Modeling Soot Emissions From Wall Films in a Direct-Injection Spark-Ignition Engine." *International Journal of Engine Research* 16, no. 8 (December): 994–1013. DOI: 10.1177/1468087414562008

7

Optimization of Direct-Injection Gasoline Engines

In an IC engine, air and fuel (and EGR sometimes) are introduced into the combustion chamber where exothermic chemical reactions take place. As a result, torque is generated, harmful pollutants such as CO, hydrocarbon, NO_x, particulates, and greenhouse-gas CO_2 are exhausted at the cost of fuel consumption. Thus, the goal of combustion development and combustion-system design optimization of an IC engine is to maximize torque (power) output with minimum fuel consumption and, at the same time, to minimize the internal formation of pollutants in a cost-effective way. Improving designs of air induction path, fuel injection strategy, and combustion-chamber geometry, exploring new methods and strategies, and reducing the cost of materials and time of development are all the tasks of combustion development and optimization. Numerical engine simulation has been proven to be of great help to these tasks.

In this chapter, methodology for engineering simulation application is introduced. Often used CFD codes and software for IC engines are briefly described. Optimization of fuel–air mixing in both homogeneous-charge and stratified-charge mode in DI gasoline engines is discussed with examples. Formation of soot and unburned hydrocarbon from fuel wall-wetting is addressed and designs for wall-wetting reduction are described in the examples. The intention of these examples is to focus on the key issues in these engines that impact combustion and performance greatly. It is hoped that readers are able to understand the fundaments of these examples and be inspired to explore new ideas and means for better solutions in their study and work.

7.1 Advanced Combustion Development Methodology

7.1.1 Modeling-Driven Approach

The phenomena in engines are highly complex and can be characterized as unsteady and 3D, turbulent and reactive, spray and gas two-phase, and under rapid volume change in complicated geometries according to physics. On the other hand, from an engineering point of view, engine processes involve complicated interactions among subsystems and components in multiple design-parameter spaces, and it is required to deliver robust function attributes (fuel economy, performance, emissions, drivability, NVH, etc.) with a cost-effective design. At the technical level of today's engines and with the complexity of the combustion processes in modern engines, traditional experiment-based trial-and-error methods can no longer be of help. Advanced R&D technology and methodology are called upon for modern engine combustion development.

A modeling-driven design approach has been developed at the FRL of Ford Motor Company that has been used in many internal engine development programs. A number of DI combustion concepts using this approach were reported publicly [1–5]. The approach systematically integrates numerical simulation, optical-engine diagnostics, and thermal-engine tests with the core connotation of upfront CFD design optimization and hardware test validation. It was proven effective and efficient in the production-intent LSC DI program with an improvement of 10.1% in vehicle fuel economy compared to the baseline PFI engine over the NEDC cycle [6] in a reduced development time period. A summary of this approach for DI gasoline engines is shown in Figure 7.1. This methodology can be also applied to other engines that involve complicated interactions of gas motion, sprays, and combustion chamber geometry.

The methodology starts with a target cascading process using vehicle simulation to identify projected engine-out, vehicle-level fuel economy, and emissions performance and to cascade the fuel consumption and emissions targets to a set of engine load and speed mini-map points [6]. The process is then followed by a design and development stage working through experimental validation. The methodology relies heavily on upfront detailed CFD simulation of the intake and in-cylinder flow, spray atomization and evaporation, mixing, and combustion processes. Various engine operating conditions are explored to determine an optimized design specification of the components that make up the combustion system: combustion chamber shape, injector position, port shape, piston bowl/crown shape, spray pattern specification, cam timings, and injection strategy, etc. At Ford FRL, an in-house CFD model MESIM has been used, which is reviewed in Chapter 1.

The analytically developed design is then replicated in hardware for an optically accessible single-cylinder engine and a thermodynamic single-cylinder engine for design verification. At this stage, the experimental components work concurrently with iterative CFD simulations of the hardware to identify and verify various performance characteristics of the combustion system relative to the targets/guidelines. Final validation of the overall emissions and fuel economy performance is evaluated across the engine performance map using vehicle simulation methods and multicylinder engine experiments. One advantage of such a methodology is that it significantly reduces the development time and cost while achieving a high-quality design with insights into the details and interactions of the in-jcylinder physical phenomena.

FIGURE 7.1   Advanced combustion system design methodology.

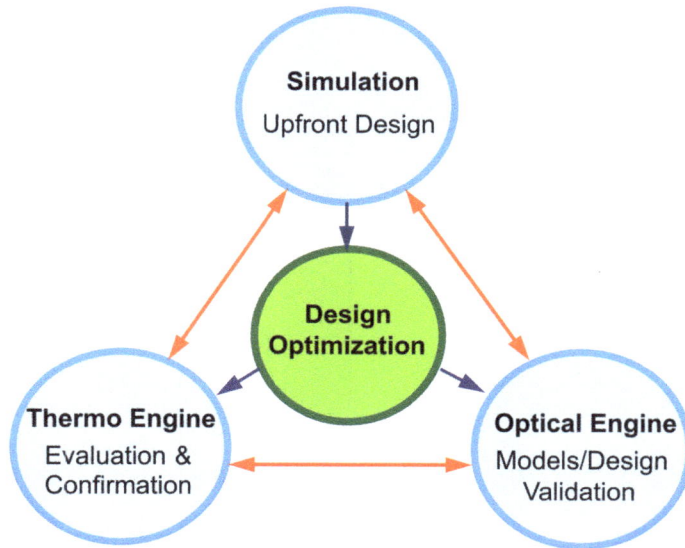

Optical-engine diagnostics provide tools to visualize fuel spray and flame behavior inside an engine temporally to gain insights into the engine combustion process. The single-cylinder optical engine should be identical in design to the thermodynamic engine regarding the combustion system features and engine architecture but is modified for optical access. Various laser-enabled optical diagnostic techniques have been used in engines with optical accesses [7–9]. Optical access arrangements can be made in an optical research engine with full optical access, and in a metal production engine with limited optical access for endoscopy. An optical research engine allows full optical access with a high-quality quartz liner and a transparent piston to the engine combustion chamber, which is used for studying the fuel–air mixing, in-cylinder flows, combustion, and flame propagation. The optical engine can run at conditions close to its metal counterparts. Engine optical diagnostics can reveal new phenomena in the combustion process, and its measured data can be used to validate CFD models. However, optical research engines can only be operated at low-to-medium loads and speeds in a highly customized test cell environment because of the added mass to the elongated piston, which causes extra inertia forces, and the limited structural strength of the optical window.

An optical engine used in Ford FRL is shown in Figure 7.2. It is modified for optical access with a piston extension to accommodate a "Bowditch" style piston and a drop-down liner that can be replaced with a full-length transparent (fused silica) liner. In addition, a separate window installed in the cylinder head provides optical access to the pent-roof combustion chamber. In-cylinder optical measurements are performed in that engine as part of the LSC combustion system development. These include fuel-tracer planar laser-induced fluorescence (PLIF) for evaluation of mixture formation, PIV for in-cylinder flow measurement, laser Mie scattering for fuel spray and soot particle imaging, and combustion luminosity [10–13].

FIGURE 7.2 An optical engine with optical extension.

Different diagnostic techniques are suitable for specific physical properties. Hotwire anemometry (HWA), laser Doppler anemometry (LDA), and PIV are widely known methods for flow measurements. Mie scattering, PDPA, and laser sheet droplet (LSD) sizing are used to measure fuel drop size. Laser Rayleigh scattering (LRS), spontaneous Raman scattering (SRS), and laser induced fluorescence (LIF) are for the measurements of fuel vapor concentrations, and laser induced exciplex fluorescence (LIEF) and dual wavelength laser extinction/absorption (LEA) are capable of simultaneous measurements of fuel vapors and liquid droplets. In-cylinder flame visualization can be carried out by a high-speed CCD camera and associated visualization equipment (e.g., LIF).

As part of the overall development methodology, a single-cylinder thermodynamic engine is used to validate the design by demonstrating engine part-load fuel consumption and emissions, and full-load performance targets [6,14]. Although multicylinder engines are used in a later phase, the single-cylinder engine can provide a number of unique advantages for combustion system development, including rapid manufacturing, low cost, and isolation of cylinder-to-cylinder variations (which could have complicated development on multicylinder engines). Nonetheless, the single-cylinder engine also presents a number of unique challenges, including unrepresentative friction and manifold tuning; these limitations can be mitigated through careful model-based load and boost setting, respectively.

Engines are required to operate at a wide speed-load range that requires quite different operation strategies. Design compromises are often made with a balanced design. In many cases, engines are operated with mode changes according to the operating conditions in which combustion is organized differently. As discussed in Chapter 2, stratified-charge gasoline combustion mode or LTC combustion mode is used at engine part-loads for fuel economy and stochiometric or enriched homogeneous-charge combustion mode is employed at high to full loads for torque output. To ensure the desired mixture stratification at stratified-charge operation and mixture homogeneity at homogeneous-charge operation, a series of simulations are usually performed at selected critical mini-map points that have unique mixture formation requirements and challenges.

Typical numerical simulation points for an stratified-charge combustion system are demonstrated in Figure 7.3 in which the focus of modeling analysis at each point is explained. For example, port design, injector specification (spray pattern), flow control, and injection strategy are optimized to ensure good mixture homogeneity at WOT homogeneous-charge operations; at the same time valve wetting must be minimized to minimize engine smoke at low-speed WOT through injector specification optimization. On the other hand, the candidate injector specification must deliver desired mixture stratification at part-load stratified-charge operation to achieve stable combustion, minimized engine smoke, and reduced NO_x formation. For engine cold-start (not shown in the diagram), particular fuel injection strategies (e.g., multiple injections) are needed for quick starts and emissions reduction, and the candidate injector and combustion chamber must be verified for those purposes.

FIGURE 7.3 Modeling points and design metrics for a SCDI system.

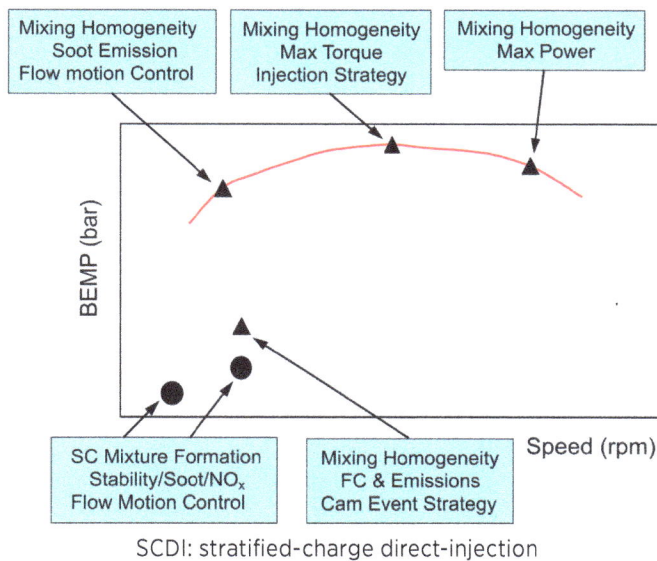

SCDI: stratified-charge direct-injection

Although the approach of multipoint simulation and the corresponding design metrics at each modeling point is applicable for other types of combustion development (e.g., diesel or LTC engines), the particular details are dependent on specific combustion concepts. Issues and design targets should be identified first in order to generate such a modeling task map as shown in Figure 7.3, which requires physical insights of the focused combustion. For example, the ratio of the dual fuels in an RCCI engine is used to control engine combustion smoothness; the ratio as well as the in-cylinder fuel injection strategy, are very important to extend the RCCI operation range. In a diesel engine, the bowl-in-piston combustion chamber geometry and its relation with spray jets are crucial for emissions formation. Since spray tip penetration is not scaled with engine speed, the design of injector specification or spray pattern and its relation with bowl-in-piston chamber must be evaluated at both high- and low-speed conditions.

For effective and rapid design iteration and sign-off, a systematic modeling-driven process needs to be implemented in the whole design process. A basic flowchart is illustrated in Figure 7.4 which outlines the main steps in the process. The major design focuses are injector specifications (spray pattern), combustion chamber shape, and intake ports. Other design and operation parameters (e.g., cam timing, injection strategy, EGR, etc.) are included in the engine CFD simulations. After simulations are made, hardware is built and experiments on the thermodynamic and optical engine are carried out. The measurements in the engine tests can be fed back to validate the CFD model and simulation results. The above steps may need to be iterated until satisfactory engine combustion results are obtained. The key point of this process is a numerical simulation upfront before hardware is built. Thus, many design iterations are performed on the computers. By doing so, development time and cost are reduced.

CFD simulation is often challenged by its accuracy and outputs. The accuracy can be improved using high-fidelity models at cost of computational cycle time. One big challenge is that CFD simulation often cannot, at the current stage, produce engine functional results that can be directly measured by engine tests, e.g., hydrocarbon emissions, particulate size, number density, combustion stability measured by standard deviation, misfire counts, and KLSA. All these parameters are critical to judge the quality of a combustion system. It is very difficult, if not impossible for current engine engineering simulations to directly generate these results due to the limits of model predictivity and computational time cost if more advanced models are used.

However, detour solutions to this issue can be obtained by the adoption of a combined deterministic-empirical approach, which is illustrated in Figure 7.5. With this approach, CFD simulation accurately predicts some in-cylinder intermediate variables I, which respond to the changes of design parameters D and then relate these variables to engine functions F using empirical correlations. Thus, engine functions influenced by design parameters can be indirectly evaluated by CFD simulation. Mathematically, in a deterministic approach, F is a function of D, or $F = f(D)$; while in a combined deterministic-empirical approach, I is a function $g(D)$ of D, and F is a correlation $h(I)$ of I, thus, $F \approx h(g(D))$. Bold typing here indicates multiple variables.

CFD can deterministically predict cylinder pressure, indicated work (or power with engine speed), and indicated thermal efficiency. However, it cannot deterministically predict engine-out soot and hydrocarbon emissions in a DI gasoline engine. Since wall-wettings are believed to be the main source of these emissions, CFD is then used to deterministically predict fuel depositions due to spray wall impingement. Fuel depositions are correlated with engine-out soot and hydrocarbon emissions qualitatively or quantitively. Thus, the effects of design and injection strategy on engine-out soot and hydrocarbon emissions can be evaluated and reduced by minimizing spray wall impingement through modeling-driven design optimization [1,15]. The combined deterministic–empirical approach is proven to be very effective, and it is the key to find out the intermediate variables and their correlations with engine functions. This calls upon fundamental research and expert knowledge. Many examples using this approach are given in the following sections.

7.1.2 Overview of Optimization Algorithms

Engine simulations can be used to guide engine design optimization. However, the current design optimization activities heavily rely on engineers' expertise and intuition. Since engine optimization is a multiobjective problem, it involves simultaneously optimizing multiple design parameters. As the complexity of engine combustion grows in modern engines, the considered design and operation parameters increase dramatically. Sound multiple parameter and multiobjective mathematical methods are needed for engine computational optimization.

FIGURE 7.4   Modeling-driven combustion development process.

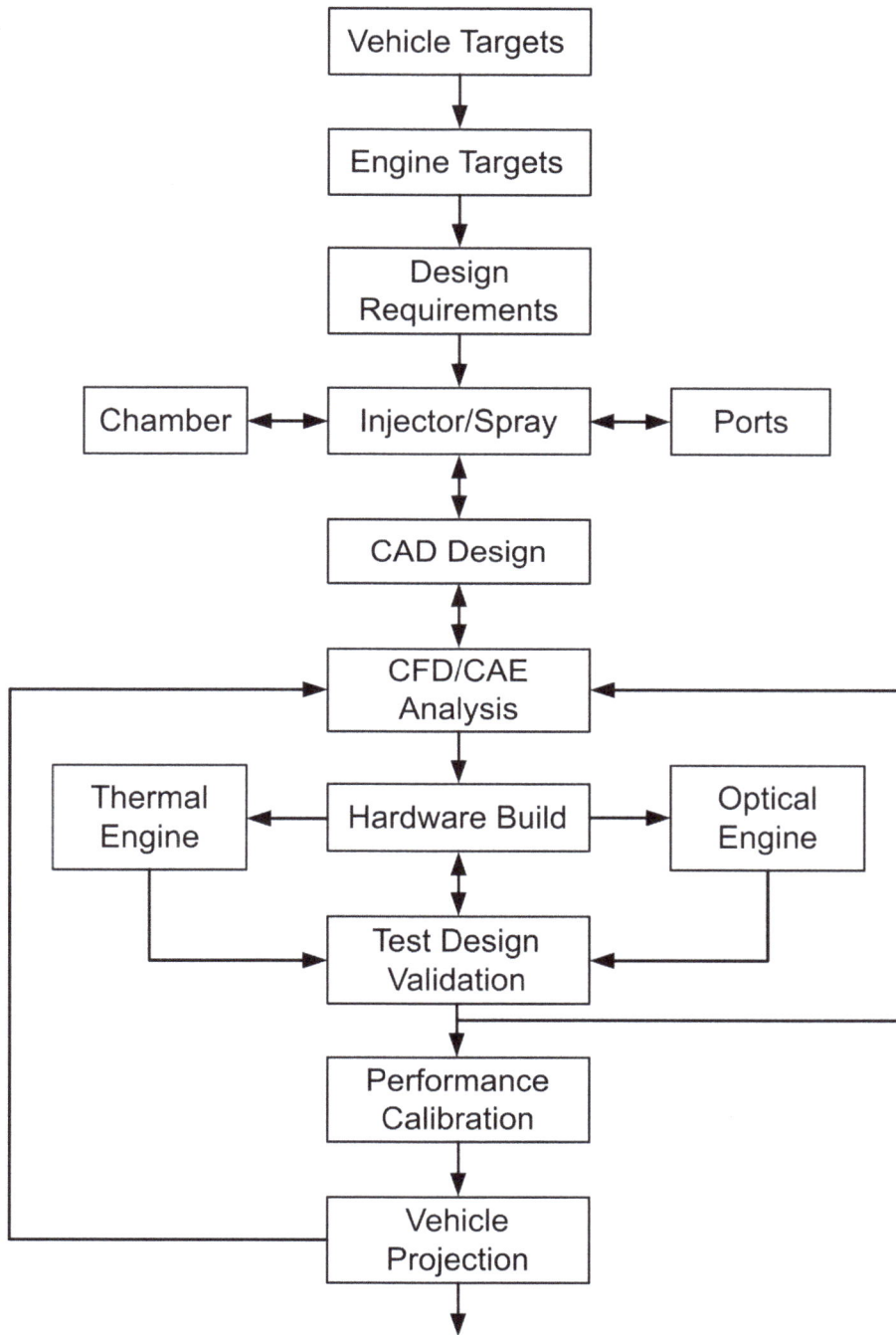

FIGURE 7.5 Conceptual diagram of a combined deterministic-empirical approach for engine simulation.

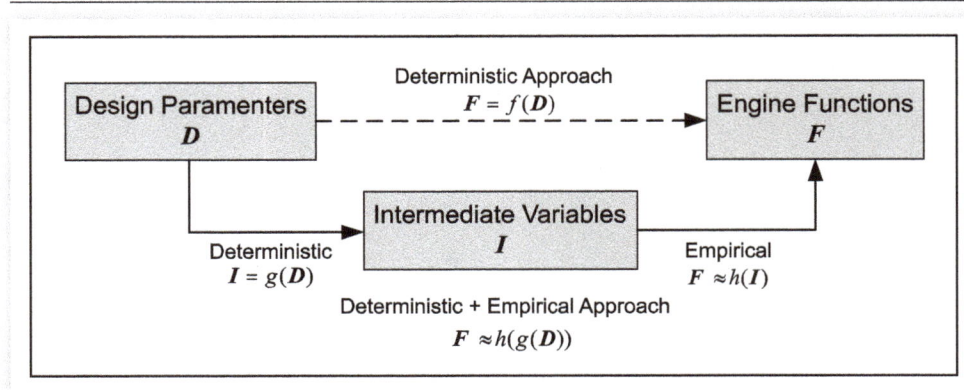

Courtesy of Zhiyu Han

Shi et al. summarizes engine computational optimization methods and applications well [16]. A brief introduction is given here. The task of computational optimization of IC engines is to identify optimal combinations of design variables that can achieve minimum or maximum objective functions of interest. The methods are generally grouped into three categories.

The one mostly used is parametric studies, which extend over the practical range of design variables using modeling tools. There are numerous examples to obtain a maximum or minimum objective through varying influential design parameters. For example, in developing a light SCDI gasoline engine, the effects of spray-cone angle and piston bowl shape on the stratified-charge formation (combustion stability) at part-loads, mixture homogeneity (power output) at full loads, and wall-wettings (smoke) were carefully evaluated and optimized through parametric studies [4,5]. However, the number of evaluations needed to achieve the optimal solutions significantly increases with the number of design variables, which limit their applications in complex design problems. In addition, the nonlinear effects among the variables complicate the problems.

The second group is the non-evolutionary methods. Tanner and Srinivasan [17] explored the conjugate gradient optimization method for a non-road DI diesel engine optimization, and Jeong et al. [18] directly adopted a response surface method to optimize the combustion chamber for a passenger car diesel engine. The performance of non-evolutionary methods relies heavily on spatial information, such as the gradient of response surfaces of objective functions to design variables. In real world optimization problems, such response surfaces can be very complicated and non-differentiable, which limits the use of non-evolutionary optimization methods.

The third group comprise evolutionary methods such as genetic algorithms (GAs) and particle swarm optimization (PSO) methods. These are more generally applicable for optimizing complex non-linear problems and, hence, have been more widely used in computational engine optimization.

The concept of GAs was first proposed by Holland [19]. They are mathematical algorithms that simulate the evolutionary processes of ecosystems. Their ease of use and global perspective are the primary reasons that they have become increasingly popular in engineering optimization. GAs are simple to search techniques that utilize the "fittest" attributes of previously created designs to generate new designs with the aim of moving the evolution process towards better solutions. In this process, the design-space is coded in mathematical expressions (representing "genes"), either with binary strings or real numbers, which are evaluated to provide the "fittest" information for the crossover of genes. GAs are usually initialized with several randomly generated designs, and the number of designs that are evaluated in each generation is called the population size. The crossover among evaluated designs gives a higher possibility to those designs (genes) with better merit to survive to the next generation. To simulate natural evolution, random changes of part of the coded design variables (gene mutation) are introduced to avoid local optimization and to maintain the diversity of solutions. Therefore, parameters that influence crossover and mutation are critical to the performance (optimality and diversity) of genetic algorithms. There are many methods for gene coding, crossover function, as well as mutation. The combination of these methods and additional ad hoc treatments for special applications result in a variety of GAs.

The advantages of GAs attracted a number of researchers at the ERC of University of Wisconsin-Madison who applied them in diesel engine optimizations around the beginning of this century [20–22] and this initiated, probably, the application of GAs in engine community. Senecal and Reitz [20] integrated a micro-GA method with KIVA and studied the simultaneous effects of six engine operational parameters on emissions and performance of a heavy-duty diesel engine at a high-speed and medium-load point. These parameters are SOI, injection pressure, EGR rate, boost pressure, and split injection rate and shape. Wickman et al. [22] optimized nine design variables, including piston geometrical parameters, injection patterns, swirl ratio, and EGR rate, for a high-speed direct-injection (HSDI) diesel engine and a heavy-duty diesel engine.

Applications of GAs have been progressed from single objective GAs to multiobjective GAs, which have become the predominant approach for computational engine optimization and design. Shi and Reitz [23] assessed three widely used multiobjective GAs which are μ-GA [24], non-dominated sorting genetic algorithm (NSGA) II [25], and adaptive range multiobjective genetic algorithm (ARMOGA) [26].

They applied the three methods to optimize the piston geometry, spray targeting, and swirl ratio for a heavy-duty diesel engine at high-load. They also defined four quantities that quantify the performance of the optimization methods in terms of the optimality and diversity of the optimal solutions. NSGA II with a large population size was found to perform the best in their study. To extend the optimization process from a single operating condition to multiple conditions, Ge et al. [27] suggested the optimization of hardware design parameters (piston geometry, number of nozzle orifices, injection angle) first under high- or full-load condition. This was followed by evaluation of the controllable operation parameters (SOI, swirl ratio, boost pressure, and injection pressure) for the fixed hardware geometry under different operating conditions. This procedure was found to be effective for the studied case in which optimal designs with simultaneous reduction of fuel consumption and pollutant emissions were obtained except for a very low load case.

Jeong et al. [28] developed a hybrid evolutionary method that includes a GA and a particle swarm optimization method. The basic idea came from the fact that GAs maintain diverse solutions, while PSO shows fast convergence to the optimal solution in multiobjective optimization problems. They tested the hybrid algorithm using two sets of mathematical functions and showed that the hybrid algorithm had better performance than either a pure GA or a pure PSO. However, due to the high computational cost, the performance of the hybrid method was not compared with other methods for engine optimization problems.

Unlike single objective optimization methods, which always lead to a single global optimal objective function, multiobjective optimization methods normally produce many optimal solutions in engine design problems. It is tedious work to analyze such a large volume of data using human intelligence. Therefore, the use of regression methods in computational engine optimization is also desirable. Since the data-mining is sometimes equally as important as the optimization simulation, comprehensive studies were carried out by Shi et al. to form and evaluate various regression methods [16].

7.2 CFD Codes and Software for IC Engines

Multidimensional modeling of IC engines started in the 1970s. The early expansion of its capability and scope was limited by computer capacity and by the availability of a general CFD code or software until the release of the KIVA code developed at LANL in 1985 [29] and the STAR-CD in 1987. These two codes had built-in features for reciprocating engines. Since KIVA opened its source code, it soon became dominant in the engine research community. After more than 30 years' evolution, engine CFD has been improved significantly in both accuracy and efficiency by enhancement in the main areas of physics models, numerical techniques, mesh generation, postprocessing, parallel computing, and ease of use. Driven by industry demands and technological advances, new CFD codes and software are emerging, and the existing ones are evolved.

Researchers at universities started to use the KIVA codes. Tremendous improvements in physical models have been mainly driven by the ERC at the University of Wisconsin-Madison, giving rise to modified versions of the KIVA codes with ERC models. These advances made by the ERC were reviewed in Chapter 1.

CONVERGE is a commercial CFD software for simulating 3D fluid flows with emphasis on IC engines that was released in 2007 [30]. One of its advantages is the cutting-edge autonomous meshing technique with adaptive mesh refinement (see Sec. 3.2). CONVERGE also features many physical models for a wide variety of phenomena, including chemistry, turbulence, multiphase fluid flow, spray, and radiation. It includes both RANS and LES turbulence models, a host of options to simulate injection, breakup, vaporization and other spray-related processes, and combustion models for premixed and non-premixed flames. A detailed chemistry solver, namely SAGE, was developed to use local conditions to calculate reaction rates based on the principles of chemical kinetics. This solver is fully coupled to the flow solver, but the chemistry and flow solvers parallelize independently of one another, which speeds up the simulation. To accelerate simulations with detailed kinetics, a number of strategies such as adaptive zoning, dynamic mechanism reduction, and load balancing are used, which allow the use of more detailed reaction mechanisms to accurately simulate kinetically controlled phenomena and emissions. Other features of CONVERGE include conjugate heat transfer, a method to fully coupled with GT-SUITE for boundary conditions transferred bidirectionally, GAs for design optimization, and convenience to add user-defined models or other functionality to CONVERGE software.

A brief update on the current status of the CFD codes and software that have specific features for IC engines are described (in alphabetical order) as follows:

1. AVL FIRE ™: A simulation package for IC engines, Lagrangian and Eulerian multiphase capabilities with automated preprocessing computational models. By AVL List GmbH (Graz, Austria), http://www.avl.com/fire.

2. CONVERGE CFD: A CFD software for simulating 3D fluid flows with emphasis on IC engines. By Convergent Sciences, Inc. (Madison, Wisconsin), http://www.convergecfd.com.

3. ANSYS Fluent: A general CFD package to predict fluid flow, heat and mass transfer, chemical reactions with modeling capability for reciprocating engines. By Ansys (Canonsburg, Pennsylvania), https://www.ansys.com/products/fluids/ansys-fluent.

4. ANSYS Forte: An IC engine simulation software. It incorporates CHEMKIN-Pro solver for detailed chemical kinetics and combines multicomponent fuel models with spray dynamics. By Ansys (Canonsburg, Pennsylvania), https://www.ansys.com/products/fluids/ansys-forte.

5. KIVA: Family codes for IC engines with open source codes. The latest version KIVA-4imp is a parallel version of KIVA-4, and the most advanced version of KIVA. KIVA-4mpi also solves chemically reacting, turbulent, multiphase viscous flows, but does this on multiple computer processors with a distributed computational domain (grid). KIVA-4mpi IC engine modeling capabilities are the same as that of KIVA-4, and are based on the KIVA-4 unstructured grid code. By Los Alamos National Laboratory (Los Alamos, New Mexico), https://www.lanl.gov/projects/feynman-center/deploying-innovation/intellectual-property/software-tools/kiva/index.php.

6. OpenFOAM: A free, open source CFD software that comprises many C^{++} modules for simulating complex fluid flows involving chemical reactions, heat transfer, liquid sprays, and films including those in IC engines with unstructured mesh. By the OpenFOAM Foundation, http://www.openfoam.org.

7. Simcenter STAR-CCM+: A commercial CFD software that replaces STAR-CD started in the late 1980s. It uses a generalized polyhedral cell formulation and physics models for IC engines. By Siemens Digital Industries Software (Plano, Texas), https://www.plm.automation.siemens.com/global/en/products/simcenter/STAR-CCM.html.

8. VECTIS: A CFD simulation tool to flows in vehicles and engines. It has structured and unstructured flow solver with any type of mesh. By Ricardo PLC (Shoreham-by-Sea West Sussex), https://software.ricardo.com/products/vectis.

7.3 Direct-Injection Spray Characterization

Fuel injection and resulting sprays directly impact on air–fuel mixing in DI gasoline engines (see Sec. 2.2). Thus, accurate modeling of sprays is the basis of good combustion simulation of DI gasoline engines. As the first step in development of DI gasoline combustion, characterization and performance of fuel sprays need to be evaluated. Measurements of spray images and drop sizes in a vessel under room or elevated pressure and temperature conditions are the common ways for such evaluations. The measured spray images, sizes, cone angles, and tip penetrations (measured from the images) are used to evaluate the characterization and performance of the sprays and to validate the spray model. Sometimes, sprays are measured in an optical engine under conditions closer to real engines.

Models of hollow-cone sprays from a pressure-swirl injector for a DI gasoline engine were described in Sec. 5.2.3. As can be observed in Figure 5.8, the cone-shaped spray tends to collapse. The reason for this phenomenon and for the vortex cloud at the spray tip is explained in Figure 7.6, in which the computed gas flow field (represented by the velocity vectors) in the spray axis plane is shown with the spray drops superimposed for reference. In this plot, the computed velocity vectors are remapped onto a coarse grid system using a postprocessor so that the velocity entrainment behavior can be shown more clearly. The spray motion causes the gas flow to recirculate through the spray. The entraining gas flows interact with the spray drops and suppress the spray cone development. The gas vortex flow also tends to carry the small spray drops and, as a result, the vortex cloud is seen at the spray tip. The air entrainment phenomenon is helpful for air–fuel mixing.

FIGURE 7.6   Spray-induced gas entrainment.

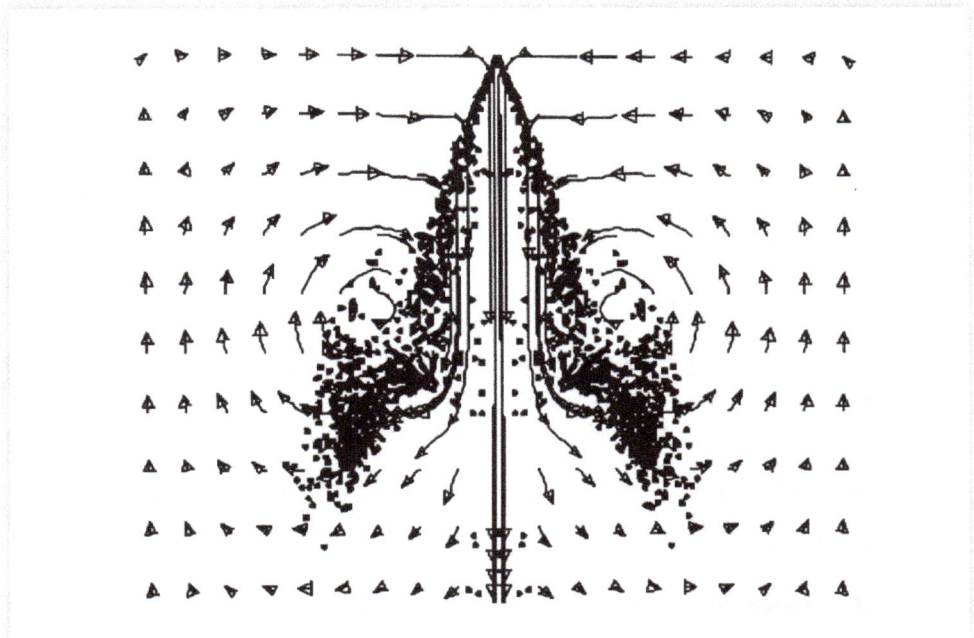

© Reprinted with permission from Ref. [31]. © Begell House Inc

The structure of an evaporating hollow-cone spray is shown in Figure 7.7. The same spray as that of Figure 7.6 was computed but the initial temperature of the environmental gas was set to be 500 K. The computed liquid and vapor density contours (at a delay time of 2.7 msec) in the plane of spray axis are shown in Figure 7.7. It is interesting to see that the liquid phase keeps the hollow-cone spray structure, although it is deformed due to the gas entrainment effects, as shown in Figure 7.6. However, the recirculating gas flow carries the vapors to the center region of the spray and the vapor phase features a "solid cone" distribution. This result has important implication on the air–fuel mixing process in DI gasoline engines.

Other types of fuel injectors are also used in DI gasoline engines as described in Figure 2.10. Multihole injectors are mostly used in recent turbo-charged DI gasoline engines. The Reitz Wave atomization and breakup models in Secs. 5.2 and 5.3 can be used for this type of sprays. An example is shown in Figure 7.8. A 10-hole injector with uneven distributed

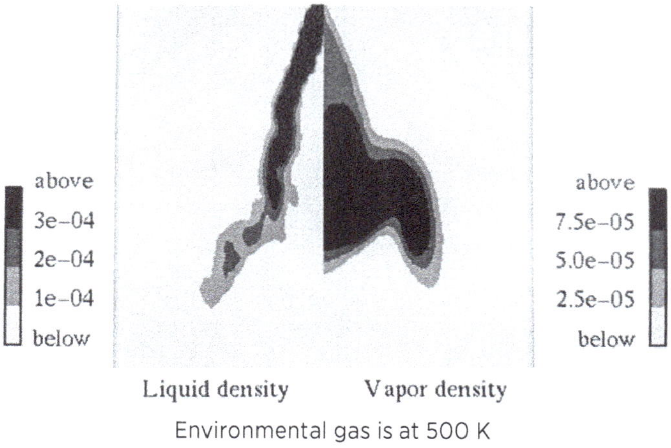

Reprinted with permission from Ref. [31]. © Begell House Inc

FIGURE 7.7 Liquid and vapor density (g/cm³) contours showing the structure of a vaporizing pressure-swirl atomized spray.

Liquid density   Vapor density

Environmental gas is at 500 K

FIGURE 7.8 Sprays from a multihole DI gasoline injector: (a) experimental; (b) computational.

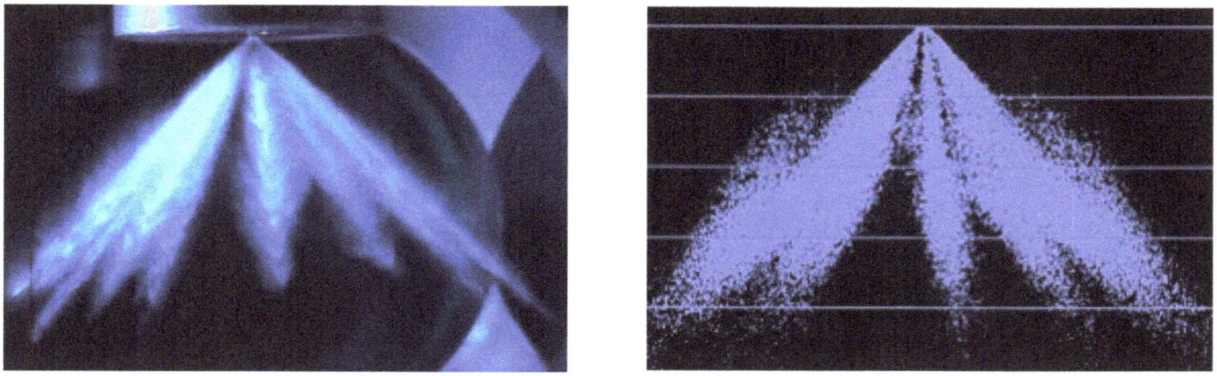

Courtesy of Zhiyu Han

(a)   (b)

spray plumes are modeled and the results at 1.0 msec delay time are shown. The injection pressure is 10 MPa and the background is at room condition. One advantage of a multihole injector is the greater flexibility to design sprays with specific orientation of each spray plume. This can improve mixing and reduce wall-wettings in an engine as discussed later.

Usually spray drop sizes are measured 30 mm downstream from an injector since this distance is typical when sprays travel in the cylinder of a DI gasoline engine before impinging on the piston or liner wall. The radial distribution of the SMD of sprays from a multihole injector measured by PDPA is given in Figure 7.9. The injection pressure is 10 MPa and the measurement environment is at room conditions. The larger droplets are at the peripheries of the sprays and smaller ones are inside the sprays. The averaged droplet size is about 20 μm.

FIGURE 7.9 Droplet size radial distribution of sprays from a multihole injector under 10 MPa injection pressure. Repeated measurements are made at 30 mm downstream of the injector.

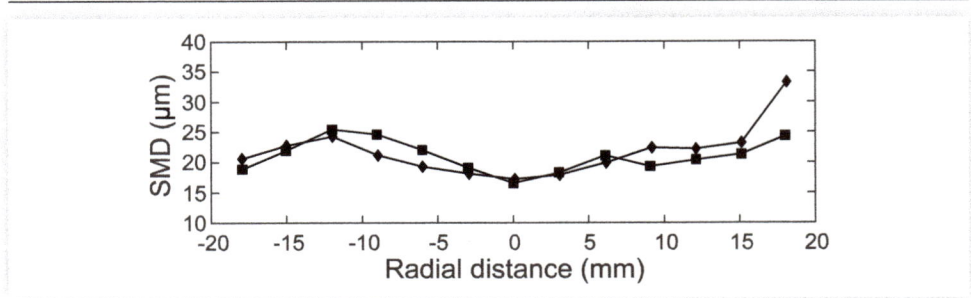

The behavior of swirl-injector produced sprays were studied in an optical engine by Han et al. [32]. Fuel flow rates of two injectors are first evaluated. Although a swirl injector can have a complicated internal geometry, it usually consists of swirl ports and a circular discharge nozzle passage. The volumetric flow rate Q versus injection pressure is assumed to be $C_D A_0 \sqrt{2(P_1 - P_2)/\rho_l}$, where P_1 is fuel injection pressure, P_2 is ambient pressure, A_0 is the nozzle orifice cross-section area, and C_D is the discharge coefficient.

The static flow rate as function of injection pressure is shown in Figure 7.10. Injector A has a nozzle hole diameter 0.505 mm and injector B 0.9 mm. Both measurement and computation suggest that the discharge coefficient C_D is independent of injection pressure within the normal working range of a DI engine injector. In these cases, C_D of Injector A is 0.38 and that of Injector B is 0.16. Therefore, when C_D is calculated from a known flow rate (usually provided by injector suppliers), it can be used to predict flow rate under other pressure conditions. It is also noticed that the discharge coefficient has a very low value; this is partially due to the presence of the air core (see Figure 5.7), which blocks off the central portion of the orifice. The calculated flow rate can be used as initial conditions in the sheet atomization model (see Sec. 5.2.3).

FIGURE 7.10   Fuel flow rates of pressure-swirl injectors.

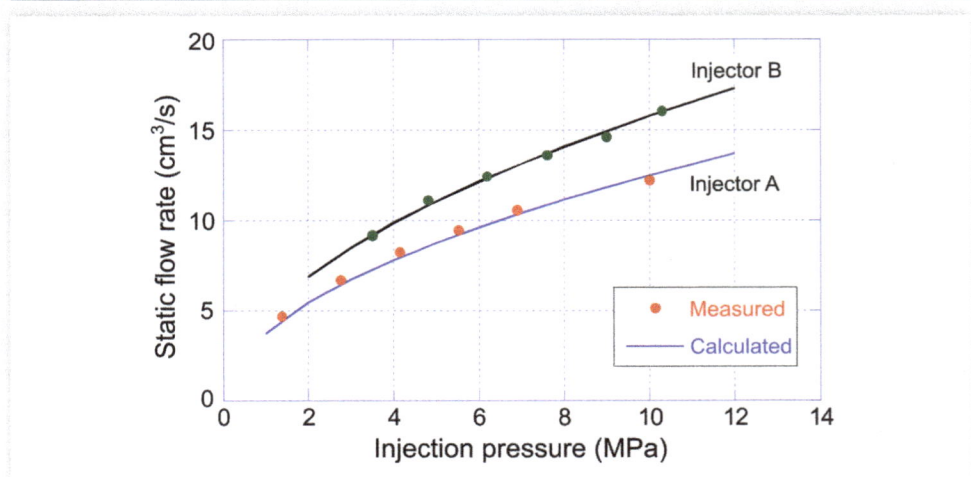

The optical engine is a single cylinder research engine designed for SCDI operations and modified for optical access. The engine features a pent-roof combustion chamber with two intake valves and two exhaust valves. It has a bowl-in-piston design. Spray images in the tumble plane orientation are acquired at selected crank angle positions throughout the injection event. Two-dimensional images of the in-cylinder fuel spray (liquid phase) are obtained using laser elastic (Mie) scattering. The imaging system is composed of a CCD camera and a laser light source. An injector is located on the cylinder symmetry plane, on the intake side of the head. The fuel injector has a 70° nominal cone angle, and operates at 10 MPa fuel pressure.

Computations were performed of the sprays in the optical engine and the results were compared with the experimental images. Comparisons were made under two sets of distinct engine operating conditions. One set of spray images was obtained under stratified-charge conditions (late injection), and the other was under homogeneous-charge conditions (early injection). In both experiments, the engine was motored at 1500 rpm. The operating condition details are listed in Table 7.1 [32]. A 3D engine mesh system with moving intake valves was used in the simulations. The mesh size is around 2 mm. Computations were started at intake valve opening (IVO) and completed at the compression TDC. The RNG k-εturbulence model was used to model gas turbulence. Drop evaporation, collision, and coalescence were included. Iso-octane was used in the computation to simulate gasoline fuel used in the optical engine test.

TABLE 7.1 Optical engine operating conditions

Operating Condition	Stratified Charge	Homogeneous Charge
Engine Speed (rpm)	1500	1500
Intake Manifold Pressure (kPa)	90	100
Swirl Ratio	0.8	0
Coolant Temperature (°C)	50	50
Oil Temperature (°C)	50	50
Fuel Type	Indolene	Indolene
Fuel Injection Pressure (MPa)	10	10
Start of Injection (SOI) (°BTDC)	66.5	249.5
Injection Duration (°CA)	11.2	34.4
Fuel Mass Injected (mg/cycle)	11.72	39.55

Reprinted with permission from Ref. [32]. © SAE International

In the early injection case, spray is injected at 249.5° BTDC for homogeneous-charge operation. Comparison of the Mie scattering spray images and the computed sprays is given in Figure 7.11. The oblique lines in the images are aligned with the injector centerline, and the horizontal lines in the computed plots correspond to the lower boundary of the quartz liner in the optical engine below which the sprays are blocked from view in the optical engine. In the first few CA after injection, an initial spray slug with an expanding cone angle can be observed. This initial phase of the spray lasts about 2CA. Then the main spray begins to form with a wider cone angle. This part of the spray features a smaller cone angle and larger droplets in comparison with the main spray. It is believed that this initial spray slug is formed during the initial stages of the injection. It was also observed in the experiments that the cone angle of the main spray underwent a development stage during a very short period (~0.1 msec), changing from a smaller angle to a steady state value. Physically, this may be related to the transient angular momentum development inside the injector as the nozzle needle was lifted. The model captures the transition from the initial spray to the main spray well.

FIGURE 7.11 Mie scattering images and computed sprays under early injection conditions in a DI gasoline engine.

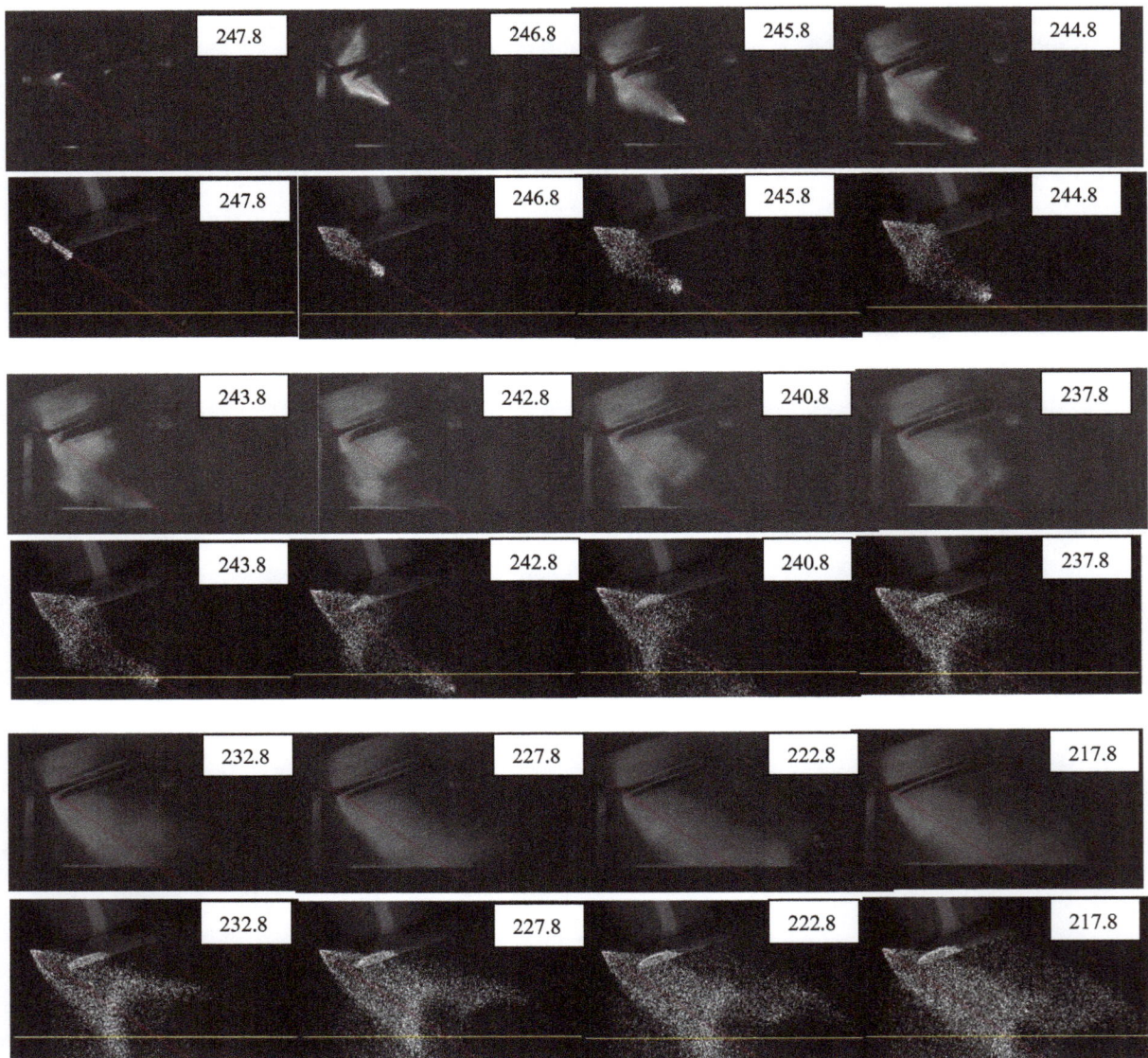

The number on each image indicates the crank angle, BTDC

Reprinted with permission from Ref. [32]. © SAE International

FIGURE 7.12 Spray tip penetrations for an early injection case in a DI gasoline engine.

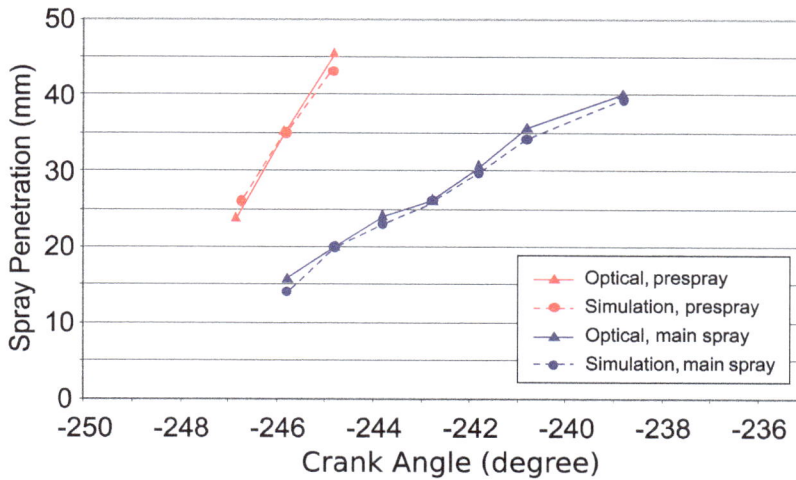

DI: direct-injection

The overall agreement is also very good in terms of spray shape and detailed structure evolution. A detailed spray tip penetration comparison is given in Figure 7.12. As can be seen, the computed penetrations of both initial slug and main spray agree well with the ones measured from the experimental images.

The most prominent characteristic of the spray structure is the deflection of the pre-spray downward (below the injector centerline) shortly after the injection starts. This feature starts to develop at 246.8° BTDC (2.7° after SOI). Two degrees later, at 244.8° BTDC, the entire pre-spray is below the centerline. The model simulation can capture this structure characteristic. The simulation indicated that at 248.8° BTDC, 0.7° before SOI, strong airflow exits from the intake ports directed radially relative to the valve. This translates to the flow directed towards the injector and downwards. At 247.8° BTDC, the spray is injected and induces a stronger gas motion along the injection direction. This trajectory is nearly perpendicular to the main gas flow direction. As a result of the interaction between the spray and intake flow, the spray-induced flow is pushed and steered downwards.

For the late injection case, a detailed comparison of spray images over a 12 CA period between the Mie scattering images and the model simulation results is shown in Figure 7.13. For the stratified-charge case, the pressure in the cylinder is higher (~0.3 MPa) at the time of injection, compared with that in the homogeneous-charge case. As indicated by the experimental images, under this relatively high ambient pressure condition, the spray collapses, resulting in a reduced spray width. The computation captures the spray collapse behavior very well. Overall, the simulated spray closely resembles the optical images in terms of shape and penetration.

From the experimental images, the main spray is observed to catch up with the initial spray slug. The boundary between the main and initial spray is unclear compared to that in the early injection case; this is also demonstrated in the computed spray. A detailed comparison of the spray tip penetration is given in Figure 7.14, and the computed penetration again agrees well with the measured value.

FIGURE 7.13 Mie scattering images and computed sprays under late injection conditions in a DI gasoline engine.

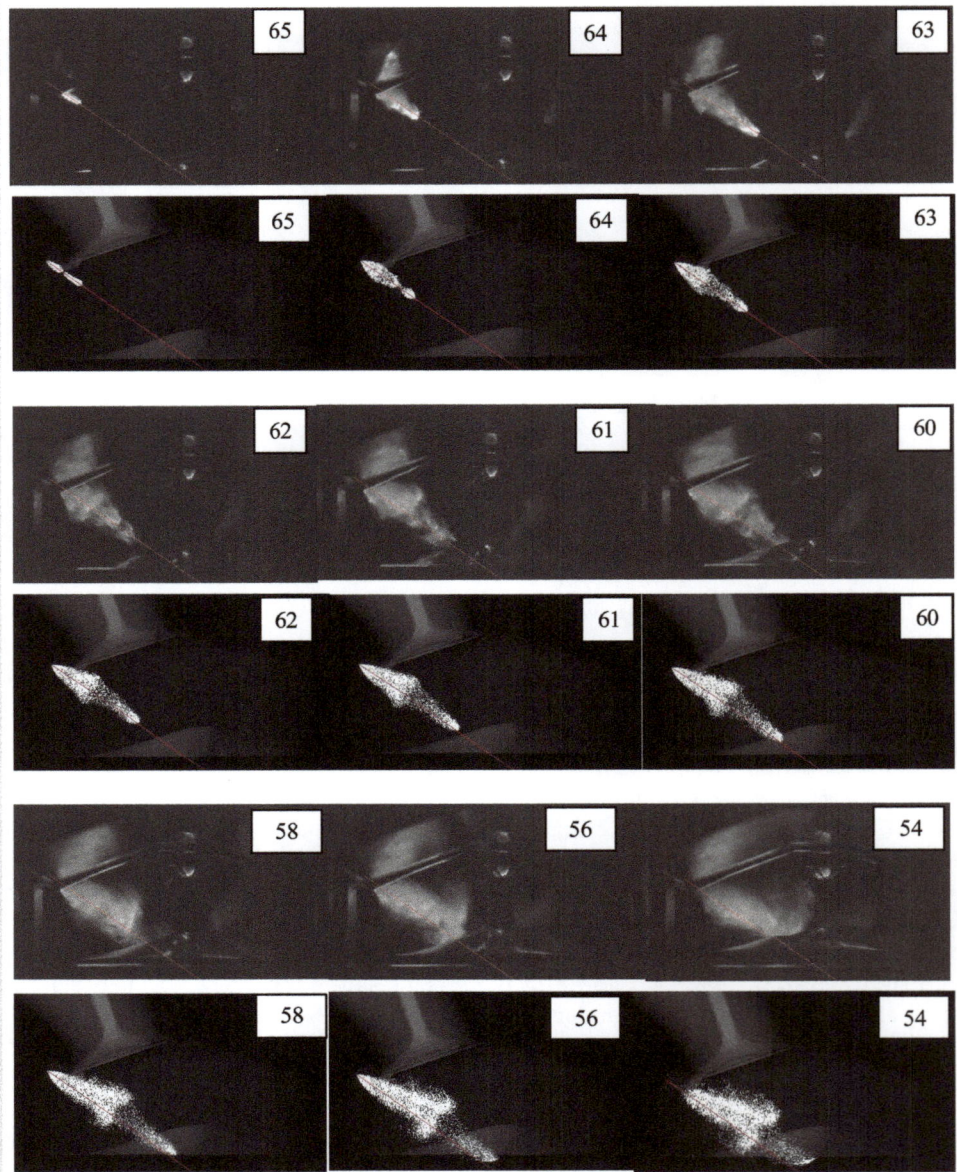

The number on each image indicates the crank angle, BTDC

FIGURE 7.14 Spray tip penetration for a late injection case of a DI gasoline engine.

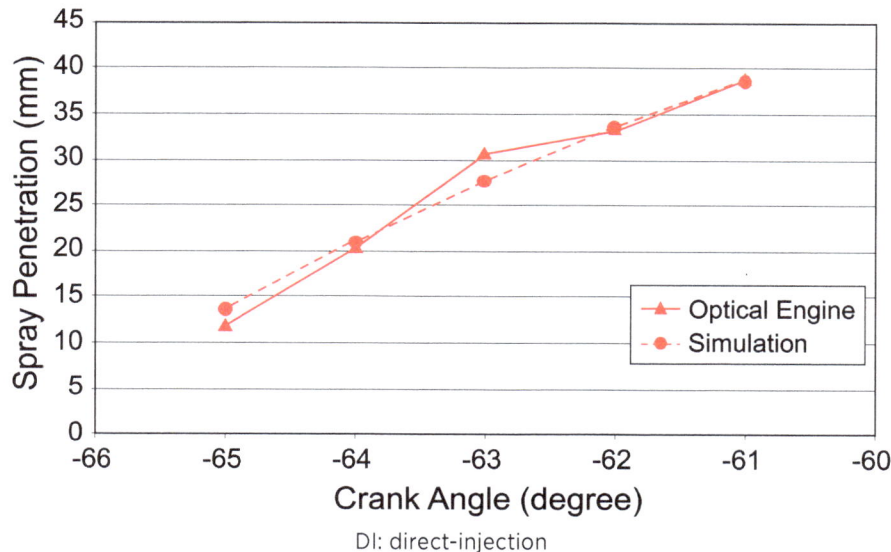

DI: direct-injection

It is very interesting to see from the Mie scattering images in Figure 7.13 that at the early stage of the injection, the spray is aligned with the injector centerline. Later in the injection event, however, it begins to deviate from this reference line and is lifted towards the chamber roof. At 58° BTDC, most of the spray is above the centerline and, at 54° BTDC, almost all the spray is above the centerline. It is also demonstrated in Figure 7.13 that the modeled sprays display the same pattern. To help explain the mechanism of this phenomenon, a vector field of the gas velocity relative to the piston on the section cutting through the engine center plane is given in Figure 7.15. This figure shows that the spray induces a gas vortex flow on its sides due to air entrainment. The vortex flow has a ring structure in 3D space but is asymmetrical. The air-entrained flow (and forming vortex) below the spray is stronger due to the less confined space; that above the spray is weaker due to the roof-wall restricted flow. The unsymmetrical flow pattern results in a stronger flow below the spray pushing the spray up. As the piston moves closer to the TDC, the flow asymmetry is enhanced and the spray is pushed even higher.

To further understand the effect of the piston-induced flow on the phenomenon, a simulation with an engine speed of 3000 rpm was performed. The injected fuel mass and SOI were kept the same as those in the above case to maintain the same spray momentum. The computed spray at 50° BTDC and the corresponding gas relative velocity on the cylinder center plane are also shown in Figure 7.15. Although the gas motion induced by the piston is stronger due to the higher engine speed, the gas entrainment induced by the spray maintains at a similar level as that in the 1500 rpm case. As a result, the spray is not seen to have a stronger tendency to be lifted up. This confirms that the interaction between the spray and the spray-induced gas flow plays a more important role.

FIGURE 7.15 Computed spray and gas relative velocity (cm/s) vector field at: (a) 56° BTDC at 1500 rpm; (b) 50° BTDC at 3000 rpm.

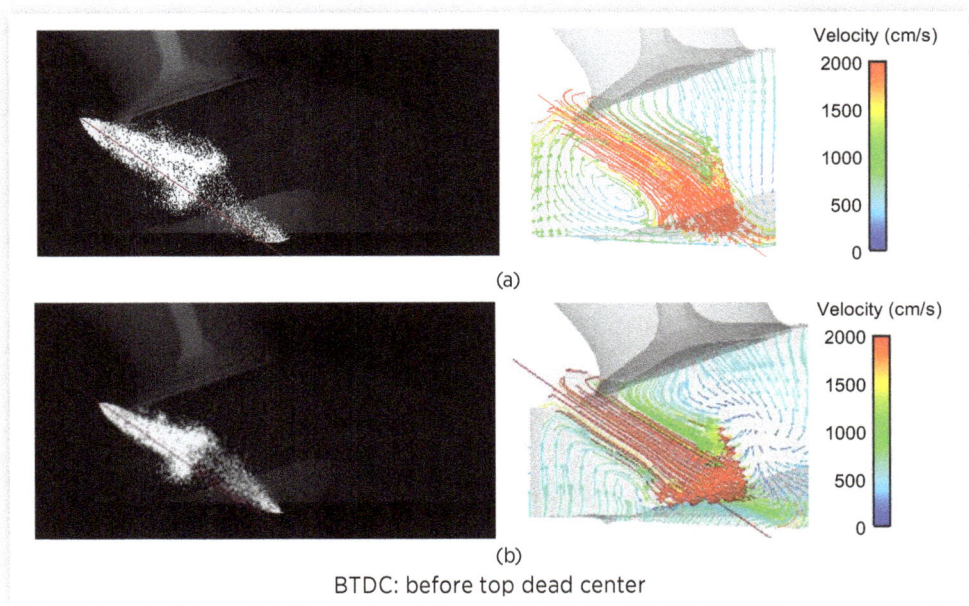

(a)

(b)

BTDC: before top dead center

The structure of DI gasoline sprays is affected by fuel temperature. Sprays tend to collapse under superheat conditions, which influences strongly mixing and combustion [33–35]. The effects of fuel temperature on sprays from a six-hole injector were studied by Wu et al. [36]. In the study, they observed that when fuel was, spray structures were almost identical, which were mainly dominated by the injector configuration and fuel properties. When fuel was heated to the boiling temperature, the fuel vapor pressure got higher than the ambient pressure and flash boiling occurred, which led to collapse of the spray and its structure changed and resulted in a more compact, shorter penetrating spray. However, multiple plumes could still be observed, which indicated that both flash boiling and injector configuration governed the spray structure. As the fuel temperature was further increased to be some tens degrees higher than the boiling temperature, all fuel plumes merged into a single plume, which resulted in a longer spray penetration and a significantly narrower spray. No obvious effect of injector configuration on spray structure could be identified, and spray structure was dominated by the flash boiling phenomenon.

The impact of flash boiling on the behavior of sprays from other injector-nozzle configurations was also studied by Wu et al. [36] and by Zhang et al. [37]. In these works, injector nozzles with a single narrow slot opening, evenly spaced three-slot openings, and four-slot openings were also tested. Different closely-spaced split injection strategies were examined. It was found that injector configuration has a great influence on spray structure at subcooled conditions, but under strong flash boiling conditions, the spray structure is nearly independent of injector configurations, and a single plume spray is identified due to spray collapse. For the six-hole injector, a closely-spaced double-injection did not alter fuel spray penetration regardless of fuel temperature. However, a significant penetration reduction was observed when a quadruple injection was used. Shortened spray penetrations were also found in other injectors when triple- and quadruple-injection strategies were employed [36].

7.4 Mixing in Wall-Guided DI Systems

7.4.1 Homogeneous Mixture Formation

In a PFI engine or a DI engine operated at homogeneous-charge mode, the key issue is to form a homogeneous mixture in the cylinder. Charge inhomogeneity may result in incomplete combustion and hence result in excessive CO and hydrocarbon emissions at part-loads and reduced engine torque output at full loads. Mixture homogeneity is affected by spray characteristics, injection strategy, large-scale flow motion, turbulence, etc. Thus, optimization of these parameters is necessary.

7.4.1.1 In-Cylinder Mixing Phenomena

Fuels are injected into the cylinder during the intake stroke in homogeneous-charge operations during which some important phenomena occur. There exists strong interactions between the intake gas motion and spray induced flows in a DI gasoline engine as discussed in Sec. 4.1.1. These interactions modify spray drop trajectories, entrain vapors, increase gas turbulence, and enhance mixing. When fuels are injected, evaporation of spray droplets takes place. Figure 7.16 shows the global evolution of fuel liquid and vapor phase normalized by the total injected fuel in a DI engine. The engine features 4-valve, pent-roof, 10.5 CR, top-mounted swirl injector. The engine is operated at 1500 rpm, WOT, and $\phi=1.18$. As can be seen in Figure 7.16, fuel continues to vaporize and over 97% of the total injected fuel is vaporized by the end of the compression stroke in the computation (combustion is not modeled in the study). It is seen that the evaporation rate decreases after about 180 CA by comparing the slope of the vapor phase curve at different times during the cycle. This is because the amount of impinged liquid droplets accumulated after this time undergo decreased heat and mass transfer.

FIGURE 7.16   Evolution of fuel vaporization in a DI gasoline engine.

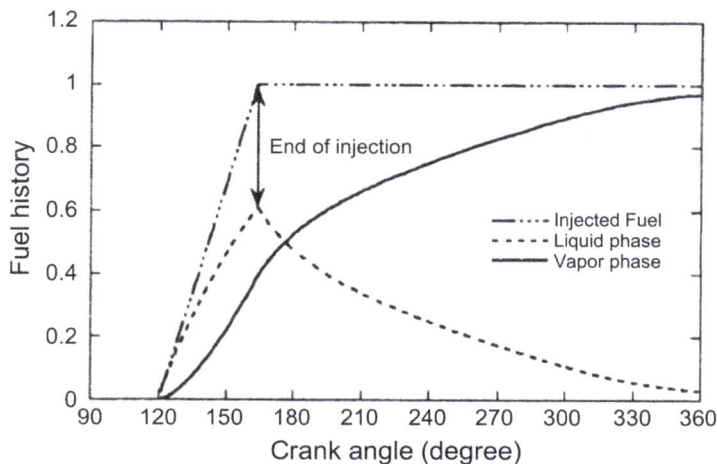

SOI: 120 degrees, 360 degrees is the compression TDC
DI: direct-injection; SOI: start of injection; TDC: top dead center

In this case, sprays impinge on the surfaces of the piston and liner. The temporal variation of the instantaneous amount of liquid fuel located on (or near) the cylinder walls is plotted in Figure 7.17, which is calculated by collecting the liquid fuel within a 0.1 mm layer immediately adjacent to the wall surface and normalized by the total injected fuel. It is seen that significant wall impingement does not occur until about 180°. The liquid fuel on the wall surfaces increases at first due to the continuous wall impingement of the droplets, and then it decreases due to vaporization. In this particular case, up to 16% and 2% of the total injected fuel are seen on the piston and liner surfaces, respectively, by 240° during the early compression period, but these surface fuels are reduced to less than 2% and 1%, respectively, by the end of the compression stroke.

FIGURE 7.17   Temporal variation of wall-wettings on the piston and liner surfaces.

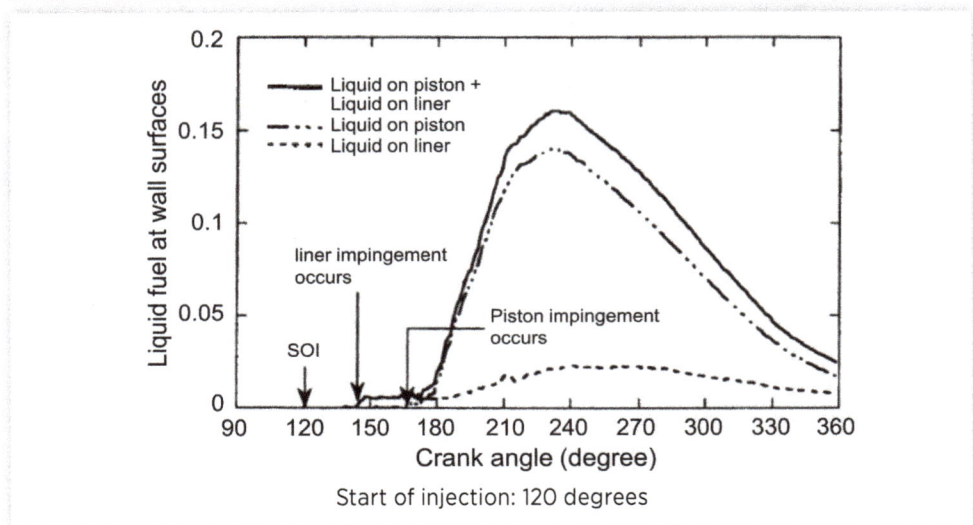

Start of injection: 120 degrees

Note that sprays are deflected by the intake flow, which brings some of the droplets to the liner. Spray droplets move downward to catch up the piston. Thus, the amount of fuel impinged on the cylinder liner is reduced substantially in the retarded injection cases as the effects of the intake flows on the spray become less. It is not difficult to understand that fuel impingement on the piston surface will increase when fuel is injected too early and too late while the piston is close to the cylinder head. The effects of injection timing on wall impingement and other parameters in the air–fuel mixing process in a DI gasoline engine were studied by Han et al. [39].

Fuel vapor mixes with air to form the air–fuel mixture. The local mixture equivalence or air/fuel (A/F) ratio varies spatially and temporally due to the varyingly distributed liquid fuel and the anisotropic gas motion. To characterize the mixture quality, the evolution with a crank angle of the local mixture volumes in different equivalence ratio regimes is given in Figure 7.18. It is seen that the amount of flammable mixture ($0.5 \leq \varnothing \leq 1.5$) increases and the extra-lean mixture ($\varnothing < 0.5$) decreases during the induction and compression processes since the fuel vapors are continuously replenished through fuel vaporization. Note that after about 270° CA, the formation rate of the flammable mixture is increased considerably,

FIGURE 7.18 Evolution of the total in-cylinder gas mixture volume (normalized by the instantaneous cylinder volume) in different equivalence ratio ranges.

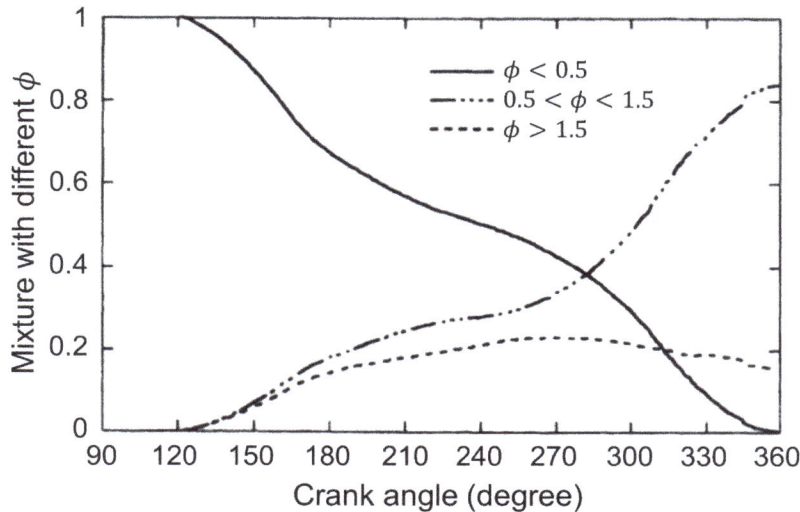

accompanied by a rapid decrease in the amount of the extra-lean mixture. This indicates that the elevated compression temperatures due to the change of the cylinder volume and diffusion effects become important to air–fuel mixing during the late compression stroke. By the end of the compression stroke, the majority of the mixture, i.e., about 84% of the total chamber volume, is in the flammable regime. While no appreciable extra-lean mixture remains by the end of the compression stroke, about 16% of the mixture is in the rich regime ($\emptyset > 1.5$). This part of the mixture accumulated by about 240 CA and then kept almost constant throughout the compression process. The major contribution to this part of the mixture is from the near piston surface region where more vapor is located during the late compression stroke.

In-cylinder fuel injection also has an impact on the thermodynamic properties of the mixture. One important phenomenon is charge cooling, which is the distinct feature of a DI engine. Figure 7.19 compares the computed average gas temperature of the non-spray case and of a spray case. Evaporating fuel drops absorb heat from the in-cylinder gases, and as a result, the in-cylinder charge temperature decreases as indicated in Figure 7.19. At IVC, the gas temperature drop in the fuel injection cases is 16.4°C. Apparently, charge cooling is affected by wall-wetting in which fuel absorbs heat from the wall for evaporation. Thus, factors that influence spray wall impingement, e.g., injection timing, will have an impact on charge cooling.

Some benefits can be gained from the charge cooling phenomenon. Specifically, the gas temperature at the end of the compression stroke is seen to be reduced significantly as in Figure 7.19. It is decreased from 812.5 K in the non-spray case to 696.5 K when fuel is injected at 120°—a decrease of 116 K. The decreased gas temperature makes an increase of the engine CR possible in a DI gasoline engine. It is known that one of the obstacles that limit an increase of the CR in an SI engine is knocking combustion. It is the result of the end-gas autoignition. Since the gas temperature is reduced in a DI engine, the CR of the engine can be increased, which is beneficial to engine fuel economy.

FIGURE 7.19 Computed in-cylinder mixture temperature reduction (charge cooling) due to fuel evaporation.

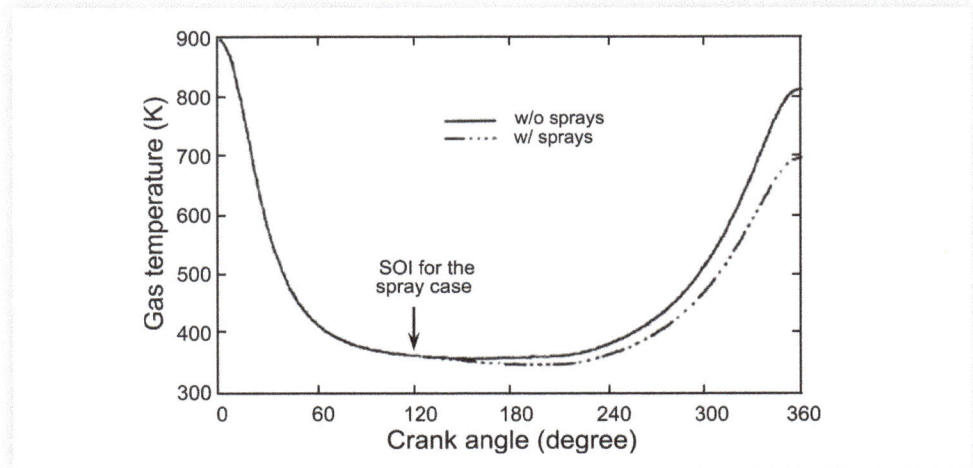

Another benefit of charge cooling is an improvement on engine volumetric efficiency. Computation indicated that the amount of trapped mass during the induction stroke increased by as much as 2.5% in the DI engine case in comparison with the non-spray case. This improved volumetric efficiency has a positive impact on engine torque output. However, the extra gain of the trapped mass generally decreases at a retarded SOI. In particular, the volumetric efficiency benefit will vanish when the fuel is injected during the compression stroke.

7.4.1.2 Mixture Homogeneity and Improvement

In the development of a DI engine, mixing is examined at the high speed (or full speed) and WOT conditions because this operation point is of importance to the rated power. However, this point is the most challenging part to form a homogeneous mixture in a naturally-aspirated DI engine because of several reasons. First, the time for induction and compression strokes is the shortest due to the highest engine revolution, which means the available time for fuel evaporation and mixing is the shortest. Second, intake flow momentum is the strongest, which results in the strongest impact on fuel sprays. Third, assisting means to enhance in-cylinder gas motion (e.g., a swirl control valve installed on one of the intake ports, see Figure 4.1) cannot be used at these high-speed conditions to avoid volumetric efficiency deterioration. Last, the flexibility of the fuel injection timing window is limited due to the long injection duration. The fuel injection window is about the same as the duration of the intake process when injection pressure is 10 MPa.

In the development of a SCDI combustion engine, researchers at Ford FRL implemented a modeling-driven design approach. Upfront CFD simulations were performed to improve air–fuel mixing at both homogeneous-charge mode and stratified-charge mode. Parametric studies of a number of design parameters and operation strategies were carried out to find optimized solutions. Figure 7.20 shows the effects of intake flow on spray droplets distribution under 6000 rpm and WOT conditions in an SCDI engine. The engine is naturally-aspirated and each cylinder has a side-mounted injector and a displacement of 0.5L. The CR is 11.3. It is seen that the strong intake flow deflects the spray droplets towards the intake side, and there is not enough cross-cylinder penetration of spray droplets. This biased distribution of the remaining droplets can be still seen 60° before the compression

FIGURE 7.20   Impact of intake flow on the distribution of fuel liquid droplets at 6000 rpm and WOT conditions: (a) droplets distribution at BDC; (b) droplets distribution at 60° BTDC; (c) gas velocity at the central plan superimposed the fuel droplets at BDC.

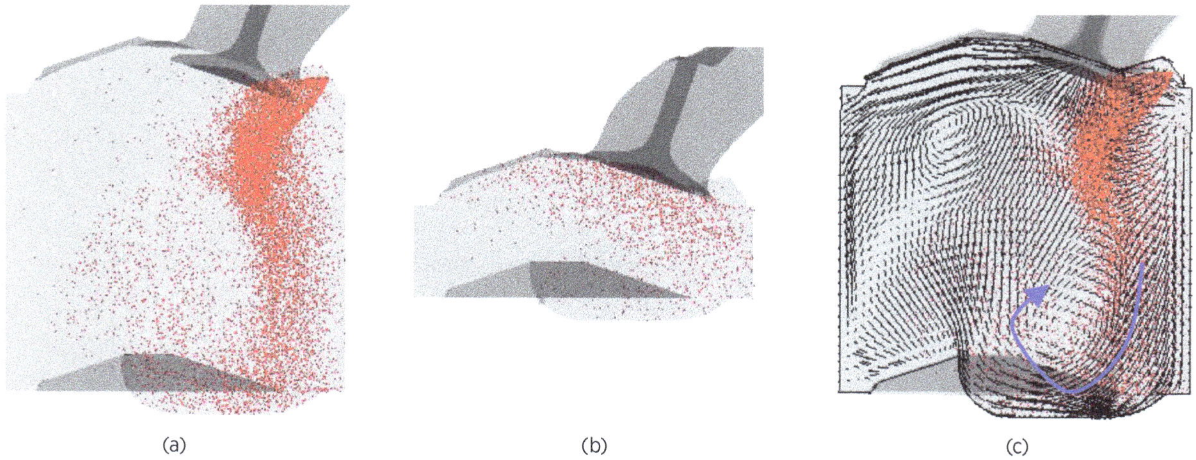

Adapted with permission from Ref. [2]. © SAE International

(a)   (b)   (c)

BDC: bottom dead center; WOT: wide-open throttle

TDC. The gas velocity distribution superimposed with fuel droplets at BDC is also given in Figure 7.20. It is seen that the intake flow pushes fuel sprays away from its axial penetration direction. In addition, the bowl-in-piston geometry inducts the airflow into the bowl and forms a vortex structure as indicated by the arrowed circle. The vortex motion confronts the penetration of the spray droplets, which helps to trap the droplets on the intake side.

As a result of the intake-flow affected liquid fuel distribution, air–fuel mixture is stratified as shown in Figure 7.21. The mixture is lean at the exhaust side and rich at the intake side as indicated by the air–fuel distribution, which is represented by the colors. This mixture inhomogeneity is very undesirable as it will cause high soot emissions and torque deterioration.

To improve the fuel spray penetration, the impact of intake flow on sprays has to be mitigated. A mask design between the two intake valves that blocks part of the intake flow on the injector side was proposed by Yi et al. [2], which can effectively alleviate the impact of intake flow on sprays. As a result, the sprays are more intact and penetrate further into the cylinder as shown in Figure 7.22. Comparing the spray droplet distribution in Figure 7.22 and that in Figure 7.21, it is seen that the remaining droplets are much more uniformly distributed in the cylinder at 60° BTDC. Consequently, the mixture is much more homogeneous as shown in Figure 7.22. The improvement of mixture homogeneity leads to about 2% increases in combustion-simulation predicted torque and reduced soot emissions in the engine tests [14].

As seen in Figure 7.20, the piston bowl has adverse effects on spray drop distribution and mixing. In an SCDI engine, the piston bowl shape has a profound impact on the formation of stratified charge. Its design ought to meet the charge stratification formation at part-load first, and then is checked for homogeneous-charge formation at high-load. The shallower piston-bowl design shown in Figure 7.23 improves stratification significantly and it also improves mixture homogeneity at 6000 rpm full load as shown in the same figure; in particular, the lean mixture packets on the exhaust side in the deeper bowl design is removed in the shallower bowl design.

FIGURE 7.21 Stratified air–fuel distribution at 6000 rpm and WOT conditions of an early DI combustion system design. The time is at 100°, 60° and 20° BTDC, respectively, from left to right.

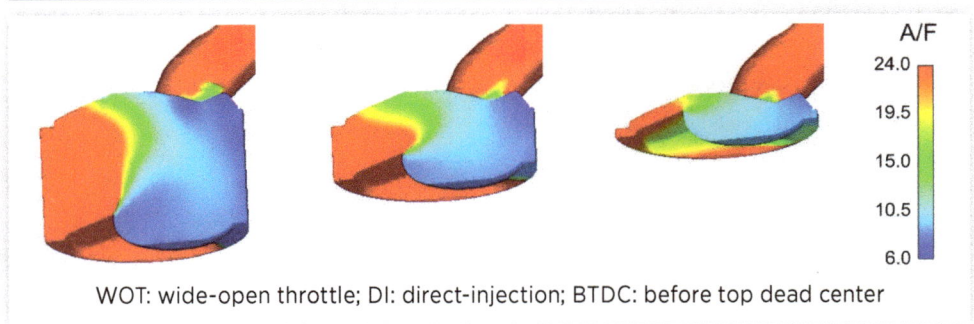

WOT: wide-open throttle; DI: direct-injection; BTDC: before top dead center

FIGURE 7.22 Improved distributions of spray droplets and of air–fuel ratio with a mask design between the two intake valves at 6000 rpm and WOT conditions. The time is at 180°, 60°, and 20° BTDC, respectively, from left to right.

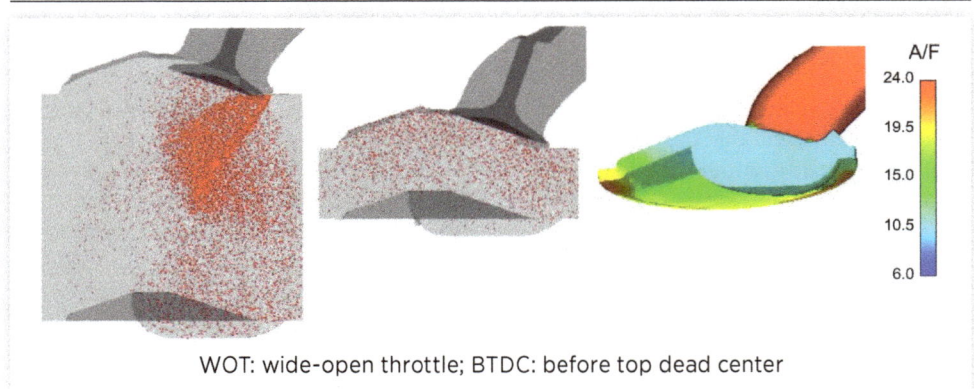

WOT: wide-open throttle; BTDC: before top dead center

FIGURE 7.23 Air–fuel distribution at 20° BTDC showing a shallower piston bowl improves mixture homogeneity at 6000 rpm and WOT conditions.

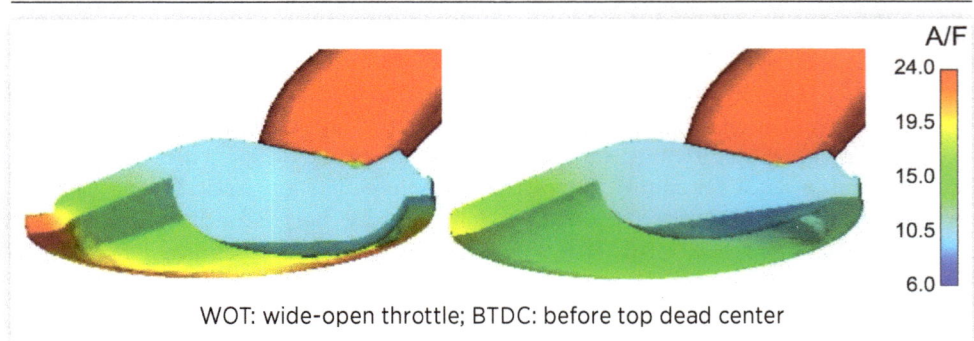

WOT: wide-open throttle; BTDC: before top dead center

At lower engine speed, the mixing issue is less severe and some operation parameters can be adjusted to further improve mixture homogeneity. For example, injection duration at 1500 rpm is only one-fourth of that at 6000 rpm for injecting the same amount of fuel, hence, simulations are often made with SOI sweep. In order to improve fuel burn rate and mixing at low-speed operations, a swirl control valve (SCV) installed in one of the intake ports (see Figure 4.1) can be used to generate enhanced large-scale inflow motion. Figure 7.24 shows the effects of SCV on mixing at 1500 rpm full load condition. In this figure, the percentage of mixture volume in a given equivalence ratio ϕ, range, or probability distribution function of ϕ is plotted at 20° BTDC. This parameter can measure the degree of mixture homogeneity at a given time. In theory, the narrower and taller the PDF of ϕ, the more homogeneous the mixture is. As seen in Figure 7.24, when the SCV is closed, the mixture becomes more homogeneous, and all the mixture gases are within $0.7 < \emptyset < 1.3$ before combustion takes place.

FIGURE 7.24   Volumetric PDF of equivalence ratio at 20° BTDC showing effects of the SCV on mixing homogeneity.

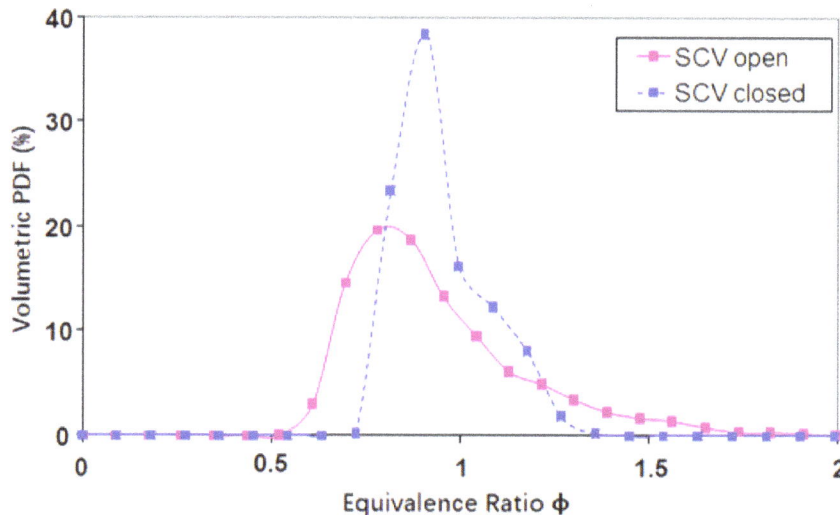

1500 rpm and WOT
PDF: probability density function; BTDC: before top dead center;
SCV: swirl control valve; WOT: wide-open throttle

Courtesy of Zhiyu Han

However, the use of SCV has contradictory effects on torque output because engine volumetric efficiency decreases and heat loss increases when the SCV is closed. Hence, in the development of an LSC DI system (see Figure 2.6), the SCV was kept open at low-speed WOT conditions. The split injection strategy was used to improve mixture homogeneity [40]. In the split injection strategy, fuel is injected in two separated pulses with a dwell in between. Figure 7.25 compares the evolution of spray droplet distribution with a single and a 50–50 split injection at 1500 rpm WOT conditions. In this case, the single-injection starts as 463° BTDC (compression TDC) with a duration of 35°, while in the split injection, the first pulse (50% fuel) starts at 437° BTDC and ends at 454° BTDC and the second one starts at 507° BTDC and ends at 524° BTDC. It is seen that in the single-injection case, spray droplets move towards the lower portion of the cylinder on the exhaust side; while

FIGURE 7.25 Evolution of spray droplet distribution with a single and a split injection at 1500 rpm and WOT conditions.

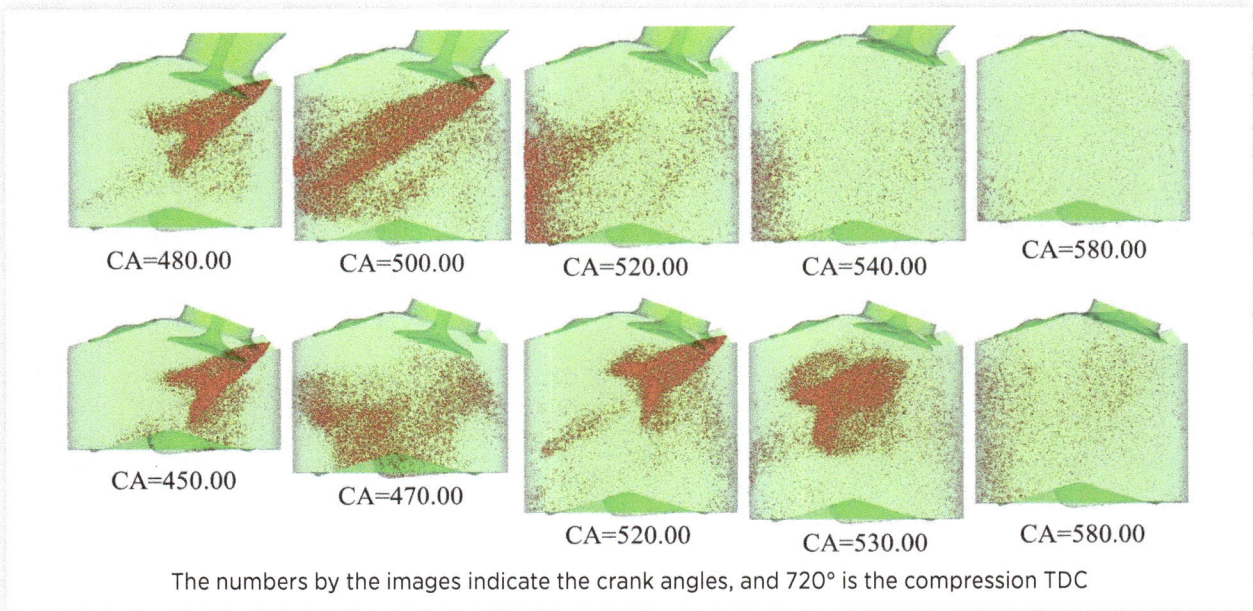

CA=480.00   CA=500.00   CA=520.00   CA=540.00   CA=580.00

CA=450.00   CA=470.00   CA=520.00   CA=530.00   CA=580.00

The numbers by the images indicate the crank angles, and 720° is the compression TDC

Reprinted with permission from Ref. [40]. © SAE International

FIGURE 7.26 Mixture distribution affected by injection strategy at 20° BTDC: (a) single-injection; (b) split injection.

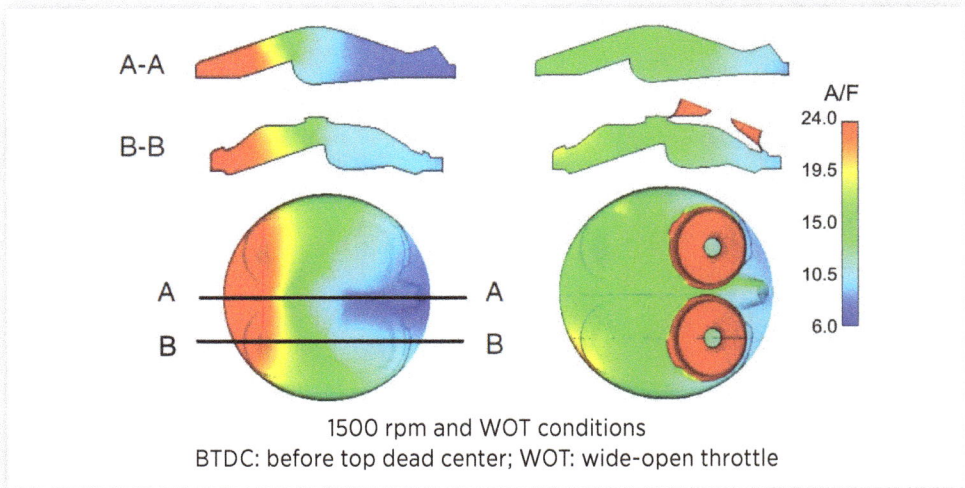

A-A

B-B

A/F
24.0
19.5
15.0
10.5
6.0

A   A
B   B

1500 rpm and WOT conditions
BTDC: before top dead center; WOT: wide-open throttle

Reprinted with permission from Ref. [40]. © SAE International

in split injection case, spray droplets are mostly located in the central region before evaporated. The first-pulse injected fuel moves towards the exhaust side of the cylinder, and the second-pulse injected fuel is influenced by the vortex motion introduced by the first injection and stays longer in the central region due to entrainment, resulting in a more uniform distribution of fuel droplets. This leads to an improved mixture homogeneity before combustion occurs as shown in Figure 7.26. The lean mixture region in the exhaust side in the single-injection case is not seen in the split injection case. Engine dyno tests indicate that the exhaust oxygen emissions with split injection were reduced, which confirmed that mixing homogeneity was improved [40].

Another advantage of split injection is to reduce liner wetting. Figure 7.27 compares the instantaneous fuel wall-wetting in these two injection schemes. The peak fuel impingement is 0.8% in the single-injection case, while it is 0.1% in the split injection case. This can be explained by comparing the evolution of the spray droplet distribution in Figure 7.25, where less fuel droplets reached the exhaust side in the split injection case.

FIGURE 7.27   Effect of split injection on liner wetting at 1500 rpm and WOT conditions.

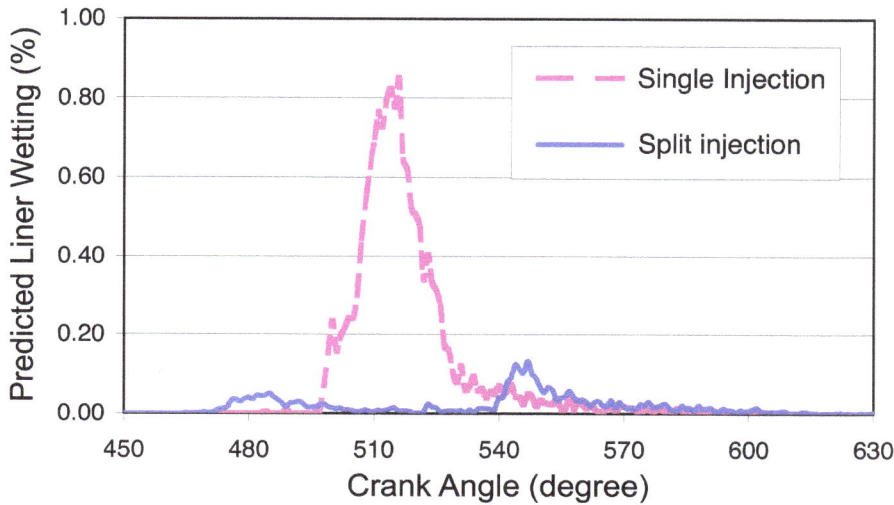

WOT: wide-open throttle

In closing of this section, it is summarized that air–fuel mixing is significantly affected by the large-scale flow motion in the cylinder. Intake flow has a strong impact on the trajectories of fuel droplets at high speeds, which should be avoided. Besides mark design, piston bowl shape, injection strategy (split injection and injection timing), intake motion control (SCV) as discussed, injector orientation, spray cone angle, and cam timing also have impacts on mixture homogeneity, which are not discussed here for brevity. It is also important to avoid or reduce wall-wetting while improving mixing.

7.4.2 Stratified-Charge Formation

As discussed in Chapter 2, fuel-lean combustion can achieve improved part-load fuel consumption in an SI engine due to reduction of pumping loss and heat transfer through the cylinder walls. In a DI gasoline engine, late fuel injection (during the latter part of compression stroke) can operate the engine with fuel-lean combustion by stratifying the fuel–vapor/air mixture within the combustion chamber so that a cloud of mixture close to stoichiometric is formed around the spark plug and the excess air is located around this contained flammable mixture cloud. Thus, the key issues in developing stratified-charge combustion include the formation of close stoichiometric mixture cloud around the spark-plug for stable combustion and minimization of piston wetting for reduction of engine smoke (soot) and hydrocarbon emissions. The influential factors need to be numerically analyzed and experimentally tested for an optimized design. These factors usually include engine design parameters such as piston-bowl shape and injector installation, spray structure parameters (injector specification), and operational parameters such as injection strategy (timing and pressure), spark timing, and gas motion control. The stratified mixture formation process in a WGDI is illustrated in Figure 2.7. The following discussion will again use the LSC concept (see Figure 2.6) as an example to describe the important aspects of stratified mixture formation and how numerical simulation helps to achieve an optimal solution.

A conventional SCDI system is commonly designed to operate under stratified-charge mode at a relatively high engine load and speed range (or window) (e.g., up to 4.0 bar BMEP and 2500 rpm for a wall-guided system) [1]. However, the stratified-charge operation load is limited by engine smoke in a wall-guided system. Although a deeper piston bowl can help extend the stratified-charge operation load range, it will penalize engine output at WOT operation due to increased piston surface area (increasing heat transfer loss) and deteriorated air–fuel mixture homogeneity as illustrated in Figure 7.23 (reducing air-utilization and combustion efficiency). Other trade-offs have to be made in a conventional SCDI system. For reference, as the stratified-charge window increases in size, drive cycle fuel consumption decreases, although at a diminishing rate due to: a) the decreasing benefit of stratified over homogeneous fuel consumption and b) the increasing aftertreatment purge frequency (requiring rich operation). Also, as the window increases in size, aftertreatment cost increases to allow reasonable duration steady-state operation without a purge. Hence the "value" of the stratified-charge window deteriorates with increasing size.

The LSC DI concept uses a reduced stratified-charge operation window ranging from the idle operation to 2000 rpm and 2 bar BMEP. The reduced stratified-charge window allows a significant relaxation in the requirements for the lean aftertreatment system but still enables significant fuel economy gains over the non-stratified-charge operation.

The LSC DI concept drives for a simple piston crown shape design. The required light stratified-charge window warrants a shallow bowl with its sides formed by the crown. The raised crown or dome shape is designed to meet the variable cam timing (VCT) range and CR requirement and to have a minimized surface area to reduce heat transfer losses. On the other hand, the shallow bowl piston design minimizes the weight of the piston for better NVH and improves mixture homogeneity for early injection operations as discussed earlier.

The details of the piston bowl shape were found to be very influential on engine functions and were optimized accordingly [4]. Figure 7.28 shows two representative designs in which exhaust is on the left and intake is on the right. The "circular" design features a wide opening at the front edge (the exhaust side), while the "converging" one reduces the width of the bowl from the back of the bowl (the intake side) to the front side. CFD modeling was carried out first to compare the mixture formation of these two designs. Figure 7.29 shows air–fuel ratio contours at 20° BTDC on the central plane between the two intake valves under engine idle conditions. The simulated engine operating conditions are 1 bar BMEP, 750 rpm. The choice of this operating condition is due to the consideration that this point is more difficult to form mixture stratification than those at higher engine loads. The EOI is 60° BTDC. It can be seen that these two designs result in similar mixture stratification in the early stage. However, the vapor distribution details become different near the spark timing; the mixture is richer around the spark plug gap (represented by a cross symbol in the plots) in the converging design.

FIGURE 7.28 Comparison of two piston bowl shapes: (a) Circular bowl; (b) Converging bowl.

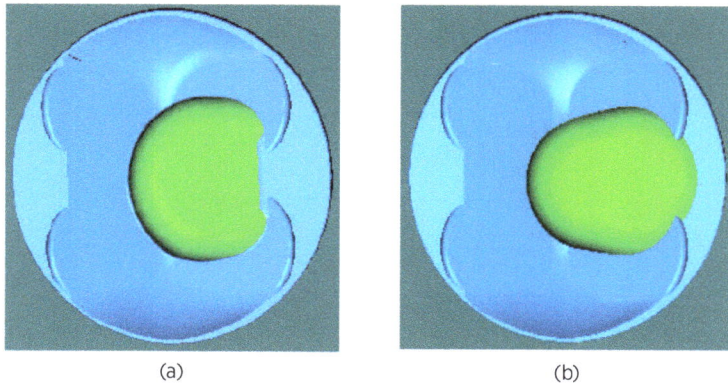

(a)   (b)

FIGURE 7.29 Computed air–fuel distribution in: (a) the Circular bowl piston and (b) the Converging bowl piston at 20° BTDC.

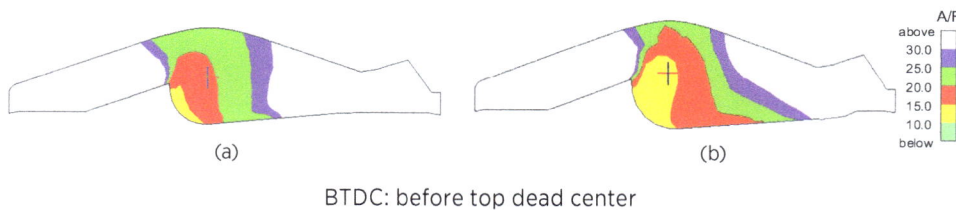

(a)   (b)

BTDC: before top dead center

Figures 7.31 and 7.32 show the computed air–fuel ratio evolution at the spark plug gap location and fuel vapor distribution in the piston bowl at 25d BTDC (near the MBT spark timing). The simulated operating conditions are 1500 rpm, 1 bar BMEP, EOI 62c BTDC. A 70° cone spray injector was simulated. Figure 7.30 indicates that the converging shape design delivers a richer mixture core around the spark plug gap, which can give slower initial combustion and overall better combustion phasing. As shown in Figure 7.31, the converging design also produces less mixture near the stoichiometric A/F ratio that reduces the total NO formation and more mixture in the rich air–fuel ratio region (equivalence ratio greater than 1.5) that increases smoke formation.

Engine dynamometer test results of the two piston-designs are given in Figure 7.32. The intake cam timing was the same for the two tests. The converging piston was run with the exhaust cam retarded 10° compared to the circular piston timing. The converging piston produced a higher smoke level over the entire EOI sweep, confirming the CFD prediction as discussed above. On the other hand, the richer mixture allows the converging piston to run with more valve overlap. Despite the increased valve overlap, the standard deviation of net mean effective pressure (NMEP) for the converging piston configuration was almost as good as or better than the circular piston configuration. As discussed earlier, the less near stoichiometric mixture in the converging piston and increased valve overlap led to improved NO_x emissions. Also, the converging piston reduces fuel consumption due to the reduced unburned hydrocarbon emissions and improved combustion phasing.

FIGURE 7.30 Effect of piston bowl shape on local air–fuel ratio at the spark plug location.

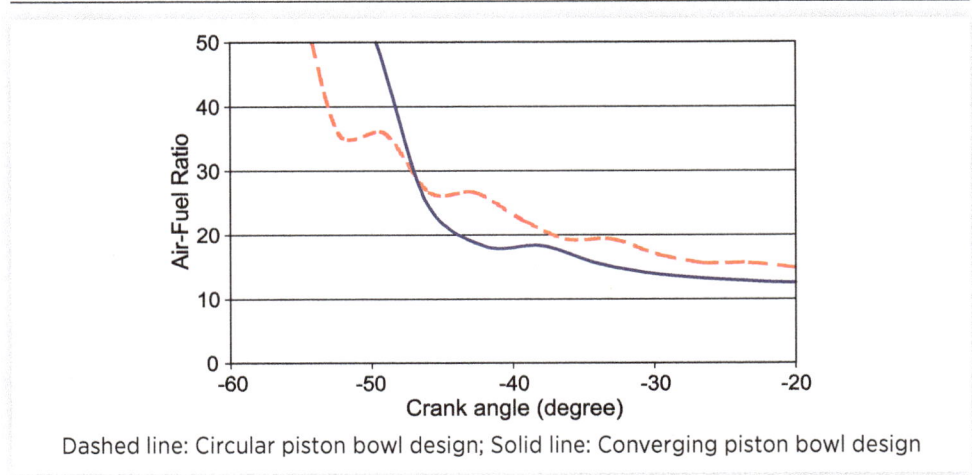

Dashed line: Circular piston bowl design; Solid line: Converging piston bowl design

FIGURE 7.31 Effect of piston bowl shape on fuel vapor distribution at spark timing (25° BTDC).

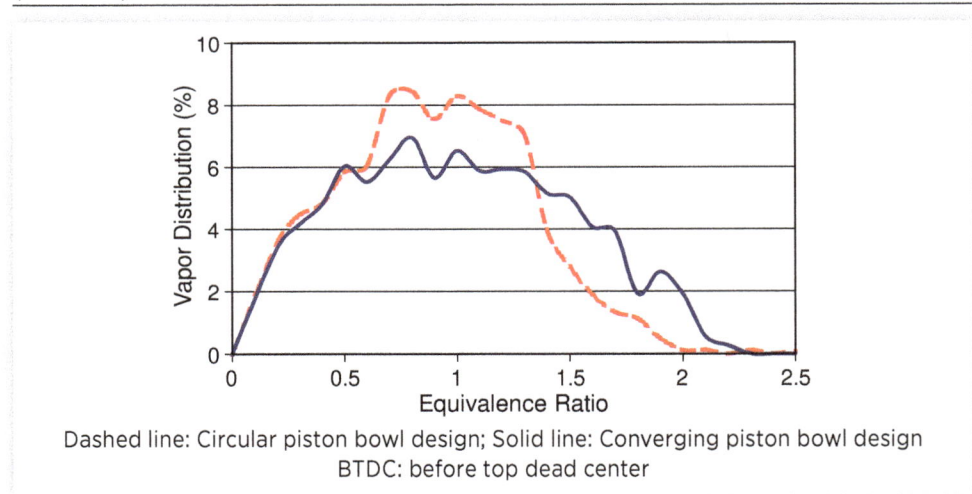

Dashed line: Circular piston bowl design; Solid line: Converging piston bowl design
BTDC: before top dead center

For stratified-charge operation, it is important to provide a robust operating window in which low fuel consumption is achieved and soot emissions are minimized. For a WGDI engine, soot emissions in stratified-charge operation are generally dominated by piston wetting occurring at late EOI. Dynamometer testing showed that lower fuel consumption and low soot emissions could be achieved by advancing the injection timing using a 60° spray injector, as seen in Figure 7.33. The combustion stability rapidly degraded beyond 70° BTDC EOI, resulting in a relatively small operating window (6° of EOI) that satisfied both the soot limit and stability constraint. Similar results were observed using a 70° spray injector (not shown). Testing with a 70° spray plus 5° offset injector (70°/5° injector) with the fuel spray directed 5° further down toward the piston, revealed a much larger stability window (20° of EOI) where soot could be minimized or essentially eliminated.

Reprinted with permission from Ref. [4]. © SAE International

FIGURE 7.32   EOI sweeps showing piston-bowl design effect.

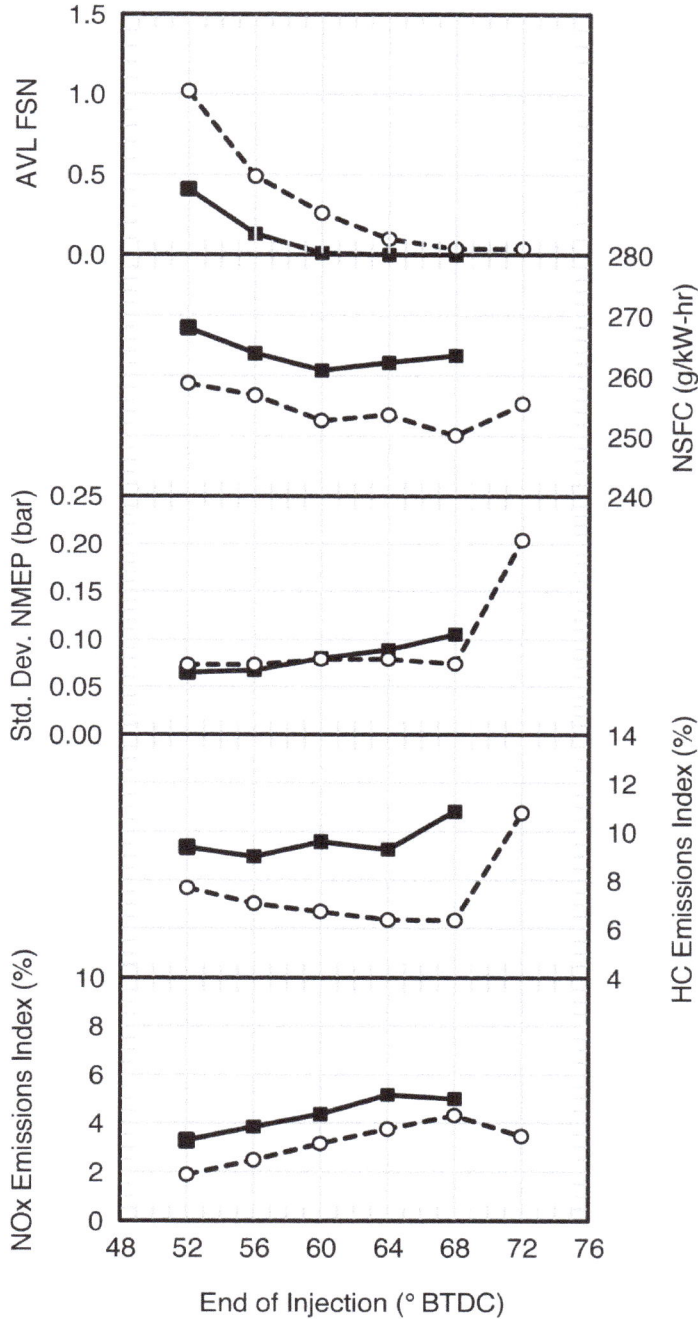

Solid symbol: Circular piston bowl design; Open symbol: Converging piston bowl design
EOI: end of injection

©2022, SAE International

FIGURE 7.33 Single-cylinder engine test data showing spray angle effect on combustion stability window.

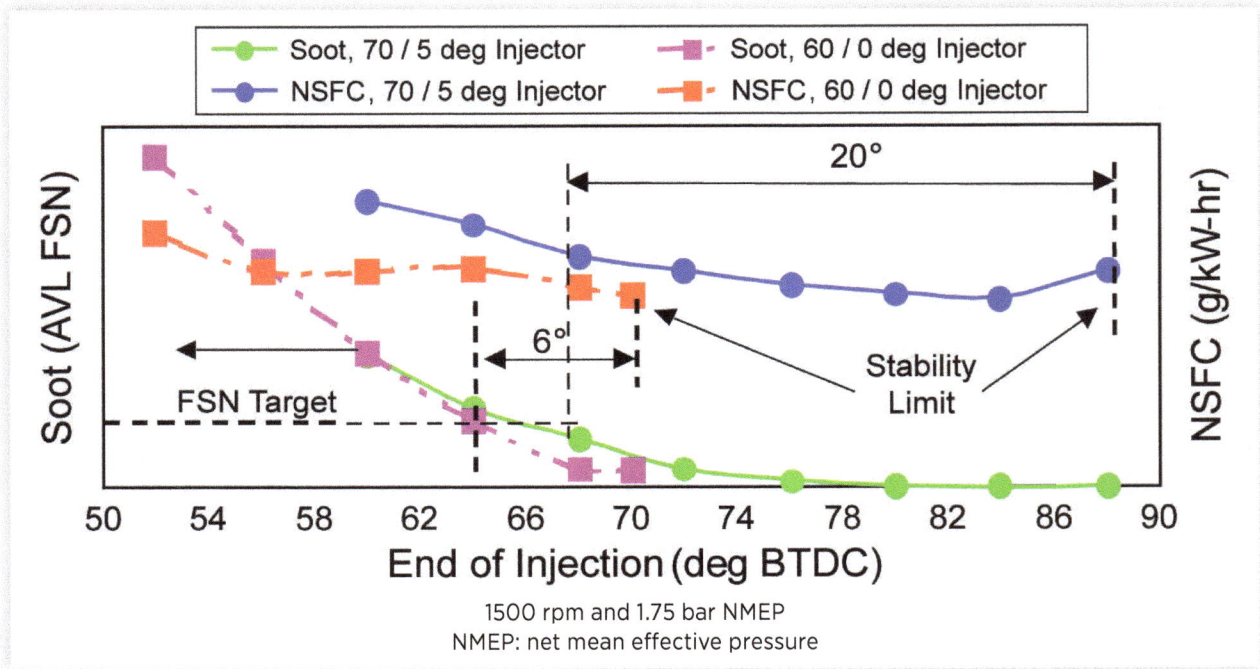

1500 rpm and 1.75 bar NMEP
NMEP: net mean effective pressure

FIGURE 7.34 Computed air–fuel ratio distribution on the central plane at 20° BTDC.

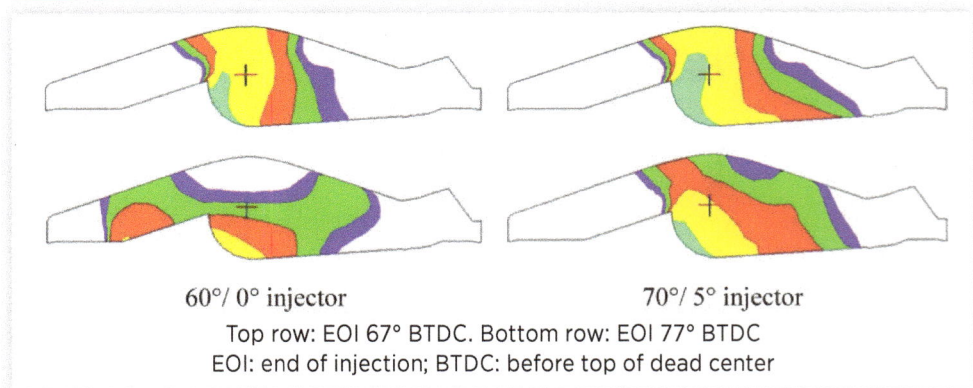

60°/ 0° injector   70°/ 5° injector
Top row: EOI 67° BTDC. Bottom row: EOI 77° BTDC
EOI: end of injection; BTDC: before top of dead center

Figure 7.34 shows the simulated fuel–air distribution created by the two injectors. The engine operating conditions are similar to those shown in Figure 7.33. In the top row, both injectors produced an ignitable mixture at the spark plug (symbol "+" location) at 67° EOI. The bottom row shows at 77° EOI, the 60°/0° injector produced mixture partially overshoots the piston bowl leaving a lean mixture at the spark plug. The 70°/5° injector continues to produce a good fuel–air distribution at early EOI.

The PLIF measurements of the stratified-charge fuel mixture distribution in Figure 7.35 show significant differences in mixture formation between the 70°/5° offset injector

FIGURE 7.35 Optical engine obtained SC mixture distribution (PLIF) for two injector specifications.

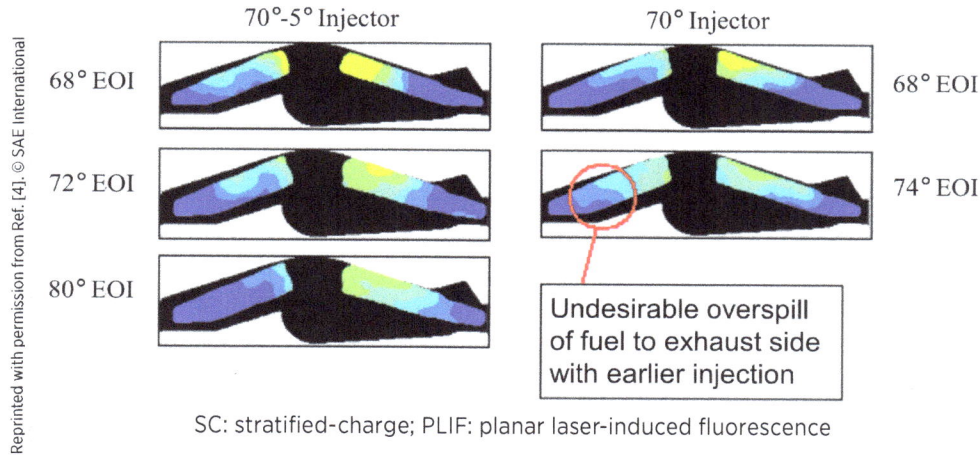

SC: stratified-charge; PLIF: planar laser-induced fluorescence

and the 70° injector. The engine operating conditions are similar to those Figure 7.33. The offset injector results in a much wider range of injection timing (range of EOI) where a well-contained F/A mixture exists in the region of the spark plug. With the non-offset injector, overshoot of the spray into the exhaust side of the combustion chamber is evident for the 74° EOI timing, which marks the limit of injection timing for this injector (COV NMEP increased sharply for further advance). The net result is a much more robust range of injection timing for the offset injector where stable combustion is possible with low soot production (generally, soot increases with retarded injection timing due to increasing piston wetting).

7.5 Soot and Hydrocarbon Emissions by Wall-Wettings

In a WGDI system, spray wall impingement cannot be avoided and this results in piston surface wetting. Wall-wettings can occur in other situations, e.g., piston and liner wetting in early injection homogeneous-charge mode, and in spray-guided DI systems. Liquid films are formed in the wall-wetting surface and, consequently, pool fires of the films take place, which generates soot (smoke) and hydrocarbon emissions. Thus, minimization of wall wettings is one of the major tasks in developing a DI combustion system. Accurate prediction of spray wall impingement is necessary. Numerical simulations can give detailed information of the location, mass deposition, and vaporization of the liquid films once spray impingement occurs. Physical models in Sec. 5.5 are used in the following examples.

In modeling of a DI engine at stratified-charge mode by Han et al. [1], it was found that spray impinged on the piston bowl soon after SOI, and most of the impinged fuel adhered to the bowl surface so that so-called piston wetting occurred. The adhering liquid fuel either vaporizes and contributes to very rich mixtures near the wall or remains as a liquid phase at the time of spark. Figure 7.36 shows the computed liquid fuel on the piston bowl surface represented by the fuel drops at 20° BTDC. The liquid fuel is concentrated inside the bowl in the region opposite the injector side.

FIGURE 7.36 Computed liquid fuel on piston surface with comparison to carbon deposits on an optical engine piston.

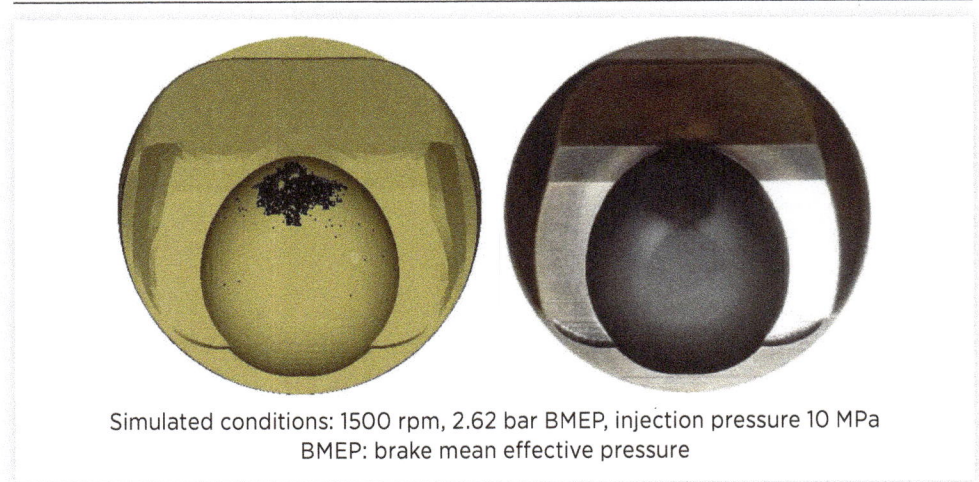

Simulated conditions: 1500 rpm, 2.62 bar BMEP, injection pressure 10 MPa
BMEP: brake mean effective pressure

Also shown in Figure 7.36 is a photo picture of the piston from the optical engine. The optical engine had the same piston design as that used in the simulation and was operated with iso-octane (the same as used in simulation) for about five minutes under similar operating conditions before the piston was pulled out for taking pictures. It is clearly seen in the picture that carbon deposits built up inside the piston bowl as a result of pool firing. The fact that the computed liquid fuel share the same location with the carbon deposits leads to the conclusion that the liquid fuel on the piston due to spray impingement results in soot formation during combustion and, hence, in engine-out soot emissions.

The conclusion was supported by the simultaneous flame and soot scattering imaging experiments in the same optical engine as shown in Figure 7.37. Two-dimensional images of in-cylinder soot particles were obtained using laser elastic (Mie and Rayleigh) scattering. Color images of combustion flames were obtained using a color video CCD camera. The engine conditions are 1500 rpm, 90 kPa MAP, 25° BTDC SA, SCV partially open, 90°C coolant temperature, 60° BTDC EOI, 1.6 msec fuel injection pulse-width, 10 MPa fuel injection pressure, indolence fuel. This image set shows that small soot particles are formed in the combustion chamber and transported to the exhaust valve and out of the combustion chamber. PLIF imaging of unburned fuel and hydrocarbons was conducted under the same engine conditions. The PLIF results show unburned fuel or partially oxidized fuel components in the regions with soot formation. Thus, the elastic scattering and fuel-tracer LIF measurements provide evidence that fuel spray wall-wetting is an important source of soot and hydrocarbon in SCDI engines [13].

Since accurate soot models were not available at that time, the combined deterministic–empirical approach (see Figure 7.5) was adopted and the remaining liquid fuel deposited on the piston surface at the time of spark was used as the intermediate variable to correlate qualitatively with the engine-out smoke level. The correlation is shown in Figure 7.38 in which the computed history of liquid fuel on the piston surface for various design configurations is plotted and the measured engine smoke number obtained from a single-cylinder engine under the same operating conditions is given. It is found that more the liquid fuel remains on the piston at the time of spark, the higher the engine smoke number.

FIGURE 7.37 Image sequence showing soot plume from the piston bowl during expansion (exhaust ports are on the right of the images). Individual image pairs correspond to 33°, 43°, and 63° ATDC.

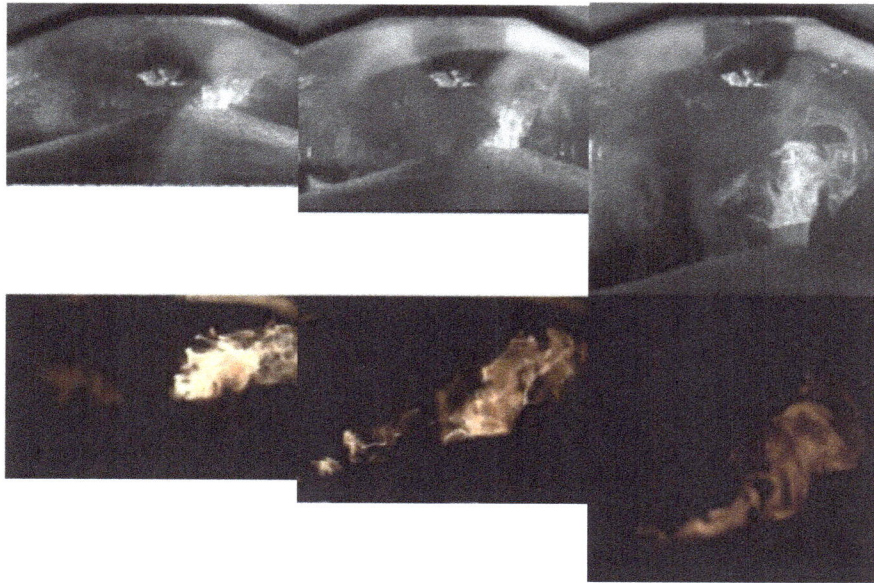

ATDC: after top dead center

FIGURE 7.38 Computed liquid fuel evolution on piston surface with its correlation with measured engine-out smoke number (FSN).

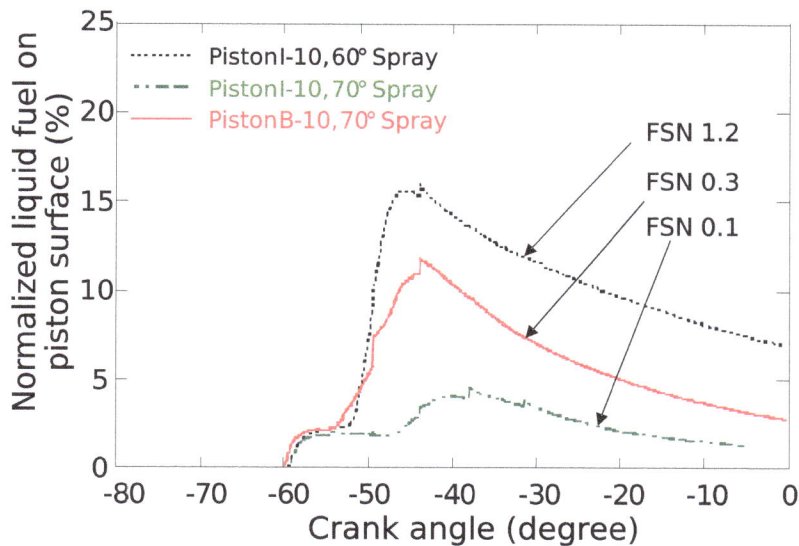

Simulated conditions: 1500 rpm, 2.62 bar BMEP, and injection pressure 10 MPa
FSN: filter smoke number; BMEP: brake mean effective pressure

It is important to point out that the engine-out smoke emissions level is an integrated result of a very complicated combustion process. During the engine combustion process, the liquid fuel continues to vaporize and mix out with other gases. Soot forms and also oxidizes. Hence, it is not surprising that the amount of liquid fuel before combustion cannot quantitatively correlate with engine smoke level.

Effects of SOI, spray cone angle, gas swirl motion, and engine load on piston wetting are illustrated in Figure 7.39. The simulated engine and engine conditions are the same as those in Figure 7.38. A delayed injection results in more spray impingement, and more liquid fuel deposition on the piston surface as indicated by the higher peak value in Figure 7.39a. Spray cone angle (60° vs 70°) has a significant effect on spray dynamics and hence mixture formation (see Figure 7.39b). A larger cone angle results in wider and more dispersed fuel distribution that helps fuel vaporization and, hence, mixing. Also, the axial velocity of a larger cone angle spray is smaller under the same injection pressure. This results in slower spray axial penetration. For these reasons, spray impingement on the piston is reduced. Swirl

FIGURE 7.39   Piston wetting affected by: (a) SOI; (b) spray cone angle; (c) gas swirl motion; and (d) engine load. 1500 rpm and 2.62 bar BMEP for (a), (b), and (c); 1500 rpm and 4.0 bar BMEP for (d).

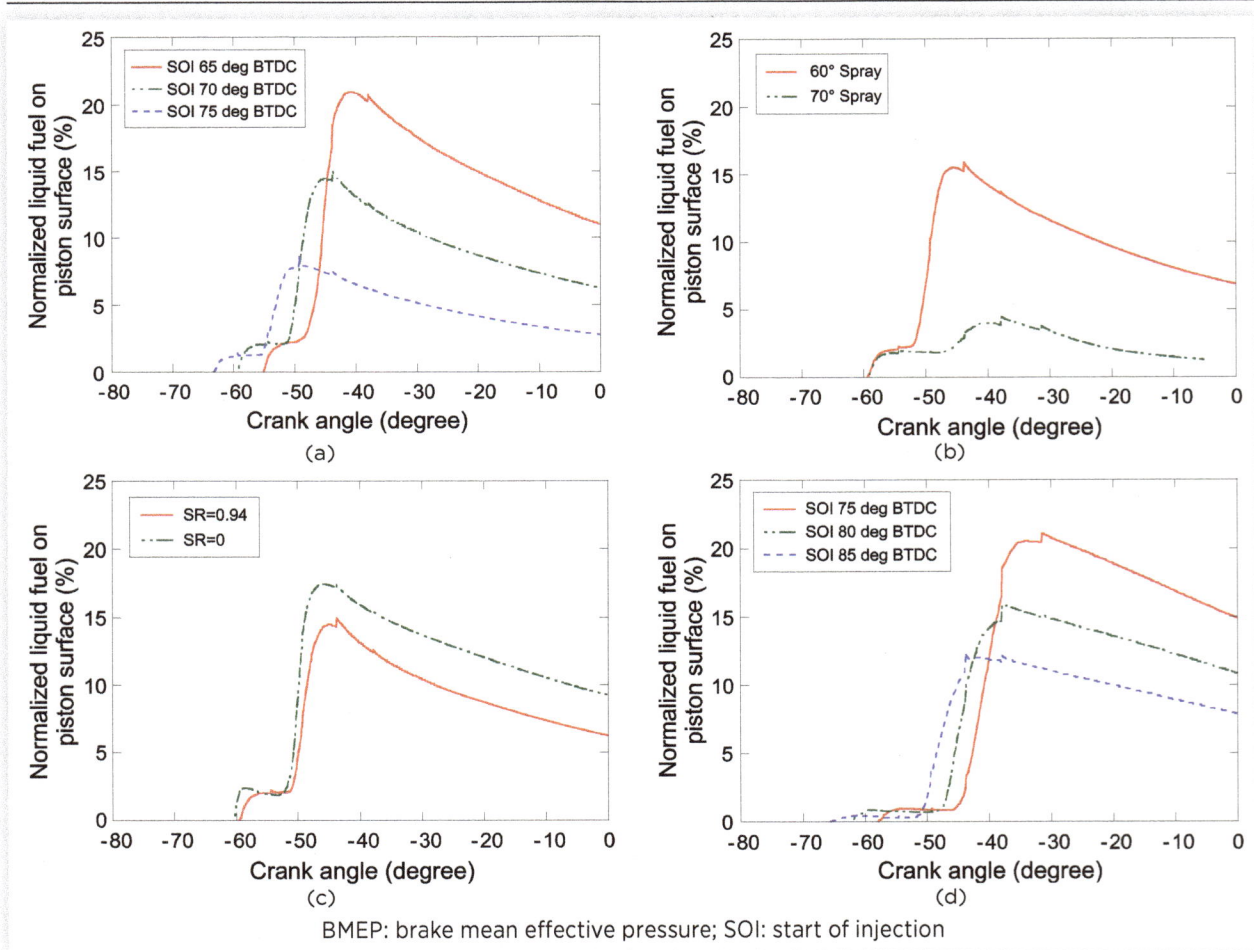

BMEP: brake mean effective pressure; SOI: start of injection

Adapted with permission from Ref. [1] SAE International

gas motion (and tumble) can be generated by closing the SCV on the intake port. In this engine, a swirl ratio of 0.94 was measured when the SCV was closed. As seen in Figure 7.39c, gas swirl motion leads to reduction of spray piston impingement and of the remaining liquid fuel on the piston in comparison with the no swirl motion (SCV open) case. This is because gas motion deflects the spray and disperses the fuel drops. Fuel drop vaporization is then enhanced, and consequently, fewer liquid drops impinge on the piston surface.

The effects of engine load are also evaluated as shown in Figure 7.39d. As the engine load is increased to 4.0 bar BMEP, the liquid fuel deposition on the piston surface increases. Although advancing SOI can help reduce piston wetting, a large number of liquid fuel depositions cannot be avoided, which indicates high-level engine soot emissions occur at high engine loads. A high level of soot emissions is often a limiting factor for wall-guided stratified-charge combustion operating at loads beyond 4 bar BMEP.

The injector location, mounting angle, and fuel spray cone angle are among the critical design parameters for a wall-guided SCDI engine. A larger spray cone angle is preferred for reduced piston wetting as seen in Figure 7.39. However, sprays with large cone angles may lead to undesired intake-valve wetting in an early injection case. In the LSC engine, the engine smoke level at low-speed WOT conditions was high with an early design of 70° spray. CFD simulation later found out that sprays impinged on the intake valves due to the impact of intake flows. Figure 7.40 shows the computed sprays during the induction stroke at 1500 rpm and WOT conditions comparing the results of the 70° and 60° cone angle injector. Modeling predicted that the sprays impinged on the intake valves when the 70° injector was used and the sprays "missed" the valves when the 60° injector was used, as further qualified valve-wetting evolution in Figure 7.41.

FIGURE 7.40 CFD computed spray images on the intake valves from: (a) a 70° spray injector and (b) a 60° spray injector at 1500 rpm and WOT conditions.

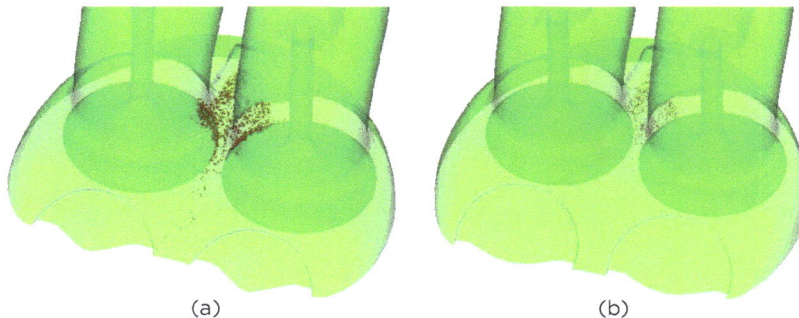

(a)   (b)

CFD: computational fluid dynamics; WOT: wide-open throttle

The effect of spray-valve impingement on engine-out smoke was verified using the optical engine. The flame images in Figure 7.42 show pool fires for the 70° injector where more fuel impingement is expected compared with the 60° spray injector. These pool fires are presumably the source of soot production for this operating condition. Comparing Figures 7.41 and 7.43, it is seen that the predicted valve impingement region coincides with the pool fire locations. The excessive valve-wetting in the 70° spray case correlated with the measured high engine smoke level as shown in Figure 7.43, while the engine-out smoke was reduced significantly when the 60° spray was used.

FIGURE 7.41 CFD predicted valve wetting affected by spray cone angles at 1500 rpm and WOT conditions.

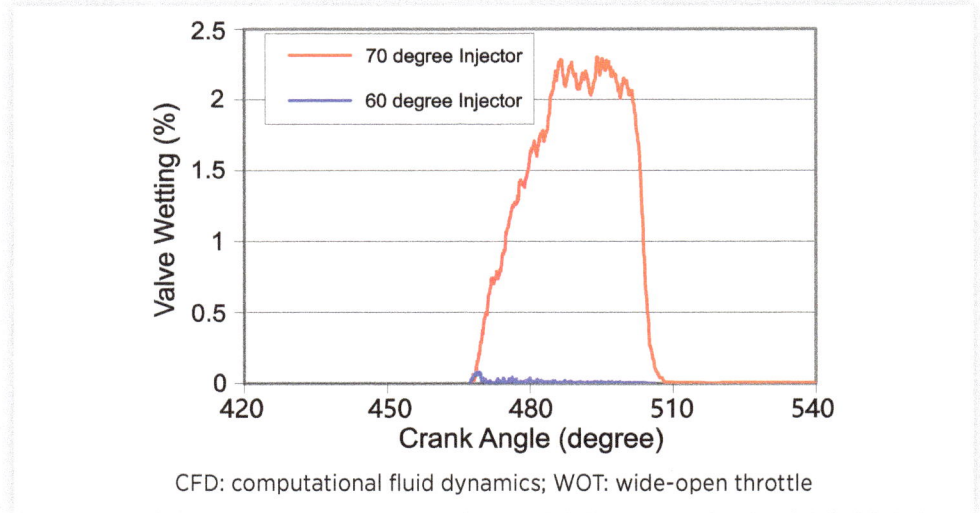

CFD: computational fluid dynamics; WOT: wide-open throttle

FIGURE 7.42 Flame images obtained in the optical engine from: (a) a 70° spray injector and (b) a 60° spray injector at 1500 rpm and WOT conditions.

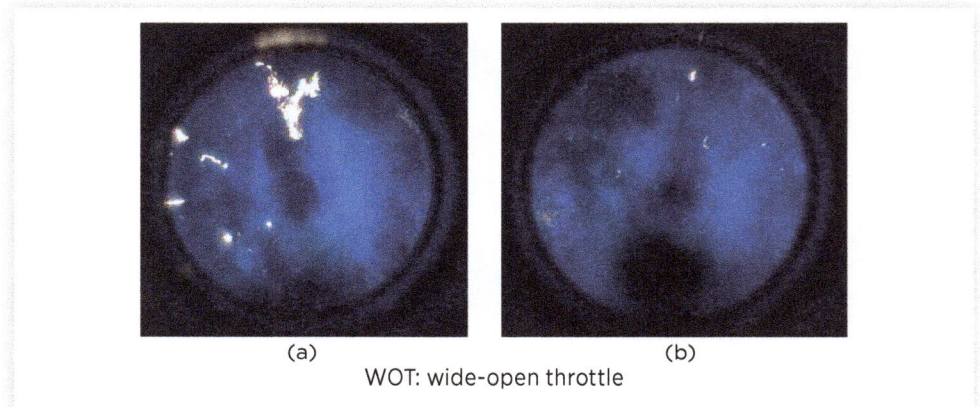

(a)   (b)

WOT: wide-open throttle

While the 60° spray injector reduced the engine-out smoke much lower than the targets, it did not deliver the satisfactory combustion robustness window under stratified-charge operation as discussed earlier. The injector specification was modified to 70° spray with a 5° offset angle (downward toward the piston) to direct the spray away from the open intake valves. This injector was verified to avoid spray-valve impingement [4]. A modeling-driven solution for a practical design issue is demonstrated in this case. A traditional trial-and-error method would take a long time to find the root cause.

High level of unburned hydrocarbon emissions is a problem associated stratified-charge combustion sometimes. Accurate prediction of unburned hydrocarbon has been a challenge

FIGURE 7.43 Engine-out smoke level showing spray cone angle effect at 1500 rpm and WOT conditions.

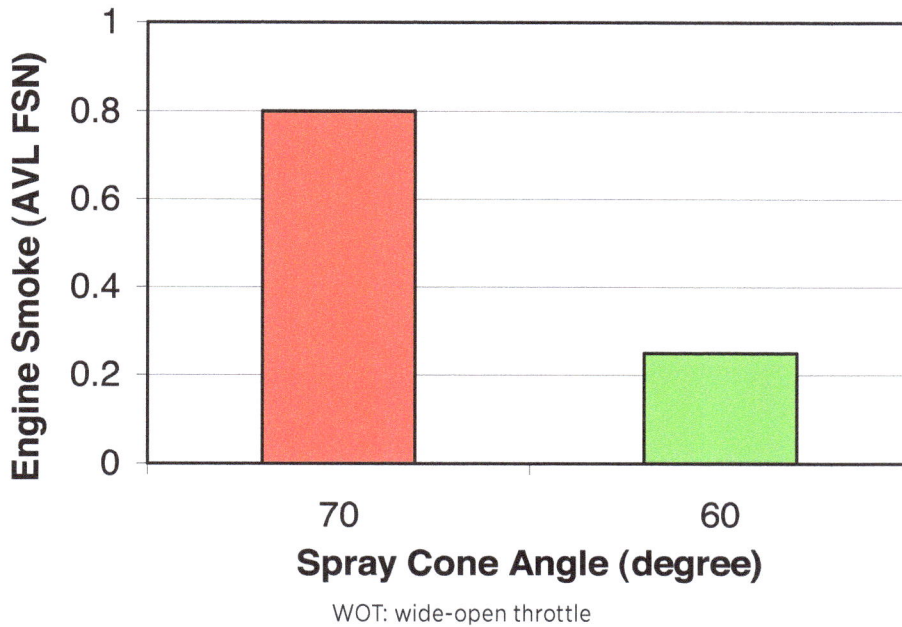

WOT: wide-open throttle

in SCDI combustion simulation due, partially to the uncertainties in hydrocarbon formation mechanism. An empirical model was proposed by Hilditch et al. [15] to correlate hydrocarbon emissions using the combined deterministic–empirical approach. In homogeneous-charge SI engines, trapped fuels in the crevice regions are recognized as the main sources of unburned hydrocarbon emissions [41]; however, this hydrocarbon mechanism does not apply for stratified-charge combustion since fuels are stratified in the piston bowl region. Instead, Hilditch et al. [15] suggested three primary sources of hydrocarbon emissions as shown in Figure 7.44. These are:

1. Very lean regions at the periphery of the stratified-charge fuel cloud.
2. Liquid fuel films on the piston (and elsewhere, potentially).
3. The spark plug crevice volume.

The following correlation is proposed for the unburned hydrocarbon emissions HC:

$$HC = 2.1(\alpha + \beta) \exp(-0.055SA) + 2.5 \qquad (7.1)$$

where α and β are the percentage of injected fuel in regions with an equivalence ratio smaller than 0.3 and the percentage of deposited fuel on the piston surface, respectively. These values are predicted using CFD models. The prediction is adjusted for combustion phasing via the exponential function of spark advance SA. It is expected that the coefficients may change for different engines or operating conditions.

FIGURE 7.44   Sources of unburned hydrocarbons in stratified-charge combustion.

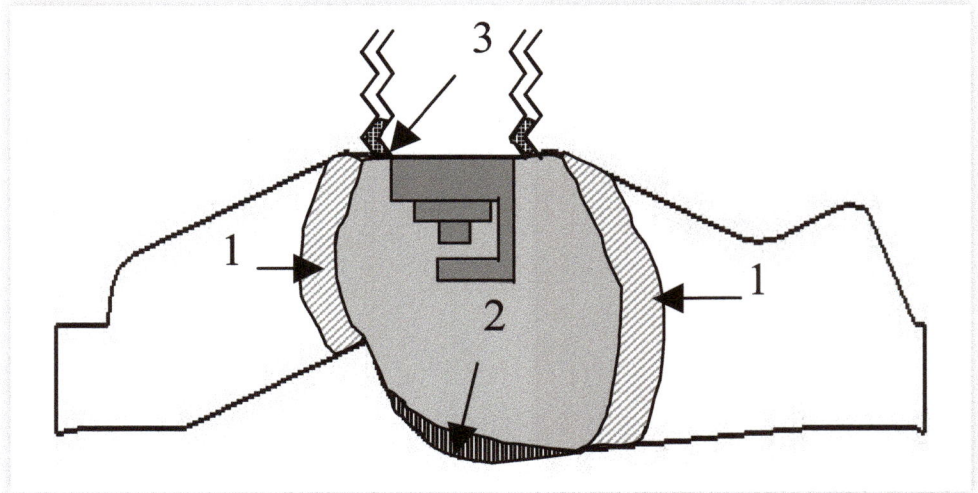

Thus, to achieve reduced hydrocarbon emissions, design parameters are evaluated to reduce α and β as suggested by Equation (7.1). As an example, two design variants were examined in a 0.38L/cylinder engine, namely "Design A" and "Design B". In Design A, the piston bowl is 1.5 mm deeper, the radius curve at the front edge of the bowl (on the exhaust side) is smaller, and the spray cone is 5° wider, all relative to Design B.

Figure 7.45 compares the evolution of the air–fuel ratio at the spark plug for the two designs at 1500 rpm, 1.75 bar NMEP conditions. There is a slight difference in mixture strength between the two designs that may be explained by the arrival time of the fuel cloud. The fuel cloud arrives at the sparking location earlier for Design B, due to the higher penetration of the Design B injector. Based on this plot, both designs can be considered to be able to deliver near stoichiometric ignitable mixture at the spark-plug in a long time-window from 35° BTDC.

However, the details in air–fuel distribution reveal clear differences in the two designs as shown in Figure 7.46. The fuel cloud in Design A is focused vertically at the spark plug location, while the fuel cloud in Design B is more dispersed with the edge of the fuel cloud moving to the exhaust side of the chamber. PLIF images obtained in an optical engine with the same engine configuration supported the CFD results as also shown in Figure 7.46. The difference in the predicted fuel vapor distribution at 25° BTDC is further illustrated in Figure 7.47. Note that Design B has significantly more fuel distributed in the leanest regions. In addition, the computed piston-wetting for Design B is substantially higher than Design A. The shallow bowl and narrow cone angle injector of Design B result in over 10% of the fuel being deposited in a liquid film on the piston. The piston-wetting is approximately 1% for Design A. With these results, Equation (7.1) was used to predict the engine hydrocarbon emissions for the two designs and the predicted results [15] are plotted in Figure 7.48. As can be seen, Design A significantly reduces unburned hydrocarbon emissions, and as a result improves overall stratified-charge fuel economy.

FIGURE 7.45 Computed evolution of air–fuel ratio at the spark plug gap.

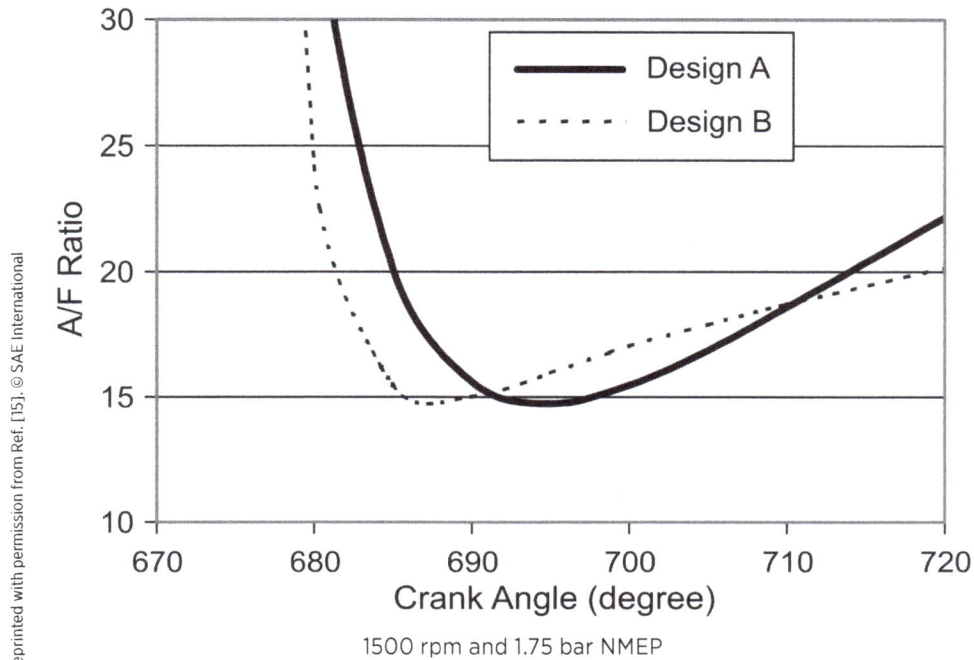

1500 rpm and 1.75 bar NMEP

FIGURE 7.46 Mixture stratification as predicted by CFD and measured in the optical engine (PLIF) showing piston bowl design effect.

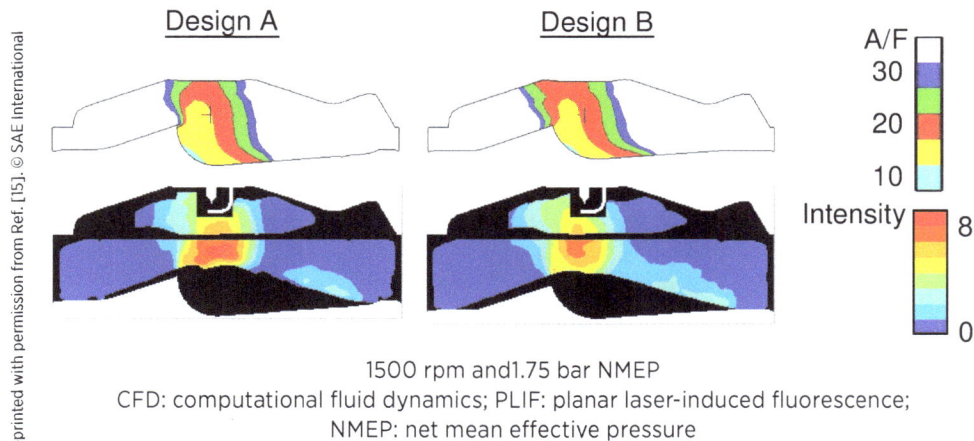

1500 rpm and1.75 bar NMEP
CFD: computational fluid dynamics; PLIF: planar laser-induced fluorescence;
NMEP: net mean effective pressure

Reprinted with permission from Ref. [15]. © SAE International

FIGURE 7.47 Equivalence ratio histogram of design variants at 25° BTDC.

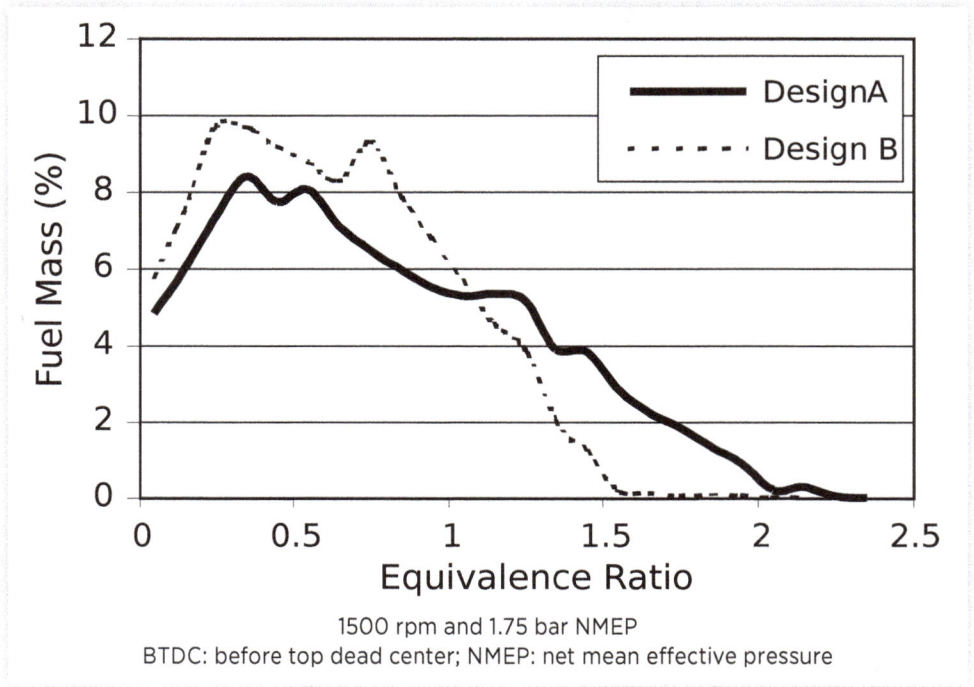

1500 rpm and 1.75 bar NMEP
BTDC: before top dead center; NMEP: net mean effective pressure

FIGURE 7.48 Reduction of hydrocarbon emissions by improved design.

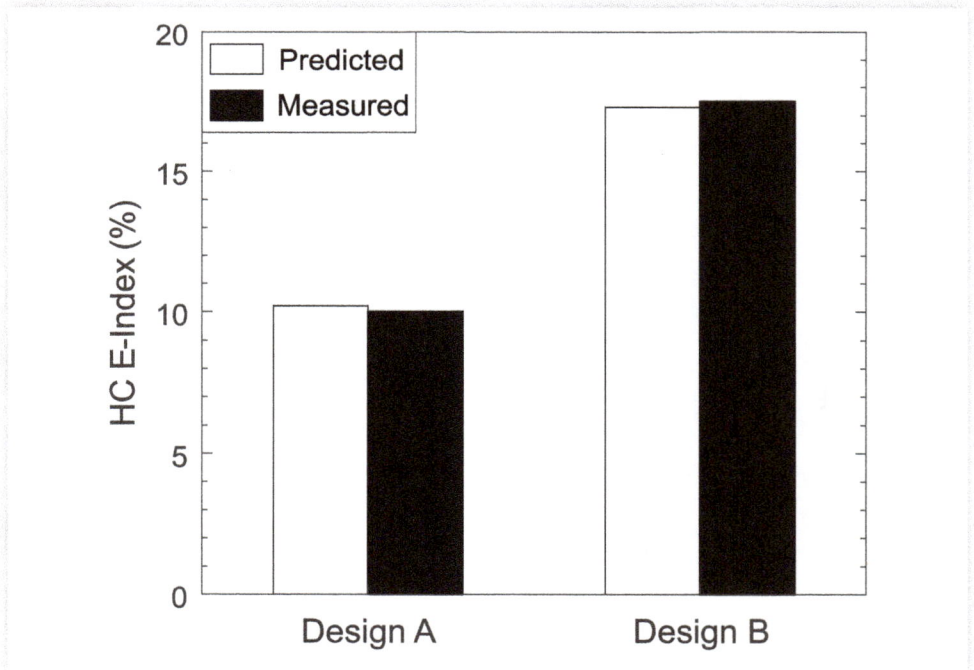

Courtesy of Zhiyu Han

7.6 Mixing in Spray-Guided and Turbocharged DI Systems

Stratified-charge combustion can be also achieved in the spray-guided direct-injection (SGDI) concepts as illustrated in Figure 2.8. As an example, features of mixing and combustion in the VISC DI concept [3] are discussed first. After that, the mixing strategy in recent turbo-charged HCDI engines is addressed.

In a wall-guided system, typically, the fuel injector is placed below the intake valves and has a relatively narrow spray cone (50°–70°) to project the fuel across the cylinder bore. Some issues were carefully addressed before. The spray tends to wet the piston and not mix sufficiently, resulting in high smoke emissions, in particular at high engine loads. Therefore, the fuel injection timing has to be relatively early in the compression stroke to avoid excessive piston wetting. However, the early injection timing may result in over-mixed regions, which causes high hydrocarbon emissions and increased fuel consumption. Although advanced spark timing can be used to reduce hydrocarbon emissions, it causes the combustion to be over-advanced, reducing thermal efficiency and increasing NO_x emissions. This trade-off of combustion efficiency and thermal efficiency has proven difficult to improve with wall-guided systems.

On the other hand, typical spray-guided combustion system concepts use a wide cone angle fuel spray and a close-coupled fuel injector and spark plug. The wide cone angle spray does not tend to collapse and so is effective at high ambient pressures, and the close coupling allows the injection timing to be near the time of spark. Together, these effects allow substantially later fuel injection timing, improving the trade-off between hydrocarbon emissions and combustion phasing. In such a close-coupled system, however, there is typically no mechanism to slow the penetration of the fuel spray, which allows the spray to quickly convect past the spark plug. The spark timing is therefore limited to a narrow window near the end of the injection event to avoid misfire. This close timing limits control flexibilities and times for the fuel droplets to evaporate and to mix. This can lead to the burning of the droplets and overly rich mixture, which tends to produce smoke emissions.

To overcome the mentioned drawbacks of spray-guided stratified-charge combustion for improved combustion efficiency and combustion stability, a spray-guided combustion system concept was developed using a mixture formation method referred to as VISC [3]. The VISC system was demonstrated to deliver reduced fuel consumption over a WGDI as shown in Figure 2.14. It provides a better tradeoff between combustion completeness and combustion phasing than typical wall-guided combustion systems, while providing robust combustion and controlling smoke emissions. This system is more typical of a spray-guided combustion system, using a close-coupled fuel injector and spark plug, and minimizing the interaction of the fuel spray with the piston surface. This, along with the use of a wide cone angle spray, allows substantially later injection timing; roughly 30–40 CA later than that in a WGDI system. The difficulty with a spray-guided combustion system is in determining how to robustly provide an ignitable mixture to the spark plug.

The spark plug gap in any stratified-charge engine must be located such that it is within an ignitable portion of the fuel–air mixture, without wetting the spark plug electrodes with fuel droplets. Spark plug wetting causes fouling and corrosion issues. In the VISC system, the fuel spray is not aimed directly at the spark plug; instead, the spark plug is located in the recirculation zone (vortex) naturally formed on the outside of the fuel spray. This vortex draws fuel vapor out of the spray to the spark location. The CFD simulation in Figure 7.49 shows how the spray vortex affects the mixture formation. Figure 7.49a shows the spray

CFD images showing the VISC concept: (a) showing the spray induced vortex; (b) showing the air–fuel ratio contours and their relations with the spark plug gap.

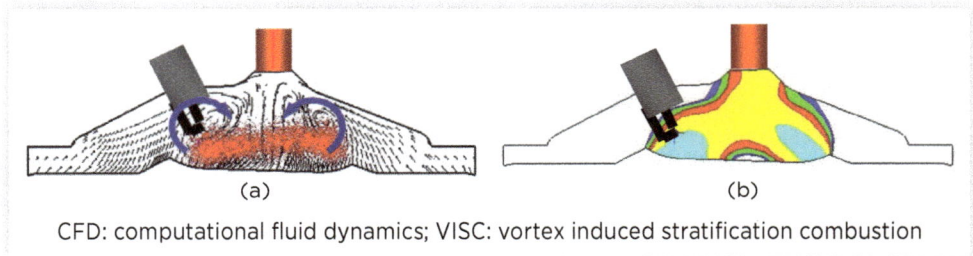

(a)   (b)

CFD: computational fluid dynamics; VISC: vortex induced stratification combustion

droplets passing below the spark plug, while the velocity vectors show the formation of the recirculation vortex (represented by large arrows) at the periphery of the spray. The air–fuel ratio contours in Figure 7.49b show how the fuel vapor surrounds the spark gap, providing an ignitable mixture.

An outward-opening pintle injector, which produces fuel sprays with a 90° cone angle was used in the VISC system [3,42]. The experimental and computed spray images are shown in Figure 7.50 and the spray tip penetration in Figure 7.51. The sheet atomization model in Sec. 5.2.3 and secondary drop breakup TAB model in Sec. 5.3.1 are used in the simulation [42]. Reasonably good predictions were obtained. Both experiment and simulation indicate:

1. Spray tip penetration length and rate are not large under pressured ambient condition, which is preferred in a spray-guided concept.

2. The spray does not collapse under the pressured ambient condition as seen in swirl-injector produced sprays (see Figure 7.13).

3. There is no initial slug in the sprays as seen in swirl-injector produced sprays (see Figure 7.11).

The measured SMD is 14.3 μm at a location 25 mm from the nozzle with a 10 MPa fuel injection pressure in a 0.1 MPa ambient pressure. The simulations predict an average SMD of 13 μm for the same conditions.

One of the major challenges in SGDI combustion is the short time window between the EOI and spark-advance (SA) for non-misfire and stable combustion. The closely-coupled EOI and SA are unfavorable for robust combustion operations. In the early stage of VISC development, SA and EOI could not be separated. After a redesign of piston top shape (combustion domain), protrusions of spark plug gap, and injector tip, sufficient separation of SA and EOI was achieved without misfire, which is a feature not common among spray-guided DI systems. This is a likely indication that SC combustion with the VISC system is realized through a very robust mixture stratification mechanism. At 1500 rpm and 2.62 bar BMEP, it was correlated that the minimum time interval between EOI and SA was about 4° after EOI, which was then used in CFD simulation at which mixing was evaluated.

A series of CFD simulations were carried out to evaluate the effects of design and operational parameters on mixing [42]. The major goal is to optimize these parameters so that a stable ignitable mixture is obtained around the spark plug gap under stratified-charge combustion at elevated engine speed and load conditions and to form a homogeneous mixture at high to full loads.

FIGURE 7.50 Comparison of experiment and simulation spray images at ambient pressures of (a) 1 bar and (b) 5 bar.

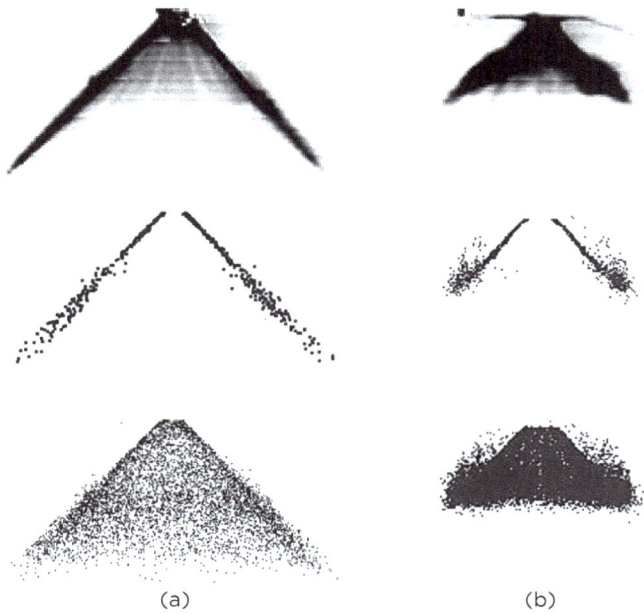

(a)   (b)

Top row: experiment; Middle row: simulation showing a slice through the spray;
Bottom row: simulation showing a front view of the spray

FIGURE 7.51 Spray tip penetration vs. time for three different ambient pressures.

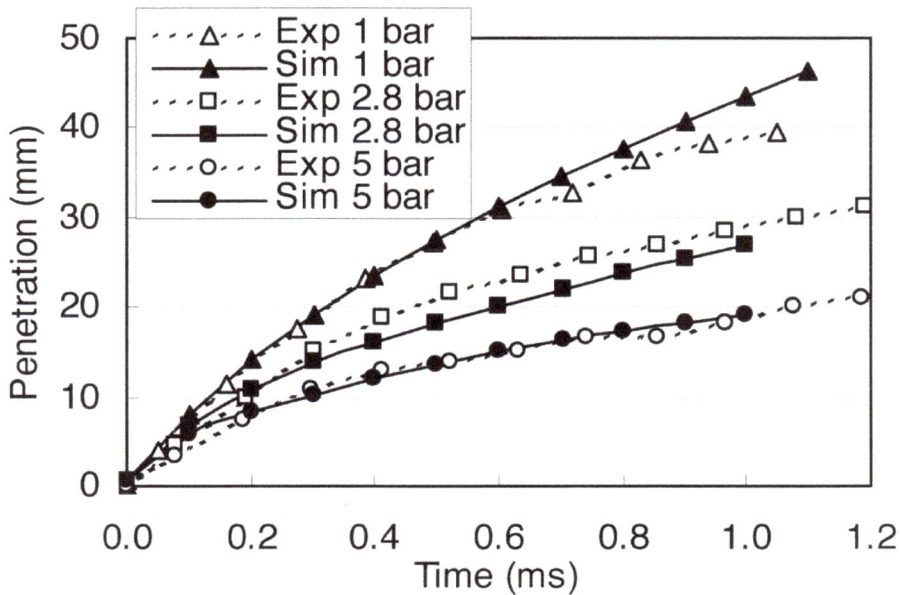

The effects of piston shape on mixture formation are shown in Figure 7.52. With the early designed dome-on-piston geometry, which had serious misfires, the squish gas motion interacts with the vortex and pushes the mixture to the sides causing the mixture cloud to separate in the middle. The cause for misfires is attributed to the poor stratification due to dome-on-piston design. In contrast, the improved bowl-in-piston design is good at containing the mixture cloud around the spark plug gap. The misfires were eliminated with this piston, when using a raised injector axial position.

FIGURE 7.52 Comparison between the mixture structure shown by air–fuel ratio contours in (a) the dome-on-piston geometry and (b) the bowl-in-piston geometry. The cross with the blue dot in the middle corresponds to the location of the spark plug.

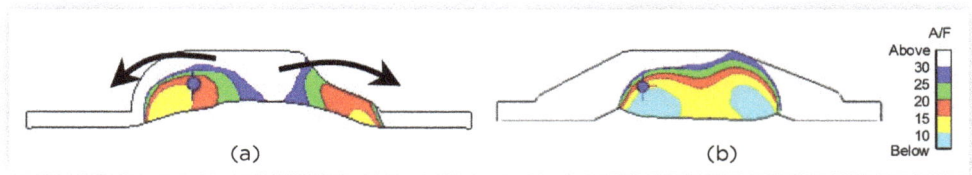

Reprinted with permission from Ref. [42]. © SAE International

It was found the injector axial position (protrusion) had a significant impact on misfire [3,42]. A reduction of 4 mm protrusion (injector raised up) eliminated misfire in a wide window of EOI and SOI. Figure 7.53 shows the CFD simulation results of two axial injector positions: injector with a 1.5 mm spacer and with a 5.5 mm spacer. Air–fuel ratio distribution superimposed with gas velocity vector field is shown.

The reasons for misfires elimination with the raised injector position are:

1. The head vortex at the periphery of the spray has more room to form before it impinges on the piston with the raised injector position, 5.5 mm lift, than with the lower injector position, 1.5 mm lift, as seen in the velocity vector plots at 22° BTDC. This way air entrainment can develop with less interference from the piston motion.

2. The spark plug has a better placement relative to the fuel vortex because the mixture cloud is higher around the spark plug for the raised injector position, as seen in the A/F ratio distribution at 16° BTDC. Also, at EOI, the spark plug is closer to the spray for the raised injector position.

3. The shapes of the mixture cloud for the two injector positions are very different. With the 1.5 mm injector position the mixture cloud is a lot lower and stays closer to the piston while with the injector position at 5.5 mm the mixture cloud covers a larger region of the chamber, as seen at 16° BTDC.

Since the spark plug has to be placed in the vortex recirculation zone, the relative position of the spark plug to the injector is very important for providing good ignition for stratified-charge operation. In the experiments, the spark-plug position was adjusted by adding spacers under it. The effect of the spark-plug position on misfires was evaluated by using three spark-plug axial positions: nominal, pulled back 0.7 mm, and pulled back 1.4 mm. Dyno data showed that by pulling the spark plug back away from its nominal position severely worsened the misfire tendency.

CFD modeling was employed to study the reason for misfires when the spark plug is pulled back. Figure 7.54 presents modeling results for the A/F ratio at the spark plug for the three spark-plug locations for EOI 26° BTDC. As described earlier it was determined that the ignitable mixture doesn't reach the spark plug until 4°–6° after EOI. So, we need to look

FIGURE 7.53 Computed air–fuel ratio distribution and superimposed spray and gas velocity vector fields in the central cutting plane with: (a) injector position 1.5 mm, and (b) injector position 5.5 mm. 1500 rpm, 2.62 bar BMEP, EOI 24° BTDC, SCV closed, MAP 95 kPa. Time sequence from top (every two are at the same time): 24°, 22°, and 16° BTDC.

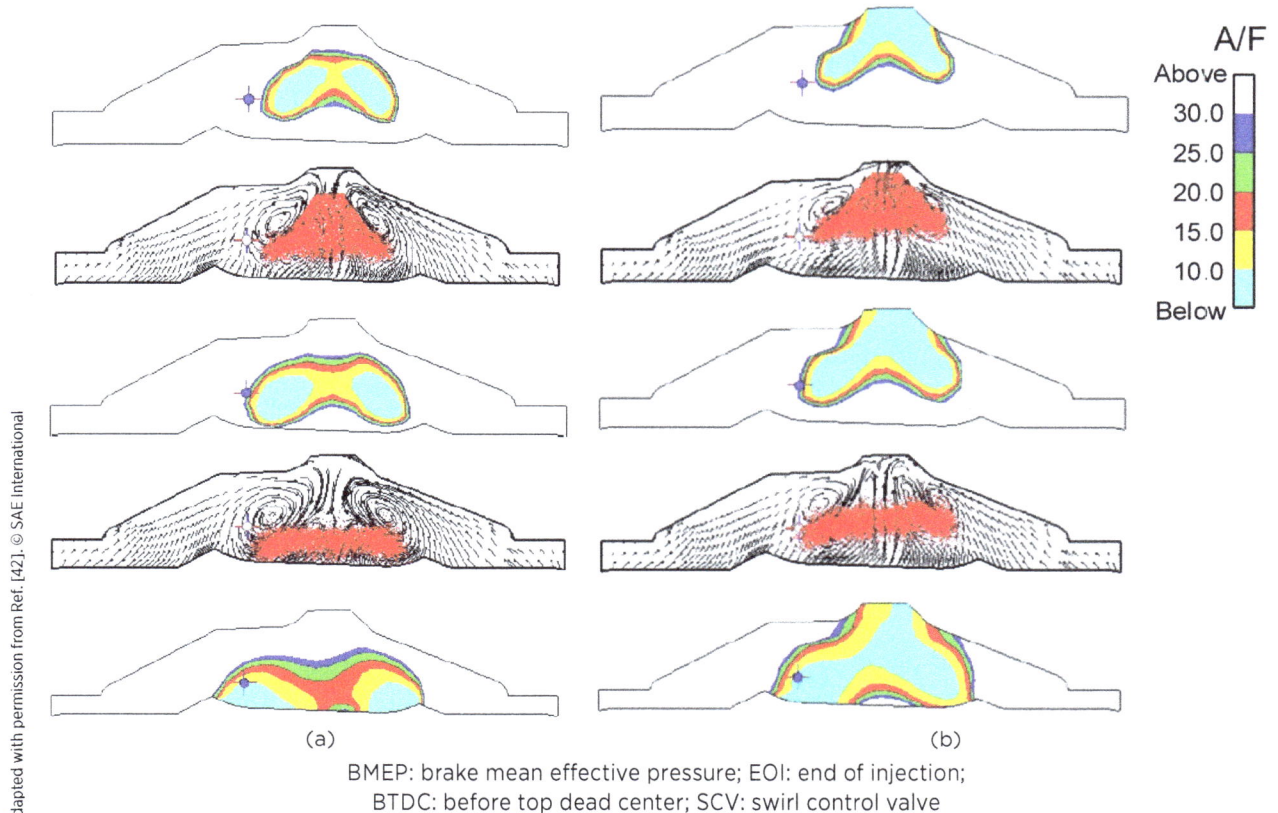

A/F
Above
30.0
25.0
20.0
15.0
10.0
Below

(a)   (b)

BMEP: brake mean effective pressure; EOI: end of injection;
BTDC: before top dead center; SCV: swirl control valve

at the A/F at spark plug at EOI+4° to determine whether the mixture is ignitable. As seen in Figure 7.54, where the EOI+4° line is marked, for the nominal spark plug position, the mixture is ignitable having an A/F ratio of 23, for the spark plug position with 0.7 mm lift the mixture has an A/F of 30, and for the spark plug position with 1.4 mm lift the mixture has an A/F of 37. Therefore, it is expected that misfires occur for the two spark plug positions lifted from the nominal position.

The effects of intake flow motion by closing the SCV were also simulated. By closing the SCV the swirl ratio (SR) increases from 0.03 to 1.47, the tumble ratio (TR) increases from 0.04 to 0.82, and the cross-tumble ratio (CTR) increases from 0.01 to 0.55. With SCV closed, the flow has a high SR and TR that indicates a spiral flow. Figure 7.55 presents the A/F ratio and the velocity distribution in the center plane. The swirl motion in the chamber causes higher velocities that lead to a higher cloud around the spark plug with the SCV closed than with SCV open. This leads to higher stability in the SCV closed case. In addition, the higher turbulence intensity in the cylinder with the SCV closed improves mixing and contributes to higher combustion stability.

FIGURE 7.54 Computed A/F ratio at the spark plug for three different spark plug positions: nominal, 0.7 mm lift, and 1.4 mm lift from the nominal.

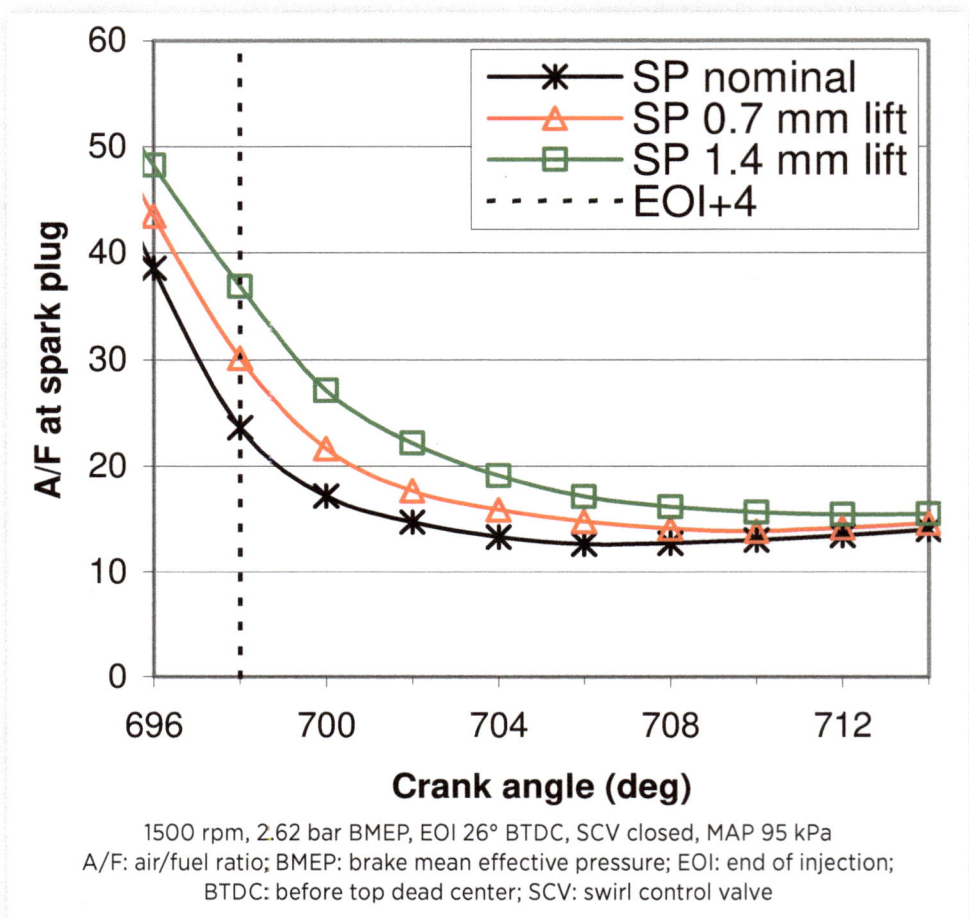

1500 rpm, 2.62 bar BMEP, EOI 26° BTDC, SCV closed, MAP 95 kPa
A/F: air/fuel ratio; BMEP: brake mean effective pressure; EOI: end of injection;
BTDC: before top dead center; SCV: swirl control valve

FIGURE 7.55 A/F ratio and velocity distribution in the center plane at 20° BTDC with: (a) SCV closed and (b) SCV open.

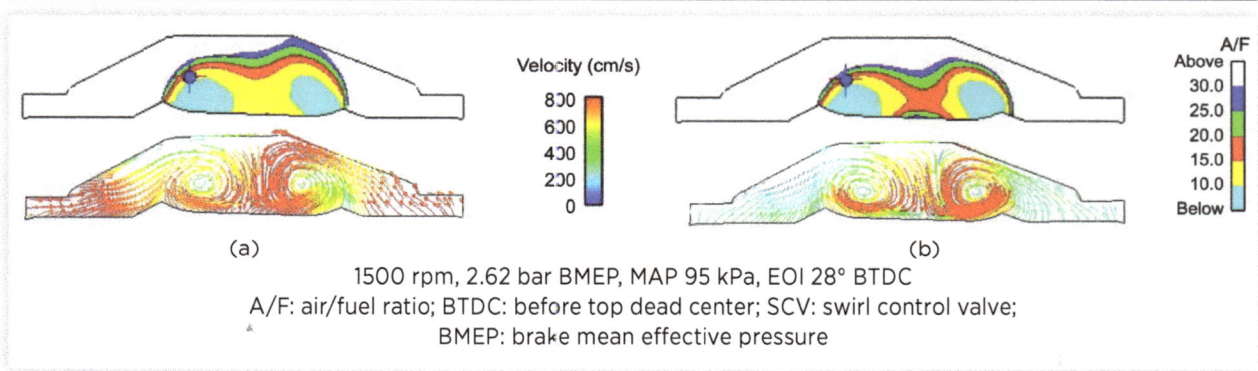

1500 rpm, 2.62 bar BMEP, MAP 95 kPa, EOI 28° BTDC
A/F: air/fuel ratio; BTDC: before top dead center; SCV: swirl control valve;
BMEP: brake mean effective pressure

The split injection was utilized to reduce fuel consumption and smoke emissions at high stratified-charge loads [3,42]. The split injection strategy was studied using CFD modeling by comparing a single-injection case with EOI=40° BTDC to a double-injection case with EOI1=50° BTDC and EOI2=26° BTDC at 1500 rpm and 5 bar BMEP. Figure 7.56 shows the evolution of the A/F ratio distribution in the central cutting plane for the two cases. In the split injection case, it is noticed that the A/F ratio becomes lean around the spark plug before the second injection, as seen at the crank angle of 28° BTDC, and the mixture becomes richer after the second injection as seen in Figure 7.57.

FIGURE 7.56 A/F ratio distribution in the central cutting plane: (a) single-injection: EOI 40° BTDC; (b) split injection: EOI1 50° and EOI2 26° BTDC. Time sequence from top: 28°, 24°, 20°, and 18° BTDC.

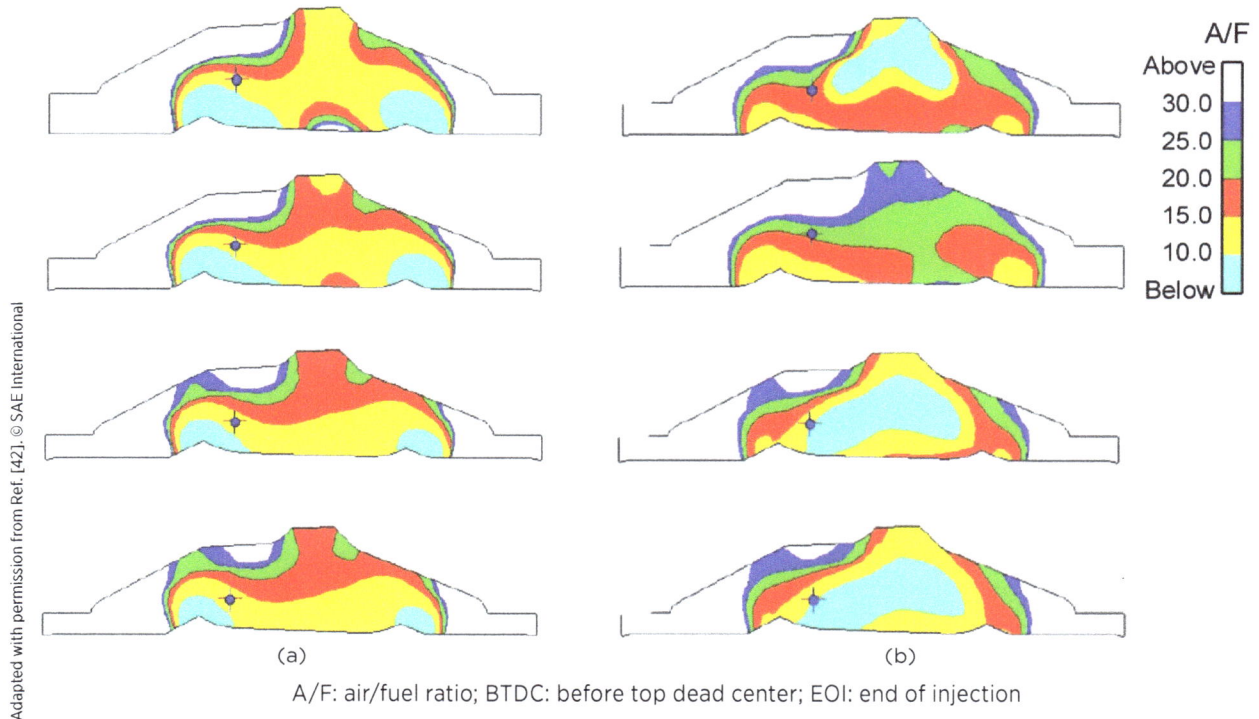

A/F: air/fuel ratio; BTDC: before top dead center; EOI: end of injection

Figure 7.58a presents the amount of liquid fuel deposited on the piston as a function of crank angle degrees. The split injection has a considerably lower liquid fuel deposition on the piston. This correlates well with the dyno measured filter smoke number (FSN), shown by the measured data next to the graphs, which drops from 0.92 for the single-injection to 0.37 for the split injection. The split injection strategy also reduces the amount of vapor in the rich mixtures compared to the single-injection strategy, as seen in Figure 7.58b, which shows the distribution of the equivalence ratio in the cylinder at the time of spark (20° BTDC). The lower amount of vapor in the rich regions leads to lower CO emissions for split injection, as shown by the measured data next to the graphs. The net specific CO (NSCO) emissions drop from 55 g/kWh for the single-injection to 28 g/kWh for the split injection.

FIGURE 7.57 A/F ratio evolution at the spark plug. 1500 rpm, 5 bar BMEP, SCV closed. Single-injection: EOI=40° BTDC; split injection: EOI1=50° and EOI2=26° BTDC.

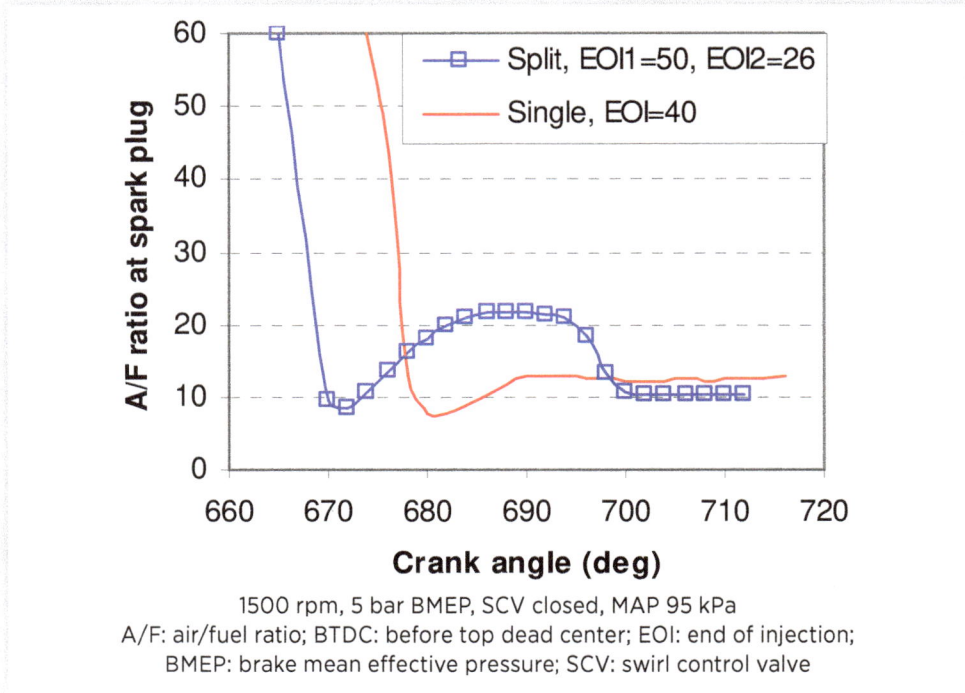

1500 rpm, 5 bar BMEP, SCV closed, MAP 95 kPa
A/F: air/fuel ratio; BTDC: before top dead center; EOI: end of injection;
BMEP: brake mean effective pressure; SCV: swirl control valve

FIGURE 7.58 Comparison of single (EOI 40° BTDC) versus split injection (EOI1 50° BTDC, EOI2 26° BTDC): (a) computed piston wetting evolution; (b) computed fuel vapor mass PDF at 20° BTDC. Measured smoke and CO emissions data are shown next to the graphs.

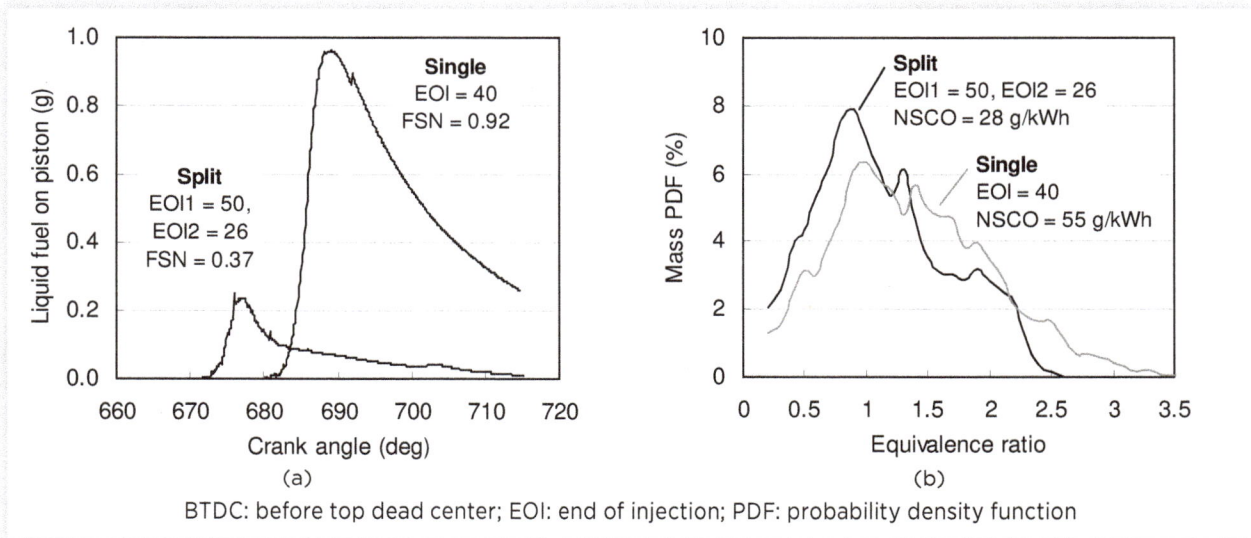

BTDC: before top dead center; EOI: end of injection; PDF: probability density function

Fuel–air mixing homogeneity at full load has been a very challenging issue for WGDI SI engines with side-mounted fuel injectors as discussed earlier. In the VISC concept, the injector is mounted in the center of the combustion chamber. The interaction between the intake flow and the fuel spray is much less significant than that in the WGDI SI engines. Good fuel–air mixing homogeneity can be achieved at both low and high engine speeds [3,42].

Mixing evolution is shown in Figure 7.59 for engine operating at 1500 rpm and WOT, with the SCV open and EOI 300° BTDC. Note that even with the SCV open, where there is less swirl and tumble motion, good mixing is achieved and there are no lean or rich packets are seen. Simulations of EOI sweep indicated mixing homogeneity was insensitive to EOI until the end of an intake stroke.

FIGURE 7.59 Mixing evolution in homogenous-charge mode of VISC at 1500 rpm and WOT conditions.

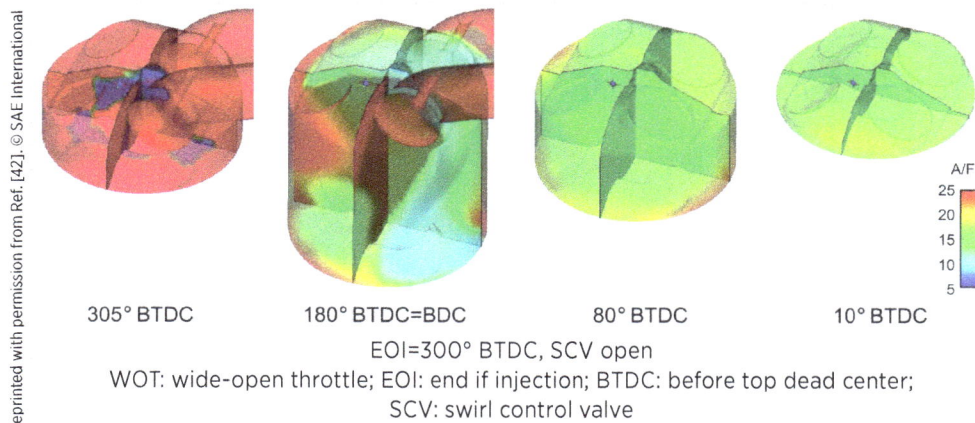

305° BTDC   180° BTDC=BDC   80° BTDC   10° BTDC

EOI=300° BTDC, SCV open

WOT: wide-open throttle; EOI: end if injection; BTDC: before top dead center; SCV: swirl control valve

Mixing at high speed is presented here. Mixing difficulty in the WGDI system is shown clearly at a high engine speed as discussed before. Figure 7.60 compares the mixture homogeneity of the VISC system with that of the WGDI system (see Figure 7.20). It is seen that mixing in the VISC engine is better than that in the WGDI engine. The VISC system has a narrower distribution of equivalence ratio and the peak of the curve is higher than the WGDI curve, which indicates better mixing.

This effect can be explained through Figure 7.61, which compares the fuel spray droplet distribution in the VISC and WGDI systems. These images show that in the WGDI system, the fuel droplets are held mainly on the intake side. The fuel spray is deflected from its axis by a strong interaction between the intake flow and the injected fuel. There is not enough cross-cylinder penetration of the spray droplets, resulting in less fuel on the exhaust side as seen in Figure 7.61a. In the VISC engine, however, the fuel spray is relatively unaffected by the intake flow, as shown in Figure 7.61b. This is primarily due to the central location of the injector, in between the relatively widely spaced intake valves. In addition, due to the higher flow rate of the VISC injector, injection is complete before the peak intake flows. The wide cone angle spray then distributes the fuel droplets more evenly throughout the cylinder. Although mixing in the WGDI can be improved by employing a mask design (see Figure 7.22) or split injection (see Figure 7.25), the centrally located injector, and wide and slowly-penetrating sprays are the inherent advantages of the VISC system.

FIGURE 7.60 Comparison of full-load mixing at 20° BTDC for the WGDI and VISC systems at 6000 rpm and WOT conditions.

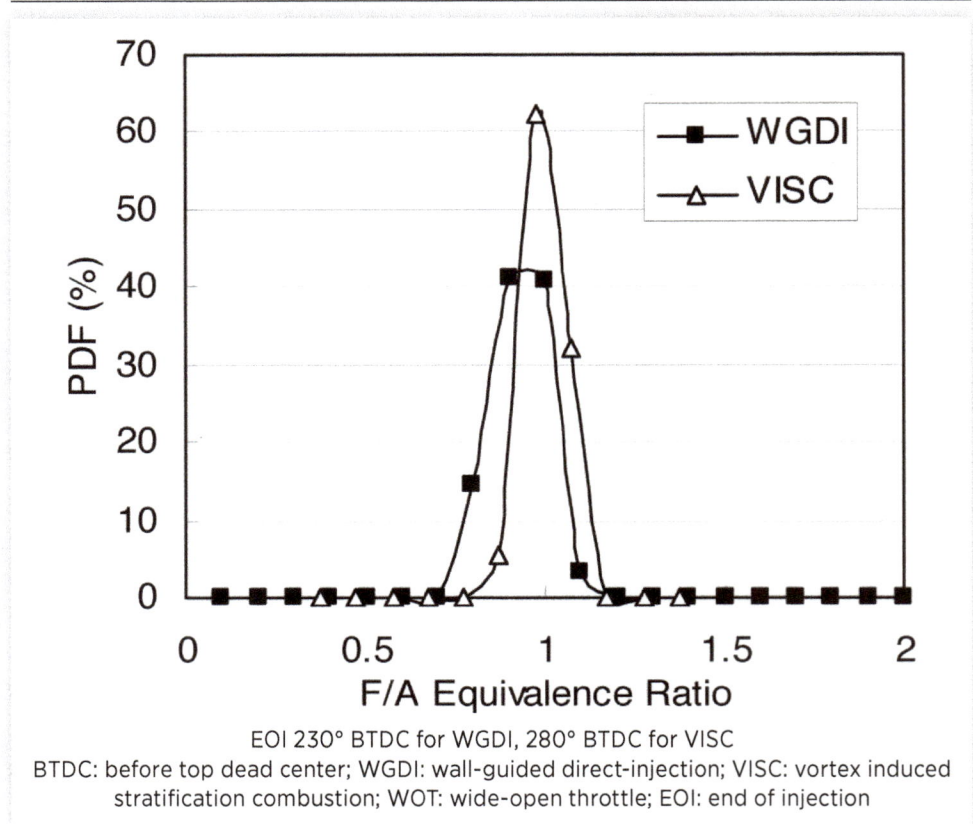

EOI 230° BTDC for WGDI, 280° BTDC for VISC
BTDC: before top dead center; WGDI: wall-guided direct-injection; VISC: vortex induced stratification combustion; WOT: wide-open throttle; EOI: end of injection

In recent downsized turbo-charging DI engines, stoichiometric homogeneous-charge operation is adopted in the entire operating range, primarily for utilization of the existing three-way catalyst emissions aftertreatment devices [5,43–46], except for stratified-charge at cold-start in some systems [47,48]. High tumbler flow ports and multihole injectors are used in these DI engines. Good mixing homogeneity can be achieved through computational optimization of spray pattern, piston shape, intake flow, and injection strategy with the use of the modeling methods discussed above. Since multihole injectors offer some degrees of flexibility in numbers and orientations of the spray plumes as shown in Figure 7.62, difficulties of wall wetting and flow-blocked spray-penetration in case of swirl injectors can be mitigated by optimizing the orientation of each spray plume [5,49].

Some freedom in reshaping intake ports is also granted by using a turbocharger so that the trade-off between volumetric efficiency and tumble ratio as in a naturally-aspirated engine becomes less important. Intake-port design optimization is then driven by achieving a large tumble ratio for which CFD can be of great help [50]. As invented in the early DI development works, split injection strategy offers many advantages over single-injection for mixing in both homogeneous-charge and stratified-charge modes [3,4] as discussed in the given examples. The split injection is also applicable at engine cold-start operations. Multiple injection strategies have been commonly employed in recent DI systems.

FIGURE 7.61 Comparison between fuel spray distribution in: (a) the WGDI system; (b) the VISC system. 6000 rpm and WOT conditions.

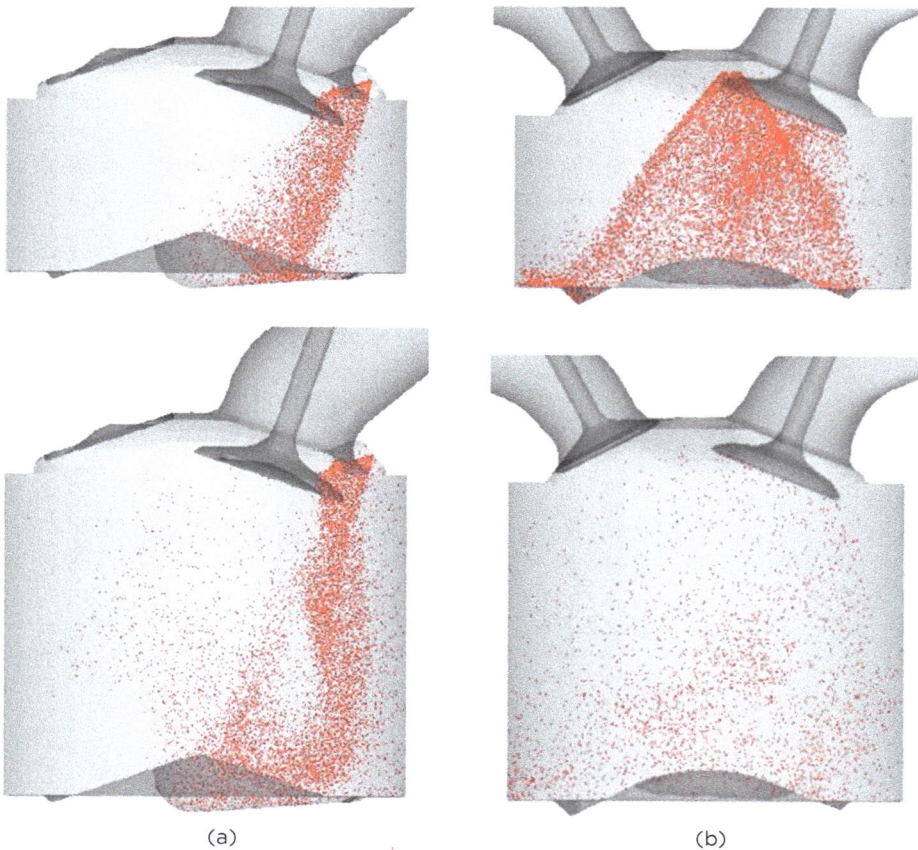

(a)   (b)

EOI 230° BTDC for WGDI, 280° BTDC for VISC.
Top row: 280° BTDC. Bottom row: 220° BTDC
BTDC: before top dead center; WGDI: wall-guided direct-injection; VISC: vortex induced stratification combustion; WOT: wide-open throttle; EOI: end of injection

FIGURE 7.62 Examples of spray patterns evaluated with CFD modeling in development of a turbo-charged DI engine.

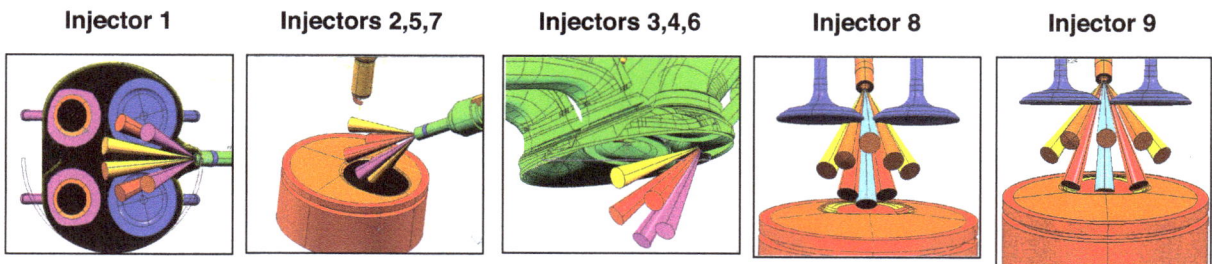

| Injector 1 | Injectors 2,5,7 | Injectors 3,4,6 | Injector 8 | Injector 9 |

CFD: computational fluid dynamics; DI: direct-injection

Combustion optimization at cold start is critical for DI gasoline engines to meet current very stringent emission regulations since the emissions during the first 20 seconds of the cold start constitute more than 80% of the hydrocarbon emissions for the entire EPA FTP75 drive cycle. Split injection generated stratified-charge can provide near stoichiometric mixture at the spark plug gap so that reliable ignition and stable combustion can be achieved with less over-fueling. Due to the dynamic flow range limit of an injector, injection pressure must be reduced to achieve a longer injection duration. It was demonstrated that by use of split injection, an optimized piston shape, and varied injection timing and pressure, engine-out hydrocarbon emission was reduced by 30% with lower fuel consumption and substantially improved combustion stability during engine cold-start in a DI engine [51].

References

1. Han, Z., Yi, J., and Trigui, N. 2002. "Stratified Mixture Formation And Piston Surface Wetting in a DISI Engine." SAE Technical Paper No. 2002-01-2655. DOI: 10.4271/2002-01-2655

2. Yi, J., Han, Z., and Trigui, N. 2002. "Fuel-Air Mixing Homogeneity And Performance Improvements of a Stratified-Charge DISI Combustion System." *SAE Transactions* 111, no. 4 (October): 965–975. DOI: 10.4271/2002-01-2656

3. VanDerWege, B.A., Han, Z., Iyer, C.O., Muñoz, R.H., and Yi, J. 2003. "Development And Analysis of a Spray-Guided DISI Combustion System Concept." *SAE Transactions* 112, no. 4 (October), 2135–2153. DOI: 10.4271/2003-01-3105

4. Han, Z., Weaver, C., Wooldridge, S., Alger, T., Hilditch, J., McGee, J., Westrate, B., et al. 2004. "Development of a New Light Stratified-Charge DISI Combustion System for a Family of Engines With Upfront CFD Coupling With Thermal And Optical Engine Experiments." *SAE Transactions* 113, no. 3 (March): 269–293. DOI: 10.4271/2004-01-0545

5. Yi, J., Wooldridge, S., Coulson, G., Hilditch, J., Iyer, C.O., Moilanen, P., Papaioannou, G., et al. 2009. "Development and optimization of the Ford 3.5 L V6 EcoBoost combustion system." *SAE International Journal of Engines* 2, no. 1 (April): 1388–1407. DOI: 10.4271/2009-01-1494

6. Westrate, B., Coulson, G., Kenney, T., Kumar, B., Rogers, M., and Weaver, C. 2004. "Dynamometer Development of a Lightly Stratified Direct Injection Combustion System." *SAE Transactions* 113, no. 3 (March): 310–323. DOI: 10.4271/2004-01-0547

7. Zhao, H., and Ladommatos, N. 1998. "Optical Diagnostics for In-Cylinder Mixture Formation Measurements in IC Engines." *Progress in Energy and Combustion Science* 24, no. 4): 297–336. DOI: 10.1016/S0360-1285(98)80026-9

8. Zhang, Y., Zhang, G., and Xu, M. 2011. "Laser Diagnostics for Spray of Spark Ignition Direct Injection (SIDI) Combustion System." *Journal of Automotive Safety and Energy* 2, no. 4): 294–307. (in Chinese)

9. Agarwal, A.K., and Singh, A.P. 2017. "Lasers And Optical Diagnostics for Next Generation IC Engine Development: Ushering New Era of Engine Development" In *Combustion for Power Generation and Transportation*, 211–259. Singapore: Springer.

10. McGee, J., Alger, T., Blobaum, E., and Wooldridge, S. 2004. "Evaluation of a Direct-Injected Stratified Charge Combustion System Using Tracer PLIF." *SAE Transactions* 113, no. 3 (March): 324–336. DOI: 10.4271/2004-01-0548

11. Alger, T., Blobaum, E., McGee, J., and Wooldridge, S. 2003. "PIV Characterization of a 4-Valve Engine With Camshaft Profile Switching (CPS) System." *SAE Transactions* 112, no. 4 (May): 1066–1078. DOI: 10.4271/2003-01-1803

12. Alger, T., McGee, J., and Wooldridge, S. 2004. "Stratified-Charge Fuel Preparation Influence on the Misfire Rate of a DISI Engine." *SAE Transactions* 113, no. 4: 229–238.

13. Wooldridge, S., Lavoie, G., and Weaver, C. 2003. "Convection Path for Soot And Hydrocarbon Emissions From the Piston Bowl of a Stratified Charge Direct Injection Engine." In *Proceedings of the Third Joint Meeting of the US Sections of the Combustion Institute, March 16–19.* Chicago: University of Illinois.

14. Westrate, B., Warren, C., VanDerWege, B., Coulson, G., and Anderson, R. 2002. "Dynamometer Development Results for a Stratified-Charge DISI Combustion System." SAE Technical Paper No. 2002-01-2657. DOI: 10.4271/2002-01-2657

15. Hilditch, J., Han, Z., and Chea, T. 2003. "Unburned Hydrocarbon Emissions From Stratified Charge Direct Injection Engines." SAE Technical Paper No. 2003-01-3099. DOI: 10.4271/2003-01-3099

16. Shi, Y., Ge, H.W., and Reitz, R.D. 2011. *Computational Optimization of Internal Combustion Engines.* London: Springer.

17. Tanner, F.X., and Srinivasan, S. 2005. "Optimization of Fuel Injection Configurations for the Reduction of Emissions And Fuel Consumption in a Diesel Engine Using a Conjugate Gradient Method." SAE Technical Paper No. 2005-01-1244. DOI: 10.4271/2005-01-1244

18. Jeong, S., Obayashi, S., and Minemura, Y. 2008. "Application of Hybrid Evolutionary Algorithms to Low Exhaust Emission Diesel Engine Design." *Engineering Optimization* 40, no. 1: 1–16. DOI: 10.1080/03052150701561155

19. Holland, J.H. 1975. *Adaptation in Natural And Artificial Systems.* Cambridge, Mass.: MIT press.

20. Senecal, P.K., and Reitz, R.D. 2000. "Simultaneous Reduction of Engine Emissions And Fuel Consumption Using Genetic Algorithms And Multi-Dimensional Spray And Combustion Modeling." *SAE Transactions* 109, no. 4 (June): 1378–1390. DOI: 10.4271/2000-01-1890

21. Shrivastava, R., Hessel, R., and Reitz, R.D. 2002. "CFD Optimization of DI Diesel Engine Performance And Emissions Using Variable Intake Valve Actuation With Boost Pressure, EGR And Multiple Injections." *SAE Transactions* 111, no. 3 (March): 1612–1629. DOI: 10.4271/2002-01-0959

22. Wickman, D.D., Senecal, P.K., and Reitz, R.D. 2001. "Diesel Engine Combustion Chamber Geometry Optimization Using Genetic Algorithms And Multi-Dimensional Spray And Combustion Modeling." *SAE Transactions* 110, no. 3 (March): 487–507. DOI: 10.4271/2001-01-0547

23. Shi, Y., and Reitz, R.D. 2008. "Assessment of Optimization Methodologies to Study the Effects of Bowl Geometry, Spray Targeting And Swirl Ratio for a Heavy-Duty Diesel Engine Operated at High-Load." *SAE International Journal of Engines* 1, no. 1 (April): 537–557. DOI: 10.4271/2008-01-0949

24. Coello, C.A.C., and Toscano Pulido, G. 2001. "A Micro-Genetic Algorithm for Multiobjective Optimization." In *Lecture Notes in Computer Science: Vol. 1993. Evolutionary Multi-Criterion Optimization*, edited by E. Zitzler, K. Deb, L. Thiele, C.A. Coello, and D. Corne, 126–140. Berlin: Springer.

25. Deb, K., Pratap, A., Agarwal, S., and Meyarivan, T. 2002. "A Fast And Elitist Multiobjective Genetic Algorithm: NSGA-II." *IEEE Transactions on Evolutionary Computation* 6, no. 2: 182–197. DOI: 10.1109/4235.996017

26. Sasaki, D., and Obayashi, S. 2005. "Efficient Search for Trade-Offs by Adaptive Range Multi-Objective Genetic Algorithms." *Journal of Aerospace Computing, Information, and Communication* 2, no. 1 (May): 44–64. DOI: 10.2514/1.12909

27. Ge, H.W., Shi, Y., Reitz, R.D., Wickman, D., and Willems, W. 2010. "Engine Development Using Multi-Dimensional CFD And Computer Optimization." SAE Technical Paper No. 2010-01-0360. DOI: 10.4271/2010-01-0360

28. Jeong, S., Minemura, Y., and Obayashi, S. 2006. "Optimization of Combustion Chamber for Diesel Engine Using Kriging Model." *Journal of Fluid Science Technology* 1, no. 2: 138–146. DOI: 10.1299/jfst.1.138

29. Amsden, A.A., Ramshaw, J.D., O'Rourke, P.J., and Dukowicz, J.K. 1985. *KIVA: A Computer Program for Two- And Three-Dimensional Fluid Flows With Chemical Reactions And Fuel Sprays* (LA-10245-MS). Los Alamos, New Mexico: Los Alamos National Laboratory.

30. Richards, K.J., Senecal, P.K., and Pomraning, E. 2017. *CONVERGE Manual (v2.4)*. Madison, Wisconsin: Convergent Science Inc.

31. Han, Z., Parrish, S.E., Farrell, P.V., and Reitz, R.D. 1997." Modeling Atomization Processes of Pressure-Swirl Hollow-Cone Fuel Sprays." *Atomization and Sprays* 7, no. 6: 663–684. DOI: 10.1615/AtomizSpr.v7.i6.70

32. Han, Z., Xu, Z., Wooldridge, S.T., Yi, J., and Lavoie, G. 2001. "Modeling of DISI Engine Sprays With Comparison to Experimental In-Cylinder Spray Images." *SAE Transactions* 110, no. 3 (September): 2376–2386. DOI: 10.4271/2001-01-3667

33. Aori, G., Hung, D.L.S., Zhang, M., Zhang, G., and Li, T. 2016. "Effect of Nozzle Configuration on Macroscopic Spray Characteristics of Multi-Hole Fuel Injectors Under Superheated Conditions." *Atomization and Sprays* 26, no. 5: 439–462. DOI: 10.1615/AtomizSpr.2015011990

34. Xu, M., Hung, D., Yang, J., and Wu, S. 2015. "Flash-Boiling Spray Behavior And Combustion in a Direct Injection Gasoline Engine." In *Proceedings of Australian Combustion Symposium 2015*, 14–23. Melbourne, Australia: University of Melbourne.

35. Xu, M., Zhang, Y., Zeng, W., Zhang, G., and Zhang, M. 2013. "Flash Boiling Easy And Better Way to Generate Ideal Sprays Than the High Injection Pressure." *SAE International Journal of Fuels and Lubricants* 6, no. 1: 137–148. DOI: 10.4271/2013-01-1614

36. Wu, S., Meinhart, M., and Yi, J. 2019. "Experimental Investigation of Spray Characteristics of Multi-Hole And Slot GDI Injectors at Various Fuel Temperatures Using Closely Spaced Split-Injection Strategies." *Atomization and Sprays* 29, no. 12: 1109–1131. DOI: 10.1615/AtomizSpr.2020033439

37. Zhang, G., Xu, M., Zhang, Y., and Hung, D.L.S. 2013. "Characteristics of Flash Boiling Fuel Sprays From Three Types of Injector for Spark Ignition Direct Injection (SIDI) Engines." In *Proceedings of the FISITA 2012 World Automotive Congress*, 443–454. Berlin: Springer.

38. Han, Z., Fan, L., and Reitz, R.D. 1997. "Multidimensional Modeling of Spray Atomization And Air–Fuel Mixing in a Direct-Injection Spark-Ignition Engine." *SAE Transactions* 106, no. 3 (February): 1423–1441. DOI: 10.4271/970884

39. Han, Z., Reitz, R.D., Yang, J., and Anderson, R.W. 1997. "Effects of Injection Timing on Air–Fuel Mixing in a Direct-Injection Spark-Ignition Engine." *SAE Transactions* 106, no. 3 (February): 848–860. DOI: 10.4271/970625

40. Yi, J., Han, Z., Xu, Z., and Stanley, L. 2004. "Combustion Improvement of a Light Stratified-Charge Direct Injection Engine." *SAE Transactions* 113, no. 3 (March): 294–309. DOI: 10.4271/2004-01-0546

41. Heywood, J.B. 2018. *Internal Combustion Engine Fundamentals*. 2nd ed. New York: McGraw-Hill Education.

42. Iyer, C.O., Han, Z., and Yi, J. 2004. "CFD Modeling of a Vortex Induced Stratification Combustion (VISC) System." SAE Technical Paper No. 2004-01-0550. DOI: 10.4271/2004-01-0550

43. Shimizu, M., Yageta, K., Matsui, Y., and Yoshida, T. 2011. "Development of New 1.6 Liter Four Cylinder Turbocharged Direct Injection Gasoline Engine With Intake And Exhaust Valve Timing Control System." SAE Technical Paper No. 2011-01-0419. DOI: 10.4271/2011-01-0419

44. Mitani, S., Hashimoto, S., Nomura, H., Shimizu, R., and Kanda, M. 2014. "New Combustion Concept for Turbocharged Gasoline Direct-Injection Engines." *SAE International Journal of Engines* 7, 2 (April): 551–559. DOI: 10.4271/2014-01-1210

45. Shibata, M., Kawamata, M., Komatsu, H., Maeyama, K., Asari, M., Hotta, N., Nakada, K., and Daicho, H. 2017. "*New 1.0L I3 Turbocharged Gasoline Direct Injection Engine*." SAE Technical Paper No. 2017-01-1029. DOI: 10.4271/2017-01-1029

46. Xu, Z., Ping, Y., Cheng, C., Zhang, X., Yin, H., Li, W., Cai, D., et al. 2020. "The New 4-Cylinder Turbocharged GDI Engine From SAIC Motor." SAE Technical Paper No. 2020-01-0836. DOI: 10.4271/2020-01-0836

47. Xu, Z., Yi, J., Curtis, E., and Wooldridge, S. 2009. "Applications of CFD Modeling in GDI Engine Piston Optimization." *SAE International Journal of Engines* 2, no. 1 (June): 1749–1763. DOI: 10.4271/2009-01-1936

48. Shinagawa, T., Kudo, M., Matsubara, W., and Kawai, T. 2015. "The New Toyota 1.2-Liter ESTEC Turbocharged Direct Injection Gasoline Engine." SAE Technical Paper No. 2015-01-1268. DOI: 10.4271/2015-01-1268

49. Iyer, C.O., and Yi, J. 2009. "Spray Pattern Optimization for the Duratec 3.5L Ecoboost Engine." *SAE International Journal of Engines* 2, no. 1 (June): 1679–1689. DOI: 10.4271/2009-01-1916

50. Iyer, C.O., and Yi, J. 2009. "3D CFD Upfront Optimization of tThe In-Cylinder Flow of the 3.5 L V6 Ecoboost Engine." SAE Technical Paper No. 2009-01-1492. DOI: 10.4271/2009-01-1492

51. Xu, Z., Yi, J., Wooldridge, S., Reiche, D., Curtis, E., and Papaioannou, G. 2009. "Modeling the Cold Start of the Ford 3.5L V6 Ecoboost Engine." *SAE International Journal of Engines* 2, no. 1 (April): 1367–1387. DOI: 10.4271/2009-01-1493

8

Optimization of Diesel and Alternative Fuel Engines

This chapter is the continuation of Chapter 7, and focuses on the simulation and optimization of diesel engines and alternative fuel engines. In the first part of this chapter, several important aspects of diesel combustion are discussed: First, the reduction mechanism of soot and NO_x emissions with the use of multiple injections is discussed; second, optimization of helical-port and combustion-chamber shapes of a high-speed automotive diesel engine is presented; third, the effects of split injection on emissions at cold-start conditions are addressed. In the second part, the modeling of alternative fuel engines is discussed. The mixture formation of natural gas in a PFI SI engine is numerically studied. RCCI combustion in a diesel-natural gas dual-fuel engine is computationally optimized. Since biodiesel is attractive due to its low soot emissions, combustion and NO_x emissions of an automotive engine fueled with diesel–biodiesel blend are simulated and the effects of biodiesel blending ratio are addressed.

8.1 Direct-Injection Diesel Engines

Combustion phenomena in a conventional diesel engine are unlike those in a SI gasoline engine since mixture formation and combustion cannot be separated as discussed in Sec. 2.3. Thus, the engine numerical simulation of a diesel engine covers the entire process of fuel injection, autoignition, combustion, and emissions formation. Some diesel combustion prediction examples are already presented in Chapter 6. We continue to discuss the simulation and optimization of diesel engines in this section. Again, the focus is on some of the key issues in diesel combustion and design parameters.

8.1.1 Emissions Reduction by Multiple Injections

Multiple injection strategies have been used in DI diesel engines for control of combustion noise and emissions. The mechanism of reduction of soot and NO_x emissions using split injection was studied through combustion simulation analysis by Han et al. [1]. In the simulations, the RNG k-ε model, Reitz Wave spray model, Shell ignition model, CTC combustion model, NO_x, and soot formation model as discussed in previous chapters are employed. The models were validated and demonstrated to be able to predict split injection combustion at a wide range of operating conditions without tuning the model constants. Examples of the predicted cylinder pressure and HRR of a Caterpillar heavy-duty diesel engine are shown in Figure 8.1. Engine specifications are listed in Table 6.1. The nomenclature of a split injection scheme represents the percentage of the total injected fuel in the first pulse, and the dwell in CA, and the percentage of the injected fuel in the second pulse. For instance, 10-8-90 represents 10% fuel injected in the first pulse, 90% in the second, and an 8° CA dwell between the two pulses. It is seen in Figure 8.1 that as the fuel in the first pulse increases, the observation of two peaks in the diffusion heat release is clearer.

FIGURE 8.1 Predicted pressure and HRR for split injections in a diesel engine.

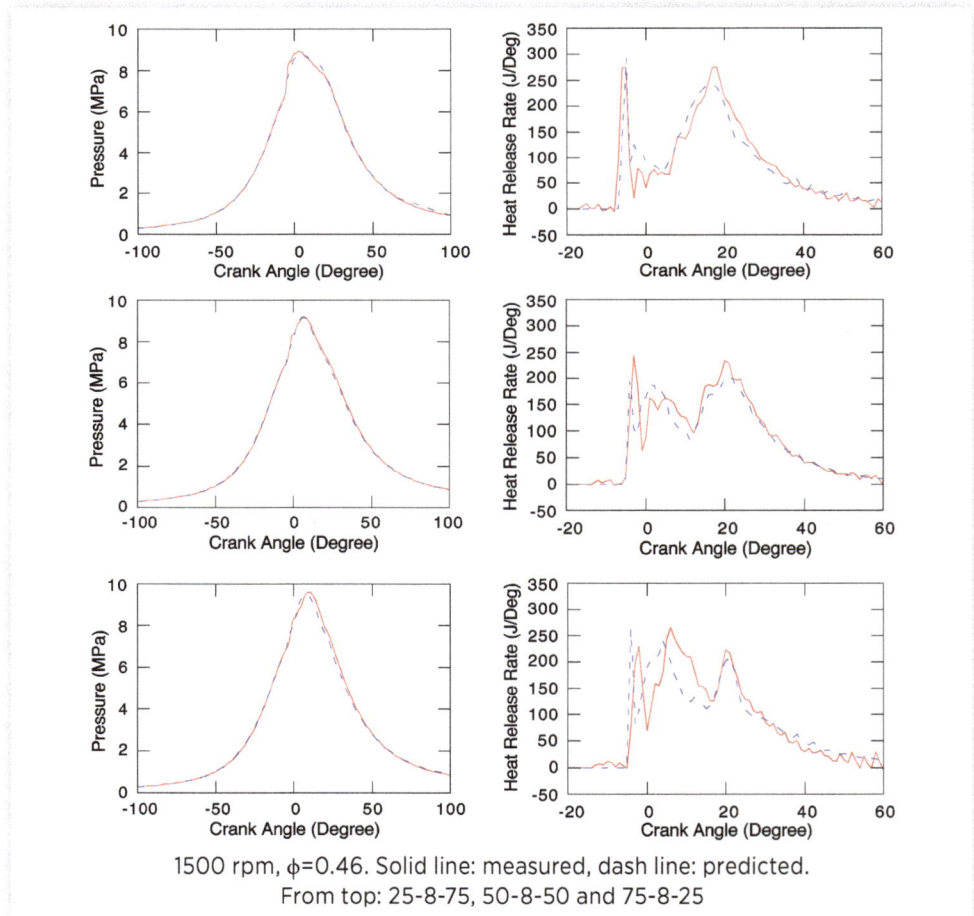

1500 rpm, φ=0.46. Solid line: measured, dash line: predicted.
From top: 25-8-75, 50-8-50 and 75-8-25

To study the mechanism of emissions reduction using split injections, a set of injection schemes was designed and computed, which included three single-injections and three double-injections. The injection schemes are shown schematically in Figure 8.2. In the nomenclature 75-8-25 (-10), the number in the brackets indicates the SOI (ATDC). The same amount of fuel is injected in all the cases considered. The injection rate shape is simply modeled by using one (or two) step functions with the same step height to keep the same injection velocity in all cases. In all the double-injections, the total injection duration is 28° CA, and in the first two single-injections, the injection duration is 20 CA. In the last case, the injection duration is 8 CA longer than the other single-injection cases, and hence the nozzle radius was reduced by 15.5% to keep the same amount of fuel injected. With the use of the designed schemes, the possible effects of differences in equivalence ratio and injection timing in experiments are eliminated. Besides, the designed schemes are not far from the experiments, hence the computed results can be confidently used, based on the good agreement with the experiments presented in Figure 8.1.

FIGURE 8.2 Schematic diagram showing model injection schemes.

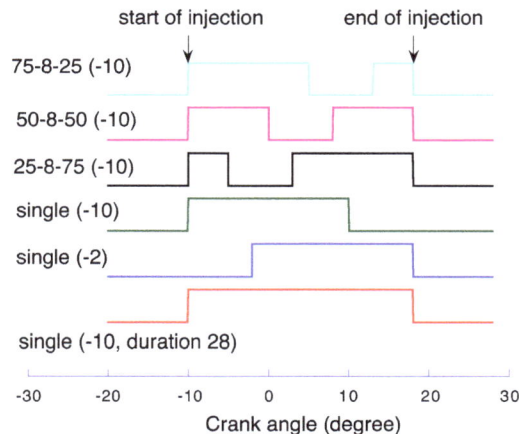

Reprinted with permission from Ref. [1]. © SAE International

Figure 8.3a shows the injected fuel history for each case and Figure 8.3b the predicted evolution of burned fuel. It is seen that the fuel-burned history curves of the double-injections are embraced by those of the two single-injections with the 20 CA injection duration. The double-injections follow the single (-10) case during the time corresponding to the first injection-pulse and then become close to the single (-2) case during the time corresponding to the second injection-pulse. Between the injection pulses, a transition occurs due to the dwell period of the injection. The time at which the transition starts depends on the amount of fuel injected in the first injection pulse. These features indicate that the double-injection combustion is partially the same as, or close to, either of the single-injections from a macroscopic viewpoint. However, the combustion process is complicated by the second injection-pulse after the dwell. For example, combustion in the 25-8-75 (-10) case differs from that in the single (-10) case soon after ignition, which is due to the non-linear effects of fuel vaporization and mixing. The fuel-burned curve of the single (-10, duration 28) case, which has a 28 CA injection duration is also included within those of the two single-injections with 20 CA injection duration. However, as can be seen, it does not follow or become close to any portion of them after ignition.

Reprinted with permission from Ref. [1]. © SAE International

FIGURE 8.3 (a) Evolution of injected fuel normalized by the total fuel; (b) Evolution of burned fuel normalized by the total fuel.

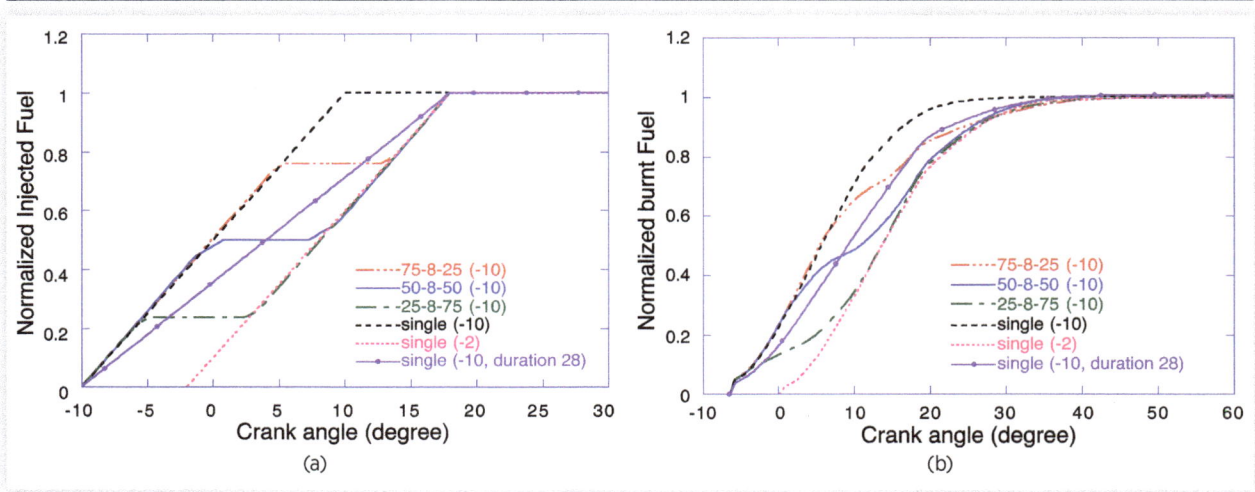

(a)

(b)

The computed soot–NO trade-off is shown in Figure 8.4 for the above cases. It is again confirmed that double-injections are effective in reducing soot and NO$_x$ emissions. For example, the soot emission of the single (-10) case is reduced by a factor of 4 using the 75-8-25 (-10) injection while the NO emission is increased very slightly, as indicated in Figure 8.4. When the 50-8-50 (-10) double-injection is used, both the soot and NO emissions are reduced below those of the single (-10) case. However, when the 25-8-75 (-10) double-injection is used, the soot emission is increased significantly at the same time that the NO emission is reduced greatly.

FIGURE 8.4 Computed soot–NO trade-off of the designed injection schemes.

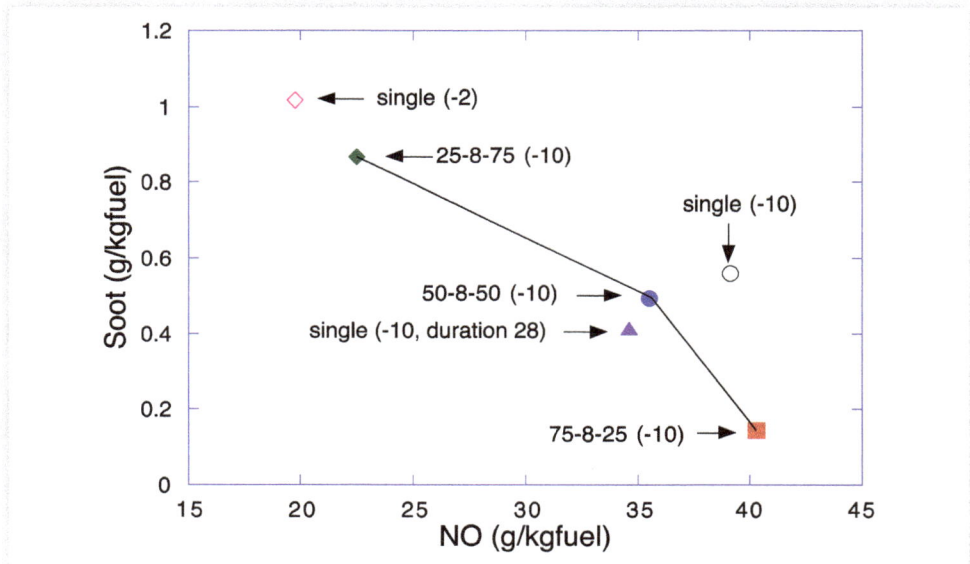

Reprinted with permission from Ref. [1]. © SAE International

It is of interest to examine the reasons for these observed emissions reduction phenomena. In general, the NO emissions are reduced by using the double-injections studied here and the soot emissions are reduced in some cases. It is noticed that soot and NO are also reduced relative to the single (-10, duration 20) case with the use of the single (-10, duration 28) scheme. Since the combustion of this case is similar to that of the 50-8-50 (-10) double-injection as indicated in Figure 8.3b, this case results in about the same level of soot and NO emissions, as seen in Figure 8.4. However, this scheme will require the use of smaller nozzle hole sizes in applications.

8.1.1.1 NO Reduction Mechanism

The predicted accumulated in-cylinder NO-formation history vs. burned fuel mass is shown in Figure 8.5. As can be seen, the NO formation history of the 75-8-25 (-10) case is very similar to that of the single (-10) case. The injection pause does not affect NO formation significantly and combustion of the 25% fuel in the second injection pulse only causes a small increase in NO production. However, the effect of the injection pause becomes more significant in the 25-8-75 (-10) case. It delays the major portion of the combustion and reduces the NO formation rate considerably compared with the single (-10) case. Comparing the 25-8-75 (-10) case with the single (-2) case in which the injection timing is retarded 8° CA, their NO formation rates become similar during the major portion of the combustion, as also seen in Figure 8.5. These phenomena are closely related to the combustion evolution histories in Figure 8.3b. Although combustion in the 25-8-75 (-10) and single (-2) cases are quite different before 10° ATDC due to their different injection timings, combustion of the fuel in the second pulse is delayed by the injection pause in the former case. As a result, the major portion of combustion in the 25-8-75 (-10) case becomes similar to that of the single (-2) case after 10° ATDC.

FIGURE 8.5 NO formation history as the function of the normalized burned fuel mass.

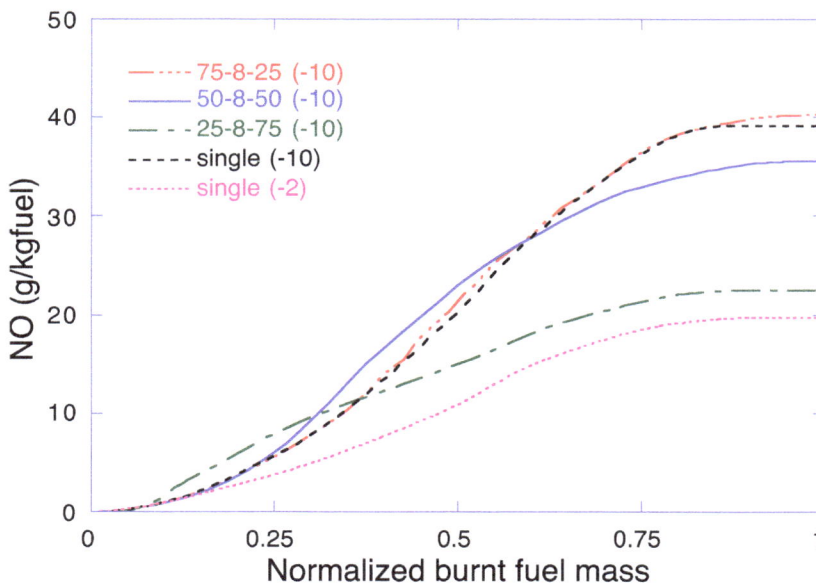

It is also noticed in Figure 8.3b that combustion in the 75-8-25 (-10) case becomes very different from that of the single (-10) case after 5° ATDC since this is the time that the injection is paused. However, the NO formation histories of the two cases are not very different. It is known that NO formation is very sensitive to the gas temperature during combustion. As can be seen in Figure 8.6, which gives the mixture mass fraction with temperatures greater than 2200 K vs. fuel mass burned, the evolution of the high-temperature gas mixture in the 25-8-75 (-10) case is similar to that of the single (-2) case after 25% of the fuel is burned. The greater amount of high-temperature gas before 25% fuel is burned in the former case results in the slightly higher NO level seen in Figure 8.5. It is also seen that the 75-8-25 (-10) and single (-10) cases have almost the same high-temperature gas histories before 80% of the fuel is burned. Hence, they have very similar NO formations.

FIGURE 8.6   In-cylinder high-temperature mixture fractions.

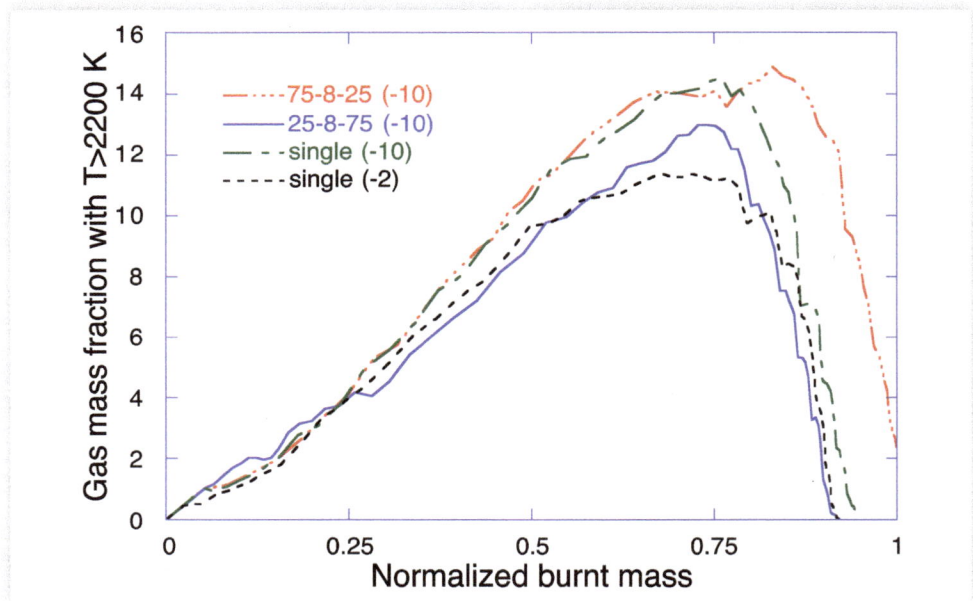

Reprinted with permission from Ref. [1]. © SAE International

Comparing Figures 8.5 with 8.6, it is interesting to note that the large gas temperature difference between the 75-8-25 (-10) and single (-10) cases seen in the later stage of combustion does not affect the NO formation significantly, while the relatively smaller temperature differences between the 25-8-75 (-10) and single (-2) cases in the earlier stage of combustion result in relatively larger changes in the NO formations. This can be explained by the fact that the NO chemistry is sensitive to the early combustion details because these combustion products stay at a high temperature for the longest time, and the combustion region is not cooled by the vaporization of the continuously injected fuel that occurs in the single-injection case. The NO chemistry is effectively frozen after about 80% of the fuel is burned in all the cases considered because the late-burning mixture has a shorter residence time at a high temperature due to the piston expansion.

Based on this, it can be said that the NO reduction mechanism of multiple injections is similar to that of retarding the injection timing. With the use of multiple injections, combustion of the second-pulse injected fuel is delayed by the injection pause. When the percentage of the first-pulse injected fuel is large, the NO formation history of the double-injection is like that at a single-injection with the same injection timing. The effect of combustion of the second-pulse injected fuel does not influence the NO formation significantly. As the percentage of the fuel in the first injection pulse becomes small, the NO formation rate of the double-injection becomes similar to that of a single-injection retarded with the dwell-time of the double-injection. In this case, combustion of the first-pulse injected fuel has an important effect on NO production, and it results in more NO being formed in the earlier combustion stage, and hence more total NO production.

8.11.2 Soot Reduction Mechanism

As indicated in Figure 8.4, some multiple injection schemes can reduce soot emissions significantly, while some can increase soot emissions. Figure 8.7 gives the soot production history during the combustion process for the various cases. It is seen that the in-cylinder soot production histories of the multiple injections are changed significantly from the original single-injection cases. The peak values of in-cylinder soot from the multiple injections are largely reduced due to the injection pause, and the net productions have different values at the end of combustion. It is known that the net soot production is the result of the competition between soot formation and soot oxidation. This is illustrated in Figure 8.8 in which the 75-8-25 (-10) and single (-10) cases are compared. It is seen that the injection pause affects both the soot formation and oxidation processes and depresses the soot chemistry. However, the soot formation is reduced more than the soot oxidation (as indicated by the numbers and arrows in the figure). Therefore, the formation–oxidation competition results in a significant reduction of soot production in the 75-8-25 (-10) case (by a factor of 4).

FIGURE 8.7 In-cylinder soot production history for single- and double-injection schemes.

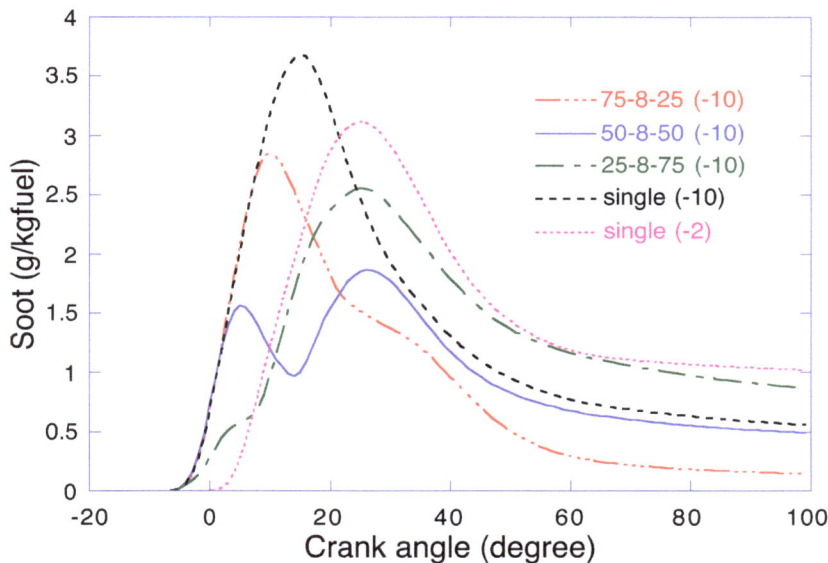

Reprinted with permission from Ref. [1]. © SAE International

FIGURE 8.8 In-cylinder histories of soot formation, oxidation, and net production.

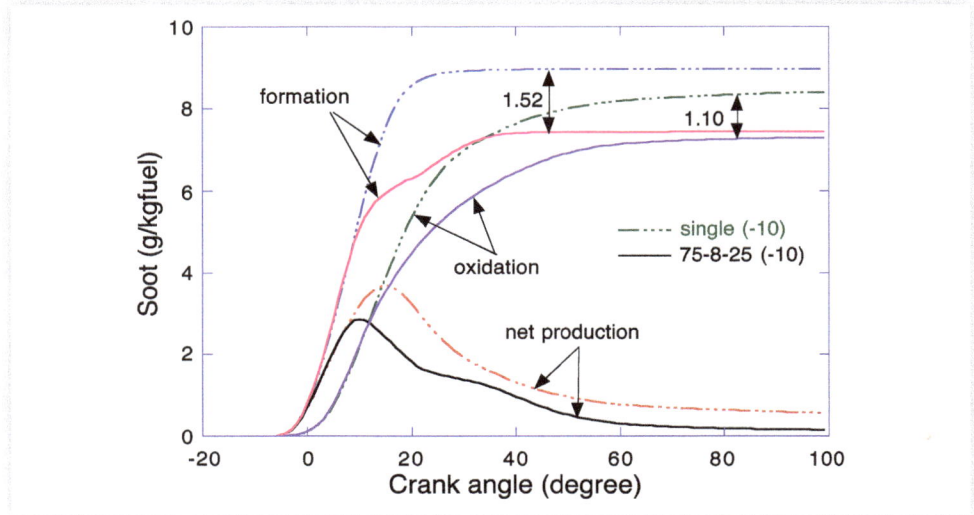

It is expected that the second-pulse injected spray also enhances fuel–air mixing. This reasoning is supported by Figure 8.9, which shows the change of the in-cylinder gas volume containing rich mixtures (equivalence ratio greater than 2.0) as a function of the burned fuel mass. As can be seen, the amount of rich mixture is reduced significantly after the injection pause in the 75-8-25 (-10) double-injection case because of the dispersion of the fuel–air mixture between the injection pulses, which is no longer maintained by the high momentum fuel jet. This process tends to lean out the mixture. Soot formation is therefore reduced.

FIGURE 8.9 In-cylinder rich mixture volume normalized by the instantaneous total volume showing the effects of the second-pulse injection on fuel–air mixing.

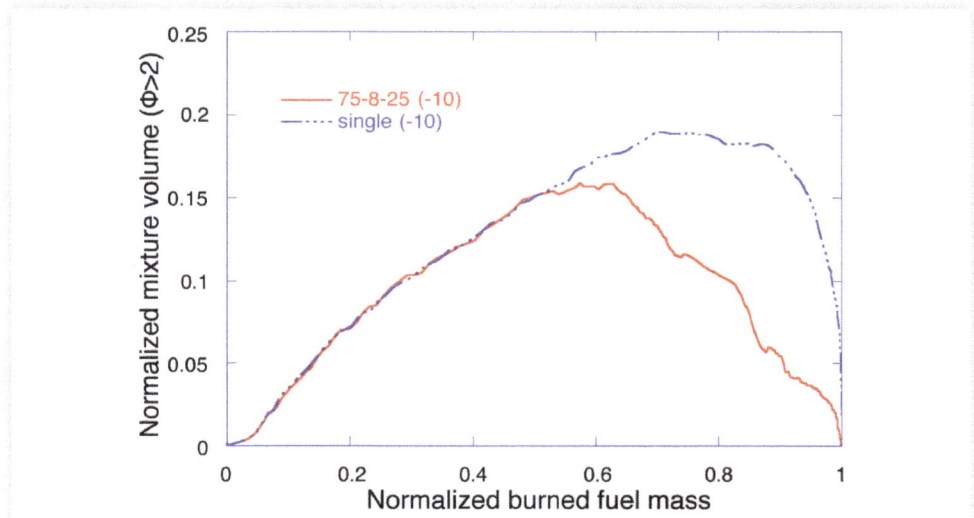

Reprinted with permission from Ref. [1]. © SAE International

The mechanisms of soot reduction using split injections are illustrated schematically in Figure 8.10. Soot is formed and accumulates in the tip region of the spray jet. This soot accumulation at the spray tip has also been observed experimentally by Dec and Espey [2] in an optically accessible DI diesel engine. In single-injection combustion, the high momentum injected fuel penetrates to the fuel-rich, relatively low-temperature region at the jet tip and continuously replenishes the rich region, producing soot. In a split injection, however, the second-pulse injected fuel enters into a relatively fuel-lean and high-temperature region, which is left over from the combustion of the first pulse. Soot formation is therefore significantly reduced because the injected fuel is rapidly consumed by combustion before a rich soot-producing region can accumulate. This can also be seen in Figure 8.9, where the last 25% of the injection is seen not to increase the amount of rich mixture in the chamber. In addition, the soot cloud of the first spray plume is not replenished with fresh fuel and, instead, continues to oxidize. As a result, the net production of soot in split injection combustion can be reduced substantially, particularly if the dwell between the two injections is optimized - long enough that the soot formation region of the first injection is not replenished with fresh fuel, but short enough that the in-cylinder gas temperature environment seen by the second pulse remains high enough to prompt fast combustion, reducing soot formation.

FIGURE 8.10 Schematic diagram showing soot-reduction mechanisms of split injections.

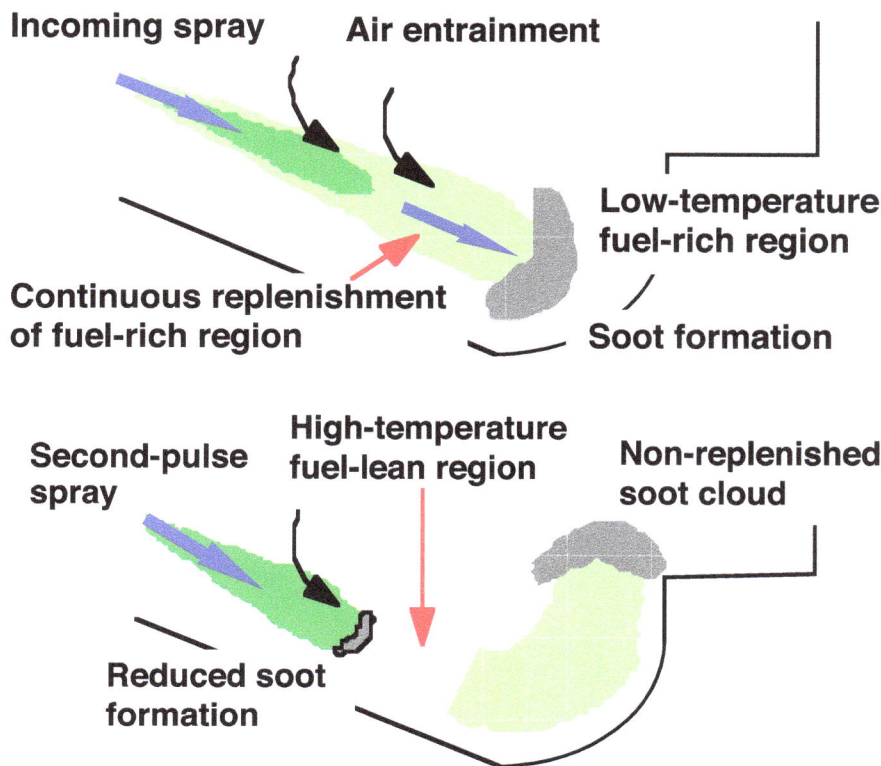

This mechanism is further demonstrated in the computed results shown in Figure 8.11. Although the computed soot distributions are complicated by the gas flow motion and wall–spray interactions, it is seen clearly that the high soot region is located in the leading portion of the spray-induced flow in both the single and split injection cases (see the middle row of Figure 8.11. at 15° ATDC). By this time, the leading edge of the first-formed soot cloud has moved up the bowl and is located on the cylinder head. In the single-injection case, the larger quantity of soot formed is oxidized (a larger quantity because this region was replenished with fresh fuel until the end of the injection). As a result, the amount of total in-cylinder soot is reduced by 25° ATDC. However, in the split injection case, the second-pulse injected fuel forms a separate burning region near the nozzle at 15° ATDC, as seen in the temperature contours of the right-top plot. The fuel–air mixture becomes relatively lean compared with the single-injection case, as is shown in Figure 8.9. Since the soot formation of the second spray plume is greatly reduced, no appreciable soot is produced in this region at this time. By 25° ATDC, however, some amount of soot is seen to have been produced from the second spray plume deep in the bowl, but its concentration is very low relative to that from the first spray plume.

FIGURE 8.11 Computed temperature (K) (top row) and soot concentration (g/cm³) (middle and bottom rows) contours in the plane of the spray axis: (a) single-injection (-10); (b) split injection 75-8-25 (-10).

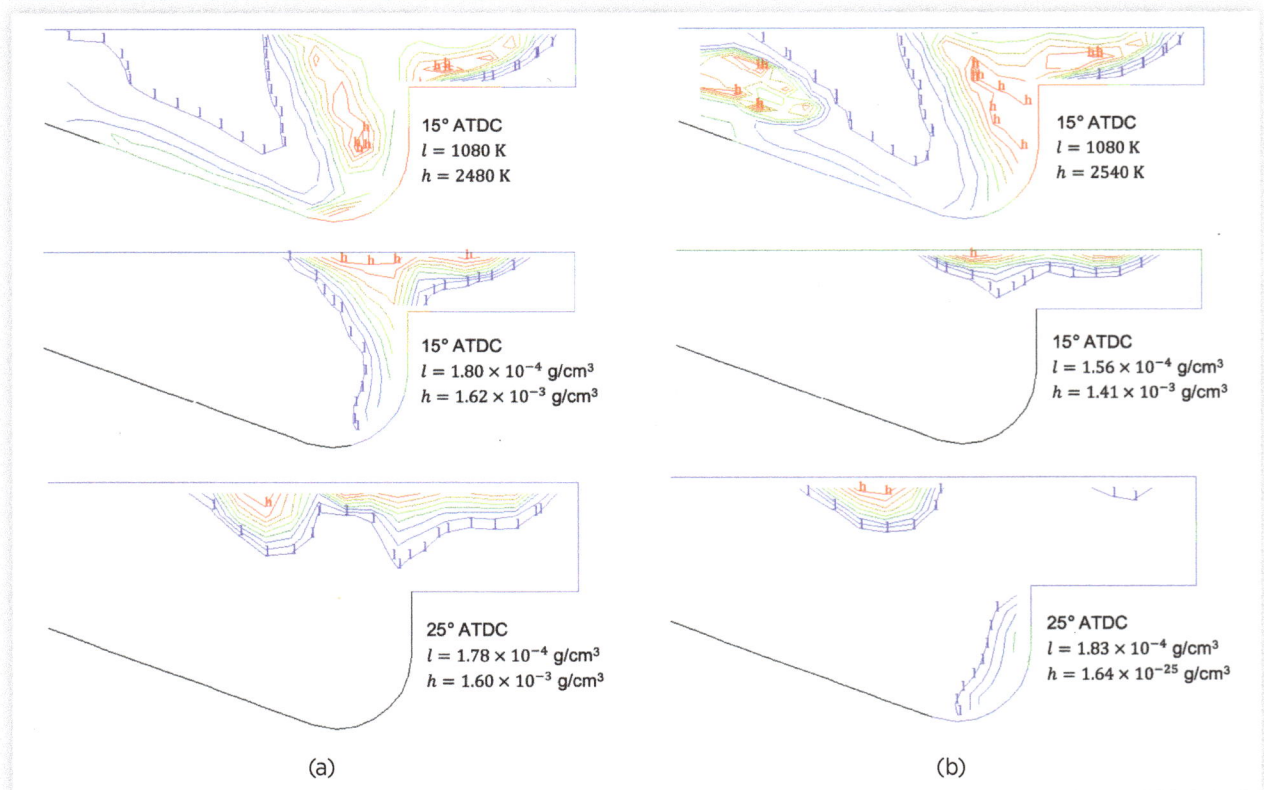

15° ATDC
$l = 1080$ K
$h = 2480$ K

15° ATDC
$l = 1080$ K
$h = 2540$ K

15° ATDC
$l = 1.80 \times 10^{-4}$ g/cm³
$h = 1.62 \times 10^{-3}$ g/cm³

15° ATDC
$l = 1.56 \times 10^{-4}$ g/cm³
$h = 1.41 \times 10^{-3}$ g/cm³

25° ATDC
$l = 1.78 \times 10^{-4}$ g/cm³
$h = 1.60 \times 10^{-3}$ g/cm³

25° ATDC
$l = 1.83 \times 10^{-4}$ g/cm³
$h = 1.64 \times 10^{-25}$ g/cm³

(a)

(b)

These results explain why multiple injections can improve the soot–NO trade-off. On the other hand, soot increased for the 25-8-75 (-10) case. As discussed before, the major portion of the combustion process in the 25-8-75 (-10) case is delayed by the injection pause, thus it is expected that the soot production mechanism, in this case, is like that of a single-injection case with retarded injection timing in which soot emissions increase due to the deteriorated soot oxidation. This is indeed seen by comparing the 25-8-75 (-10) case with the single (-2) case in Figure 8.7.

It is expected that further emissions reductions could be obtained when the injection timing was also varied. For example, the predicted soot–NO trade-off trend using different injection schemes is illustrated in Figure 8.12. It is seen that the 75-8-25 double-injection shifts the soot–NO trade-off to a lower level of soot emission relative to that of the single-injection with almost no penalty in the NO emission in the timing range studied (from −14° to 2° ATDC) when the same injection timing is used for the two injection schemes. In the single-injection cases, NO can be reduced with a corresponding penalty of increased soot with retarding injection timings. This phenomenon is also seen in the 75-8-25 split injection case. However, with the combination of double injections and retarded injection timings, significant reductions of NO and soot can be achieved simultaneously. These predicted results are consistent with the experimental observations of Tow et al. [3], which demonstrates the dramatic reduction of particulate that has been observed with a 75-10-25 double injection.

FIGURE 8.12 Predicted soot–NO trade-off showing emission reduction using split injections.

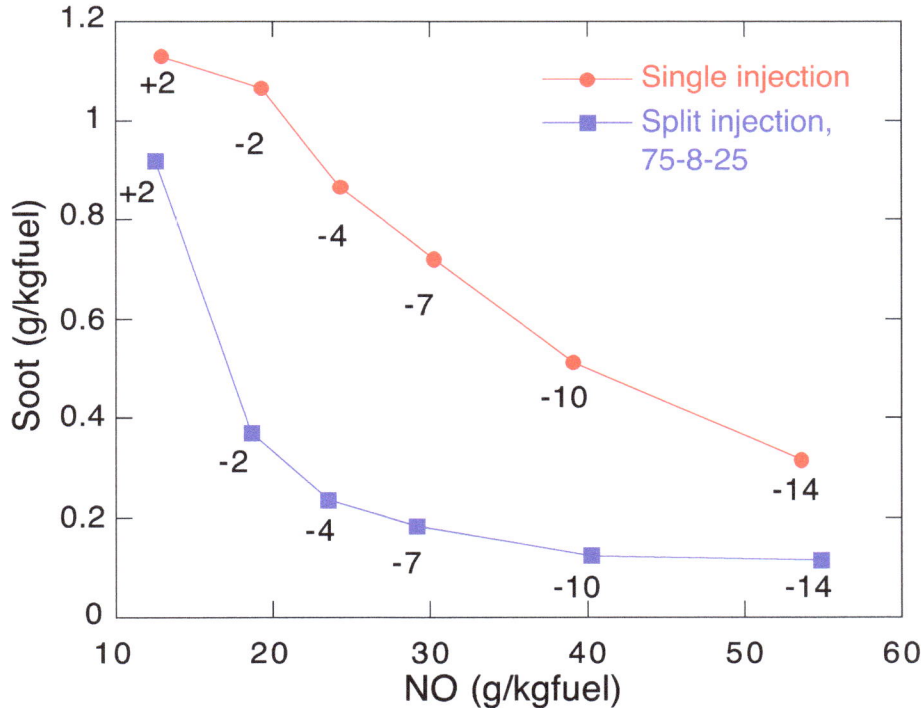

Numbers in the figure are the injection timings (ATDC)

8.1.2 Geometry of Helical-Port and Combustion Chamber

It is well known that combustion chamber and fuel injection strategy are extremely important to diesel combustion and emissions formation. In a high-speed small-bore diesel engine, in-cylinder swirl motion generated by the helical intake port is also crucial in promotion of fast combustion to control soot emissions. Design optimization of intake port and combustion chamber, as well as injector specifications (spray characteristics) for engine function improvements, is a continuous task as government regulations of fuel economy and emissions are increasingly tightened. Use of engine simulation to help improve diesel combustion started in the mid-1990s in a few leading engine companies, and is now practiced globally.

In the development of a 1.6-liter high-speed automotive diesel engine, simulations were carried out to help optimize helical port, combustion chamber, and injector specifications [4,5]. The goal of helical port design is to achieve maximum swirl ratio and to minimize compromise in volumetric efficiency at the same time. CFD calculations were applied to achieve such a goal by Wang et al. [4]. The engine is a turbocharged, 4-valve I4 engine and its compression ratio is 18.1. Its rated power and max torque are 80 kW at 4000 rpm and 230 Nm at 2000 rpm.

Figure 8.13 shows the engine and helical port geometry. There are five design parameters of the helical port to consider, which are labeled in the port graph as in Figure 8.13b. CONVERGE [6] was used for flow CFD simulation. The simulated engine conditions are 2000 rpm WOT, and turbocharging was not included for swirl ratio calculation. A sensitivity study was performed first to examine the effects of design parameters on the swirl ratio and in-cylinder trapped mass. The results are given in Table 8.1. It is seen that by the end of compression, the change of swirl ratio is 18.81%, 17.0%, 16.54%, 13.54%, and 6.26%, respectively, by varying θ, η, Δa, H, and μ independently within engineering ranges. However, their effects on the trapped mass are within 1.06% because the tangential port design was not alternated.

FIGURE 8.13 (a) Geometry of a high-speed diesel engine; (b) Design parameters (labelled by in the graph) of the helical port (b).

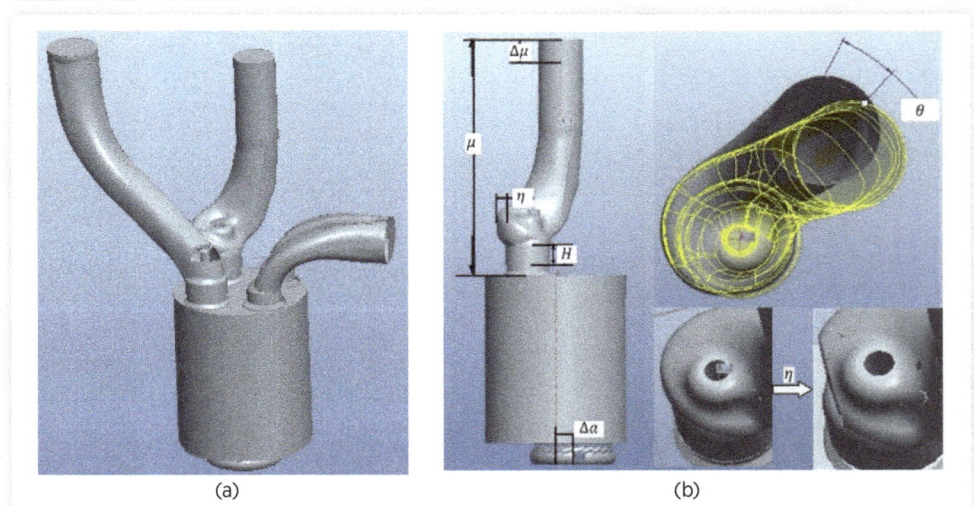

(a) (b)

Adapted with permission from Ref. [4]. © CICEE

TABLE 8.1 Effects of individual parameter on swirl ratio (%)

Crank angle (ATDC)	H (mm)	η (mm)	μ (mm)	θ (°)	Δa (mm)
−20°	16.37	17.94	5.84	18.87	13.46
0°	13.28	17.00	6.26	18.81	16.54
20°	10.64	17.09	6.36	17.92	12.49

Adapted with permission from Ref. [4]. © CICEE

It is noticed that all the studied parameters demonstrated considerable effects on the swirl ratio, and their interactive influences are expected. To further evaluate the interactions, if 11 levels of change are considered in each parameter, the problem becomes a five-factor and 11-level test set. The number of calculations will be $5^{11}=48828125$, which is, apparently, not acceptable. Thus, the Uniform Design method [7] was applied to simplify the test cases. Uniform Design method is a factorial experiment design method that can significantly reduce the number of test cases. With the use of the Uniform Design method, only 11 cases were calculated and the results are given in Table 8.2.

TABLE 8.2 CFD simulation test cases suggested by Uniform Design method and the results

Test	H (mm)	η (mm)	Δμ (mm)	Δa (mm)	θ (°)	SR (TDC)	Trapped Mass (g)
1	6.5	0.30	4	4	20	0.9601	0.5435
2	7.0	0.90	10	9	3	0.7099	0.5433
3	7.5	1.27	16	3	29	0.9482	0.5430
4	8.0	1.80	0	8	15	0.8530	0.5434
5	8.5	2.40	6	2	1	0.9029	0.5434
6	10.5	0	12	7	27	0.8249	0.5432
7	12.5	0.60	18	1	10	0.8981	0.5429
8	13.0	1.10	2	6	0	0.7834	0.5436
9	13.5	1.50	8	0	25	0.9250	0.5432
10	14.0	2.10	14	5	5	0.7979	0.5431
11	14.5	2.54	20	10	30	0.7289	0.5428

Adapted with permission from Ref. [4]. © CICEE

Applying the multiple nonlinear regression method, a regression equation is obtained, and an optimal value of each parameter can be predicted based on the regression equation; the maximum and minimum value of the swirl ratio is obtained as 1.0358 and 0.6708, respectively, and the corresponding values of the design parameters are determined [4]. To confirm the optimization suggested parameters, simulations were performed of Case A, which should give maximum swirl ratio, and Case B, which should give minimum swirl ratio, and the results are shown in Figure 8.14. It is seen that the two sets of design parameters result in a very different evolution of swirl ratio, and the simulated peak swirl ratios at TDC are consistent with the predicted values using the above optimization method.

FIGURE 8.14 Simulated evolution of swirl ratio of Case A (solid line) and Case B (dashed line).

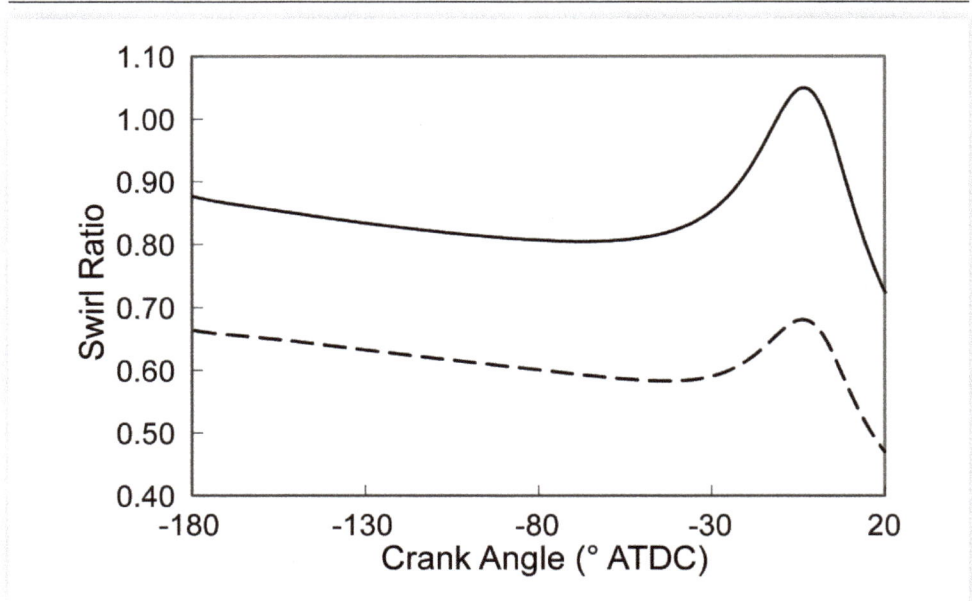

Adapted with permission from Ref. [4]. © CICEE

Combustion simulations of the same engine were made by Li et al. [5]. A total of 24 cases were studied in a combination of three chamber geometries as shown in Figure 8.15, two spray included angles (149° and 153°), and four injector nozzle axial positions (1.1, 1.6, 2.1, and 2.6 mm protrusion). The parameter set of Chamber A, 149° spray angle and 1.6 mm, was the baseline and tested in experiments. The injector is centrally located with six spray plumes evenly distributed and, thus, a 60° sector mesh was used. The simulated condition is 2000 rpm and 2 bar BMEP, which is a typical part-load condition for passenger vehicle applications. The KIVA code with the improved physical models, as described earlier, was used.

FIGURE 8.15 Combustion chamber geometries of a high speed 1.6-liter diesel engine.

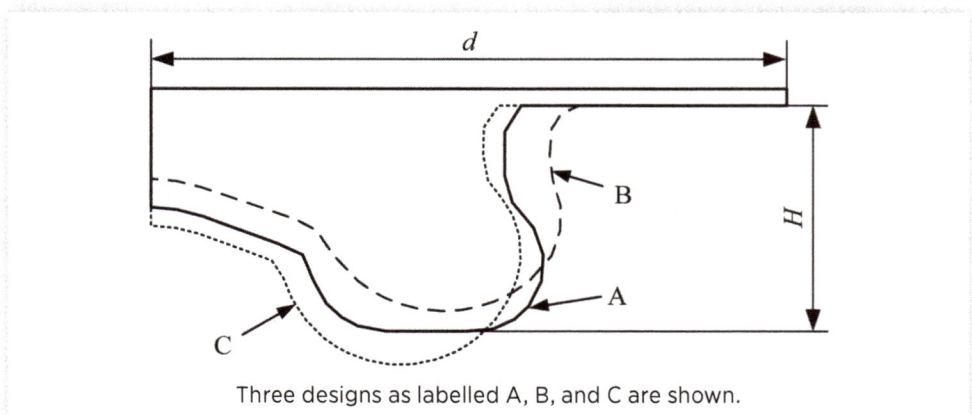

Three designs as labelled A, B, and C are shown.

Courtesy of Zhiyu Han

The model was first validated with existing experimental data of pressure trace, HRR, and soot and NO_x emissions under different operating conditions. Figure 8.16 shows the comparison of predicted and measured emissions in the baseline case. Prediction of NO_x is in good agreement with experiments, but a prediction of soot only agrees with experiments quantitively.

FIGURE 8.16   Comparison of predicted and measured NO_x and soot emissions for the baseline case.

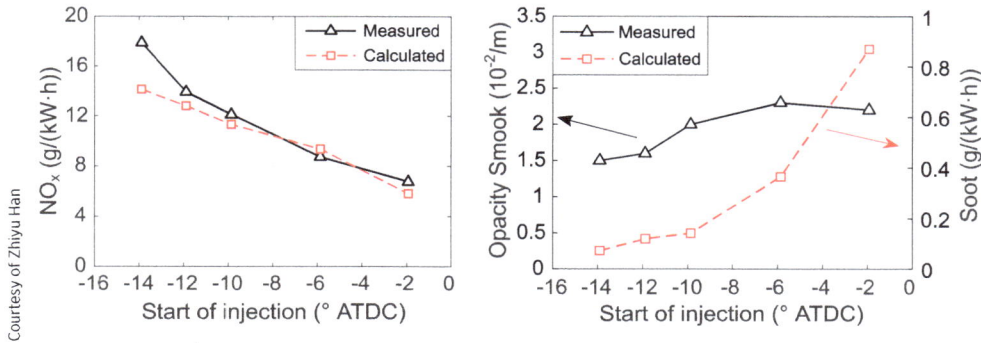

Courtesy of Zhiyu Han

Figure 8.17 shows the predicted NO_x and soot emissions as the parameters are changed. Generally, a declined trend of soot emissions is seen as the injector protrusion becomes larger. A shallower piston bowl results in reduced soot emissions, and Chamber B gives the least soot emissions. However, the effects on NO_x emissions are nonlinear. NO_x emissions increase first and then reduce as the protrusion becomes larger except in one case. Chamber B leads to higher NO_x emissions in general, which is the typical trade-off of soot and NO_x emissions. A small increase of the spray included angle (4°) results in considerable deterioration in soot emissions and a minor increase in NO_x emissions. It is seen that at injector tip protrusion of 1.6 mm, Chamber B with a 149° spray reduces soot emissions by a factor of 2 and NO_x emissions slightly from the baseline case (Chamber A and 149° spray) at the part-load conditions.

FIGURE 8.17   Predicted effects of design parameters on NO_x and soot emissions at 2000 rpm and 2 bar BMEP.

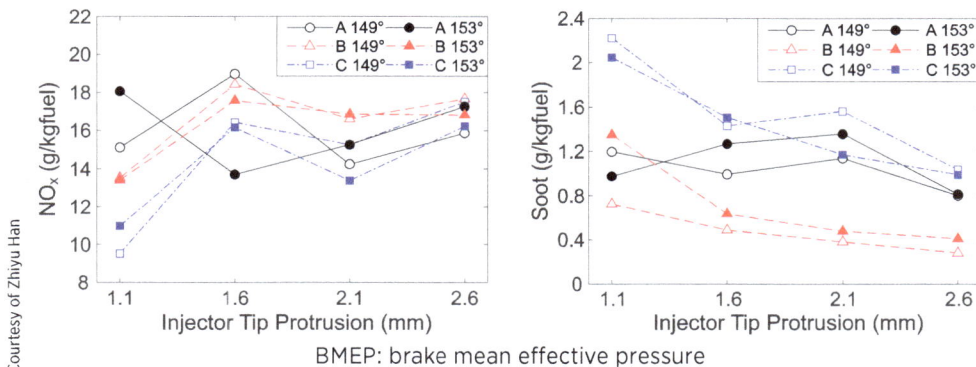

Courtesy of Zhiyu Han

BMEP: brake mean effective pressure

8.1.3 Emissions at Cold Start

The effects of fuel injection strategy on cold start emissions were studied numerically by Deng [8]. The same diesel engine shown in Figure 8.13 was modeled. In order to predict combustion and detailed emissions formation, the KIVA–CHEMKIN model with a reduced n-heptane kinetics mechanism [9], a 12-reaction NO_x mechanism [10], and the Hiroyasu–NSC soot model (see Sec. 6.7) were used. Hydrodynamics and heat transfer of spray wall impingement were simulated with the models by Deng et al. [11,12]. A 60° sector mesh was used and simulations were started at the IVC time.

Validations of these models were made under two engine operating conditions and comparisons of experiments and simulations are given in Figures 8.18 and 8.19. Under Condition 1, the engine runs at 1000 rpm and 2 bar BMEP using a double fuel injection with SOI1 −30° ATDC, 1.8 mg fuel, and SOI2 −10° ATDC, 5.7 mg fuel. While under Condition 2, the engine runs at 2000 rpm and 2 bar BMEP with SOI1 −30° ATDC, 1.8 mg fuel, and SOI2 −10° ATDC, 7.8 mg fuel. Figure 8.18 indicates the simulation can reproduce engine cylinder pressure trace and HRRs well, but some discrepancies between the simulated and measured emissions are seen in Figure 8.19. Note that soot prediction was not compared since soot experimental data were unavailable.

FIGURE 8.18 Comparison of predicted and measured cylinder pressure and HRR with double injection.

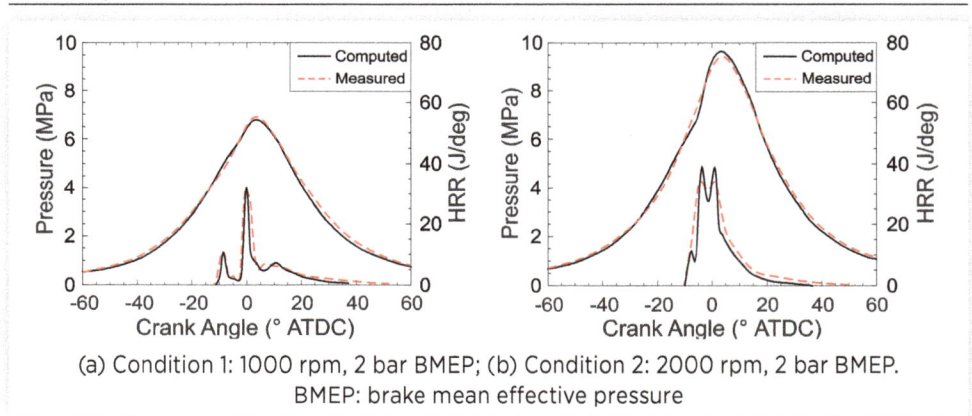

(a) Condition 1: 1000 rpm, 2 bar BMEP; (b) Condition 2: 2000 rpm, 2 bar BMEP.
BMEP: brake mean effective pressure

Courtesy of Deng Peng

FIGURE 8.19 Comparison of predicted and measured hydrocarbon, CO, and NO_x at Conditions 1 and 2.

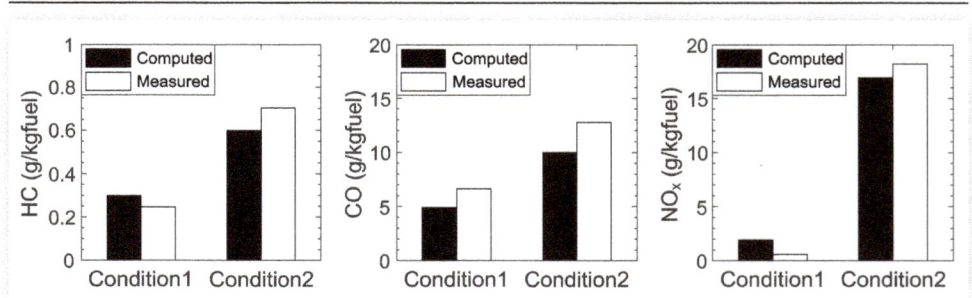

Courtesy of Deng Peng

©2022, SAE International

Engine cranking process during cold-start is a highly complicated and transient process. Simulation of many cycles with cycle-by-cycle varied fueling and thermal conditions is required, which is limited by computer capacity and data availability. To simplify the problem, four engine speeds are selected to represent the cranking process: 200, 400, 600, and 800 rpm. After searching the misfiring boundary at each engine speed, the minimum fuel amount is determined in each engine speed in the simulation. Considering warm-up effects while the engine speeds up, the injected fuel mass corresponding to the above engine speeds is set to be 50, 30, 24, and 20 mg, respectively, and the initial gas temperature at IVC is 273, 278, 283, and 293 K, respectively. The engine surface temperatures are assumed to be 273 K (0°C) and kept the same during the starting process. More details about the simulation conditions are given by Deng [8].

Various injection schemes are used in the simulation to study their effects, which are given in Figure 8.20; there is a label by each scheme in which the first number means the percentage of fuel in the first injection pulse and the second the percentage in the second pulse. For instance, 50-50 means 50% fuel is injected in the first pulse and 50% in the second. SOI and EOI are set to be −15° ATDC and TDC, respectively, for all the studied cases.

FIGURE 8.20 Diagram to show injection schemes.

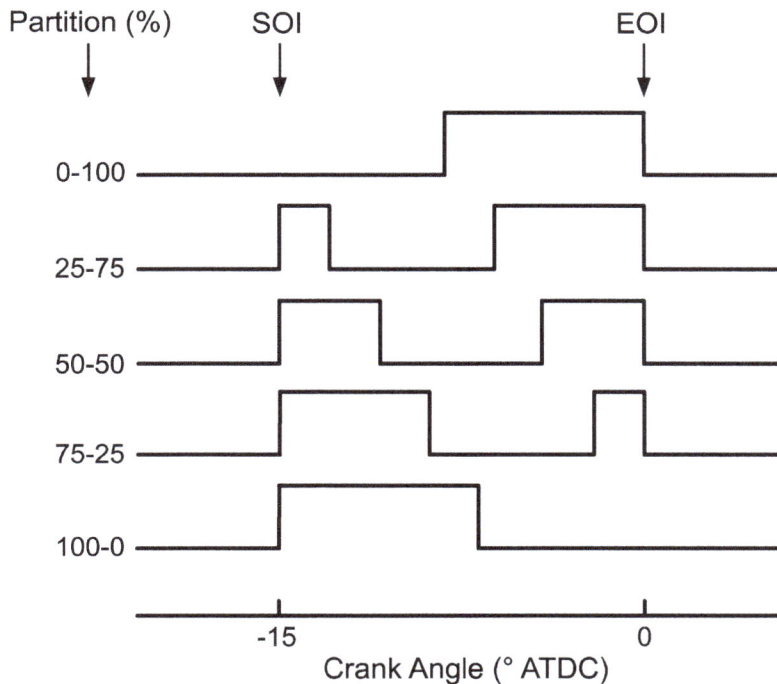

It was interesting to notice that a double-injection could have a shorter ignition delay than a single-injection with the same SOI as illustrated in Figure 8.21. The reason is that the charge cooling phenomenon results in lower gas temperature in the single-injection case due to more fuel vaporization as seen in the figure.

FIGURE 8.21 In-cylinder gas-temperature evolution at 400 rpm.

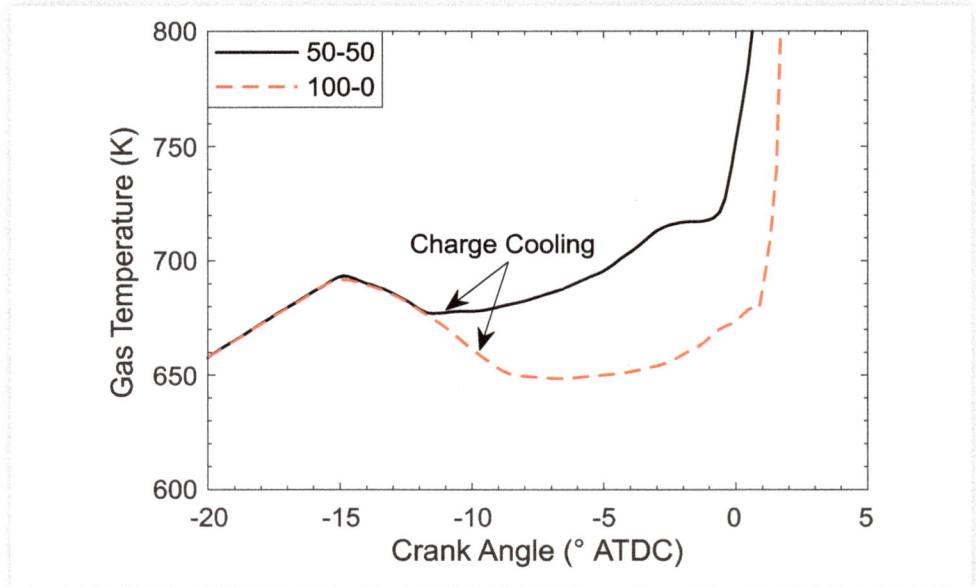

Courtesy of Deng Peng

Injection strategy also impacts fuel depositions or liquid film formation on the piston bowl surface. The evolution of the liquid film mass on the piston surface is shown in Figure 8.22 for two engine speed cases. It is observed that an ample amount of liquid fuel deposits on the piston surface and remains until combustion takes place, during which the liquid film vaporizes rapidly due to increased gas temperature. However, part of the deposited fuel remained after combustion, and the double-injections reduce both the impinged fuel and remained fuel. As the engine speed is increased from 400 rpm to 800 rpm, the deposited and remained fuel are all reduced significantly. Figure 8.23 summarizes the post-combustion remained liquid fuels for all the cases studied. The results are consistent with the available experiment data by Tsunemoto et al. [13].

FIGURE 8.22 Evolution of liquid film mass on the piston surface.

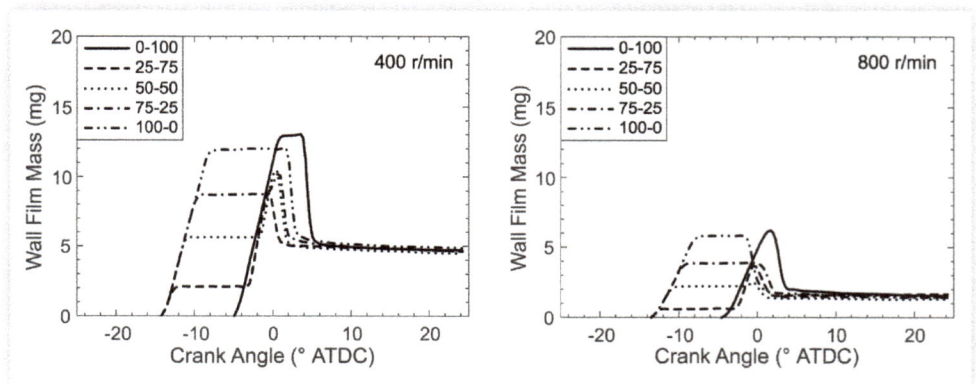

Courtesy of Deng Peng

Courtesy of Deng Peng

FIGURE 8.23 Influence of injection schemes on post combustion remained fuel normalized by the total fuel.

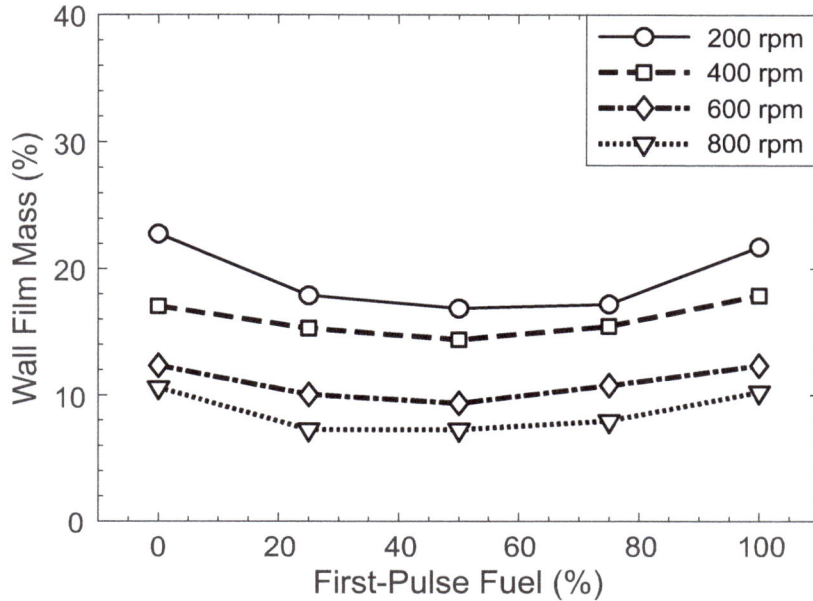

Figure 8.24 gives the computed cylinder pressure traces, HRRs, and gas temperatures at 400 rpm with various injection schemes. Diesel engine cold-start featuring long ignition delays are shown. The long ignition delays result in rapid combustion and heat release and, hence, noisy operation at cold-start is evident. Double-injection helps mitigate the combustion rate, but not much. Similar behaviors are seen at other speeds. Note that the peak temperatures are generally less than 2000 K, which have negative effects on hydrocarbon, CO, and soot emissions as will be discussed next.

FIGURE 8.24 Computed cylinder pressure, heat release rate, and gas temperature at 400 rpm.

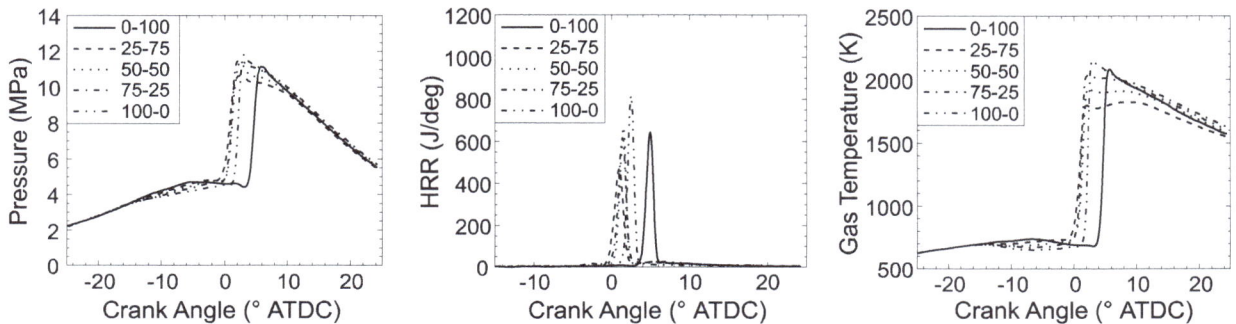

Courtesy of Deng Peng

Predictions of hydrocarbon, CO, NO$_x$, and soot emissions at 200 rpm are given in Figure 8.25. High levels of hydrocarbon and CO emissions are generated due to a rich mixture burning near the piston bowl surface where liquid fuels have deposited. On the other hand, oxidation of hydrocarbon and CO is not sufficient because of the relatively low combustion temperature and inadequate oxygen concentration locally inside the piston bowl due to poor mixing at low engine speeds. Due to similar reasons, soot emissions are high as well. However, NO$_x$ emissions are low primarily due to the low combustion temperature. The simulation indicated that the peak gas temperature was lower than 2000 K in all the cases (see Figure 8.24), which are favorable to NO$_x$ emissions and unfavorable to soot emissions as suggested by Figure 2.23 in Sec. 2.4. Double-injections can reduce hydrocarbon and CO emissions due to improved mixing and less fuel deposition. However, their effects on NO$_x$ and soot emissions are seen to be insignificant, which suggests that global temperature and gas motion play more important roles. As the engine speed increases, e.g., to 800 rpm, hydrocarbon and CO emissions reduce significantly as also seen in Figure 8.25 due to improved mixing and thermal conditions. Thus, insights into the cold-start during which the first few cycles at low engine speeds produce much higher hydrocarbon and CO emissions are given by CFD simulations.

FIGURE 8.25   Predicted emissions during cold-start with varied injection strategies.

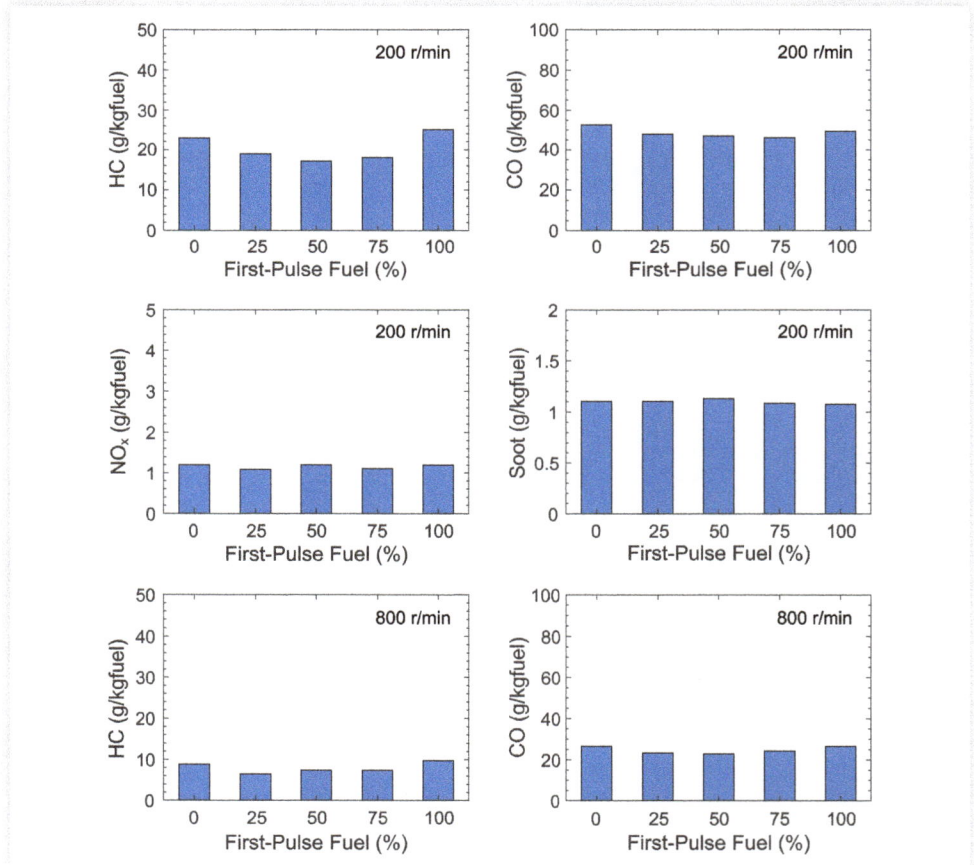

8.2 Alternative Fuel Engines

Alternative fuels, such as natural gas, alcohol (methanol, ethanol), biodiesel, dimethyl ether (DME), hydrogen, etc., have been used in IC engines to substitute conventional petroleum fuels (gasoline and diesel) in consideration of the preservation of petroleum resources. On the other hand, some kinds of alternative fuels can reduce CO_2 emissions as well. In gasoline and diesel engines, reduction of fuel consumption will lead to a reduction of CO_2 emissions. Combustion of low-carbon fuels (e.g., natural gas, alcohol, hydrogen, etc.) result in much less CO_2 production at the same fuel conversion efficiency. For example, burning natural gas produces about 20% less CO_2 emissions than burning gasoline under stoichiometric conditions. Thus, low-carbon fuels should be used when resources are available.

Natural gas has been widely used due to its attractive chemical and physical properties [14]. It is primarily composed of methane, and burning methane can lead to reduced CO_2 and PM emissions than burning gasoline or diesel because of its lowest C/H ratio among hydrocarbon fuels. Over the period 2005 to 2018, the proportion of natural gas vehicles (NGVs) in the total vehicle population increased from approximately 0.5% to 1.52%, and the total number of NGVs in use reached more than twenty-six million by 2018 as seen in Figure 8.26. Natural gas can be burned in both SI engines and compression ignition engines by using various combustion strategies. Some recent research works were reported by the author's group on SI natural gas (SING) engines [16,17] and on diesel and natural gas dual-fuel (DNGDF) engines [18–21]. CFD simulations to help understand and optimize combustion in these engines are discussed in this section.

FIGURE 8.26   Global NGV numbers by region from 2000 to 2018.

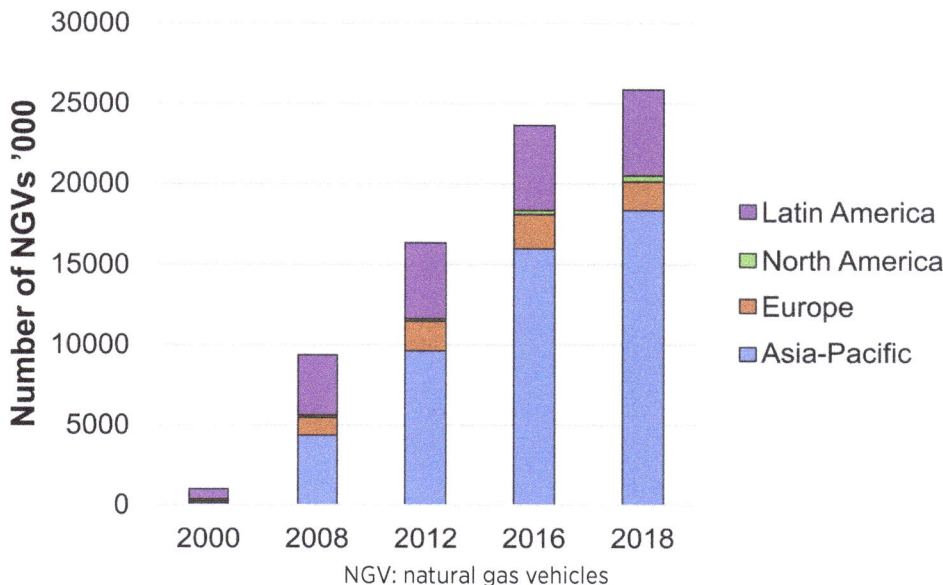

Biodiesel fuels can be produced from many biomasses, such as soybean, grapeseed, linseed, sunflower seed, algae, animal fat, etc. The raw materials of biodiesel are renewable and, therefore, burning biodiesel in IC engines can reduce the dependence on nonrenewable energy resources or fossil oil. Biodiesel can mix with conventional diesel well and is compatible with most modern diesel engines without major engine modifications. Due to the oxygen content in biodiesel molecules, biodiesel fueled engines can simultaneously reduce emissions of CO, PM, and hydrocarbon [22]. However, there are some negative effects of using biodiesel [23]. Burning biodiesel results in higher NO_x emissions than burning petroleum diesel by 10–20%. This higher NO_x emission tendency has become one of the main obstacles in biodiesel utilization, especially pure biodiesel utilization. Research on the reduction of NO_x emissions has been intensively carried out. CFD simulations coupled with chemical kinetics of biodiesel-blends combustion have been reported to evaluate the performance and emissions in addition to experimental investigations [24]. A detailed biodiesel reaction mechanism has been proposed [25] and reduced mechanisms have been used in CFD combustion simulations. In Sec. 8.2.3, CFD simulation research on biodiesel combustion by Li et al. [26,27] is presented as examples.

8.2.1 Spark-Ignition Natural Gas Engines

Similar to gasoline premixed combustion, good mixing of natural gas with air to form a homogeneous mixture in the cylinder is important. Directly injecting natural gas into the cylinder faces mixture homogeneity challenges due to the limitation of the injectors, which are only capable of delivering low radial and axial fuel penetrations because of the low gaseous mass density of natural gas. Diffusion processes have a minor influence on mixture formation due to the very small timescales available in engines. Therefore, natural gas mixing primarily depends on large-scale charge motion in the cylinder. Due to the non-availability of commercial DI injectors, injecting natural gas into the intake ports (PFI) is the primary method in light-duty automotive SING engines. Torque deterioration in PFI SING engines due to reduced volumetric efficiency can be compensated by turbocharging [28,29].

Multiple cycle simulations were carried out to study mixture formation in a turbocharged PFI SING engine [16]. The engine is a 1.2-liter I4 engine and the computational domain is shown in Figure 8.27 in which the positive direction of cylinder flow motion (swirl and tumble ratio) is also illustrated. An electronically controlled injector is installed at the upstream location of the junction of the two intake ports in each cylinder, and natural gas is injected under an injection pressure of 0.7 MPa. The engine specifications are summarized in Table 8.3 and the simulated engine conditions are given in Table 8.4.

FIGURE 8.27 A 1.2-liter I4 turbocharged PFI SI natural gas engine and its computational domain at BDC.

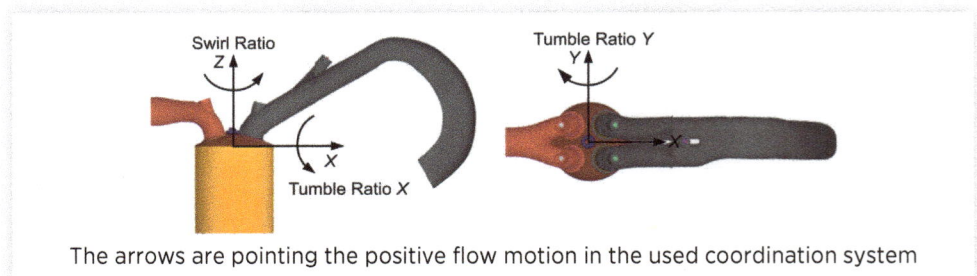

The arrows are pointing the positive flow motion in the used coordination system

TABLE 8.3 Specifications of a 1.2-liter PFI turbocharged natural gas engine

Parameters	Value
Bore × Stroke (mm × mm)	69.7 × 79
Displacement (L)	1.2
Connecting rod length (mm)	121
Compression ratio	12
Intake valve open (° ATDC)	343
Intake valve closure (° ATDC)	619
Exhaust valve open (° ATDC)	130
Exhaust valve closure (° ATDC)	363
Injection pressure (bar)	7
Fuel	Natural gas

TABLE 8.4 Operating conditions in the simulations

Parameters	Condition 1	Condition 2	Condition 3
Engine speed	5500	2000	2000
Engine load	WOT	WOT	2-bar
BMEP (bar)	16.73	13.78	2
Intake pressure (bar)	1.576	1.425	0.41
Intake temperature (K)	306	297	294
Injection fuel mass (mg)	27.27	23.8	4.82
Injection duration (°CA)	473	150	30
Injection timing (°ATDC)	−103	180	180
Spark timing (°ATDC)	−12	−12	−23.6

CH_4 is used to represent the natural gas used in the engine experiments since it comprises more than 96% of the compositions. The CONVERGE code [30] with the RNG k-ε turbulence model, compressible heat transfer model, and G-equation combustion model discussed previously is used in the simulation. To obtain an in-depth understanding of mixture formation, three typical engine operating conditions: middle speed and low load, middle speed and high load, and high speed and high load were simulated. All the calculations started from −420° ATDC, close to the exhaust TDC. Since gas injection and mixing inside the intake port significantly affected by the backflow from the cylinder and pre-existing conditions are difficult to estimate, multiple cycle simulations had to be carried out and the results of the previous cycle were used to be the boundary and initial conditions in the current cycle. Computation indicated that at least three cycles were needed to obtain converged results as shown in Figure 3.1 in Sec. 3.3.1. In the following discussion, all the results were taken from the third simulation cycle.

Figures 8.28 and 8.29 show the evolution of fuel mass fraction spatial distributions for the three operating conditions during the intake stroke and compression stroke, respectively. Different maximum values of legend scale in Figures 8.28 and 8.29 were selected to better exhibit the differences. Due to the CVI and the pushback of the fuel at the early stage of the compression stroke, the intake port is full of fuel at the onset of intake stroke

(−380° ADTC), as shown in Figure 8.28. The rich fuel–air mixture accumulates near the intake valves, which blocks fresh air entering the cylinder at the early intake phase, leading to in-cylinder nonuniformity, as can be seen at −320° and −300° in Figure 8.28. At −340°, the fuel begins to enter into the cylinder under Conditions 2 and 3, while a significant charge pushback is shown in Condition 1. The pushback motion results from the pumping pressure difference between in-cylinder and intake port (negative pressure), improving intake port fuel–air mixing. This leads to a better homogeneity during the intake stroke than that in the other two full load conditions. It should be noted that the intake runner must be long enough to ensure that the pushback fuel–air mixture does not flow into the plenum (not modeling in this study); else it would cause poor injection control and high cylinder to cylinder variations. This undesirable phenomenon does not occur in the current study, as is illustrated at −340° in Figure 8.28, indicating the rational design of the intake manifold.

FIGURE 8.28 Fuel mass fraction distribution in two cut planes for a turbocharged SING engine during the intake stroke. Time sequence from the top row: −380°, −340°, −320°, −300°, −270°, and −240° ATDC.

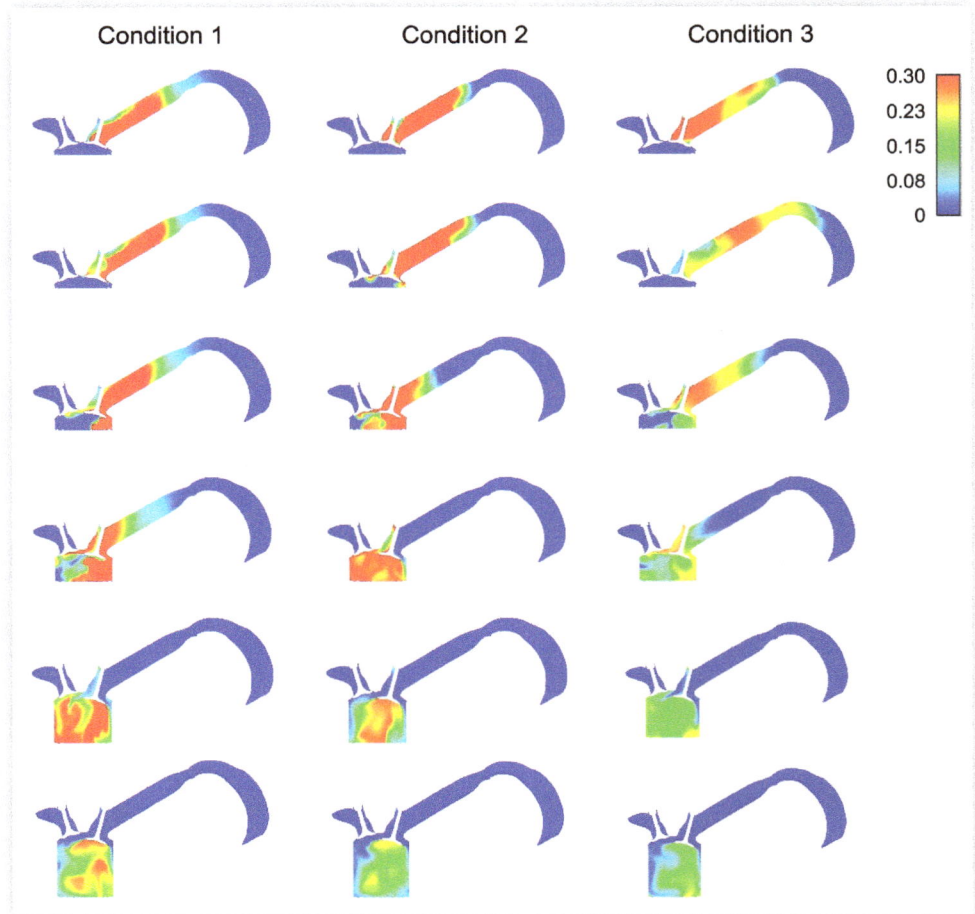

Compared to Condition 2, Condition 3 shows a lagging intake phase due to the fact that the used monotonous valve-phasing cannot cover the entire engine operating range. Overall, the mixing processes for the three operating conditions at the late stage of intake stroke are similar, and adequate fresh air flows into the cylinder and mixes with the rich mixture with the strong flow motion. As for the compression stroke in Figure 8.29, similar results are observed for the three conditions. Late intake-valve-closing results in part of the fuel flowing back into the intake port, which is a challenge for precise fuel mass control. The in-cylinder fuel and air start to form a homogenous mixture as the piston moves toward the TDC. However, the mixture in the exhaust-valve side of the cylinder still tends to be leaner than that in intake-valve side.

FIGURE 8.29 Fuel mass fraction distribution in two cut planes for a turbocharged SING engine during the compression stroke. Time sequence from the top row: −180°, −120°, −60°, and −30° ATDC.

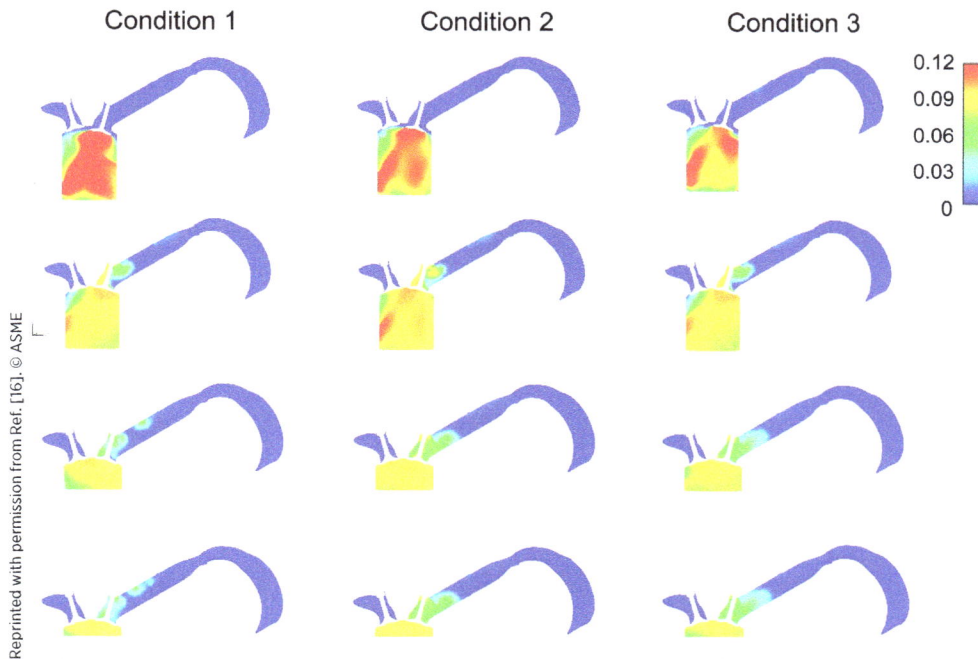

The equivalence ratio (∅) spatial distribution at the spark timing is shown in Figure 8.30. It is seen that the maximum homogenous mixture distribution was obtained at Condition 1, while the worst mixing was observed at Condition 3. It demonstrates that even with a short available mixing time, strong flow motion (high velocity) at high engine speed enables the engine to produce more uniform mixing. By contrast, longer available mixing time at Condition 3, which is almost three times as that at Condition 1 cannot ensure a better mixing due to its weak gas flow motion. It is observed in the simulations that there are no rich mixtures with an equivalence ratio greater than 1.25 for all the cases studied. And mixture throughout the cylinder is flammable (0.7 < ∅ < 1.25) for Condition 1 and Condition 2. At Condition 3, some lean mixtures were seen; thus, unburned fuel and hydrocarbon emissions are expected.

FIGURE 8.30   Equivalence ratio distribution at the spark timing.

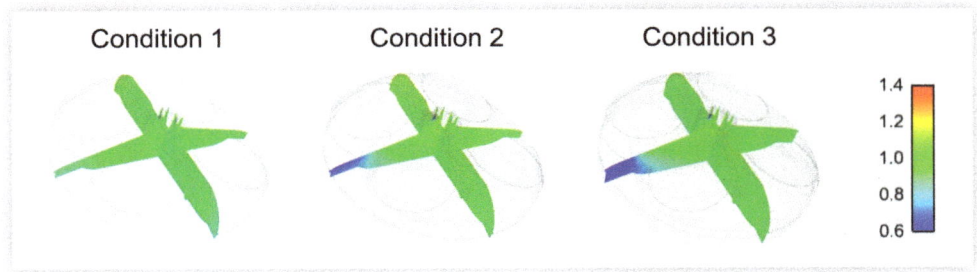

Now, we define a parameter, $\varnothing_{nstd}$, to measure mixture inhomogeneity as:

$$\varnothing_{nstd} = \frac{\varnothing_{std}}{\varnothing_{mean}} \tag{8.1}$$

where the standard deviation of the in-cylinder mixture equivalence ratio is given as:

$$\varnothing_{std} = \sqrt{\frac{\Sigma_{cell} m_{cell} (\varnothing_{cell} - \varnothing_{mean})^2}{\varnothing_{mean}}} \tag{8.2}$$

and $\varnothing_{mean}$ is the mean equivalence ratio of the entire in-cylinder mixture, m_{cell} is the mixture mass in a computational cell, and $\varnothing_{cell}$ is the equivalence ratio of in that cell. Physically, the smaller the $\varnothing_{nstd}$, the more homogeneous.

Effects of injection timing on mixture preparation were evaluated. As illustrated in Table 8.4, the injection duration in Condition 1 is 473°, with a duration of approximately three quarters of the engine work cycle. Since this engine uses PFI, injection timing is more flexible than that in a DI system in which all fuels must be injected within the intake stroke. Mixture preparation with different injection timing strategies is assessed, which are CVI, half open valve injection (HOVI), and open valve injection (OVI). Table 8.5 gives the intake-valve timing schemes of the three injection strategies under the three operating conditions.

TABLE 8.5   Intake-valve timing schemes with CVI, HOVI, and OVI

Parameters	Value		
Operating conditions	2000 rpm 2bar	2000 rpm WOT	5500 rpm WOT
Injection fuel mass (mg)	4.86	20.5	27.27
Injection duration (°CA)	30	129	473
Start to end of CVI (°ATDC)	180 to 210	180 to 309	−103 to 370
Start to end of HOVI (°ATDC)	330 to 360	290 to 419	−3 to 470
Start to end of OVI (°ATDC)	480 to 510	400 to 529	−303 to 170

In-cylinder mixture inhomogeneity at the spark timing under three injection strategies is shown in Table 8.6. Compared to the CVI strategy, HOVI and OVI strategies have the lower inhomogeneity for all the conditions, which indicates that better mixing is obtained when the injection takes place during the intake stroke. However, these differences among the three strategies at the two full load conditions are small, while the difference between CVI and OVI is relatively bigger at 2000 rpm, 2 bar condition (around 0.03). In addition, it was found that under the two full load conditions, all the injection timings lead to flammable air–fuel mixture $(0.7 < \varnothing < 1.25)$ at spark timing. At 2000 rpm and 2 bar BMEP, the completely flammable mixture is obtained only with the OVI strategy, while both CVI and HOVI strategies resulted in a small quantity of air–fuel mixture beyond the flammable range. Nevertheless, the nonflammable fraction is very small, between 0.01 and 0.03. Thus, it is concluded that injection timing has a negligible impact on mixture flammability at the spark timing. In this context, flexible injection timing can be used in engine calibrations.

TABLE 8.6 Mixture inhomogeneity of three injection strategies at the spark timing

Operating conditions	CVI	HOVI	OVI
5500 rpm, WOT	0.050	0.041	0.033
2000 rpm, WOT	0.070	0.058	0.062
2000 rpm, 2-bar BMEP	0.110	0.104	0.081

Reprinted with permission from Ref. [16]. © ASME

In the conceptual design stage of this SING engine, a unique intake-valve opening scheme, the so-called "asynchronous valve-opining scheme", was proposed [31]. In this new design, the cam profiles for the two intake valves are not the same so that the two valves do not open synchronously. By using this valve-opening strategy, skewed intake flows are generated so that swirl motion, in addition to tumble motion, is generated and the in-cylinder turbulence is enhanced [16]. As evidenced by the mixture equivalence ratio distribution shown in Figure 8.31, the mixture homogeneity is improved with the use of an asynchronous valve-opening scheme. In particular, the lean mixture packets seen in Figure 8.30 at Condition 3 are removed. Quantitively, at 2000 rpm and 2 bar BMEP, the inhomogeneity parameter $\varnothing_{nstd}$ is reduced to 0.075 from 0.11 when the conventional synchronous valve-opening scheme is used as listed in Table 8.6.

FIGURE 8.31 Improved mixing at the spark timing by using an asynchronous valve-opening scheme.

Condition 1   Condition 2   Condition 3

Reprinted with permission from Ref. [16]. © ASME

An asynchronous valve-opening scheme was also found to result in increased trapped mass, which benefits engine torque output. For this engine, about a 4% increase in the trapped mass was found in simulation, which was confirmed by measurements. The improved mixing and increased trapped mass lead to an increase in engine torque output by 7.7% at 2000 rpm WOT without tuning of the turbocharger [16]. A similar asynchronous valve-opening scheme was also applied to a 2.4-liter turbocharged SING engine and the test results are shown in Figure 8.32. Configuration 1 carried over the turbocharger and intake-valve timing of the base gasoline engine, Configuration 2 adopted a reduced flow-rate turbo, Configurations 3 and 4 used the same turbo as that in Configuration 2, but with a shortened intake duration and valve lift. These last two configurations increased the low-end torque significantly. With the use of an asynchronous valve-opening scheme in Configuration 4, the low-end torque was improved even more as shown in the figure.

FIGURE 8.32   Improved low-end torque by using an asynchronous valve-opining scheme demonstrated in a 2.4-liter turbocharged SING engine.

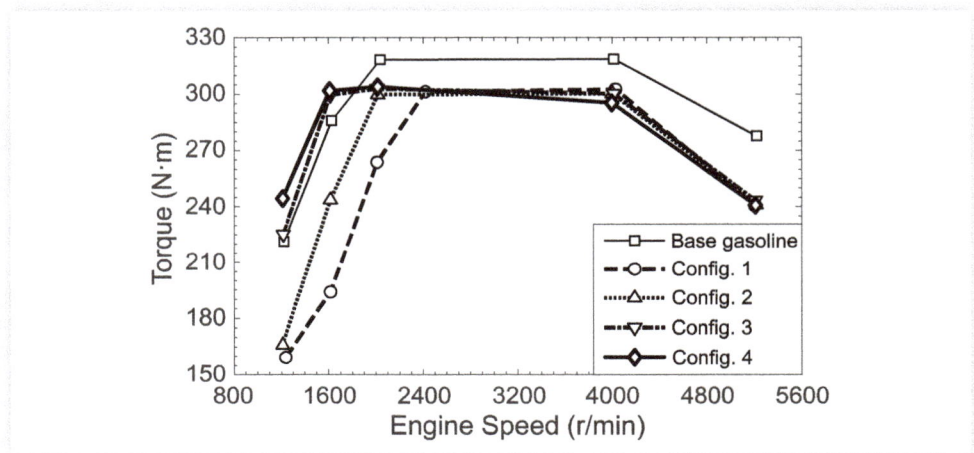

Courtesy of Zhiyu Han

8.2.2 RCCI in Diesel–Natural Gas Dual-Fuel Combustion

As discussed in Sec. 2.4, in the RCCI combustion concept, in-cylinder fuel blending can be arranged using PFI of a lower reactivity fuel, coupled with in-cylinder injection of more reactive fuel. This novel concept has been demonstrated to deliver superior fuel economy and extra low emissions over conventional combustion methods [33].

In a DNGDF engine, natural gas can be regarded as the lower reactive fuel, and diesel the more reactive fuel. Thus, RCCI combustion can be realized in a DNGDF engine by premixing natural gas with PFI and injecting diesel with in-cylinder DI, in which the diesel fuel autoignites, promoting fast kinetics-controlled combustion of the blended fuel [34]. On the other hand, diesel pilot ignition (DPI) combustion has been traditionally used in DNGDF engines in which natural gas is premixed with air in the intake ports, and the mixture is then introduced into the cylinder and ignited by the direct injected diesel fuel near the end of compression [35]. The main difference between DPI and RCCI is that in DPI, diesel fuel is injected near the TDC, and in RCCI, diesel fuel is injected much earlier in the compression stroke. Consequently, there is ample time in RCCI for diesel fuel to mix with the air–natural gas mixture, resulting in near homogeneous blended-fuel mixture and multiple site autoignited low-temperature combustion. In DPI, diesel fuel soon auto-ignites once injected into the air–natural gas mixture, leading to high-temperature local flame propagation in the dual-fuel mixture. Thus, in the RCCI combustion, very low NO_x emissions (due to low-temperature) and improved fuel consumption (due to lower heat transfer loss) are achieved.

One way to control the start of combustion or combustion phasing in RCCI is to vary the amount of premixed low reactive fuel as shown in Figure 2.20 in Sec. 2.4. Other effects on RCCI combustion phasing in a DNGDF engine were numerically studied by Wu et al. [18]. The simulated engine is a heavy-duty diesel-converted DNGDF engine that has a 2.44-liter displacement with a compression ratio of 16.0. The CONVERGE code with its SAGE chemistry solver [30] was used for combustion modeling. The Reitz Wave model is adopted for modeling the in-cylinder injected diesel sprays. A reduced fuel mechanism of 45 species and 142 reactions developed by Ra and Reitz [9] was used for the diesel and natural gas reaction chemistry. This also includes a reduced NO_x mechanism [10]. For an affordable computational time, the simulation is simplified by using a 51.43° sector-mesh (for a 7-hole injector) and a well premixed air–natural gas mixture at IVC.

Effects of diesel SOI on combustion were evaluated. In this case, the engine was operated at 1300 rpm and in a medium load condition. There was consumption of 115 mg equivalent diesel fuel mass in which 90% of the total heat was from the premixed natural gas and 10% was from the injected diesel. Figure 8.33 shows the SOC variation with the diesel SOI where the SOC is defined as the crank angle, CA10, at which 10% of the cumulative heat release occurs. As seen, when SOI is advanced, SOC (or CA10) exhibits a nonmonotonic trend. In the DPI regime, as SOI is advanced from TDC, SOC advances as well. As SOI is advanced beyond −30° ATDC, the trend reverses and SOC starts to retard, at which time combustion starts to exhibit RCCI characteristics. When SOI is further advanced, apparent RCCI combustion occurs, which features separated two-stage combustion including a low-temperature heat release of the diesel fuel and a single-peak fast heat release of the blended fuel as indicated in Figure 8.34.

FIGURE 8.33 Illustration of computed combustion regime transition with pilot-diesel SOI in a DNGDF engine.

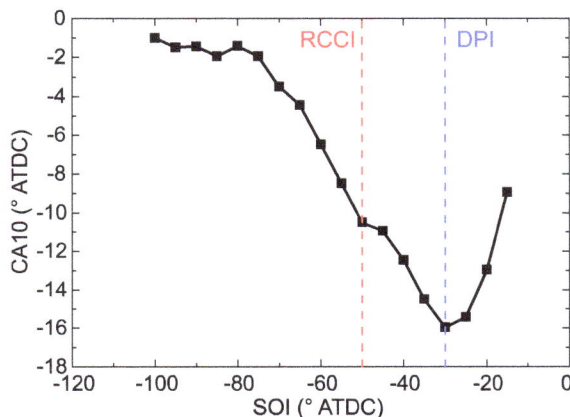

SOI: start of injection; DNGDF: diesel and natural gas dual-fuel

Figure 8.34 shows the computed HRR at three SOIs. When SOI is −85° ATDC, separated two-stage combustion is clearly seen, which is the typical RCCI combustion characteristic. As SOI is retarded to −45° ATDC, the first peak in the HRR becomes larger due to a shortened ignition delay. As SOI is delayed further to −15° ATDC, a typical DPI combustion with closely-coupled two-stage high-temperature heat release can be observed. In this case, the first stage of heat release is from the pilot-diesel premixed combustion as a result of diesel ignition delay, and the second one is from the flame propagation in the mixture of the blended-fuel and air.

Combustion phasing versus SOI is shown in Figure 8.35. Here, CA50, the crank angle at which 50% of cumulative heat release occurs, is used to define the combustion phasing. The

FIGURE 8.34 Comparison of computed heat release rates as SOI varies. Wu et al. (2017) [18].

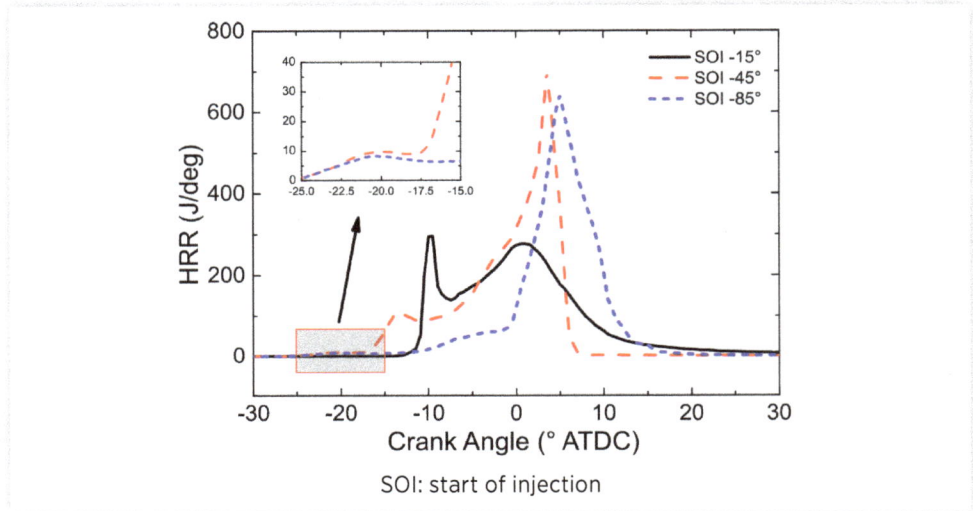

SOI: start of injection

FIGURE 8.35 Effects of SOI on combustion phasing as indicated by CA50.

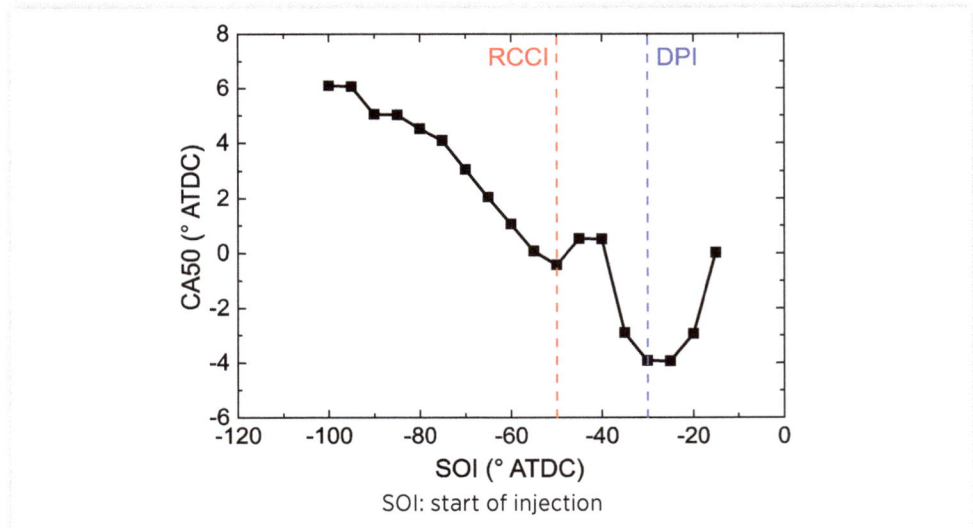

SOI: start of injection

well-known dependence of combustion phasing on SOI is apparently seen for DPI because combustion is closely coupled with fuel injection here. However, it is also clearly observed that SOI can alternate, near monotonously, the combustion phasing in the RCCI regime, which means SOI can be used as an additional parameter to control RCCI combustion in a DNGDF engine.

The computed gross indicated specific fuel consumption (GISFC) versus CA50 is shown in Figure 8.36 in which GISFC is calculated with the work integrated from the cylinder pressure from IVC to EVO. RCCI combustion gives better fuel efficiency than DPI does. Higher combustion efficiency due to lower hydrocarbon and CO emissions (see Figure 8.37), faster combustion, and more appropriate combustion phasing in the RCCI mode have attributed to this advantage. Furthermore, RCCI combustion results in low emissions as illustrated in Figure 8.37. Although relatively high CO and hydrocarbon emissions can be reduced in DPI combustion by an earlier injection, a penalty of NO_x emission follows. These results of low fuel consumption and emissions in RCCI combustion demonstrate its advantages again over the regular DPI combustion.

A comprehensive evaluation was also conducted to optimize the diesel injection strategy in the above DNGDF engine using the micro-GA optimization by Wu and Han [21]. Multiple operating parameters including injected diesel quantity, single-injection, double-injection, injection pressure, and timing for each injection pulse, EGR rate were considered and the goal was to achieve the maximum thermal efficiency with the constraints of peak pressure, peak pressure rise rate, and NO_x emission at three (low, medium, and high) engine loads.

FIGURE 8.36 Effects of CA50 (varying on SOI) on fuel consumption in RCCI and DPI.

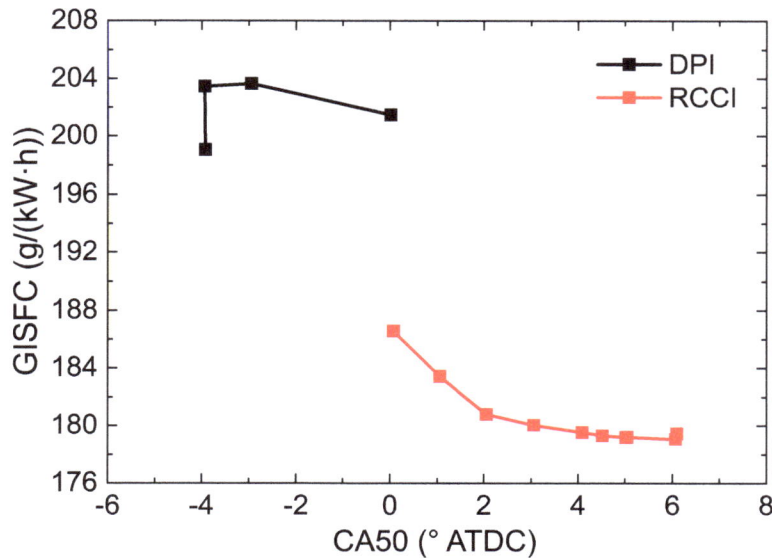

SOI: start of injection; RCCI: reactivity-controlled compression ignition; DPI: diesel pilot ignition

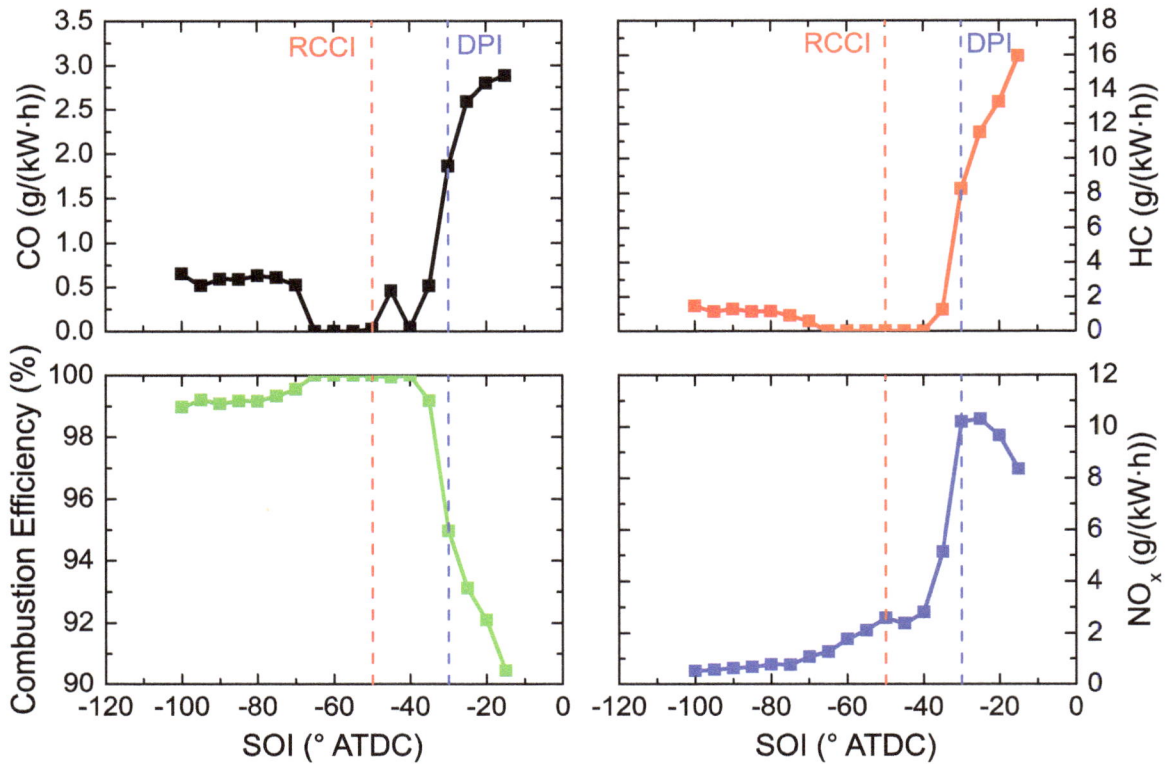

SOI: start of injection; RCCI: reactivity-controlled compression ignition; DPI: diesel pilot ignition

In Sec. 7.1.2, GAs for optimization of complex nonlinear problems in engines are introduced briefly. Among the GAs, the micro-GA allows for using a small population size (eight in this study) in one generation, which benefits reduction in computational time. A population means a group of individuals, and an individual is a combination of the considered parameters, or a solution of the optimization problem. During the optimization, all the design parameters for each individual are translated to a binary string, which is referred to as a chromosome. This is convenient for the code to perform the crossover operation in the reproduction process [21 36]. The details of micro-GA optimization process are given by Wu and Han [21].

The CONVERGE code was used in the study implementing the micro-GA solver CONGO. Three operating conditions with different loads at the same engine speed were selected for this study. The details of these conditions are shown in Table 8.7. OP1, OP2, and OP3 refer to a low, medium, and high load condition, respectively. The equivalent diesel fuel mass (EDM) in Table 8.7 is defined as:

$$EDM = m_d + m_{ng} \frac{LHV_{ng}}{LHV_d} \tag{8.3}$$

where m_{ng} and m_d represent the natural gas and diesel mass, respectively, and LHV_{ng} and LHV_d represent the lower heating value of natural gas and diesel fuel, respectively.

TABLE 8.7 Operating conditions for optimization

Operating condition number (/)	OP1	OP2	OP3
Engine speed (rpm)	1300	1300	1300
Nominal load (bar)	~5	~10	~20
Pressure @IVC (kPa)	135	190	342
Temperature @IVC (K)	363	363	363
Equivalent diesel mass (g)	0.06	0.115	0.245
Compression ratio (/)	16.1	16.1	16.1

Reprinted with permission from Ref. [21]. © Elsevier

Seven parameters were evaluated and their variation ranges are given in Table 8.8. The percent energy substitution (PES) is defined as:

$$PES = \frac{m_{ng} LHV_{ng}}{m_{ng} LHV_{ng} + m_d LHV_d} \tag{8.4}$$

and EGR rate is defined as:

$$EGR = \frac{(CO_2)_{in}}{(CO_2)_{ex}} \times 100\% \tag{8.5}$$

where $(CO_2)_{in}$ and $(CO_2)_{ex}$ is the CO_2 concentration in the intake flow and in the exhaust flow, respectively.

TABLE 8.8 Parameters and their variation ranges for optimization

Injection strategy	Double	Single
PES (%)	50–100	50–100
EGR (%)	0–60	0–60
SOI1 (°ATDC)	−130 to −50	−130 to 20
SOI2 (°ATDC)	−40 to 20	\
1st Injection pressure (bar)	300–1500	300–1500
2nd Injection pressure (bar)	300–1500	\
Split ratio (/)	0–1	\

Reprinted with permission from Ref. [21]. © Elsevier

A merit function MF is defined considering the constraints for peak pressure, peak pressure rise rate, and NO_x emission as:

$$MF = 100 \times \left[\frac{\eta_i}{\eta_{i,tg}} + \alpha \left(\left(\frac{NOx}{NOx,c} \right)^{0.2} - 1 \right) \right] + \beta \left[\left(\frac{P_m}{P_{m,c}} \right)^5 - 1 \right]$$
$$+ \gamma \left[\left(\frac{dp_m}{dp_{m,c}} \right)^5 - 1 \right] + \delta \left[\frac{\eta_{c,c}}{\eta_c} - 1 \right] \tag{8.6}$$

In Equation (8.6), η_i is the indicated thermal efficiency (ITE), $\eta_{i,tg}$ is the targeted thermal efficiency, NO_x is the NO_x emissions, p_m is the peak pressure, dp_m is the peak pressure rise rate, and η_c represents the combustion efficiency. Other parameters with the subscript c represent the constraint values. Table 8.9 shows the target and constraints of the parameters used in the merit function. In Equation (8.6), α, β, γ, and δ are the multipliers. When the response values of NO_x, p_m, or dp_m exceed their constraint values, α, β, or γ will be set to be -1, which otherwise will be set to be 0. δ will be set to -1 when the response value of η_c is smaller than the constraint value, which otherwise will be set to 0. Based on the merit function, it can be seen that the individual that has high thermal efficiency and acceptable NO_x emissions, peak pressure, peak pressure rise rate, and combustion efficiency would obtain a high merit value. During the optimization process, the micro-GA code maximizes the merit value as the evolution of the individuals. By doing this, the optimization solution moves towards high thermal efficiency and clean combustion.

TABLE 8.9 Target and constraints of the parameters used in the merit calculation

ITE (%)	50
NO_x (g/kW.h)	0.4
Peak pressure (bar)	170
Peak pressure rise rate (bar/°)	15
Combustion efficiency (%)	80

After the calculations, the optimal performance results are obtained as listed in Table 8.10, and the corresponding operational parameters are given in Table 8.11. As can be seen, thermal efficiency higher than 45% is achieved while keeping the peak pressure, peak pressure rise rate, and NO_x emissions within their constraints under all the three operating conditions for the double-injection strategy. In particular, thermal efficiency is near 50% at the medium load. For the single-injection, thermal efficiency higher than 45% is obtained at the low and medium loads, but only 35.5% of thermal efficiency yields at the high load, which is mainly caused by the low combustion efficiency (85.03%).

TABLE 8.10 Optimal performance results

Cases	OP1		OP2		OP3	
Injection strategy	Double	Single	Double	Single	Double	Single
ITE (%)	47.0	45.6	48.4	47.6	45.5	35.5
IMEP (bar)	4.97	4.82	9.80	9.64	19.64	15.31
CA50 (°ATDC)	0.0	−1.4	4.5	3.0	12.6	12.0
Combustion duration (°)	14.95	14.02	13.0	14.5	23.45	41.05
Peak pressure (bar)	88.3	90.5	134.9	132.4	168.4	167.1
Peak pressure rise (bar/°)	4.4	4.9	11.3	8.6	5.5	4.8
Combustion efficiency (%)	99.1	97.32	99.8	99.02	97.1	85.03
NO_x (g/kW.h)	0.157	0.108	0.231	0.272	0.400	0.355
CO (g/kW.h)	2.392	5.112	0.196	1.21	1.875	17.329
HC (g/kW.h)	0.860	3.215	0.247	1.237	4.180	26.928

TABLE 8.11 Settings of the operational parameters to achieve the optimal engine result

Cases	OP1		OP2		OP3	
Injection strategy	**Double**	**Single**	**Double**	**Single**	**Double**	**Single**
PES (%)	64.3	58.69	84.9	82.85	89.0	54.58
EGR (%)	3.2	4.07	6.2	5.78	39.0	57.35
SOI1 (°ATDC)	−127.6	−124.2	−129.5	−87.5	−86.6	−10.3
SOI2 (°ATDC)	−19.4	\	−22.8	\	−6.2	\
1st Injection pressure (bar)	401.0	1321.0	1048.6	1415.4	802.9	874.8
2nd Injection pressure (bar)	523.9	\	703.0	\	1084.3	\
Split ratio (/)	0.938	\	0.915	\	0.610	\

Figure 8.38 compares the cylinder pressure and HRR for the optimized double-injection and single-injection cases under different operating conditions. It is seen that at OP1 and OP2, the SOC are almost the same for both the injection strategies. At OP3, the SOC for the double-injection case starts from around −20° ATDC due to low temperature reaction of the diesel fuel, while combustion in the single-injection case is delayed because of a later injection time. It can be also seen that the initial HRRs in the single-injection cases are stronger than those in the double-injection cases, which causes earlier combustion phases leading to advances in the occurrences of the peak pressures and peak HRRs.

FIGURE 8.38 Cylinder pressure and heat release rates with different injection strategies.

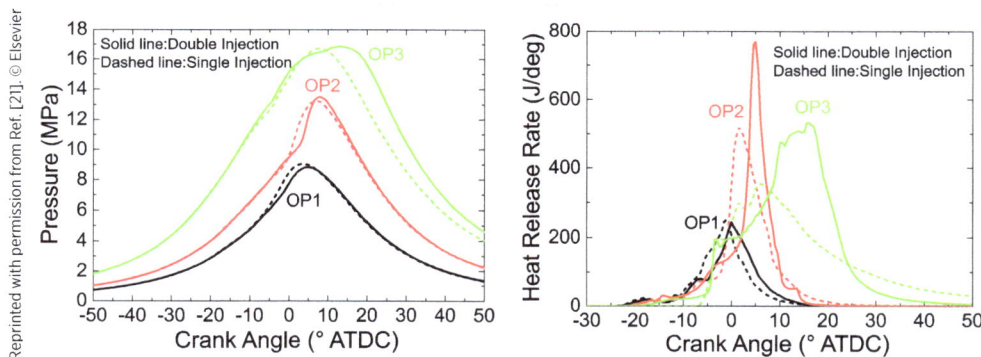

To better understand the differences in thermal efficiency among the injection strategies, energy balances are compared in Figure 8.39. It is seen that compared to the double-injection, the deterioration of thermal efficiency for the single-injection is mainly caused by the high combustion loss. And the heat transfer loss is also higher for the single-injection, which is because of its earlier combustion phasing. It is also observed that despite a higher thermal efficiency obtained by the double-injection, its exhaust loss is slightly higher than that of the single-injection. The combustion and heat transfer losses decrease for the double-injection, which increases the amount of energy produced. However, the increased energy cannot be completely converted to useful work and a portion of the increased energy is wasted in the exhaust.

FIGURE 8.39   Energy balance for different injection strategies.

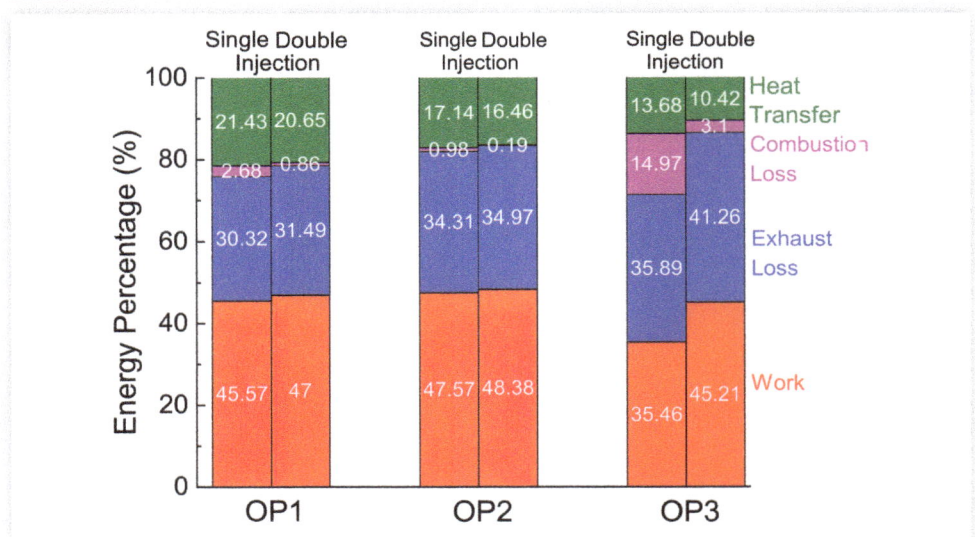

This example demonstrated the usefulness of engine combustion simulation and optimization to help search operating parameters in a complicated and interactive space to achieve the optimal targets. It will be, if possible, a very time-consuming and costly process in searching the targets by trial-and-error experiments. Furthermore, the above optimization is based on a given combustion chamber design; simulation with the same optimization process was also applied in searching for a better chamber-shape design and additional improvements in the thermal efficiency were obtained [21].

8.2.3 Combustion and NO_x Emissions of Biodiesel Fuels

Biodiesel is an attractive alternative fuel for IC engines due to its merits of renewability and low soot emissions. However, biodiesel combustion results in higher NO_x emissions than diesel combustion does. Modeling studies on combustion and formation of NO_x emissions in biodiesel engines were made [26,27]. As an example, the numerical study on NO_x emissions characteristics of biodiesel–diesel blends in a light-duty diesel engine by Li et al. [26] is presented in this section. Li and co-workers investigated cases with pure diesel (B00), 20% (in volume) biodiesel–80% diesel (B20) and 50% biodiesel–50% diesel (B50). The characteristics of combustion and NO_x emissions for different fuels were compared. The effects of injection timing and EGR rate on NO_x emissions were also evaluated.

The CFD code, KIVA3V2 [37] coupled with the CHEMKIN-II chemistry solver [38] was extended to work with a two-component chemical kinetics mechanism. The blended fuels are the mixtures of soybean methyl ester (SME) biodiesel and premier diesel. To model the fuels, n-heptane (C_7H_{16}) combined with methyl butanoate ($C_5H_{10}O_2$) in a molar ratio of 2:1 to represent SME ($C_{19}H_{35}O_2$). A biodiesel surrogate mechanism [39,40] with a reduced methyl butanoate and n-heptane mechanism was used in the present simulations to represent SME. A detailed description of the mechanism and chemical/physical properties of the blended fuels is given by Li et al. [26]. The modeled engine is a Ford Lion 2.7-liter V6 DI diesel engine. A 60° sector mesh was modeled, which contained one of the six sprays evenly arranged in the combustion chamber. The simulation began at 220 CA (IVC time), and ended at 490 CA (EVO time), where 360 CA is the compression TDC.

Simulations were first conducted for B00, B20, and B50 with EGR of about 20%. The engine was operated at 1000 rpm and BMEP of 0.5 MPa. A multiple injection strategy was used for both the B20 and B50 blends. The SOI was swept with a 2 CA interval while the dwell between the pilot injection and the main injection was unchanged. Figure 8.40 shows the comparison of predicted and measured NO_x emissions. The NO_x emissions for the B20 cases are over-predicted by a maximum relative error of 37.0%. The NO_x emissions for the B50 cases are over-predicted except for the case with the main SOI of 366° CA. The maximum relative error in the B50 cases is 8.2%.

FIGURE 8.40 Comparison of the predicted and measured NO_x emissions.

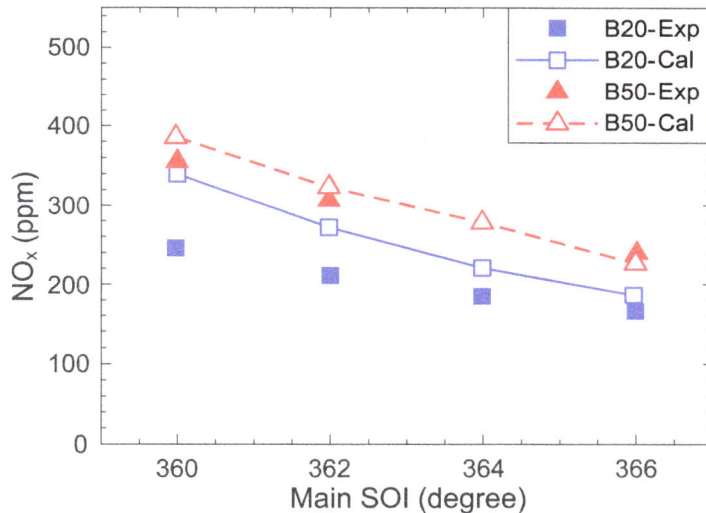

In order to address the influence of the biodiesel blending ratio on the formation of NO_x emissions, the cases with the main SOI of 362 CA were analyzed in detail. The comparisons between the experimental pressures and the apparent HRRs (AHRRs) for different blends are presented in Figure 8.41, which shows that the ignition delays of the pilot injection for B20 and B00 are almost the same. For the main injection, B20 shows a slightly longer ignition delay, which may be caused by the slightly lower compression pressure. As shown in Figure 8.40, the experimental NO_x emissions for B50 are much higher than for B20. For the B50 case, the pressure curve is similar to that for B00, but the AHRR curve for B50 shows that the ignition delays of both injections are slightly shorter. The peaks of the AHRR are lower than those for B00. The significant increase in the NO_x emissions for the B50 cases may be partially caused by the slightly lower EGR rate. The earlier SOC for B50 may also be one of the key factors that cause the significant increase in the NO_x emissions.

Reprinted with permission from Ref. [26]. © SAGE Publications

FIGURE 8.41 Experimental pressures and AHRRs for different fuels with a pilot SOI of 344 CA and a main SOI of 362 CA.

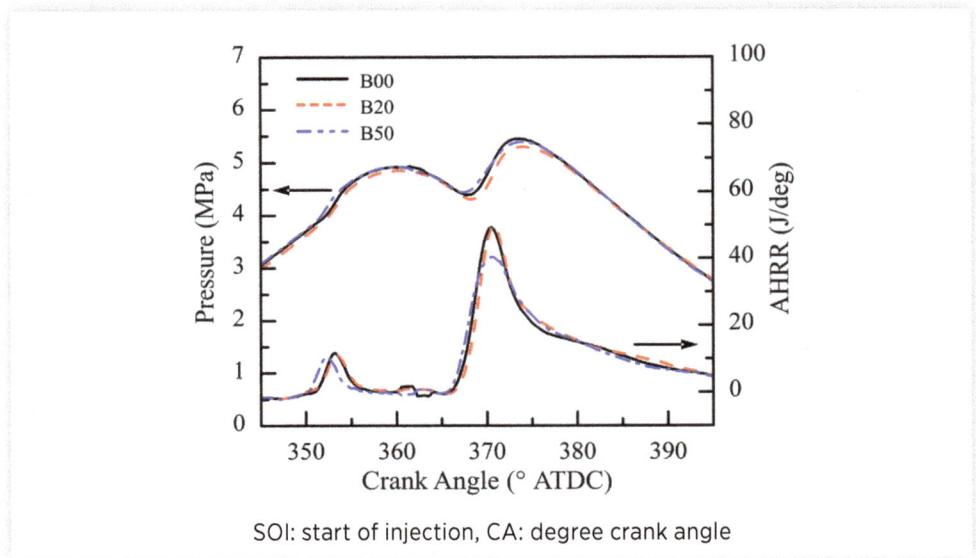

SOI: start of injection, CA: degree crank angle

FIGURE 8.42 (a) Evolution of NO_x volume fractions; (b) Mass fractions of the cells at a temperature greater than 2200 K.

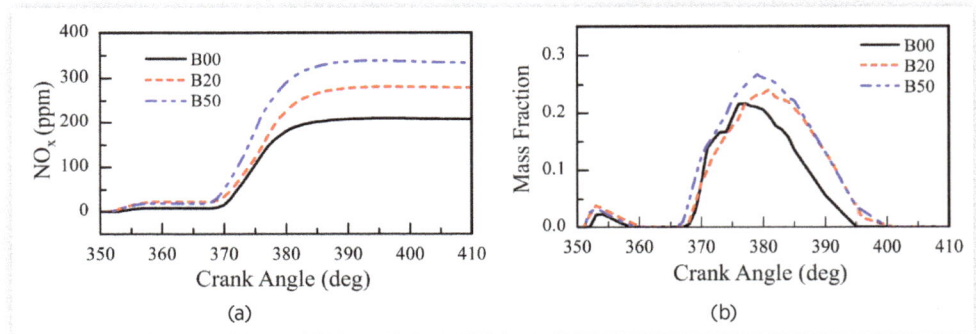

Reprinted with permission from Ref. [26]. © SAGE Publications

The mass fractions of the computational cells at temperatures greater than 2200 K and the history curves of NO_x emissions are shown in Figure 8.42. The mass fractions represent the quantities of local high-temperature gaseous mixture, which favors NO_x formation. As shown in Figure 8.42, the majority of the NO_x is formed during the combustion of the main injection. Focusing on the combustion of the main injection, the quantity of high-temperature mixture for B50 is greater than those for B00 and B20. This is the result of the earlier SOC and higher oxygen concentrations in the B50 case. The quantity of high-temperature mixture for B20 is similar to that for B00 before 378 CA and more than that for B00 after 378 CA.

The spatial distributions of the temperature and the NO_x mass fraction are shown in Figures 8.43 and 8.44, respectively. Results at three crank angles are shown, which are the representative timings of major NO_x formation. The temperature distribution for B50 shows the largest high temperature region. The high-NO_x-concentration region follows a pattern similar to that of the high temperature region. As a result, B50 produces the highest NO_x emissions.

FIGURE 8.43 Temperature distributions for different fuels.

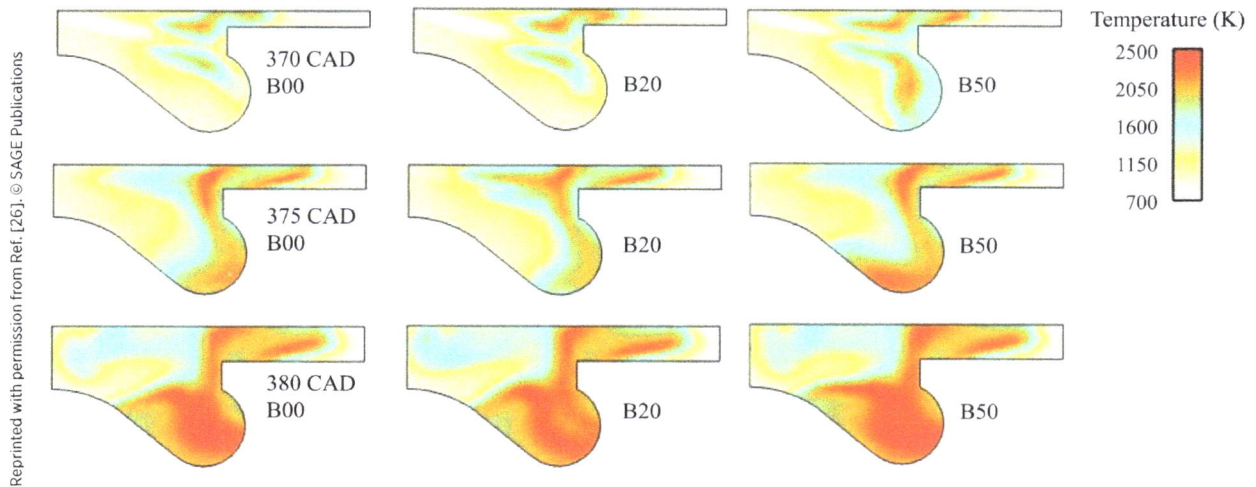

370 CAD
B00

B20

B50

375 CAD
B00

B20

B50

380 CAD
B00

B20

B50

Temperature (K)

2500
2050
1600
1150
700

FIGURE 8.44 Distributions of the NO$_x$ mass fraction for different fuels.

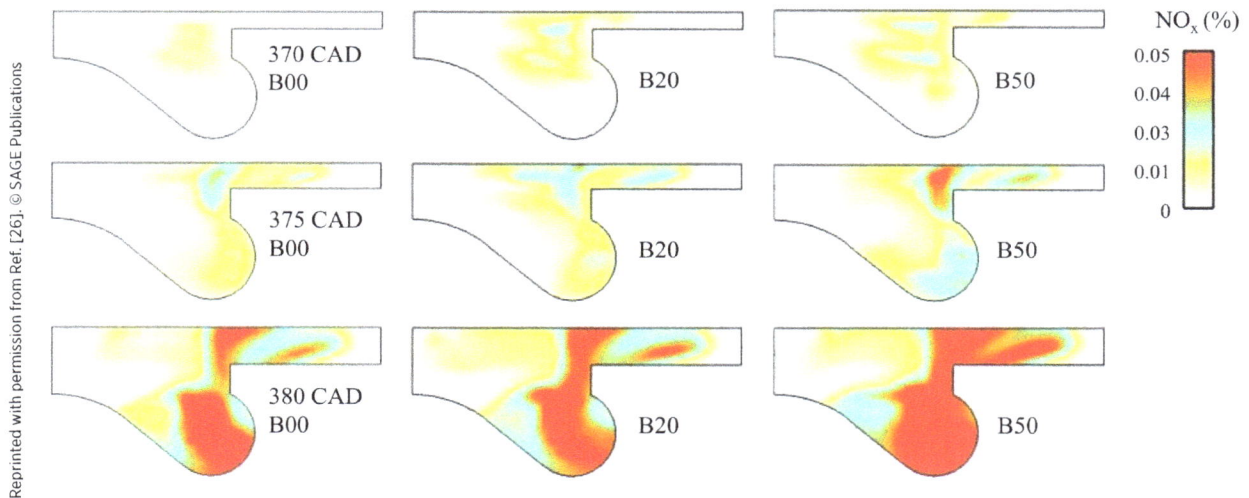

370 CAD
B00

B20

B50

375 CAD
B00

B20

B50

380 CAD
B00

B20

B50

NO$_x$ (%)

0.05
0.04
0.03
0.01
0

In a separate study by Li et al. [27], combinations of four SOIs and three EGR rates are studied to reduce NO$_x$ emissions. The predicted NO$_x$ emissions can be reduced by 20.3% and 32.9% when SOIs are delayed by 2 and 4 crank angles, respectively. NO$_x$ emissions of the cases with 24.0% and 28.0% EGR are reduced by 38.4% and 62.8%, respectively. A soot–NO$_x$ trade-off is also seen in biodiesel combustion. When EGR rate is 28.0% and SOI is delayed by 2°, NO$_x$ emissions can be reduced by 55.1%, and soot emissions are controlled as that of the case with 24% EGR. Thus, NO$_x$ emissions of biodiesel combustion can be effectively improved by SOI retardation or increasing EGR rate.

These studies demonstrate again that engine CFD simulation together with chemical reaction mechanism can be used to predict NO_x and soot emissions for alternative fuels and to evaluate means to reduce emissions. Thus, there is no doubt that the same modeling approach can be applied to assess and design better fuels for future engines.

References

1. Han, Z., Uludogan, A., Hampson, G.J., and Reitz, R.D. 1996. Mechanism of soot and NO_x emission reduction using multiple-injection in a diesel engine. *SAE Transactions* 105, no. 3: 837–852. DOI: 10.4271/960633

2. Dec, J.E., and Espey, C. 1995. "Ignition And Early Soot Formation in a DI Diesel Engine Using Multiple 2-D Imaging Diagnostics." *SAE Transactions* 104, no. 3 (February): 853–875. DOI: 10.4271/950456

3. Tow, T.C., Pierpont, D., and Reitz, R.D. 1994. "Reducing Particulate And No_x Emissions by Using Multiple Injections in a Heavy Duty D.I. Diesel Engine." *SAE Transactions* 103, no. 3 (March): 1403–1417. DOI: 10.4271/940897

4. Wang, Y., Han, Z., Deng, P., and Chen, Z. 2012. "Numerical Simulation of Design Parameter Optimization of Diesel Engine Helical Port." [In Chinese]. *Chinese Internal Combustion Engine Engineering* 33, no. 5: 79–86. DOI: 10.13949/j.cnki.nrjgc.2012.05.012.

5. Li, J., Han, Z., Chen, Z., and Liu, Y. 2013. "A Numerical Simulation Study on the Combustion System Parameters of a Passenger Car Diesel Engine." [In Chinese]. *Automotive Engineering* 35, no. 4: 47–353, 384. DOI: 10.19562/j.chinasae.qcgc.2013.04.012.

6. Richards, K.J., Senecal, P.K., and Pomraning, E. 2008. *CONVERGE (v1.3)*. Madison, Wisconsin: Convergent Science Inc.

7. Fang, K.T., Lin, D.K.J., Winker, P., and Zhang, Y. 2000. "Uniform Design: Theory And Application.: *Technometrics* 42, no. 3 (March): 237–248. DOI: 10.1080/00401706.2000.10486045

8. Deng, P. 2016. Numerical study on spray/wall interaction models and cold start of a diesel engine (Publication No. U464.172). [In Chinese]. Changsha, China: Hunan University. China Doctoral Dissertations Full-text Database.

9. Ra, Y., and Reitz, R.D. 2008. "A Reduced Chemical Kinetic Model for IC Engine Combustion Simulations With Primary Reference Fuels." *Combustion and Flame* 155, no. 4: 713–738. DOI: 10.1016/j.combustflame.2008.05.002

10. Sun, Y. (2007). Diesel Combustion Optimization And Emissions Reduction Using Adaptive Injection Strategies (AIS) With Improved Numerical Models (Publication No. 3278986). Madison, Wisconsin: University of Wisconsin-Madison. ProQuest Dissertations and Theses.

11. Deng, P., Jiao, Q., Reitz, R.D., and Han, Z. 2015. "Development of an Improved Spray/Wall Interaction Model for Diesel-Like Spray Impingement Simulations." *Atomization and Sprays* 25, no. 7: 587–615. DOI: 10.1615/AtomizSpr.2015011000

12. Deng, P., Han, Z., and Reitz, R.D. 2015. "Modeling Heat Transfer in Spray Impingement Under Direct-Injection Engine Conditions." *Proceedings of the Institution of Mechanical Engineers, Part D: Journal of Automobile Engineering* 230, no. 7 (August): 885–898. DOI: 10.1177/0954407015596284

13. Tsunemoto, H., Yamada, T., and Ishitani, H. 1986. "Behavior of Adhering Fuel on Cold Combustion Chamber Wall in Direct Injection Diesel Engines." *SAE Transactions* 95, no. 4 (September): 1017–1024. DOI: 10.4271/861235

14. Korakianitis, T., Namasivayam, A.M., and Crookes, R.J. 2011. "Natural-Gas Fueled Spark-Ignition (SI) And Compression-Ignition (CI) Engine Performance And Emissions." *Progress in Energy and Combustion Science* 37, no. 1: 89–112. DOI: 10.1016/j.pecs.2010.04.002

15. Le Fevre, C. 2019. *A Review of Prospects for Natural Gas as a Fuel In Road Transport*. Oxford, UK: The Oxford Institute for Energy Studies.

16. Wu, Z., and Han, Z. 2018. "Numerical Investigation on Mixture Formation in a Turbocharged Port-Injection Natural Gas Engine Using Multiple Cycle Simulation." *Journal of Engineering for Gas Turbines and Power* 140, no. 5: article 051704. DOI: 10.1115/1.4039106

17. Han, Z., Wu, Z., Huang, Y., Shi, Y., and Liu, W. 2021. "Impact of Natural Gas Fuel Characteristics on the Design And Combustion Performance of a New Light-Duty CNG Engine." *International Journal of Automotive Technology* 22, no. 6.

18. Wu, Z., Rutland, C., and Han, Z. 2017. "Numerical Study on Controllability of Natural Gas And Diesel Dual Fuel Combustion in a Heavy-Duty Engine." SAE Technical Paper No. 2017-01-0756. DOI: 10.4271/2017-01-0756

19. Wu, Z., Rutland, C.J., and Han, Z. 2018. "Numerical Optimization of Natural Gas And Diesel Dual-Fuel Combustion for a Heavy-Duty Engine Operated at a Medium Load." *International Journal of Engine Research* 19, no. 6: 682–696. DOI: 10.1177/1468087417729255

20. Wu, Z., Rutland, C.J., and Han, Z. 2019. "Numerical Evaluation of the Effect Of Methane Number on Natural Gas And Diesel Dual-Fuel Combustion." *International Journal of Engine Research* 20, no. 4 (February): 405–423. DOI: 10.1177/1468087418758114

21. Wu, Z., and Han, Z. 2020. "Micro-GA Optimization Analysis of the Effect of Diesel Injection Strategy on Natural Gas-Diesel Dual-Fuel Combustion." *Fuel* 259: article 116288. DOI: 10.1016/j.fuel.2019.116288

22. Fazal, M.A., Haseeb, A., and Masjuki, H.H. 2011. "Biodiesel Feasibility Study: An Evaluation of Material Compatibility; Performance; Emission And Engine Durability." *Renewable and Sustainable Energy Reviews* 15, no. 2 (February): 1314–1324. DOI: 10.1016/j.rser.2010.10.004

23. Atabani, A.E., Silitonga, A.S., Badruddin, I.A., Mahlia, T.M.I., Masjuki, H.H., and Mekhilef, S. 2012. "A Comprehensive Review on Biodiesel as an Alternative Energy Resource And its Characteristics." *Renewable and Sustainable Energy Reviews* 16, no. 4: 2070–2093. DOI: 10.1016/j.rser.2012.01.003

24. Brakora, J.L., Ra, Y., and Reitz, R.D. 2011. "Combustion Model for Biodiesel-Fueled Engine Simulations Using Realistic Chemistry And Physical Properties." *SAE International Journal of Engines* 4, no. 1 (April): 931–947. DOI: 10.4271/2011-01-0831

25. Herbinet, O., Pitz, W.J., and Westbrook, C.K. 2008. "Detailed Chemical Kinetic Oxidation Mechanism for a Biodiesel Surrogate." *Combustion and Flame* 154, no. 3; 507–528. DOI: 10.1016/j.combustflame.2008.03.003

26. Li, J., Han, Z., Shen, C., and Lee, C.-f. 2014. "Numerical Study on the Nitrogen Oxide Emissions of Biodiesel–Diesel Blends in a Light-Duty Diesel Engine." *Proceedings of the Institution of Mechanical Engineers, Part D: Journal of Automobile Engineering* 228, no. 7 (February): 734–746. DOI: 10.1177/0954407014521174

27. Li, J., Han, Z., Shen, C., and Lee, C.-f. 2014. "A Study on Biodiesel No$_x$ Emission Control With the Reduced Chemical Kinetics Model." *Journal of Engineering for Gas Turbines and Power* 136, no. 10: article 101505. DOI: 10.1115/1.4027358

28. Han, Z. 2014. *A Combustion Method for Natural Gas Engines And Turbocharged Natural Gas Engines With Such a Method*, China Patent No. 201410408370.9, China National Intellectual Property Administration.

29. Mendl, G., Mangold, R., Rosenberger, S., Langa, Z., and Czuczor, B. 2017. "The New Audi 2.0 L G-Tron—Another Step Towards Future Sustainable Mobility." Paper presented at the *38th International Vienna Motor Symposium, Vienna, Austria, April 27–28*.

30. Richards, K.J., Senecal, P.K., and Pomraning, E. 2017. *CONVERGE Manual (v2.4)*. Madison, Wisconsin: Convergent Science Inc.

31. Han, Z., Wu, D.Y., Shi, Y., and Wu, Z. 2015. *A Device to Enhance In-Cylinder Gas Motion for Natural Gas Engines*, China Patent No. 201520122531.8, China National Intellectual Property Administration.

32. Guo, Z., Huang, Y., Shi, Y., Liu, W., and Han, Z. 2019. "Low-Speed Performance Improvement of Natural Gas Engine by Asynchronous Intake Timing And Turbocharger Matching." *Journal of Xi'an Jiaotong University* 53, no. 9: 55–60. (in Chinese)

33. Reitz, R.D. 2013. "Directions in Internal Combustion Engine Research." *Combustion and Flame* 160, no. 1 (January): 1–8. DOI: 10.1016/j.combustflame.2012.11.002

34. Walker, N.R., Wissink, M.L., DelVescovo, D.A., and Reitz, R.D. 2015. "Natural Gas for High Load Dual-Fuel Reactivity Controlled Compression Ignition in Heavy-Duty Engines." *Journal of Energy Resources Technology* 137, no. 4: article 042202. DOI: 10.1115/1.4030110

35. Wei, L., and Geng, P. 2016. "A Review on Natural Gas/Diesel Dual Fuel Combustion, Emissions And Performance." *Fuel Processing Technology* 142 (February): 264–278. DOI: 10.1016/j.fuproc.2015.09.018

36. Senecal, P.K., and Reitz, R.D. 2000. "Simultaneous Reduction of Engine Emissions And Fuel Consumption Using Genetic Algorithms And Multi-Dimensional Spray And Combustion Modeling." *SAE Transactions* 109, no. 4 (June): 1378–1390. DOI: 10.4271/2000-01-1890

37. Amsden, A.A. 1999. *KIVA-3V, Release 2: Improvements to KIVA-3V* (LA-13608-MS). Los Alamos, New Mexico: Los Alamos National Laboratory.

38. Kee, R.J., Rupley, F.M., and Miller, J.A. 1989. *Chemkin-II: A Fortran Chemical Kinetics Package for the Analysis of Gas-Phase Chemical Kinetics* (SAND89-8009). Livemore, California: Sandia National Laboratory. DOI: 10.2172/5681118

39. Brakora, J.L., Ra, Y., Reitz, R.D., McFarlane, J., and Daw, C.S. 2008. "Development And Validation of a Reduced Reaction Mechanism for Biodiesel-Fueled Engine Simulations." SAE Technical Paper No. 2008-01-1378. DOI: 10.4271/2008-01-1378

40. Brakora, J.L., and Reitz, R.D. 2010. "Investigation of No_x Predictions From Biodiesel-Fueled HCCI Engine Simulations Using a Reduced Kinetic Mechanism." SAE Technical Paper No. 2010-01-0577. DOI: 10.4271/2010-01-0577

Index

About the Author

Dr. Zhiyu Han is currently a professor in the School of Automotive Studies of Tongji University in China. He is a recognized automotive engine expert. Dr. Han started research on multidimensional engine simulation in the early 1990s. He proposed variable-density turbulence and heat transfer models to improve engine simulation accuracy. He also developed practical engineering methods that enable combustion system optimization at the pre-prototype stage, making pioneering contributions in the use of advanced modeling techniques for rapid and cost-effective engine design and development. Dr. Han is a Fellow of the Society of Automotive Engineering International. He has a Ph.D. from the University of Wisconsin-Madison, Wisconsin, USA. His career spanned auto industry and universities with various job positions including Staff Technical Specialist in Ford Motor Company in Dearborn, USA; General Manager of the Technical R&D Center of Great Wall Motor Company in Baoding, China; General Manager of Changfeng Engine Company in Changsha; and Director of Advanced Powertrain Research Center of Hunan University in Changsha.

www.ingramcontent.com/pod-product-compliance
Lightning Source LLC
Chambersburg PA
CBHW051928190326
41458CB00026B/6446